B

Progress in Nonlinear Differential Equations and Their Applications

Volume 1

Editor
Haim Brezis
Rutgers University
New Brunswick
and
Université Pierre et Marie Curie
Paris

Partial Differential Equations and the Calculus of Variations

Essays in Honor of Ennio De Giorgi

Volume I

Edited by
F. Colombini
A. Marino
L. Modica
S. Spagnolo

1989

Birkhäuser
Boston · Basel · Berlin

Editors
Ferruccio Colombini, Antonio Marino,
Luciano Modica, Sergio Spagnolo
Dipartimento di Matematica
Universita di Pisa
Via F. Buonarroti 2
56100 Pisa
Italy

Library of Congress Cataloging–in–Publication Data
Partial differential equations and the calculus of variations : essays
in honor of Ennio De Giorgi / Ferruccio Colombini . . . [et al.].
p. cm. — (Progress in nonlinear differential equations and
their applications)
ISBN 0-8176-3424-X (v. 1 : alk. paper). — ISBN 0-8176-3425-8 (v.
2 : alk. paper). — ISBN 0-8176-3426-6 (set : alk. paper)
1. Differential equations, Partial. 2. Calculus of variations.
3. Giorgi, Ennio De. I. Giorgi, Ennio De. II. Colombini, F.
(Ferruccio) III. Series.
QA377.P297 1989
515′.353—dc20   89-9746

Printed on acid-free paper.

Volume I   *Volume II*
ISBN 0-8176-3424-X ISBN 0-8176-3425-8
ISBN 3-7643-3424-X ISBN 3-7643-3425-8

2-Volume set
ISBN 0-8176-3426-6
ISBN 3-7643-3426-6

Camera-ready copy prepared by the editors using T_EX.
Printed and bound by Edwards Brothers, Ann Arbor, Michigan.
Printed in the U.S.A.

9 8 7 6 5 4 3 2 1

A Ennio De Giorgi
i suoi allievi ed amici
con affetto

Courtesy of Foto Frassi, Pisa.

Preface

The Italian school of Mathematical Analysis has long and glorious traditions. In the last thirty years it owes very much to the scientific pre-eminence of Ennio De Giorgi, Professor of Mathematical Analysis at the Scuola Normale Superiore di Pisa.

His fundamental theorems in Calculus of Variations, in Minimal Surfaces Theory, in Partial Differential Equations, in Axiomatic Set Theory as well as the fertility of his mind to discover both general mathematical structures and techniques which frame many different problems, and profound and meaningful examples which show the limits of a theory and give origin to new results and theories, makes him an absolute reference point for all Italian mathematicians, and a well-known and valued personage in the international mathematical world.

We have been students of Ennio de Giorgi. Now, we are glad to present to him, together with all his collegues, friends and former students, these Essays of Mathematical Analysis written in his honour on the occasion of his sixtieth birthday (February 8th, 1988), with our best wishes and our thanks for all he gave in the past and will give us in the future.

We have added to the research papers of this book the text of a conversation with Ennio De Giorgi about the diffusion and the communication of science and, in particular, of Mathematics.

We wish to thank all the authors of the papers collected here, and all those who helped us in the preparation of this book, in particular: the students of the Scuola Normale and of the Department of Mathematics of Pisa University L.Ambrosio, S. Baldo, P. Biscari, P. D'Ancona, D. Del Santo, G. Ferraro, N. Orrù, C. Saccon, F. Serra Cassano and V. M. Tortorelli who revised with extreme care the typescripts; Miss Giuliana Cai and Mrs. Margaret Mencacci who did an excellent job in typing the all papers.

Finally we thank our friend Professor H.Brezis who proposed the publication of these volumes in the Birkhäuser series "Progress in Nonlinear Differential Equations and their Applications", and Birkhäuser for accepting to publish this book and supporting us in organizational matters.

Pisa, March 1989

Ferruccio Colombini
Antonio Marino
Luciano Modica
Sergio Spagnolo

A conversation with Ennio De Giorgi

The communication of science. The relationship and dialogue between lovers of scientific disciplines and people that live in the same period and for this reason share with them many human problems and those crucial dilemmas that seem to have a determining influence on the future of humanity.

And again, the ability of scientists and scholars to make themselves reciprocally share at least the substance of their research.

And the desire to question even themselves about the ultimate meaning and importance of their own studies.

These topics are very dear to De Giorgi and have often formed the basis of our talks with him.

It seemed to us a good idea to use some things he had told us during a recent chat on those themes as the introduction to this collection of articles which so many of us dedicate to him with enthusiasm and friendship.

Also because, from the rich discussion we had with him, sometimes with impromptu exchanges of ideas or co-operating (and improvising!) in some social, humanitarian undertakings, many of us feel we are indebted to the man no less for these reasons than his exceptional qualities as a mathematician.

Furthermore, and as those that know Ennio well know, his own scientific activity of research and teaching is stamped with an extraordinary breadth of vision in which a stimulating, open and cordial dialogue with everyone develops.

We were beginning to speak about science, when Ennio referred us to a wider concept that is dear to him: that of "wisdom" according to the Bible and invited us to read a passage, that we quote here, taken from the Book of Proverbs, to show us how in the Bible wisdom is of a "convivial" nature, in that it is serene and has a community character at the same time.

> *True wisdom has built its house.*
> *It has hewn out its seven pillars.*
> *It has organized its meat slaughtering;*
> *It has mixed its wine;*
> *More than that, it has set in order its table.*
> *It has sent forth its lady attendants,*
> *That it may call on top of the heights of the town:*
> *"..*
> *Come, feed yourselves with my bread*
> *And share in drinking the wine that I have mixed."*

Proverbs 9 $(1-5)$

De Giorgi: In the life of a mathematician the problem of communicating knowledge presents many different aspects: from the relationship with those studying the same topics to that with other mathematicians, working in different branches, from dialogue with scholars of experimental sciences and different branches of the techniques mathematicians use in various forms, to teaching at all levels and the popularization of science.

Question. *Let's begin with scientific popularization. Doesn't it seem to you that in the field of math there is very little of it? And is it really possible to achieve it to a good degree? In Italy, for example, you can't say math is very popular.*

De Giorgi: The popularization of math appears to be very difficult. First of all, even many well-educated people are convinced they are unable to understand certain aspects of math. Then we have to admit that many mathematicians have no faith at all in the possibility of communicating to non-experts the problems and results of their own work. I think we have to react against this distrust.

An even limited paper on mathematical thinking could be very important both for society and mathematicians themselves.

Question. *In what way?*

De Giorgi: On the one hand a mathematician trying to explain to others the nature of math and its problems and also the reasons for his own work, ends up himself by understanding its meaning better.

On the other hand a certain understanding of mathematical thinking seems to me necessary if all of society is to be able to grasp the meaning of all scientific thinking and to therefore give reasoned judgments on the development and applications of science and technology that characterize the period we live in.

You have to remember public opinion can have a decisive influence on both good and evil applications of scientific discoveries and technological innovations.

But you've also to remember public opinion already has a remarkable influence on the development of every scientific discipline: this development can be greatly encouraged by the consideration and esteem encountered in society at large.

After all the balanced development of society depends to a large extent on the harmonious integration of the various forms of human thought: scientific, artistic, literary, religious etc.

Question. *We are often asked what the use is of mathematical research. Rather, to tell the truth, we are asked what the devil we mathematicians study. Questions of this kind, we suppose, are not put to physicists, biologists or philosophers.*

De Giorgi: True, the aim of math is less definable than that of other sciences. We can say that math examines structures i.e. collections of objects and relationships between them but in this study math is interested in the properties of the structure rather than in the nature of the individual objects.

Let's take for example a row of a hundred houses, or a row of a hundred people, or a row of a hundred trees and let us think in each row of the order relation that exists between the object that precedes and the one that follows: these rows, each one as far as its own order relation is concerned, are "equivalent" structures to the mathematician even if a house is clearly different from a tree or a person.

This concept of equivalence between structures is precisely stated in the various branches of math by the concepts of similarity, isomorphism etc.

If we want to limit ourselves to examples of an elementary nature, we can think of equilateral triangles. Their properties are interesting, independently of whether they are small or large, drawn on a blackboard or on a sheet of paper, modelled in wood or in metal etc.

Question. *Among examples that are more abstract and yet easily understandable, we could quote the partial order relations represented by the inclusion of sets and the divisibility of natural numbers.*

However, it's a question of finding examples and problems that arouse a certain interest.

De Giorgi: These varieties of possible models of the one same mathematical theory, on one hand, can partly explain the flexibility of math and the variety of its applications, on the other it can create some confusion in the listener. We have to be aware of this difficulty, remembering that the choice of problems and more interesting, suggestive examples is essential for the understanding of the mathematical structures we want to illustrate.

For example, it's one thing to propose a certain number of ingenious problems resolvable by the use of Algebra, but it's another to furnish a simple row of equations, inequalities and algebraic expressions.

On the other hand, the choice of examples and meaningful models is valid not only for popularization and elementary teaching but also in all phases of scientific research. A suggestive model can help us to guess at some possible developments that another model would never suggest.

Question. *Many people find it strange that mathematical problems exist that have not been solved and have difficulty in imagining what mathematical research consists of.*

De Giorgi: While it is easy to answer that in math, as in every living science, the number of open questions is much greater than the number of solved ones, you have to also honestly admit it is difficult for a single mathematician to give an idea that is not inadequate of the most important open questions that math is working on today.

In fact you have to recognize that a mathematician today knows only to a small degree his own subject and to a lesser degree the multiple relationships with other branches of the discipline.

This honest admission must not however take away our faith in communicating some aspects of the life and development of math that might enrich the mind of the listener nor must it make us lose hope of being enriched in our turn by dialogue with any interlocutor that has a sincere love for learning.

Question. *Perhaps for the popularization of math one must presuppose a certain degree of mathematical knowledge in the interlocutor, whereas certain facts of physics or biology, to give two common examples, can be given even if only roughly resorting to a language, concepts and interests that are much more widespread.*

De Giorgi: My impression is that the interlocutor must have if not a mathematical background, at least a certain interest and curiosity for mathematical problems and a certain faith in his own capacity to at least partly understand what meaning certain problems that occur might have even if he doesn't have the technical means of solving them.

To encourage this interest and faith it is no bad thing to recall that even for the most experienced mathematician it can be important and difficult to reach a clear, expressive formulation of a problem that will perhaps only be solved a long time after him.

You have to remember that in general an open question is much more interesting than a result already reached. (This is also true, I think, of the listener with wide mathematical knowledge.)

It's enough to think of Fermat's theorem or the problem of four colors: they are certainly much more popular than an ordinary arithmetical theorem.

Just like even the problem of squaring the circle: it was for a long time the object of curiosity and attention far beyond what it deserved. Perhaps even today a lot of people might be interested to know in what way such a problem could be considered solved.

Question. *That's true. Even the meaning of the word "solution" is somewhat abstract and is linked for example to the introduction of the concept of function.*

De Giorgi: Beginning with "popular" problems one can take the opportunity to show how with the passage of time the very idea of the solution of a mathematical problem has gained more ground: from research into an integer number that satisfied certain conditions to later research into a number expressible in fractions or roots, to that of a real or complex number, up to the most general concepts of today for which you first try to establish if one or more solutions to a problem exist and then you try and give information more or less qualitatively and quantitatively complete about the solutions themselves.

Furthermore this could be a good opportunity to refute the old prejudice that says math is a quantitative science: the results of modern math are for the most part qualitative results.

Question. *Naturally it's the field of the applications of math that concerns numerous disciplines. Some information can be given, even if incomplete.*

One can call attention to the fact that math does not only serve to quantify problems but often it is actually necessary to express and organize them in a theoretical context.

De Giorgi: It's important, talking of this, to try and re-examine the reasons for the successes and difficulties that are met with from time to time in these applications.

Particular attention must be paid to the idea of the mathematical model, an idea that is taking on an ever more important role in the science and technology of our time.

It's a question of a series of perspectives that is difficult to illustrate, one has to recognize, also because of the scarcity of information that an individual mathematician has, compared with the breadth of the themes involved.

Question. *Again, talking about communication, reference was made at the beginning to the link between the often very important choices, that society has to make, tied to the practical applications of science (for example the choices that concern energy resources) and the power of dialogue between scholars and public opinion.*

De Giorgi: It is actually these limits I am referring to now that must not be hidden but clarified rather .

This could be a good opportunity to give the public a more exact idea of what an "expert" is: it's a person that has accumulated good, actual experience in a certain field of study, but who is also fully aware he doesn't know all the aspects of the problems under consideration to the same degree. He is prepared to honestly expose his own certainties, doubts and problems confronting them objectively with those of his own interlocutors, in the hope that in the end everyone will be able to be enriched by the coming together of different experiences and knowledge.

Question. *If we now pass on from the popularization of math to the problem of the relationships of mathematicians with the world of science, with scholars of those disciplines to which math can be applied, cannot we perhaps say that, strange as it may seem, completely analogous problems are come up against?*

De Giorgi: In this field the experience of attempts at collaboration, more or less successful, shows that a necessary condition for success is a certain real interest on the part of each of the interlocutors in the problems of the other.

If for example, the mathematician doesn't have a sincere interest in the problems of the physicist and he for those of the mathematician, it will be difficult to find that understanding necessary for the collaboration to bear fruit.

Also in this case, you have to solve problems of intelligent information to succeed in bringing the interlocutor quickly enough to an at least qualitative understanding of the problems without going into long complicated technical details.

Question. *And among mathematicians? Even among them discussion doesn't seem to be brilliant.*

De Giorgi: You mustn't denigrate efforts to arrive at good communication among lovers of different branches of math: Algebra, Analysis, Theory of Probability etc.

Perhaps you have to invent some type of communication that

allows you to inform the entire mathematical community of the most interesting innovations registered in different sectors of research. For example, thinking only of Calculus of Variations, I think I can see, over the last few years an increase in attention given to generalized solutions to problems for which it was not easy to establish a priori the existence of classical solutions, along with the elaboration of ever more refined methods for the regularization of generalized solutions.

In particular a lot of research has been dedicated to cases where regularity is achieved everywhere except for a set having a small Hausdorff-dimension.

Besides interest has grown in the study of stationary points that are neither minimum nor maximum and research has greatly developed of topological operations performable in functional spaces (relaxation, Γ-limits, homogenization etc.).

Question. *Certainly it would be interesting to know the roots of so many ideas and so many modern techniques.*

Also the history of science, and in particular of math is a fascinating if difficult subject. It would be good above all to know better the history of certain ideas and concepts that later proved to be essential.

De Giorgi: Certainly, reflecting on the history of science can help us a lot to understand the meaning of many ideas.

On the other hand, attention to history should not go as far, in my opinion, as those that think that reflecting on the value and importance of every idea and discovery ends up in analysis, although certainly useful, of the historical context (cultural, political, economic background etc.) in which such an idea has grown.

Indeed you must remember that a good idea has within it a power that perhaps fully reveals itself only many centuries after its discovery and in answer to problems very different from those from which it sprang.

Question. *For example?*

De Giorgi: A typical example is given by the theory of conics studied by Apollonius in the 3rd century B.C. and then used by Kepler to express the laws of the motion of planets at the beginning of the 17th century.

Question. *Also the relationships between math and philosophy are fascinating. They have very ancient origins.*

De Giorgi: As we have already observed, the problem of the nature of math is already a philosophical problem of great importance (you need only think of Platonism, of nominalism etc.). Besides reflecting on math offers the philosopher innumerable opportunities for consideration.

For example it has always struck me that a great deal of applied math has its theoretical roots in sectors of pure math in which ideas of infinity and the infinitesimal dominate.

The re-emergence of some mathematical structures in the most diverse sectors of the natural sciences continues also to surprise me, almost a motif that comes up again in various parts of a symphony.

These are facts that remind us at the same time of Pythagoras' ideas on the harmony of celestial spheres, of the passage of Psalms *"the heavens tell of the glory of God"* and Einstein's saying: *"God is subtle but be is not malicious."*

The ultimate meaning of mathematical thought, according to me, lies in the last analysis in the idea of a subtle, complex harmony among all visible and invisible realities.

Translated from the Italian
by Robert Learmonth.

Contents

Volume I

Preface vii

A Conversation with Ennio de Giorgi ix

W. K. Allard — *An improvement of Cartan's test for ordinarity* 1

F. J. Almgren and M. E. Gurtin — *A mathematical contribution to Gibbs's analysis of fluid phases in equilibrium* 9

H. Attouch and H. Riahi — *The epi-continuation method for minimization problems. Relation with the degree theory of F.Browder for maximal monotone operators* 29

C. Baiocchi — *Discretization of evolution variational inequalities* 59

A. Bensoussan — *Homogenization for non linear elliptic equations with random highly oscillatory coefficients* 93

L. Boccardo — L^∞ *and* L^1 *variations on a theme of* Γ-*convergence* 135

H. Brezis and L. Peletier — *Asymptotics for elliptic equations involving critical growth* 149

G. Buttazzo, G. Dal Maso and U. Mosco — *Asymptotic behaviour for Dirichlet problems in domains bounded by thin layers* 193

S. Campanato — *Fundamental interior estimates for a class of second order elliptic operators* 251

L. Carbone and R. De Arcangelis — *Γ-Convergence of integral functionals defined on vector-valued functions* 261

M. Carriero, G. Dal Maso, A. Leaci and E. Pascali — *Limits of obstacle problems for the area functional* 285

L. Cattabriga — *Some remarks on the well-posedness of the Cauchy problem in Gevrey spaces* 311

A. Chiffi — *Approximating measures and rectifiable curves* 321

F. Colombini and S. Spagnolo — *A non-uniqueness result for the operators with principal part* $\partial_t^2 + a(t)\partial_x^2$ 331

U. D'Ambrosio — *A note on duality and the calculus of variations* 355

G. Da Prato — *Some results on periodic solutions of Hamilton-Jacobi equations in Hilbert spaces* 359

Z. Denkowska and Z. Denkowski — *Generalized solutions to ordinary differential equations with discontinuous right-hand sides via Γ-convergence* 371

M. J. Esteban and P. L. Lions — *Stationary solutions of nonlinear Schrödinger equations with an external magnetic field* 401

R. Finn and E. Giusti — *On the touching principle* 451

W. H. Fleming — *Generalized solutions and convex duality in optimal control* 461

M. Forti and F. Honsell — *Models of self-descriptive set theories* 473

Volume II

J. Frehse — *On a class of nonlinear diagonal elliptic systems with critical growth and C^α-regularity* . . . . — 519

N. Fusco and C. Sbordone — *Higher integrability from reverse Jensen inequalities with different supports* 541

M. Giaquinta, G. Modica and J. Souček — *Partial regularity of cartesian currents which minimize certain variational integrals* 563

G. H. Greco — *Théorème des minimax locaux et fonctions topologiquement fermées* 589

R. Hardt and D. Kinderlehrer — *Variational principles with linear growth* 633

H. Kacimi and F. Murat — *Estimation de l'erreur dans des problèmes de Dirichlet où apparaît un terme étrange* 661

H. Lewy — *On atypical variational problems* 697

J. L. Lions — *Sur la contrôlabilité exacte élargie* 703

M. Lobo-Hidalgo and E. Sanchez-Palencia — *Low and high frequency vibration in stiff problems* 729

E. Magenes — *A time-discretization scheme approximating the non-linear evolution equation $u_t + ABu = 0$* 743

P. Marcellini — *The stored-energy for some discontinuous deformations in nonlinear elasticity* 767

A. Marino — *The calculus of variations and some semilinear variational inequalities of elliptic and parabolic type* 787

S. Mizohata — *Some remarks on the dependence domain for weakly hyperbolic equations with constant multiplicity* 823

L. Modica — *Monotonicity of the energy for entire solutions of semilinear elliptic equations* 843

M. K. V. Murthy — *Pseudo-differential operators of Volterra type on spaces of ultra-distributions and parabolic mixed problems* 851

O. A. Oleinik, A. S. Shamaev and G. A. Yosifian — *The Neumann problem for second order elliptic equations with rapidly oscillating periodic coefficients in a perforated domain* 879

L. C. Piccinini — *Discrete exterior measures and their meaning in applications* 905

G. Talenti — *An embedding theorem* 919

L. Tartar — *Nonlocal effects induced by homogenization* 925

N. S. Trudinger — *On regularity and existence of viscosity solutions of nonlinear second order, elliptic equations* 939

J. Vaillant — *Etude d'un système en multiplicité 4, lorsque le degrée du polinôme minimal est petit* 959

C. Vinti — *On the Weierstrass integrals of the calculus of variations over BV varieties: recent results of the mathematical seminar in Perugia* 983

T. Zolezzi — *Variable structures control of semilinear evolution equations* 997

AN IMPROVEMENT OF CARTAN'S TEST
FOR ORDINARITY

WILLIAM K. ALLARD

Dedicated to Ennio De Giorgi on his sixtieth birthday

Introduction. Let Ω be an open subset of $\mathbf{R}^n$. For each non negative integer let $\mathcal{A}^\ell(\Omega)$ be the vector space of real analytic differential ℓ-forms on Ω and let $\mathcal{A}^*(\Omega) = \bigoplus_{\ell=0}^n \mathcal{A}^\ell(\Omega)$; $\mathcal{A}^*(\Omega)$ is an algebra with respect to exterior multiplication. Let $\mathcal{I}$ be an ideal in $\mathcal{A}^*(\Omega)$ such that

$$(1) \qquad \mathcal{I} = \bigoplus_{\ell=1}^n \mathcal{I}^\ell \quad where \quad \mathcal{I}^\ell = \mathcal{I} \cap \mathcal{A}^\ell(\Omega) \quad for \quad \ell \in \{0, ..., n\};$$

$$(2) \qquad d\omega \in \mathcal{I} \ whenever \ \omega \in \mathcal{I}; here \ d \ is \ exterior \ differentiation.$$

Consider the following problem: *Let $o \in \Omega$ and let $m \in \{1, ..., n\}$. Determine those open subsets U of $\mathbf{R}^m$ and real analytic embeddings $f: U \to \Omega$ such that $o \in U, f(0) = o$ and*

$$(3) \qquad\qquad f^*\varphi = 0 \ whenever \ \varphi \in \mathcal{I}.$$

As is well known and elementary, there corresponds to each system of real analytic partial differential equations Ω, m and $\mathcal{I}$ as above such that the local solutions of the system correspond to the embeddings f as above which satisfy a transversality condition. The Cartan-Kähler Theorem (see [C] or [BCG]) gives a construction of a family of embeddings f as above provided a certain natural condition is satisfied; this condition is that there exist an m-dimensional ordinary integral flag $(o; E_0, \ldots, E_m)$, the definition of which will be given below. E.Cartan has given a useful criterion for ordinarity. It is the purpose of this paper to give a similiar criterion for the ordinarity which is apparently weaker than Cartan's and which appears easier to verify.

Much of what is done here comes from [BCG].

Preliminaries. In order to proceed further we need to recall some of the basic definitions from exterior differential systems. Given a finite dimensional vector space X we let $G_*(X) = \{S : S \text{ is a linear subspace of } X\}$ and for each nonnegative integer ℓ we let $G_\ell(X) = \{S \in G_*(X) : \dim S = \ell\}$. For each $\ell \in \{0, ..., n\}$ we let

$$V_\ell = \{(p, X) \in \Omega \times G_\ell(\mathbf{R}^n) : \varphi(p)(v_1, \ldots, v_k) = 0 \text{ whenever}$$
$$k \in \{1, \ldots, \ell\}, \varphi \in \mathcal{I}^k \text{ and } \{v_1, \ldots, v_k\} \subset X\} \,.$$

Note that $V_0 = \Omega \times \{0\}$. We let $V_* = \cup_{\ell=0}^n V_\ell$. The members of V_ℓ are called $\ell - dimensional\ integral\ elements\ of\ \mathcal{I}$. Evidently, (3) is equivalent to

(4) $(f(u), \text{range}\ \partial f(u)) \in V_m$ for each $u \in U$

where $\partial f(u)$ is the differential of f at u.

For each $(p, X) \in V_*$ we let

$$P(p, X) = \{v \in \mathbf{R}^n : (p, \text{span}(X \cup \{v\})) \in V_*\}$$

and call $P(p, X)$ the *polar space of* (p, X); evidently, $X \subset P(p, X) \in G_*(\mathbf{R}^n)$. We define

$$c : V_* \to \{0, ..., n\}$$

at $(p, X) \in V_*$ by $c(p, X) = n - \dim P(p, X)$. We define

$$\tilde{c} : V_* \to \{0, \ldots, n\}$$

at $(p, X) \in V_*$ by

$$\tilde{c}(p, X) = \min\{\max\{c(q, Y) : (q, Y) \in G\} :$$
$$G \text{ is a neighborhood of } (p, X) \text{ in } V_*\}.$$

We say $(o; E_0, \ldots, E_\ell)$ is an ℓ-*dimensional integral flag* if $\ell \in \{0, \ldots, n\}$, $(o, E_k) \in V_k$ for $k \in \{0, \ldots, \ell\}$ and

$$E_0 \subset E_1 \subset \ldots \subset E_\ell;$$

we say it is *ordinary* if

$$c(o, E_k) = \tilde{c}(o, E_k) \text{ for } k < \ell.$$

We need to recall some facts about real analytic sets. So suppose A is a real analytic subvariety of the real analytic manifold B. We let reg A, the *set of regular points of* A, be the set of those $a \in A$ such that A meets some open neighborhood of a in B in a properly embedded real analytic submanifold; for $a \in \text{reg} A$ we let $\dim(A, a)$ be the dimension of the component containing a of any such submanifold. We let sing A, the *set of singular points of* A, equal $A \sim (\text{reg} A)$; for $a \in \text{sing} A$ we let

$$\dim(A, a) = \min\{\max\{\dim(A, x) : x \in U \cap \text{reg} A\} :$$
$$U \text{ is a neighborhood of a in } A\}.$$

We call $\dim(A, a)$ the *dimension of A at a*. It is well known that

(5) reg A *is a dense open subset of* A;

(6) dim (A, a) *depends uppersemicontinuously on* $a \in A$.

It is easily seen that, for each $\ell \in \{0, \ldots, n\}$,

(7) V_ℓ *is a real analytic subvariety of* $\Omega \times G_\ell(\mathbf{R}^n)$;

(8) $c(p, X)$ *depends lowersemicontinuously on* (p, X) *in* V_ℓ;

(9) $\{(p, X) \in V_\ell : c(p, X) \le t\}$ *is a real analytic subvariety of* $\Omega \times G_\ell(\mathbf{R}^n)$ *for each* $t \in \mathbf{R}^n$.

It is elementary but perhaps not so obvious that

(10) $(0, E_\ell) \in \mathrm{reg}\, V_\ell$ *whenever* $(o; E_0, E_1, \ldots, E_\ell)$ *is an ordinary* ℓ-*dimensional integral flag; see* [BGC].

The test for ordinarity. *Suppose* $(o; E_0, \ldots, E_m)$ *is an* m-*dimensional integral flag. Then*

(11) $\dim(V_m, (o, E_m)) \le n + m(n - m) - \displaystyle\sum_{\ell < m} c(o, E_\ell)$

with equality only if the flag is ordinary.

PROOF. Induct on m. The statement holds trivially for $m = o$ so suppose $m > 0$ and that the statement holds with m replaced by $m - 1$.
     We let

$$S = \{(p, W, X) : (p, X) \in V_m \text{ and } W \in G_{m-1}(X)\},$$
$$T = \{(p, W) \in V_{m-1} : c(p, W) \le n - m\},$$

and we let

$$\pi : S \to \Omega \times G_{m-1}(\mathbf{R}^n)$$

be such that $\pi(p, W, X) = (p, W)$ for $(p, W, X) \in S$. Keeping in mind (7) and (9) we infer that

(12) *S is a real analytic subvariety of* $\Omega \times G_{m-1}(\mathbf{R}^n) \times G_m(\mathbf{R}^n)$;

(13)   *T is a real analytic subvariety of* $\Omega \times G_{m-1}(\mathbf{R}^n)$;

It is evident that

(14) $\pi[S] = T$

(15) $\pi^{-1}[\{(p,W)\}]$ *is an* $[n-m-c(p,W)]$ *− dimensional real analytic submanifold of* $\Omega \times G_{m-1}(\mathbf{R}^n) \times G_m(\mathbf{R}^n)$ *for any* $(p,W) \in T.$

Let

$$G_1 = \{(p,W,X) \in S : \dim(V_{m-1},(p,W)) \\ \leq \dim(V_{m-1},(o,E_{m-1}))\},$$

$$G_2 = \{(p,W,X) \in S : c(o,E_{m-1}) \leq c(p,W)\}$$

and let

$$G_3 = \{(p,W,X) \in S : (p,W,X) \text{ is a regular point of } S \text{ and the rank of the differential of } \pi \text{ is constant near } (p,W,X)\}.$$

Keeping in mind (6) and (8) we see that G_1 and G_2 are open neighborhoods of (o, E_{m-1}, E_m) in S. Keeping in mind (5) one sees that G_3 is an open dense subset of S. Set $G = G_1 \cap G_2 \cap G_3$. Let

$$Q = \{(p,W,X) \in S : \dim(V_m,(p,X)) = \dim(V_m,(o,E_m))\}.$$

We claim that

(16) $Q \cap G$ *meets every neighborhood* H *of* (o, E_{m-1}, E_m) *in* S.

Indeed, $\{(p,X) : (p,W,X) \in H\}$ is a neighborhood of (o,E_m) in V_m so that, by (5), it has a subset I which is open in V_m such that $\dim(V_m,(p,X)) = \dim(V_m,(o,E_m))$ for $(p,X) \in I$. Thus

$$\{(p,W,X) \in H : (p,X) \in I\} \cap G_1 \cap G_2$$

is a nonempty open set which must intersect G_3 since G_3 is dense.

Suppose $(p,W,X) \in Q \cap G$. Since $(p,W,X) \in G_3$ we infer that

$$\dim(S,(p,W,X)) = \dim(V_m,(p,X)) + (m-1),$$
$$\dim(S,(p,W,X)) \leq [n-m-c(p,W)] + \dim(T,(p,W)).$$

Since $T \subset V_{m-1}$ we have

$$\dim(T,(p,W)) \leq \dim(V_{m-1},(p,W)).$$

Since $(p, W, X) \in G_1 \cap G_2 \cap Q$ we may use our inductive hypothesis to infer that

$$
\begin{aligned}
\dim(V_m, (o, E_m)) &= \\
&= \dim(V_m, (p, X)) \\
&= \dim(S, (p, W, X)) - (m - 1) \\
&\leq [n - m - c(p, W)] + \dim(T, (p, W)) - (m - 1) \\
&\leq [n - m - c(o, E_{m-1})] + \dim(V_{m-1}, (o, E_{m-1})) - (m - 1) \\
&\leq [n - m - c(o, E_{m-1})] + n + (m - 1)(n - (m - 1)) \\
&\qquad - \sum_{\ell < m-1} c(o, E_\ell) - (m - 1) \\
&= n + m(n - m) - \sum_{\ell < m} c(o, E_\ell)
\end{aligned}
$$

(17)

with equality if and only if

(18)
$$
\dim(V_{m-1}, (o, E_{m-1})) = n + (m - 1)(n - (m - 1) + \\
- \sum_{\ell < m-1} c(o, E_\ell);
$$

(19)
$$
c(p, W) = c(o, E_{m-1});
$$

(20)
$$
\dim(T, (p, W)) = \dim(V_{m-1}, (o, E_{m-1}));
$$

(21)
$$
\dim(S, (p, W, X)) = [n - m - c(o, E_{m-1})] + \\
+ \dim(V_{m-1}, (o, E_{m-1}))
$$

In particular, (11) holds since $G \cap Q \neq \emptyset$ by (16).

Suppose equality holds in (11). Then (18) holds and (19), (20), (21) hold for $(p, W, X) \in G \cap Q$. Since (18) holds our inductive hypothesis gives

(22)
$$
c(o, E_\ell) = \tilde{c}(o, E_\ell) \text{ for } \ell < m - 1.
$$

It follows from (22) and (10)

(23)
$$
(o, E_{m-1}) \in \operatorname{reg} V_{m-1}.
$$

Since (20) holds for (p, W) arbitrarily close to (o, E_{m-1}) in V_{m-1} by (16), it follows from (23) and the unique continuation property of real analytic functions that

(24) T *is a neighborhood of* (o, E_{m-1}) *in* V_{m-1}.

We may therefore choose an open neighborhood J of (o, E_{m-1}) in V_{m-1} such that

(25) $J \subset T$

and such that

(26) $K = \{(p, W) \in V_{m-1} : c(p, W) = \tilde{c}(o, E_{m-1})\}$ *is an open dense subset of* J.

Use (16) to choose $(\tilde{p}, \tilde{W}, \tilde{X}) \in G \cap Q \cap \pi^{-1}[J]$. Because $(\tilde{p}, \tilde{W}, \tilde{X}) \in G_3$ there is an open neighborhood N of $(\tilde{p}, \tilde{W}, \tilde{X})$ in $G \cap Q \cap \pi^{-1}[J]$ such that

(27) $\pi[N]$ *is a real analytic submanifold of* $\Omega \times G_{m-1}(\mathbf{R}^n)$;

(28) $\dim(S, (\tilde{p}, \tilde{W}, \tilde{X})) = [n - m - c(\tilde{p}, \tilde{W})] + \dim(\pi[N], (\tilde{p}, \tilde{W}))$.

Since (19) and (21) hold with (p, W, X) replaced by $(\tilde{p}, \tilde{W}, \tilde{X})$ we infer from (28) that

(29) $\dim(\pi[N], (\tilde{p}, \tilde{W})) = \dim(V_{m-1}, (o, E_{m-1}))$.

It follows from (27) and (29) that $\pi[N]$ is an open neighborhood of $(\tilde{p}, \tilde{W})$ in J. From (26) we infer that $\pi[N] \cap K \neq \emptyset$. Choose $(p, W, X) \in N$ such that $(p, W) \in K$. By (19) and the definition of K, $c(o, E_{m-1}) = c(p, W) = \tilde{c}(o, E_{m-1})$.

W.Allard

References

[BCG] R.L.Bryant, S.S.Chern, P.A.Griffiths, *Proceedings of the 1980 Beijing Symposium on Differential Geometry and Differential Equations,* Gordon and Breach, New York (1982).

[C] E.Cartan, *Les systèmes differentiels exterieurs et leur applications géométriques,* Hermann, Paris (1945).

Department of Mathematics

Duke University

DURHAM, NC 27707

A MATHEMATICAL CONTRIBUTION
TO GIBBS'S ANALYSIS
OF FLUID PHASES IN EQUILIBRIUM

FREDERICK J.ALMGREN   MORTON E.GURTIN

Dedicated to Ennio De Giorgi on his sixtieth birthday

1. Introduction. The pioneering work of Ennio De Giorgi and Herbert Federer in the 1950's on the structure of sets of finite perimeter laid a mathematical foundation for much of modern geometric measure theory. This contribution to continuum thermodynamics is formulated in this tradition. We here wish to examine possible interpretations of Gibbs's phase rules for equilibrium in fluid systems of several phases in which surface energies are a dominant component of the relevant free energy. We initially consider a simple model problem in some detail and then briefly discuss more elaborate situations.

2. The model problem. As a container we consider a closed surface Γ enclosing a region Ω in space. Inside this container is an idealized weightless fluid of prescribed volume whose possible positions correspond to subsets A of Ω. We assume that the relevant free energy of this system equals the surface area of the part of the surface of the fluid which is not in contact with Γ, i.e. the energy is the surface area of $\partial A \cap \Omega$ which hereafter denote a(A). An equilibrium configuration for such a system will occur whenever this free energy is a minimum among all possible nearby positions of the fluid within the container having the same volume. According to Gibbs, at such an equilibrium the pressure **p** within A plays the role of a Lagrange multiplier, viz. for an equilibrium A the expression

$$\text{area}(\partial A \cap \Omega) \text{ - } \mathbf{p}\,\text{volume}(A)$$

will be a stationary point among possible configurations, i.e.

$$(*) \qquad 0 = d\Big[\text{area}(\partial A \cap \Omega) - \mathbf{p}\,\text{volume}(A)\Big].$$

A main purpose of this note is to suggest mathematical interpretations of the expression ($*$) for our model problem.

The basic reason for subtlety interpreting ($*$) is that natural mathematical spaces corresponding to possible fluid positions are not known to admit tangent vector spaces; in particular, they do not seem representable as, say, Banach or Hilbert manifolds. One more or less "tangent vector space" approach, however, is discussed briefly in § 8.1 below in the context of smoothly deformed images of locally energy minimizing fluid positions. The equation ($*$) does hold in this context.

The primary new interpretation of ($*$) in this note can be regarded intuitively as the statement that ($*$) holds at almost all locally energy minimizing fluid position in comparison with nearby fluid positions of nearly the same volume *provided these nearby fluids have been permitted to adapt themselves to become locally energy minimizing.*

Since, for a given volume, there may be a number of different locally energy minimizing positions of the fluid, some with different energies, it seems useful (for fixed nearness parameter τ—defined below) to consider the collection **E** of ordered pairs (v, e) of positive numbers in which $0 < v < \text{volume}(\Omega)$ is a possible fluid volume and $0 < e = \text{a}(A) < \infty$ is a value of free energy assumed by at least one A which has volume v and which minimizes energy among τ nearby fluids of the same volume v. Even though there may be infinitely many distinct nearby minimizing A's in Ω having the same volume

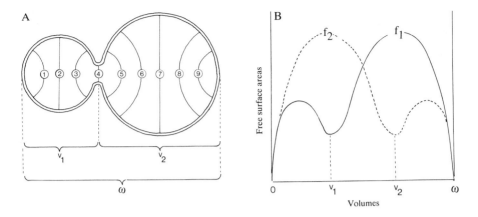

Figure 1. The function f_1 illustrated in Figure 1B is the presumed (volume, free energy) plot of various locally free energy minimizing fluid positions A in a container Ω illustrated in Figure 1A. The function $f_2(v)=f_1(\omega-v)$ is the corresponding plot when A is replaced by its complement in Ω. The infimum energy function $\epsilon_0=\inf\{f_1,f_2\}$ is not differentiable at one point between the volumes v_1 and v_2.

v and same energy e, we record only one point (v, e) in **E**. See Figure 1.

For a *fixed* value of our nearness parameter τ the number of distinct values of e for each v turns out to be a *priori* bounded as is the maximum value of such distinct e's. For volumes sufficiently small or sufficiently filling e is a (single valued) function of v with

$$(18\pi)^{1/3} = \lim_{v\to 0+} \frac{e(v)}{v^{2/3}}$$

(in case Γ is smooth). Additionally, it turns out that, for uniformly positive and nonfilling volumes, **E** is contained in the union of graphs of finitely many Lipschitz functions. We conclude that our variable e is "approximately differentiable" in an appropriate sense as a function of the variable v at almost every (v, e) in **E** in the sense of 1 dimensional measure in the plane.

Also, at almost every point (v, e) of such differentiability we show that the derivative $\frac{de}{dv}$ equals minus two times the mean curvature of the boundary of *any* A which is energy minimizing for that volume v and whose energy is e. The usual interpretation of the pressure $\mathbf{p}$ within A as minus two times the mean curvature applies so that

$$\frac{de}{dv} - \mathbf{p} = 0 \qquad \text{or} \qquad de - \mathbf{p}\,dv = 0.$$

We emphasize that the number of distinct values of e for fixed v may well depend on the size of our nearness parameter τ; in § 8.2 we show how to construct a region Ω with smooth boundary such that for many fixed v's the number of possible e's goes to infinity as τ goes to zero.

The absolute minimum energy function $e_0(A)$ is by definition the *infimum* of the numbers $\mathbf{a}(A)$ corresponding to all $A \subset \Omega$ having volume equal to v. It turns out that $e_0(v)$ is defined for all v and is locally Lipschitz; the same interpretation of pressure applies here to almost every v giving

$$\frac{de_0}{dv} - \mathbf{p} = 0.$$

Our model problem seems most naturally formulated and studied in the context of geometric measure theory which we do below. The main mathematical tools are the compactness theorems for spaces of solids having bounded surface areas and the uniform regularity theory for problems of this type.

It is a pleasure to acknowledge the benefit of discussions with J. Taylor in preparation of this note.

3. Terminology and assumptions for the mathematical model.

3.1. We denote by Ω a fixed bounded open region in 3 dimensional space $\mathbf{R}^3$ whose bounday Γ is a Hölder continuously differentiable 2 dimensional submanifold of $\mathbf{R}^3$.

3.2. We denote by $\mathcal{L}^3$ Lebesgue's 3 dimensional volume measure and by $\mathcal{H}^2$ Hausdorff's 2 dimensional surface area measure over $\mathbf{R}^3$. We typically abbreviate $|A| = \mathcal{L}^3(A)$ for $A \subset \mathbf{R}^3$ and also set $\omega = |\Omega|$. Also $\mathcal{H}^1$ denotes Hausdorff's 1 dimensional length measure over $\mathbf{R}^2$.

3.3. In order to specify the mathematical sets with boundaries which will correspond to possible fluid positions in Ω we basically follow the devices of [GWZ] for such purposes.[1] We recall that each $\mathcal{L}^3$ measurable set A in $\mathbf{R}^3$ differs at most in a set of measure 0 from the associated set A_* consisting of all the points of density of A in $\mathbf{R}^3$; in order to avoid ambiguities about sets of measure 0 we will consider only sets for which A equals A_*—if $A \subset \Omega$ then $A_* \subset \Omega$. Each $\mathcal{L}^3$ measurable subset A of $\mathbf{R}^3$ has a *measure theoretic boundary*, denoted ∂A, and A is said to have *finite perimeter* provided $\mathcal{H}^2(\partial A) < \infty$. If such A does have finite perimeter then the set of points p in $\mathbf{R}^3$ at which A has a measure theoretic unit *exterior normal vector* $\mathbf{n}(A, p)$ is denoted $\partial^* A$; in particular, $\partial^* A \subset \partial A$ and $\mathcal{H}^2(\partial A \sim \partial^* A) = 0$.

With these ideas in mind we denote by $\mathbf{A}$ the collection of all $\mathcal{L}^3$ measurable subsets A of Ω having finite perimeter for which $A = A_*$ and $0 < |A| < \omega$. The members of $\mathbf{A}$ correspond to (mathematically) possible fluids in Ω which have positive, nonfilling volumes. Also, for each $A \in \mathbf{A}$ we set

$$\mathbf{a}(A) = \mathcal{H}^2(\partial^* A \cap \Omega), \qquad \mathbf{b}(A) = \mathcal{H}^2(\partial^* A \cap \Gamma).$$

Whenever A and A' are $\mathcal{L}^3$ measurable subsets of $\mathbf{R}^3$ we set

$$\text{dist}(A, A') = |A \sim A'| + |A' \sim A|.$$

On $\mathbf{A}$ this distance function is a metric and $\mathbf{A}$ carries the metric topology.

3.4. We say that $A \in \mathbf{A}$ is τ *locally free energy minimizing* if and only if

$$\mathbf{a}(A) = \inf \Big\{ \mathbf{a}(A') : A' \in \mathbf{A} \text{ with } |A'| = |A| \text{ and } \text{dist}(A', A) < \tau \Big\}.$$

We fix τ and denote by $\mathbf{A}_0$ the collection of A's in $\mathbf{A}$ which are τ locally free energy minimizing and also set

$$\mathbf{E} = \Big\{ (|A|, \mathbf{a}(A)) : A \in \mathbf{A} \Big\}.$$

---

[1] This approach essentially is mathematically equivalent to working with Caccioppoli sets as used by De Giorgi and others in their studies of minimal surfaces; note [GE].

Finally, we set

$$e_0(v) = \inf\left\{e\colon (v, e) \in \mathbf{E}\right\}$$

for each $0 < v < \omega$ as our absolute minimum free energy function; we note that $e_0(\omega - v) = e_0(v)$ for each v.

3.5. Suppose $A \in \mathbf{A}$ is an open subset of Ω and $\partial A \cap \Omega$ is a smooth 2 dimensional submanifold of Ω. For each $p \in \partial A \cap \Omega$, $\mathbf{n}(A, p)$ is the usual unit exterior normal vector to A at p; we also let $H(A, p)$ denote the mean curvature vector of $\partial A \cap \Omega$ at p, which is normal to $\partial \Omega$ at p. If $\mathrm{mc}(A, p) = H(A, p) \bullet \mathbf{n}(A, p)$ (so that $H(A, p) = \mathrm{mc}(A, p)\,\mathbf{n}(A, p)$) then $\mathrm{mc}(A, p)$ equals one half of the algebraic sum of the two principal curvatures of $\partial A \cap \Omega$ at p in the direction $\mathbf{n}(A, p)$.

3.6. We denote by $\mathcal{X}(\Omega)$ the vector space of all smooth (initial velocity) vector fields $g\colon \Omega \to \mathbf{R}^3$ on Ω which are zero near Γ. For fixed $g \in \mathcal{X}(\Omega)$ and small times t we let G_t be the deformation of Ω given by setting $G_t(x) = x + t\,g(x)$ for $x \in \Omega$. Under the assumptions of § 3.5 two important formulas are the following
(3.6.1)

$$\frac{d}{dt}|G_t(A)|\bigg|_{t=0} = \int_{p\in\partial A\cap\Omega} g(p) \bullet \mathbf{n}(A, p)\,d\mathcal{H}^2 p = \int_A \mathrm{div}(g)\,d\mathcal{L}^3.$$

This formula gives the initial rate of change of volume of deformed sets $G_t(A)$.
(3.6.2)

$$\frac{d}{dt}\mathbf{a}\left(G_t(A)\right)\bigg|_{t=0} = -2\int_{p\in\partial A\cap\Omega} g(p) \bullet H(A, p)\,d\mathcal{H}^2 p$$

$$= -2\int_{p\in\partial A\cap\Omega} \mathrm{mc}(A, p)\,g(p) \bullet \mathbf{n}(A, p)\,d\mathcal{H}^2 p.$$

This formula gives the initial rate of change of area of deformed free boundaries $G_t(\partial A \cap \Omega)$; it can, alternatively, be used as a definition of mean curvature vectors from which the principal curvature expression in § 3.5 above can be computed.

4. Mathematical facts about the mathematical model and further terminology.

4.1. EXISTENCE OF ABSOLUTE ENERGY MINIMA. For each $0 < v < \omega$ there is at least one set $A \in \mathbf{A}$ with $|A| = v$ and $\mathbf{a}(A) = e_0(v)$. Furthermore, $\mathbf{E}$ is closed relative to $(0, \omega) \times \mathbf{R}$.

4.2. INTERIOR AND BOUNDARY REGULARITY OF τ LOCAL MINIMA. If $A \in \mathbf{A}_0$ then A is open and $\partial A \cap \Omega$ is a real analytic 2 dimensional submanifold of Ω with constant mean curvature in the sense that $\mathrm{mc}(A, p) = \mathrm{mc}(A, q)$ whenever $p, q \in \partial A \cap \Omega$; we hereafter abbreviate $\mathrm{mc}(A) = \mathrm{mc}(A, p)$ for such A and p. Also the boundary of $\partial A \cap \Omega$ is a Hölder continuously differentiable 1 dimensional submanifold C of Γ consisting of one or more simple closed curves. Additionally, $(\partial A \cap \Omega) \cup C$ is a compact Hölder continuously differentiable submanifold of $\mathbf{R}^3$ which is perpendicular to Γ along C.

4.3. ENERGY AND MULTIPLICITY BOUNDS FOR τ LOCAL MINIMA. There is a finite number L such that $0 < e \leq L$ for each $(v, e) \in \mathbf{E}$. As additional terminology we set for each $0 < \epsilon < \omega/2$,

$$\mathbf{A}_\epsilon = \mathbf{A}_0 \cap \{A: \epsilon \leq |A| \leq \omega - \epsilon\}.$$

Then, corresponding to each $0 \leq \epsilon < \omega/2$ and each $0 < \delta \leq \infty$, there is a positive integer M together with sets $A_1, \ldots, A_M \in \mathbf{A}_\epsilon$ such that

$$\mathbf{A}_\epsilon \subset \bigcup \{A: \mathrm{dist}(A, A_i) \leq \delta \text{ for some } i = 1, \ldots, M\}.$$

In particular, there is a positive integer N such that, for each v between 0 and ω, there are at most N distinct (v, e)'s in $\mathbf{E}$.

4.4. CONTINUITY OF MEAN CURVATURES. The mean curvature function mc is continuous on $\mathbf{A}_0$. Also, for each $0 < \epsilon < \omega/2$, the collection of sets $\mathbf{A}_\epsilon$ is compact; in particular, mc is bounded above and below.

4.5. RECTIFIABILITY OF τ LOCAL MINIMA VOLUME–ENERGY SPACE AND GIBBS'S RELATION. For each $0 < \epsilon < \omega/2$ there exist a nonnegative integer N and Lipschitz functions $e_0 \leq e_1 \leq \cdots \leq e_N: [\epsilon, \omega - \epsilon] \to \mathbf{R}$ such that

$$\mathbf{E} \cap [\epsilon, \omega - \epsilon] \times \mathbf{R} \subset \mathrm{graph}(e_0) \cup \ldots \cup \mathrm{graph}(e_N).$$

Additionally, for $\mathcal{H}^1$ almost every pair $(v_*, e_*) \in \mathbf{E}$, the following two properties hold:

4.5.1.   The number v_* is a point of density of the set

$$\left\{ v \colon (v, e) \in \mathbf{E} \cap U \right\}$$

for every neighborhood U of (v_*, e_*) in $\mathbf{R}^2$;

4.5.2.   For each $A \in \mathbf{A}_0$ with $|A| = v_*$ and $\mathbf{a}(A) = e_*$,

$$0 = \lim_{\mathbf{E} \ni (v, e) \to (v_*, e_*)} \frac{|(e - e_*) - (-2)\mathbf{mc}(A)(v - v_*)|}{|v - v_*|}.$$

With the derivative $\frac{de}{dv}(v_*, e_*)$ understood in the sense of § 4.5.2 the usual interpretation of pressure $\mathbf{p}$ as $-2\mathbf{mc}(A)$ is applicable so that for $\mathcal{H}^1$ almost every (v_*, e_*) in $\mathbf{E}$,

$$\frac{de}{dv} - \mathbf{p} = 0 \qquad or, \ equivalently \qquad de - \mathbf{p}\,dv = 0.$$

4.6. STRUCTURE OF e_0.   For the absolute minimum function e_0 a bit more can be said. For each v between 0 and ω we set

$$\mathbf{h}(v, +) = -2 \sup \left\{ \mathbf{mc}(A) \colon A \in \mathbf{A}_0 \text{ with } |A| = v \text{ and } \mathbf{a}(A) = e_0(v) \right\},$$

$$\mathbf{h}(v, -) = -2 \inf \left\{ \mathbf{mc}(A) \colon A \in \mathbf{A}_0 \text{ with } |A| = v \text{ and } \mathbf{a}(A) = e_0(v) \right\}.$$

Then the following five properties hold:

4.6.1.   The function e_0 is defined for each v between 0 and ω;

4.6.2.   Near 0 and ω our relation e is single valued in the sense that $e = e_0(v)$ whenever $(v, e) \in \mathbf{E}$ with $0 < v < \tau/2$ or $\omega - \tau/2 < v < \omega$;

4.6.3.   At 0 and ω our function $e_0(v)$ has the limiting behavior

$$(18\pi)^{\frac{1}{3}} = \lim_{v \to 0+} \frac{e_0(v)}{v^{\frac{2}{3}}} = \lim_{v \to \omega-} \frac{e_0(v)}{(\omega - v)^{\frac{2}{3}}};$$

4.6.4. The function e_0 is differentiable at almost every point v_* between 0 and ω with

$$\frac{de_0}{dv}(v_*) = \mathbf{h}(v_*, +) = \mathbf{h}(v_*, -) = -2\,\mathrm{mc}(A)$$

for each $A \in \mathbf{A}_0$ with $|A| = v_*$ and $\mathbf{a}(A) = e_0(v_*)$;

4.6.5. For *each* v between 0 and ω,

$$\mathbf{h}(v, +) = \lim_{t \to 0+} \frac{e_0(v+t) - e_0(v)}{t},$$

$$\mathbf{h}(v, -) = \lim_{t \to 0-} \frac{e_0(v+t) - e_0(v)}{t}.$$

The example illustrated in Figure 1 shows why these two limits can be different.

These mathematical facts are justified in § 5 below. More elaborate fluid free energy problems are discussed in § 6 and § 7. In § 8 we give several remarks and examples.

5. Justification from the mathematical facts. The mathematical facts of the previous section are largely a consequence of the analysis of general elliptic partitioning problems which appears in Chapter VI of [AF1] together with the a priori uniformity of the regularity estimates obtained in earlier chapters. The main mathematical contribution of the present paper is the use of such estimates to show that the local minimum free energies are essentially related by Lipschitz continuity conditions as a function of volumes with derivatives expressible in terms of mean curvatures of minimizing boundaries. The geometric measure theory used in the general proofs is quite heavy going for nonexperts (and for many experts as well—[AF1] is 207 pages long while [FH1] is 690 pages).

5.1. The general fact used in proving 4.1 is the Compactness Theorem for Integral Currents of [FH1 4.2.17(2)] or, alternatively, of [AF2 4.3] which we apply in the present context. The metric associated with suitable $\mathcal{F}_K$ norms in [FH1] coincides with the metric of § 3.3. The following is true.

5.1.1.   For each $0 < \epsilon < \omega/2$ and each $M < \infty$,

$$\mathbf{A} \cap \left\{ A : \epsilon \leq |A| < \omega - \epsilon \text{ and } \mathcal{H}^2(\partial A) \leq M \right\}$$

is compact in the topology of § 3.3 above.

5.1.2.   If $A, A_1, A_2, A_3, \ldots, \in \mathbf{A}$ with $0 = \lim_{i \to \infty} \text{dist}(A, A_i)$, then

$$|A| = \lim_{i \to \infty} |A_i|,$$

$$\mathcal{H}^2(\partial A) \leq \lim_{i \to \infty} \inf \mathcal{H}^2(\partial A_i),$$

$$\mathbf{a}(A) \leq \lim_{i \to \infty} \inf \mathbf{a}(A_i),$$

and, for each $0 \leq \sigma \leq 1$,

$$\mathbf{a}(A) + \sigma \mathbf{b}(A) \leq \lim_{i \to \infty} \inf \left[\mathbf{a}(A_i) + \sigma \mathbf{b}(A_i) \right].$$

Specific interpretation and verification here is left to the reader who may wish to consult sections 4.1.7, 4.1.24, 4.5.1–4.5.6 of the treatise [FH1].

5.2.   Almost everywhere regularity of $\partial_* A \cap \Omega$ for A's in $\mathbf{A}_0$ follows from [A1 VI.2] with only slight adaptation to the present context; this is because the area integrand is elliptic and the essential arguments there are entirely local. As also discussed in [AF4] one uses the fact that volume and area scale differently under homothetic dilations to check that whenever $A \in \mathbf{A}_0$ and $p \in \partial A \cap \Omega$ then each oriented tangent cone [FH1 4.3.16] to $\partial^* A$ at p must be area minimizing. It then follows from [FH2] (since we are in low dimensions) that each such tangent cone must be a tangent plane. The everywhere regularity asserted in § 4.2 then is a consequence of [AF1 II.3(11), III.2(6)(7), IV.13(5)] or one can use [AW 8.19] to replace [AF1 IV.13(5)]. The boundary regularity asserted in § 4.2 is proved in [TJ].

5.3.   With regard to the first assertion of § 4.3 we check the following. Suppose $U(r)$ is an open ball in $\mathbf{R}^3$ of radius r, $A \in \mathbf{A}$, and $t \in \mathbf{R}$ is chosen so that, with

$$V_t = U(r) \cap \left\{ (x, y, z) : z < t \right\},$$

one has $\mathcal{L}^3\Big(U(r) \cap \Omega \cap V_t\Big) = \mathcal{L}^3\Big(U(r) \cap A\Big)$. Suppose also $A' = \Big[A \sim U(r)\Big] \cup V_t$. Then $|A'| = |A|$,

$$\mathcal{H}^2\Big(\partial A' \cap \Omega\Big) \le \mathcal{H}^2\Big(\partial A \cap [\Omega \sim U(r)]\Big) + 5\pi r^2,$$

and for sufficiently small r's dist$(A, A') < \tau$. Such an A' is a comparison fluid position for the energy minimization; the area bounds of § 4.3 follow readily since Ω is bounded.

The remaining assertions of § 4.3 follow readily from the compactness estimates of § 5.1 and this area bound.

5.4. For $0 < t \le 1$ we set

$$C_t = \Big\{(x, y, z) \colon x^2 + y^2 < t,\, z^2 < t\Big\}.$$

The following propositions are the basic facts used in establishing 4.3.

Proposition 5.4.1. *For each $\epsilon > 0$ there is $\delta > 0$ with the following property. Suppose $A \in \mathbf{A}_0$ and $\varphi \colon \mathbf{R}^3 \to \mathbf{R}^3$ is a similarity transformation (the composition of an isometry with a homothety) which does not decrease distances such that $\mathcal{L}^3\left[\varphi^{-1}(C_1)\right] < \tau$, $C_1 \subset \varphi(\Omega)$, and*

$$\mathrm{dist}\Big(\varphi(A) \cap C_1,\, C_1 \cap \{(x, y, z) \colon z < 0\}\Big) < \delta.$$

Then there is a real analytic function

$$f \colon \big\{(x, y) \colon x^2 + y^2 < 1 - \epsilon\big\} \to \big\{z \colon z^2 < \epsilon\big\}$$

such that $|Df(x, y)| < \epsilon$, $|D^2 f(x, y)| < \epsilon$, and $|D^3 f(x, y)| < \epsilon$ for each (x, y) and also

$$\varphi(\partial A \cap \Omega) \cap C_{1-\epsilon} = \Big\{(x, y, z) \colon x^2 + y^2 < 1 - \epsilon \text{ and } z = f(x, y)\Big\}.$$

The proof of this proposition is basically a straightforward application of the regularity theory referenced in § 5.2 above to the φ transformation of our energy minimization.

Clearly, if $p \in \partial A' \cap \Omega$ for some $A' \in \mathbf{A}_0$ then one can choose a similarity transformation φ which does not decrease distances so that $\varphi(p) = (0, 0, 0)$, $\mathcal{L}^3 \left[\varphi^{-1}(C_1) \right] < \tau$, $C_1 \subset \varphi(\Omega)$, and

$$\mathrm{dist}\Big(\varphi(A') \cap C_1 \,,\, C_1 \cap \{(x,\, y,\, z) : z < 0\} \Big) << \delta,$$

and one obtains uniform regularity estimates for A's in a neighborhood of A'. This fact together with the compactness discussed in § 5.1 readily gives

Proposition 5.4.2. *Corresponding to each $0 < \epsilon < \omega/2$ and each $\delta > 0$ there is $M < \infty$ with the following property. Whenever $A \in \mathbf{A}_\epsilon$ there will exist a similarity transformation $\varphi \colon \mathbf{R}^3 \to \mathbf{R}^3$ which does not decrease distances and which increases distances by a factor no more than M such that $\mathcal{L}^3 \left[\varphi^{-1}(C_1) \right] < \tau$, $C_1 \subset \varphi(\Omega)$, and*

$$\mathrm{dist}\Big(\varphi(A) \cap C_1 \,,\, C_1 \cap \{(x,\, y,\, z) : z < 0\} \Big) < \delta.$$

The assertions of § 4.4 readily follow.

We note that § 4.4 implies that the supremum and the infimum in the definitions of § 4.6 are both attained.

5.5.   We assert the following.

Proposition 5.5.1. *Corresponding to each $0 < \epsilon < \omega/2$ there are numbers $0 < \delta < \epsilon$, $1 < \lambda < \infty$, and $1 < \mu < \infty$ such that whenever $A \in \mathbf{A}_\epsilon$ and $-\delta \leq \Delta v \leq \delta$ then A can be deformed to $A' \in \mathbf{A}$ such that $|A'| - |A| = \Delta v$, $\mathrm{dist}(A',\, A) < \tau/3$,*

$$\mathbf{a}(A') - \mathbf{a}(A) = -2\mathbf{mc}(A)\,\Delta v \pm \mu(\Delta v)^2,$$

$$|\mathbf{a}(A') - \mathbf{a}(A)| \leq \lambda\,\Delta v.$$

The proof of Proposition 5.5.1 is based on the uniform representations of Proposition 5.4.2 (e.g. the dilations of our similarity transformations are bounded, in particular) together with formulas 3.6.1 and 3.6.2 and the estimates of the following propositions.

Proposition 5.5.2. *Suppose*

$$f, g: \left\{(x, y): x^2 + y^2 \leq 1/4\right\} \to \left\{z: z^2 < 1/4\right\}$$

are smooth functions, and, for $-1/2 < t < 1/2$,

$$\text{volume}(t) \equiv \mathcal{L}^3 \left\{(x, y, z): -1 < f(x, y) + t\,g(x, y)\right\}$$
$$= \int \left[1 + f(x, y) + t\,g(x, y)\right] dx\,dy,$$

$$\text{area}(t) \equiv \mathcal{H}^2 \left\{(x, y, z): z = f(x, y) + t\,g(x, y)\right\}$$
$$= \int \left[1 + |\nabla(f + t\,g)|^2\right]^{\frac{1}{2}} dx\,dy.$$

Then, for each t,

$$\frac{d^2}{dt^2}\text{volume}(t) = 0$$

while the second area derivative

$$\frac{d^2}{dt^2}\text{area}(t)$$

equals

$$\int \left[1 + |\nabla(f + t\,g)|^2\right]^{\frac{1}{2}} |\nabla g|^2 \, dx\,dy -$$
$$- \int \left[1 + |\nabla(f + t\,g)|^2\right]^{-\frac{3}{2}} \left[\nabla f \bullet \nabla g + t\,|\nabla g|^2\right] dx\,dy.$$

In applying this proposition we take g either nonnegative or nonpositive with $g(x, y) = 0$ for $x^2 + y^2 = 1/4$ and transform our deformations back within Ω by the inverse similarities.

For given ϵ we chose δ is accordance with § 5.5.1 and use § 4.3 to choose a positive integer N and $A_1, \ldots, A_n \in \mathbf{A}_\epsilon$ such that

$$\mathbf{A}_\epsilon \subset \mathbf{A}_0 \cap \left\{A: \text{dist}(A, A_i) \leq \inf\{\delta, \tau/3\} \text{ for some } i = 1, \ldots, N\right\}.$$

For each i we then set

$$E_i = \mathbf{E} \cap \left\{ (|A|, \mathbf{a}(A)) : A \in \mathbf{A}_\epsilon \text{ with } \operatorname{dist}(A, A_i) \le \inf\{\delta, \tau/3\} \right\}$$

and conclude from § 5.5.1 and Kirszbraun's theorem that E_i is contained within the graph of a function $[\epsilon, \omega - \epsilon] \to \mathbf{R}$ with Lipschitz constant not exceeding λ. The assertions of § 4.5 now follow readily from § 5.5.1 and standard facts in real function theory.

We consider the assertions of § 4.6.

5.6.1.   Assertion § 4.6.1 is basically a consequence of § 5.5.1.

5.6.2.   Assertion § 4.6.2 follows from the observations that if $0 < v < \tau/2$ or $\omega - \tau/2 < v < \omega$ and $A, A' \in \mathbf{A}$ with $v = |A| = |A'|$ then $\operatorname{dist}(A, A') < \tau$.

5.6.3.   The basic fact use in establishing § 4.6.3 is that (with the obvious meanings) any absolutely energy minimizing region in $\{(x, y, z) : z > 0\}$ having volume $2\pi/3$ must be of the form $\{(x, y, z) : (x - x_0)^2 + (y - y_0)^2 + z^2 < 1, z > 0\}$ for some x_0, y_0; doubling such a region by reflection in the $z = 0$ plane reduces the assertion to well known area-volume extremal properties of spheres (compare [AF3][AF4]). One applies the methods of Chapter VI of [AF1] together with the compactness noted in § 5.1.2 to a sequence of energy minimizing A's (with actual volumes converging to 0) which have been suitably translated, rotated, and dilated to have volume $2\pi/3$ to conclude convergence to $\{(x, y, z) : x^2 + y^2 + z^2 < 1, z > 0\}$. Further details are left to the reader.

5.6.4.   Assertion § 4.6.4 is basically a consequence of § 5.5.1 and Rademacher's theorem.

5.6.5.   We verify assertion § 4.6.5 by contradiction. Suppose, for example, $0 < v < \omega$, $0 < \epsilon < 1$, and

$$\liminf_{t \to 0+} \frac{e_0(v + t) - e_0(v)}{t} < \mathbf{h}(v, +) - \epsilon.$$

There will then exist $t_1 > t_2 > t_3 > \ldots > 0$ with $\lim_{i \to \infty} t_i = 0$ so that, for each i,

$$e_0(v + t_i) - e_0(v) = \int_0^{t_i} \left(\frac{de_0}{dv} \right) ds < t_i[\mathbf{h}(v, +) - \epsilon],$$

and we infer the existence of $v < v_i < v + t_i$ such that e_0 is differentiable at v_i with $\frac{de_0}{dv}(v_i) < h(v, +) - \epsilon$. We use § 4.1, § 4.3 to infer the existence of $A_i \in \mathbf{A}_0$ with $|A_i| = v_i$ such that $\frac{de_0}{dv}(v_i) = -2\mathbf{mc}(A_i) < h(v, +) - \epsilon$ and finally the existence of $A \in \mathbf{A}_0$ such that A is the limit of a subsequence of the A_i's and $-2\mathbf{mc}(A) \leq h(v, +) - \epsilon$, which is not in accordance with our definitions. The first equality of § 4.6.5 follows. A similar argument gives the second.

6. A problem with boundary interactions and gravity. Suppose we retain the terminology of § 3.1, § 3.2, § 3.3, § 3.5 but assume the free energy of our fluid represented by A in Ω equals

$$\mathbf{a}(A) + \sigma \, \mathbf{b}(A) + \gamma \int_A z \, d\mathcal{L}^3$$

for some fixed $0 \leq \sigma \leq 1$ and some fixed $0 \leq \gamma < \infty$. In particular, there can be energy associated with the interface between the fluid and the container Γ and there can be a gravitational potential energy represented as a volume integral of the height coordinate z. The nondegeneracy assumption that $\sigma \leq 1$ lets us continue to apply § 5.1 and *the analysis of our model problem applies to this case with only the obvious changes.* The main difference is that, if A is τ locally energy minimizing, then it is not mean curvature that is constant along $\partial A \cap \Omega$ but rather the sum of minus two times the mean curvature plus γ times the height so that one obtains formulas of the form

$$\frac{de}{dv} = -2\,\mathbf{mc}(A, (x, y, z)) + \gamma z,$$

etc. for τ locally minimizing A and $(x, y, z) \in \partial A \cap \Omega$ replacing those of § 4.5.2.

It is worth noting that smoothness of the container Γ both here and earlier is used in the analysis only in showing boundary regularity of the free surfaces; here, however, we do require that Γ have finite area.

7. Problems with several immiscible fluids of varying densities with distinct nondegenerate interface energies and gravitational and other bulk energies. The mathematical setup of [AF1 VI] is quite general and applies in a straightforward way to configurations of N immiscible fluids which fill the container Γ and which locally minimize general weighted surface integrals (which need not be isotropic) plus weighted volume integrals. The bulk of the analysis of our model problem generalizes to such situations— the main difficulty is the complexity of organizing what needs to be said. In particular, the single volume variable v is replaced by an N-tuple $(v_1, \ldots, v_N)$ of volumes with $v_1 + \ldots + v_N = \omega$, and the single derviatives $\frac{de}{dv}$ are replaced by appropriate partial derivatives. The general existence, compactness, and regularity estimates on which the model problem analysis is based hold in appropriately adapted form. There are, however, no corresponding boundary regularity estimates.

We note further that, in order to apply the general theory of [AF1] the fluids need not be constrained by a container nor (mathematically) do they need to be restricted to three space dimensions.

8. Remarks and examples. We retain the terminology of § 3 and § 4.

8.1. One setup in which one might be tempted to interpret Gibbs's equation (∗) in § 2 would be to regard the "tangent space" to $A \in \mathbf{A}_0$ as all sets of the from $G_t(A)$ for small times t corresponding to various deformations G_t associated with initial velocity vector fields $g \in \mathcal{X}(\Omega)$ as in § 3.6. The obvious interpretation of the equation (∗) does indeed hold in this context. There has been an extensive mathematical study of surfaces having controlled first variations of area (i.e. having reasonable first variation distributions or generalized mean curvatures); see [AW] or Appendix C of [AF3] in particular. Smooth first variation techniques alone do not give as much regularity information as the more general Lipschitz deformations of [AF1]. In particular, with $\Omega = \{(x, y, z): x^2 + y^2 + z^2 < 1\}$, the set $A = \{(x, y, z): xy > 0\}$ has minimizing free boundary area as far as nearby smooth deformation images are concerned. The interior interface $\partial A \cap \Omega$ however contains a singular line; see Figure 2.

Second variation techniques are somewhat more subtle and can, for example, detect the nonminimality of this example. These techniques, however, are not well understood, especially in connection with several immiscible fluids as in § 7. *Caution: in general, lo-*

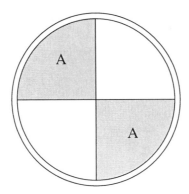

Figure 2. The shaded region A denotes a fluid within container Ω in the shape of a disk. The interior boundary $\partial A \cap \Omega$ has a singularity of codimension one at the origin. This interior boundary length (area) of A is stationary with respect to smooth variations within Ω which preserve the area (volume) of A so long as $\partial \Omega$ is carried into itself. There are however smooth area preserving deformations of Ω with respect to which this boundary length is unstable. Such a singularity also cannot occur if the fluid configuration is required to minimize boundary length locally in competition with small Lipschitz deformations of A having the same volume.

cal minima have nonnegative second variations of energy only with respect to volume preserving deformations.

8.2. For certain regions Ω and certain volumes v the number of distinct (v, e)'s in $\mathbf{E}$ can depend strongly on the size of our nearness parameter τ used in defining "local" minimality. To illustrate such a case we let $r(z)$ be a smooth nondecreasing function such that $r(z) = 1$ for $-\infty < z \leq 0$, $r(z) = 2$ for $1 \leq z < \infty$, $r\|\left(2^{-1}\right) > r\|\left(2^{-2}\right) > r\|\left(2^{-3}\right) > \ldots$, and $r(z)$ is constant in an interval about 2^{-k} for each $k = 1, 2, 3, \ldots$. With $8\pi \leq v < 12\pi$ fixed and $z_1, z_2, z_3, \ldots$ defined by requiring

$$v = \int_{2^{-k}}^{z_k} \pi\, r(z)^2\, dz$$

for each k, we choose Ω so that

$$\Omega \cap \left\{(x, y, z)\colon -1 < z < 2\right\} \subset \left\{(x, y, z)\colon x^2 + y^2 = r(z)^2\right\}$$

(see Figure 3), and set $A_k = \Omega \cap \left\{(x, y, z)\colon 2^{-k} < z < z_k\right\}$ for each k. Then $v = |A_k|$ and $\mathbf{a}(A_k) = \pi\left(4 + r\|\left(2^{-k}\right)^2\right)$ for each k so that

$\mathbf{a}(A_1) > \mathbf{a}(A_2) > \mathbf{a}(A_3) > \ldots$. It is not difficult to confirm the existence of $\tau_1 > \tau_2 > \tau_3 \ldots$ with $\lim_{k \to \infty} \tau_k = 0$ so that $A_1, A_2, \ldots, A_N$ are each τ_N locally free energy minimizing for each $N = 1, 2, 3, \ldots$

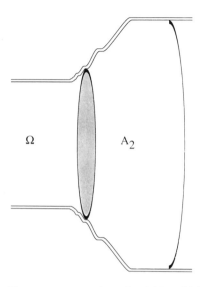

Figure 3.   Figure 3 illustrates a container Ω within which the possible distinct values of free energy associated with locally free energy minimizing fluid configurations of fixed volume becomes arbitrarily large as admissible comparison configurations become more tightly constrained. The fluids all lie between pairs of coaxial disks.

     8.3.   One might be tempted in defining τ nearness for sets A, A' to require in addition to the condition that $\text{dist}(A, A') < \tau$ that also $\text{diam}\,[(A \sim A') \cup (A' \sim A)] < \tau$. Although such a definition would be adequate for the applicability of the regularity theory of [AF1], it is not associated with a metric and most of the results of our initial problem analysis would no longer hold. In particular, mean curvature vectors of boundaries of minimizing A's would necessarily be of constant length only on connected boundary components, and typically there could be infinitely many distinct (v, e)'s for a given volume v in the corresponding $\mathbf{E}$. Various of the problems of this alternative notion of nearness are illustrated by the fluid of Figure 4 which is minimizing in this new sense.

     8.4.   It is perhaps worth pointing out that, in the terminology of § 3.4 and § 4.2, there generally will not exist and $A \in \mathbf{A}_0$ such that $\mathbf{a}(A) = e_0(|A|)$ and $\mathbf{mc}(A) = 0$. A somewhat more elaborate

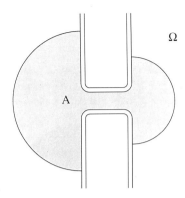

Figure 4. If comparison fluid configurations must arise from deformations having small diameters, then only connected components of the interior boundary of locally minimizing configurations need have the same mean curvatures, as illustrated by the "minimizing configuration" in Figure 4. Many of the properties listed in §4 would then not necessarily hold.

variational calculus in the large has been developed, however, by F. Almgren, E. Cook, and J. Pitts which does produce such minimal surfaces.

References

[AW] W.Allard, *On the first variation of a varifold*, Ann. of Math. **95** (1972), 417-491.

[AF1] F.Almgren, *Existence and regularity almost everywhere of solutions to elliptic variational problems with constraints*, Mem. Amer. Math. Soc. No. 165 (1976).

[AF2] F.Almgren, *Deformations and multiple valued functions*, Geometric Measure Theory and the Calculus of Variations, Proceedings of Symposia in Pure Mathematics 44 (1986), 29-130.

[AF3] F.Almgren, *Optimal isoperimetric inequalities*, Indiana Univ. Math. J. **13** (1986), 451-547.

[AF4] F.Almgren, *Spherical symmetrization*, Integral Functionals in the Calculus of Variations, Editrice Tecnico Scientifica, Pisa (to appear).

[FH1] H.Federer, *Geometric Measure Theory*, Springer-Verlag, New York (1969).

[FH2] H.Federer, *The singular sets of area minimizing rectifiable currents with codimension one and of area minimizing flat chains modulo two with arbitrary codimension*, Bull. Amer. Math. Soc. **76** (1970), 767-771.

[GE] E.Giusti, *Minimal Surfaces and Functions of Bounded Variation*, Monographs in Mathematics **80**, Birkhäuser, Boston (1984).

[GWZ] M.Gurtin, W.Williams, W.Ziemer, *Geometric measure theory and the axioms of continuum thermodynamics*, Arch. Rational Mech. Anal. **92** (1986), 1-22.

[TJ] J.Taylor, *Boundary regularity for solutions of various capillary and free boundary problems*, Comm. Partial Differential Equations **2** (1977), 323-358.

Princeton University
Departement of Mathematics
Fine Hall, Washington Road
PRINCETON, NJ 08544

Carnegie - Mellon University
Departement of Mathematics
PITTSBURGH, PA 15213

THE EPI-CONTINUATION METHOD
FOR MINIMIZATION PROBLEMS.
RELATION WITH THE DEGREE THEORY OF F. BROWDER
FOR MAXIMAL MONOTONE OPERATORS

HEDY ATTOUCH   HASSAN RIAHI

Dedicated to Ennio De Giorgi on his sixtieth birthday

Introduction. The continuation method, initiated by H.Poincaré, then systematically developped in the context of the degree theory by Kronecker and Brouwer, Leray and Schauder ... consists of imbedding the problem in a parametrized family of problems and consider its solvability as the parameter varies. The homotopy invariance is a decisive property of the topological degree of mappings.

In this paper we make a break with the classical approach in at least two ways.

We restrict ourself to variational problem of a given kind, namely minimization problems or saddle value problems... This allows us to work directly on the variational formulation of the problem

$$\begin{cases} F(u,\lambda) \le F(v,\lambda) & \forall v \in X \\ u \in X \end{cases}$$

where λ is a parameter, and formulate the (homotopy) continuity property directly on the mapping

$$\lambda \longrightarrow F_\lambda$$

where

$$F_\lambda : v \longrightarrow F(v, \lambda).$$

A major advantage of this approach is that the corresponding Euler equation

$$DF(u, \lambda) \ni 0$$

may be very difficult to handle with (think to nonconvex problem, or when X is non reflexive...).

The second feature of our approach which differs from the classical one is that the (homotopy) continuity property is taken in terms of epi-continuity. Indeed it follows from the works of De Giorgi [D.G] , Attouch and Wets [A-W]... that it is a natural concept when considering stability properties of minimization problems. This allows us to consider a large class of functions F (with possibly $+\infty$ values) and of admissible deformations.

So doing, when restricting our attention to the class of functions with connected level sets, we obtain that solvability of such problems is invariant under epi-continuous deformation (cf. theorem 1.6).

The link with the degree theory of F.Browder is obtained by proving a similar kind of result for non linear problems governed by maximal monotone operators (theorem 2.7)

$$A(u, \lambda) \ni 0.$$

The homotopy continuity property is taken in the sense of graph-convergence i.e.

$$\lambda_n \longrightarrow \lambda \quad \Rightarrow A_{\lambda_n} \longrightarrow \quad A_\lambda \qquad \text{in graph sense} \qquad (\text{in } X \times X^*)$$

where

$$A_\lambda v := A(v, \lambda).$$

When A_λ is the subdifferential of a convex lower semicontinuous function φ_λ this notion of continuity turns to be equivalent (cf. H.Attouch [A_1], [A_2]) to the epi-continuity of the map $\lambda \longrightarrow \varphi_\lambda$ which is an other justification of the above approach.

1. **Epi-continuous deformation for minimization problems.** All along this chapter we use the following notations: we consider mappings from a given topological space (X, τ) into $\overline{\mathbf{R}}$ which we denote by:

$$F : X \to \overline{\mathbf{R}}$$

$$x \to F(x).$$

The set of parameters is $\Lambda = [0, 1]$, we write $\lambda \in \Lambda$. Given a parametrized family $\{F_\lambda; \lambda \in \Lambda\}$ of mapping from X into $\overline{\mathbf{R}}$, for every $\lambda \in \Lambda$ we denote by

$$S_\lambda = \{x \in X; F_\lambda(x) = \inf_x F_\lambda\}$$

the set of global minimizers of F_λ over X which is possibly empty. We denote $inf_x F_\lambda = inf_{x \in X} F_\lambda = inf\{F_\lambda(x); x \in X\}$.

1.1. Topological results. The notion of epi-convergence will play a crucial role in this section. For convenience of the reader we recall its definition and basic variational properties (see [A$_1$] and [D.G] for further details).

Definition 1.1. *Let (X, τ) be a metrizable topological space and $\{F_n, F; n \in \mathbf{N}\}$ a sequence of functions from X into $\overline{\mathbf{R}}$. The sequence $\{F_n; n \in \mathbf{N}\}$ is said to be τ-epi-convergent to F if for every $x \in X$, the following properties are satisfied:*

(i) there exists a sequence $\{x_n; n \in \mathbf{N}\}$ τ-converging to x such that
$$F(x) \geq \limsup F_n(x_n)$$

(ii) for every sequence $\{\xi_n; n \in \mathbf{N}\}$ τ-converging to x,

$$F(x) \leq \liminf F_n(\xi_n).$$

We then write $F = \tau - \lim_e F_n$ (or $F = \lim_e F_n$ when there is no ambiguity on the choice of the topology τ).

Symmetrically we say that $\{F_n; n \in \mathbf{N}\}$ is τ-hypo-convergent to F, and denote $F = \tau - \lim_h F_n$, iff:

$$(-F) = \tau - \lim_e(-F_n).$$

Epi-convergence is well fitted to the study of stability of global minimizers:

Proposition 1.2. *Let us consider a sequence of minimizing problems*

$$\inf\{F_n(x); x \in X\} \qquad n = 1, 2, \ldots$$

and for every $n \in \mathbf{N}$, x_n an ϵ_n-minimizer of F_n, that is

$$F_n(x_n) \leq \sup\{\inf_x F_x + \epsilon_n; \quad -\frac{1}{\epsilon_n}\}$$

where ϵ_n is a sequence of positive numbers which is converging to zero. Let us assume that the sequence $\{x_n; n \in \mathbf{N}\}$ is τ-relatively compact for some topology τ. Then the following implication holds:

(i) $F = \tau - \lim_e F_n$

$$\Downarrow$$

(ii) $\inf F_n \longrightarrow \inf F$ *and every τ-limit point of some subsequence of $\{x_n; n \in \mathbf{N}\}$ does minimize F on X.*

Remark 1.3. *(a)* It is for simplicity of the exposition we have assumed (X, τ) to be metrizable in definition 1.1 of the τ-epiconvergence. Indeed the concept works in general topological spaces (see references quoted above). When $X = \mathbf{R}^k$ we shall take τ equal to the usual topology of $\mathbf{R}^k$.

(b) An equivalent terminology is Γ-convergence (see for instance [D.G]). The epi-convergence terminology carries its geometrical interpretation as set-convergence of the sequence of corresponding epigraphs.

The concepts of epi-continuity and hypo-continuity follow in a natural way from epi-convergence and hypo-convergence.

Definition 1.4. *A parametrized family $\{F_\lambda; \lambda \in \Lambda\}$ of functions from X into $\overline{\mathbf{R}}$ is said to be τ-epi-continuous (respectively τ-hypo-continuous) where τ is a topology on X, if for every converging sequence $\lambda_n \longrightarrow \lambda$ we have:*

$$F_\lambda = \tau - \lim_e F_{\lambda_n} \quad (\text{respectively } F_\lambda = \tau - \lim_h F_{\lambda_n}).$$

We shall use the following property of the epi-continuous parametrized families which are indeed direct and easy consequences of the theory of epi-convergence.

Proposition 1.5. *Let $\{F_\lambda; \lambda \in \Lambda\}$ be a τ-epi-continuous family. Then for every $\lambda \in \Lambda$, F_λ is τ-lower semicontinuous and the mapping*

$$\lambda \longrightarrow \inf_x F_\lambda$$

is upper semicontinuous.

PROOF. The τ-lower semicontinuity of F_λ is a clear consequence of the following equivalences

$$F_\lambda \text{ is } \tau - \text{l.s.c.} \Leftrightarrow \text{epi } F_\lambda \text{ is closed in } (X, \tau) \times \mathbf{R}$$
$$F_\lambda = \tau - \lim_e F_{\lambda_n} \Leftrightarrow \text{epi } F_{\lambda_n} \longrightarrow \text{ epi } F_\lambda \text{ in } (X, \tau) \times \mathbf{R}$$

and as a limit in Kuratowski sense, epiF_λ is closed (see [A$_1$] ch.2.1 for detailed statements and proofs). The upper semicontinuity of $\lambda \longrightarrow \inf F_\lambda$ is an easy consequence of definition of τ-epi-convergence. Indeed let $\{\lambda_n; n \in \mathbf{N}\}$ be a sequence converging to λ. Then for every $x \in X$, there exists $x_n \xrightarrow{\tau} x$ such that

$$F_\lambda(x) \geq \lim F_{\lambda_n}(x_n).$$

Hence

$$F_\lambda(x) \geq \lim \sup(\inf_x F_{\lambda_n}).$$

This being true for every $x \in X$,

$$\inf_x F_\lambda \geq \lim \sup(\inf_x F_{\lambda_n}),$$

see [A$_1$], prop. 2.9.

We can now state the main result of this section.

Theorem 1.6 *Let $\{F_\lambda; \lambda \in \Lambda\}$ be a parametrized family of functions from a connected metrizable topological space (X, τ) into $\overline{\mathbf{R}}$ which satisfies the following properties:*

(i) *for every $\lambda \in \Lambda$ and every $t \in \mathbf{R}$ the lower level set $\{F_\lambda \leq t\}$ is a closed connected subset of (X, τ).*

(ii) *the family $\{F_\lambda; \lambda \in \Lambda\}$ is τ-epi-continuous with respect to $\lambda \in \Lambda$.*

(iii) *there exists an open subset Ω of X such that:*

$$S_o \neq \emptyset \quad and \quad S_o \subset \Omega.$$

(iv) *for every $\lambda \in \Lambda$, $S_\lambda \cap \partial\Omega = \emptyset$.*

(v) *for every converging sequence $\{\lambda_n; n \in \mathbf{N}\}$ in Λ, the following implication holds:*

$$x_n \in \overline{\Omega} \text{ for every } n \in \mathbf{N} \text{ and } \limsup F_{\lambda_n}(x_n) < +\infty$$
$$\Rightarrow \{x_n; n \in \mathbf{N}\} \text{ possesses a } \tau - \text{ converging subsequence.}$$

Then the following conclusions holds:

$$\forall \lambda \in \Lambda \quad S_\lambda \neq \emptyset \quad and \quad S_\lambda \subset \Omega$$

and the infimal value $\inf_x F_\lambda$ depends in a continuous way on the parameter λ.

PROOF. Let us consider the case $\overline{\Omega} = X$. From (ii) for every $\lambda \in \Lambda$, F_λ is lower semicontinuous on (X, τ). From (v) it is τ-inf-compact. Thus F_λ achieves its minimum on X,

$$S_\lambda \neq \emptyset \text{ and clearly } S_\lambda \subset \Omega.$$

So the conclusion of Theorem 1.6 holds in that case.

Let us now consider the case $\overline{\Omega} \neq X$. Since X is connected, $\partial\Omega$ is non empty. As a consequence, for every $\lambda \in \Lambda$ $\inf_x F_\lambda < +\infty$. Otherwise S_λ would be equal to X which is contradictory to (iv).

Let us now introduce the set

$$I = \{\lambda \in \Lambda; S_\lambda \neq \emptyset \quad and \quad S_\lambda \subset \Omega\}.$$

By assumption (iii), $\lambda = 0$ belongs to I. In order to prove the equality $I = \Lambda$ we rely on a connectedness argument and show that I is both closed and open in Λ.

a) I is closed. Given $\lambda_n \longrightarrow \lambda$ with λ_n belonging to I for every $n \in \mathbf{N}$, let us prove that λ still belongs to I. By Proposition 1.5

$$\limsup_x (\inf_x F_{\lambda_n}) \leq \inf_x F_\lambda < +\infty.$$

For every $n \in \mathbf{N}$, let us pick up a point x_n belonging to S_{λ_n}. Then

$$F_{\lambda_n}(x_n) = \inf F_{\lambda_n}$$
$$\limsup F_{\lambda_n}(x_n) < +\infty$$
$$x_n \in \overline{\Omega} \text{ for every } n \in \mathbf{N}.$$

By assumption (v) the sequence $\{x_n; n \in \mathbf{N}\}$ is τ-relatively compact. It follows from Proposition 1.2 and from

$$F_\lambda = \tau - \lim_e F_{\lambda_n} \text{ that } x \in S_\lambda \cap \overline{\Omega}.$$

Since $S_\lambda \cap \partial\Omega = \emptyset$, it follows that $x \in S_\lambda \cap \Omega$ that is $S_\lambda \cap \Omega \neq \emptyset$. Take $t = \inf_x F_\lambda$, the level set $\{F_\lambda \leq t\}$ is precisely the set S_λ of global minimizers of F_λ over X. By assumption (i) S_λ is connected. Since $S_\lambda \cap \partial\Omega = \emptyset$, we can write

$$S_\lambda = [S_\lambda \cap \Omega] \cup [S_\lambda \cap (\overline{\Omega})^C]$$

that is S_λ is the union of two disjoint open sets (for the induced topology on S_λ). S_λ being connected, one of these two sets is empty, and since $S_\lambda \cap \Omega \neq \emptyset$ we have that $S_\lambda = S_\lambda \cap \Omega$ that is $\emptyset \neq S_\lambda \subset \Omega$, which expresses that $\lambda \in I$.

b) I is open. Let us argue by contradiction and assume that there exists some sequence $\{\lambda_n; n \in \mathbf{N}\}$ and some λ in I satisfying

$$\forall n \in \mathbf{N} \quad \lambda_n \notin I \quad \text{and} \quad \lambda_n \to \lambda \quad \text{as} \quad n \to +\infty.$$

Then there exists an infinite number of indexes n such that either

$$b.1) \qquad S_{\lambda_n} = \emptyset$$
$$b.2) \qquad S_{\lambda_n} \cap \Omega^C \neq \emptyset$$

Let us still denote λ_n such a subsequence.

In case b.1) take a minimizing sequence $\{x_k; k \in \mathbf{N}\}$, i.e.

$$\inf_x F_{\lambda_n} = \lim_{k \to +\infty} F_{\lambda_n}(x_k).$$

For k sufficiently large one has necessarily $x_k \notin \overline{\Omega}$. Otherwise, since F_{λ_n} is lower semicontinuous and inf-compact on Ω, F_{λ_n} achieves its minimum, which is contradictory to the fact $S_{\lambda_n} = \emptyset$.
Thus for every $n \in \mathbf{N}$, one can find some $x_n \notin \overline{\Omega}$ such that

$$F_{\lambda_n}(x_n) \leq \inf_x F_{\lambda_n} + 1/n.$$

In case b.2) we have the same property by taking $x_n \in S_{\lambda_n} \cap (\overline{\Omega})^C$ (since $S_{\lambda_n} \cap \partial\Omega = \emptyset$). So in both cases b.1) and b.2) we have the same type of situation.

Up to now we have only expressed that $\lambda_n \notin I$. Let us now express that $\lambda \in I$ and introduce some $\overline{x} \in S_\lambda \cap \Omega$. Since $F_\lambda = \lim_e F_{\lambda_n}$, there exists some sequence $\{\xi_n; \ n \in \mathbf{N}\}$ with $\xi_n \in \Omega$ for every $n \in \mathbf{N}$ (since Ω is open) and

(1.1) $$\limsup F_{\lambda_n}(\xi_n) \leq F_\lambda(\overline{x}).$$

Let us introduce

$$t_n := \max\{F_{\lambda_n}(\xi_n); \inf F_{\lambda_n} + \frac{1}{n}\}.$$

By definition of x_n and ξ_n we have $x_n, \xi_n \in \{F_{\lambda_n} \leq t_n\}$. Since $\xi_n \in \Omega$ and $x_n \notin \overline{\Omega}$ and the level set $\{F_{\lambda_n} \leq t_n\}$ is connected, we have necessarily $\{F_{\lambda_n} \leq t_n\} \cap \partial\Omega \neq \emptyset$. Otherwise we have a partition of $\{F_{\lambda_n} \leq t_n\}$ into two disjoint non void open subsets, which is impossible since it is connected. So let us introduce for every $n \in \mathbf{N}$

$$z_n \in \{F_{\lambda_n} \leq t_n\} \cap \partial\Omega.$$

(1.2) $$F_{\lambda_n}(z_n) \leq \max\{F_{\lambda_n}(\xi_n); \inf F_{\lambda_n} + \frac{1}{n}\} \text{ and } z_n \in \partial\Omega.$$

Noticing that for any sequence $\{a_n; n \in \mathbf{N}\}$ and $\{b_n; n \in \mathbf{N}\}$ of real numbers:

$$\limsup(\max\{a_n, b_n\}) = \max\{\limsup a_n, \limsup b_n\},$$

we derive from (1.2):

$$(1.3) \qquad \limsup F_{\lambda_n}(z_n) \leq \max\{\limsup F_{\lambda_n}(\xi_n), \limsup \inf_x F_{\lambda_n}\}$$

Moreover since $F_\lambda = \lim_e F_{\lambda_n}$, from Prop. 1.5

$$(1.4) \qquad \inf_x F_\lambda \geq \limsup(\inf_x F_{\lambda_n}).$$

From (1.1), (1.3) and (1.4) we obtain

$$(1.5) \qquad \limsup F_{\lambda_n}(z_n) \leq \max\{F_\lambda(\overline{x}), \inf_x F_\lambda\} < +\infty.$$

On the other hand, using that $F_\lambda = \lim_e F_{\lambda_n}$ and assumption (v), there exists z, some τ-limit point of the sequence $\{z_n; n \in \mathbf{N}\}$, in $\partial\Omega$ (which is τ-closed) such that:

$$(1.6) \qquad F_\lambda(z) \leq \liminf F_{\lambda_n}(z_n).$$

Combining (1.5) and (1.6) we obtain $F_\lambda(z) \leq \inf_x F_\lambda$ that is $z \in S_\lambda \cap \partial\Omega$, which is a contradiction.

Remark 1.7. *1.* We point out that the parametrized space Λ can be taken in general as a closed connected subset of a metrizable space. It allows us to consider a parametrized family of single variable $\{F_\lambda; \lambda \in \Lambda\}$ as a bivariate function F from $X \times \Lambda$ into $\overline{\mathbf{R}}$

$$F(x, \lambda) := F_\lambda(x).$$

2. The class of real functions F on topological spaces X having connected lower level sets plays an important role in optimization theory. This notion naturally occurs when studying uniqueness of local minimizers and under what condition a local minimizer should be global, see for instance [Ma], [Z-C-A], [V].

3. It is readily seen that the conclusion of theorem 1.6 holds if instead of S_λ one takes the ϵ-minimizer's subset of F_λ which is given, for some fixed $\epsilon > 0$, by:

$$S_{\lambda,\epsilon} := \{x \in X; F_\lambda(x) \leq \inf_x F_\lambda + \epsilon\}.$$

Let us notice that for every $\epsilon > 0$ and every $\lambda \in \Lambda$ the set $S_{\lambda,\epsilon}$ is non empty.

4. We have the symmetrical result of the above theorem for the family of upper semicontinuous functions $G_\lambda (\lambda \in \Lambda)$ when we replace the epi-convergence and lower level sets by hypo-convergence and upper level sets. It is sufficient to take in Theorem 1.6 $F_\lambda = -G_\lambda$.

1.2 Convex case. An important case of the above theorem is the convex one. In the preceding section, the only structure was topological. We now assume that X is a Banach space.

Let us recall, a function F from X into $\mathbf{R}$ is said to be convex (respectively quasi-convex) whenever the epigraph set $\{(x,t) \in X \times \mathbf{R}; F(x) \le t\})$ (respectively for every $t \in \overline{\mathbf{R}}$ the lower level set $\{x \in X; \ F(x) \le t\})$ is convex.

Let us first derive from Theorem 1.6 the following

Corollary 1.8. *Let us suppose that the parametrized family $\{F_\lambda; \lambda \in \Lambda\}$ satisfies assumptions (ii), (iii), (iv) and (v) of Theorem 1.6 and for every $\lambda \in \Lambda$, epi F_λ (respectively for every $\lambda \in \Lambda$ and every $t \in \mathbf{R}$ the subset $\{F_\lambda \le t\}$) is a closed convex subset of $X \times \mathbf{R}$ (respectively of X).*

Then for every $\lambda \in \Lambda$ $S_\lambda \ne \emptyset$ and $S_\lambda \subset \Omega$.

Let us now examine the compactness assumption of Theorem 1.6. We need the following notions and definitions (see [A-E], [C] and [S-K] for further details).

Definition 1.9. *Let X be a Banach space, K a non empty subset of X and F a function from X into $\mathbf{R}$.*

1. We say that $v \in T_k(x)$, the contingent cone to K at x, if and only if there exists a sequence of strictly positive numbers h_n and of $v_n \in X$ satisfying

$$\lim_{n \to +\infty} v_n = v, \quad \lim_{n \to +\infty} h_n = 0 \quad and \quad \forall n \ge 0 \quad x + h_n v_n \in K.$$

2. Let $x \in K$, the contingent derivative of F at x along a direction h in $T_K(x)$ is defined by

$$(1.7) \qquad D_c^K F(x;h) := \lim_{\substack{h' \to h \\ x+th' \in K \\ t \downarrow 0}} \inf \frac{1}{t}\{F(x + th') - F(x)\}.$$

3. A parametrized family $\{F_\lambda; \lambda \in \Lambda\}$ satisfies the generalized Palais-Smale condition in K (denoted by G.P.S.) if for every converging sequence $\{\lambda_n; n \in \mathbf{N}\} \subset \Lambda$, any sequence $\{x_n; n \in \mathbf{N}\} \subset K$, along which

$$(1.8) \qquad \limsup F_{\lambda_n}(x_n) < +\infty$$

and

$$(1.9) \qquad \limsup \left(\inf_{h \in T_K(x_n)} D_c^K F_{\lambda_n}(x_n; \frac{h}{\|h\|}) \right) \geq 0,$$

possesses a strong-convergent subsequence.

Lemma 1.10. *Let F be a lower semicontinuous function from X into $\overline{\mathbf{R}}$, which is bounded from below.*

If F satisfies the G.P.S. condition in X, then it achieves its minimum on X.

PROOF. Take $\{x_n; n \in \mathbf{N}\}$ a minimizing sequence of F, i.e.

$$\inf_x F = \lim_{n \to +\infty} F(x_n).$$

There exists a sequence of non negative numbers $\epsilon_n > 0$, converging to zero, such that, for n sufficiently large one has

$$F(x_n) \leq \inf_x F + \epsilon_n.$$

According to the Ekeland's variational principle, cf. [E], there exists $\{u_n; n \in \mathbf{N}\}$ satisfying

(i) $F(u_n) \leq F(x_n)$

(ii) $F(u_n) < F(u) + \sqrt{\epsilon_n}.\|u_n - u\|$ for every $u \neq u_n$.

By taking in (ii), $u = u_n + \theta h'$ for some $\theta > 0$ and $h' \in X$, we obtain

$$D_c^K F(u_n; \frac{h}{\|h\|}) \geq -\sqrt{\epsilon_n}.$$

Now from G.P.S. we can extract a subsequence $\{u_{n_\nu}; \nu \in \mathbf{N}\}$ converging to some x. Then, since F is lower semicontinuous, we obtain

$$F(x) \leq \liminf_{\nu \to +\infty} F(u_{n_\nu}) \leq \lim_{n \to +\infty} F(x_n) \leq \inf_x F.$$

We can now state the following result with a weakened compactness condition, namely the generalized Palais-Smale one.

Theorem 1.11. *Let E be a closed connected subset of a Banach space, $\{F_\lambda : E \to \mathbf{R}; \lambda \in \Lambda\}$ be a parametrized family of proper functions which are bounded from below and satisfy assumptions (i), (ii), (iii), (iv) of Theorem 1.6 and the generalized Palais-Smale condition in $\overline{\Omega}$ (see Definition 1.9,3.) where Ω is given by (iii).*
Then the following conclusion holds:

$$\forall \lambda \in \Lambda \qquad \text{argmin } F_\lambda \neq \emptyset \quad and \quad \text{argmin } F_\lambda \subset \Omega.$$

PROOF.

First step: Suppose $\overline{\Omega} = E$. Let us give some λ in Λ. From Lemma 1.10, (i) and G.P.S. condition it follows that F_λ achieves its minimum on E.

Second step: We suppose that $\overline{\Omega} \neq E$, and consider the set

$$I = \{\lambda \in \Lambda; S_\lambda \neq \emptyset \quad \text{and} \quad S_\lambda \subset \Omega\}.$$

As in the proof of Theorem 1.6, it is readily seen that the set I is closed. Hence, to show that I is the whole of Λ it is sufficient to prove that it is open.

Indeed, by contradiction, let us assume that there exists some sequence $\{\lambda_n; n \in \mathbf{N}\} \subset \Lambda$ and some λ in I satisfying

$$\lambda_n \notin I \text{ for each } n \geq 0 \text{ and } \lambda_n \to \lambda.$$

Necessarily for each $n \geq 0$ we have, either

$$S_{\lambda_n} = \emptyset \text{ or } S_{\lambda_n} \cap (\overline{\Omega})^C \neq \emptyset.$$

Hence, for every $n \geq 0$ we can find some $x_n \in (\overline{\Omega})^C$ such that

$$F_{\lambda_n}(x_n) \leq \inf_x F_{\lambda_n} + \frac{1}{n},$$

see the proof of the theorem 1.6. Now, since λ belongs to I and $F_\lambda = \lim_e F_{\lambda_n}$, there exists $\overline{x} \in \Omega$ and $\{y_n; n \in \mathbf{N}\}$ which is converging to $\overline{x}$ such that

$$\lim F_{\lambda_n}(y_n) \leq F_\lambda(\overline{x}) < +\infty.$$

Then $-\infty < \inf_\Omega F_{\lambda_n} < +\infty$ for $n \geq 0$ sufficiently large. We deduce, for $n \geq 0$ sufficiently large, the existence of some ξ_n in Ω satisfying

$$F_{\lambda_n}(\xi_n) \leq \inf_\Omega F_{\lambda_n} + \frac{1}{n}.$$

Take $t_n = \inf_{(\overline{\Omega})} F_{\lambda_n} + 1/n$, and as in the proof of theorem 1.6 we can find some point $z_n \in \{F_{\lambda_n} \leq t_n\} \cap \partial\Omega$. We now assert that the sequence $\{z_n; \ n \in \mathbf{N}\}$ possesses a strong converging subsequence. Indeed, since $F_{\lambda_n}(z_n) \leq t_n$, we use the Ekeland's variational principle (see [E]) to find some $u_n \in \overline{\Omega}$ such that

(i) $F_{\lambda_n}(u_n) \leq F_{\lambda_n}(z_n)$

(ii) $F_{\lambda_n}(u_n) < F_{\lambda_n}(u) + \frac{1}{\sqrt{n}}\|u - u_n\|$ for every $u \in \overline{\Omega}$ and $u \neq u_n$

(iii) $\|z_n - u_n\| \leq \frac{1}{\sqrt{n}}.$

Then for any $\theta > 0$ and $h' \in E$ with $u_n + \theta h' \in \overline{\Omega}$ one has

$$F_{\lambda_n}(u_n + \theta h') - F_{\lambda_n}(u_n) \geq -\frac{\theta}{\sqrt{n}}\|h'\|.$$

Hence

$$D_c^{\overline{\Omega}} F_{\lambda_n}(u_n; \frac{h}{\|h\|}) \geq -\frac{1}{\sqrt{n}}$$

and

$$\limsup \left\{ \inf_{h \in T_{\overline{\Omega}}(x_n)} D_c^{\overline{\Omega}} F_{\lambda_n} \left(u; \frac{h}{\|h\|} \right) \right\} \geq O.$$

On the other hand, from (i)

$$\limsup F_{\lambda_n}(u_n) < +\infty.$$

By the compactness assumption G.P.S. we can extract a subsequence of $\{u_n; \; n \in \mathbf{N}\}$ (still denoted u_n) converging to $\overline{z}$. From (iii) and above we derive the strong convergence of z_n to $\overline{z}$. Thus $\overline{z} \in \partial\Omega$, which yields a contradiction as in the proof of Theorem 1.6.

Remark 1.12. *1.* When $X = \mathbf{R}^k$ and Ω is bounded, then the compactness assumption of Theorems 1.6 and 1.11 are automatically satisfied.

2. If the connectedness assumption on level sets is dropped, the following situation can occur (see Fig 1), and the conclusion of the above Theorems fails to be true.

Take $\Omega =\;]-2,2[$ and for every $\lambda \in \Lambda$,

$$F_\lambda(x) = \begin{cases} x^2 & \text{if } x \in [0,1[\\ 2 - x & \text{if } x \in [1, 1+\lambda[\\ 1 - \lambda & \text{if } x \geq 1 + \lambda \end{cases}$$

and

$$F_\lambda(-x) = F_\lambda(x), \text{ for every } x \in \Omega.$$

One has $S_o \neq \emptyset$ and $S_o \subset \Omega$ (since $S_o = \{0\}$). But $S_1 \cap \Omega^C \neq \emptyset$.

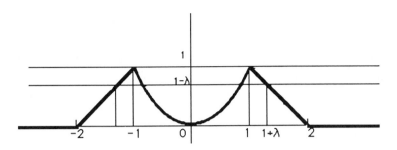

Figure 1

1.3 Bivariate functions case. Let us consider (X, τ) and (Y, ω) two topological spaces and mappings $\{F_\lambda; \lambda \in \Lambda\}$ from $X \times Y$ into $\mathbf{R}$.

We recall the following definitions and notations, see Guillerme [G]. The *lower marginal function* and the *upper marginal function* of F_λ, for a given $\lambda \in \Lambda$, are respectively defined by:

$$(1.11) \qquad m_\lambda(y) := \inf_{x \in X} F_\lambda(x,y) \qquad \text{for each } y \in Y$$

$$(1.12) \qquad M_\lambda(x) := \sup_{y \in Y} F_\lambda(x,y) \qquad \text{for each } x \in X.$$

A point (x_o, y_o) is said to be a *saddle point* of F_λ if:
for every $(x,y) \in X \times Y$ $\quad F_\lambda(x_o,y) \le F_\lambda(x_o,y_o) \le F_\lambda(x,y_o)$
which is equivalent to:

$$(1.13) \qquad M_\lambda(x_o) \le m_\lambda(y_o).$$

Likewise, we define and ϵ-*saddle point* (x_ϵ, y_ϵ) of F_λ, for some $\epsilon > 0$, by:

$$(1.14) \qquad \min\{M_\lambda(x_\epsilon) - \epsilon, \frac{1}{\epsilon}\} \le \max\{m_\lambda(y_\epsilon) + \epsilon, -\frac{1}{\epsilon}\}.$$

We respectively denote by S_λ and $S_{\epsilon,\lambda}$ the sets of saddle points and ϵ-saddle points of F_λ.

Theorem 1.13. *Let us consider* (X,τ) *and* (Y,ω) *two connected topological spaces and* $F_\lambda : X \times Y \to \overline{\mathbf{R}}$ *a parametrized bivariate function with* m_λ *and* M_λ *the corresponding lower marginal and upper marginal functions. Let us suppose moreover that:*

$$(1.15) \qquad for\ every\ \lambda \in \Lambda \quad \sup_Y m_\lambda = \inf_X M_\lambda.$$

We assume that the following hypothesis are fulfilled:

(i) For every $\lambda \in \Lambda$, *every* $s,t \in \mathbf{R}$, *the level sets* $\{x; M_\lambda(x) \le t\}$ *and* $\{y; m_\lambda(y) \ge s\}$ *are connected.*

(ii) The maps $\lambda \to M_\lambda$ *and* $\lambda \to m_\lambda$ *are respectively* τ-*epi-continuous and* ω -*hypo-continuous.*

(iii) There exists an open subset Ω *of* $X \times Y$ *such that*

$$\emptyset \ne S_o \subset \Omega\ and\ S_\lambda \cap \partial\Omega = \emptyset\ for\ every\ \lambda \in \Lambda.$$

(iv) For every converging sequence $\{\lambda_n; n \in \mathbf{N}\}$ *in* Λ, *when* $(x_n, y_n) \in \overline{\Omega}$ *for each* $n \in \mathbf{N}, \liminf m_{\lambda_n}(y_n) > -\infty$ *and* $\limsup M_{\lambda_n}(x_n) < +\infty$, *then* $\{(x_n, y_n); n \in \mathbf{N}\}$ *is* $\tau \times \omega$-*relatively compact.*

Then the following conclusion holds:

$$\forall\lambda \in \Lambda, \quad \emptyset \ne S_\lambda \subset \Omega.$$

Remark 1.14. The proof of Theorem 1.13 is quite similar to that of Theorem 1.6. So we indicate just how to adapt it to the bivariate case. If the open subset Ω is equal to $\Omega_1 \times \Omega_2$ where $\Omega_1 \subset X$ and $\Omega_2 \subset Y$, then the above bivariate problem can be reduced to the situation of Theorem 1.6 and its symmetrical one (see Remark 1.7,2). In fact, it is sufficient to see that (1.15) implies:

$$S_\lambda = \{x; M_\lambda(x) = \inf_X M_\lambda\} \times \{y; m_\lambda(y) = \sup_Y m_\lambda\}.$$

For a general Ω we follow the lines of the proof of Theorem 1.6, replacing (1.1) by the following approximation result:

Proposition 1.15. *Let (X, τ) and (Y, ω) two metrizable spaces and F_n, F saddle functions from $X \times Y$ into $\overline{\mathbf{R}}$. We respectively denote by M_n, M and m_n, m their upper marginal functions and lower marginal functions. Then the following implication holds:*

(i) $M = \tau - \lim_e M_n$ *and* $m = \omega - \lim_h m_n$.

$\Downarrow$

(ii) for every saddle point $(\overline{x}, \overline{y})$ of F, there exists some sequence $\{\epsilon_n ; n \in \mathbf{N}\}$ converging to zero such that $(\overline{x}, \overline{y}) = \lim_{n \to +\infty} (x_n, y_n)$ with (x_n, y_n) an ϵ_n-saddle point of F_n.

PROOF. Let us give some saddle point $(\overline{x}, \overline{y})$ of F. From (i) there exists sequences $\{x_n; n \in \mathbf{N}\}$ and $\{y_n; n \in \mathbf{N}\}$ converging respectively to $\overline{x}$ and $\overline{y}$, such that:

$$\limsup M_n(x_n) \leq M(\overline{x}) \text{ and } \liminf m_n(y_n) \geq m(\overline{y}).$$

Using that $(\overline{x}, \overline{y})$ is a saddle point of F i.e. $M(\overline{x}) = m(\overline{y})$, we obtain

$$\limsup\{M_n(x_n) - m_n(y_n)\} \leq 0.$$

Hence we can find a sequence of real numbers $\epsilon_n > 0$, converging to zero such that for n sufficiently large

$$M_n(x_n) - m_n(y_n) \leq 2\epsilon_n.$$

It follows immediately

$$M_n(x_n) - \epsilon_n \leq m_n(y_n) + \epsilon_n,$$

i.e. (x_n, y_n) is an ϵ-saddle point of F_n. When $M(\overline{x})$ or $m(\overline{y})$ is equal to $\pm\infty$, it is sufficient to use (1.14).

2. Relation with the degree theory for maximal monotone operators of F.Browder. Let us review some definitions. We adopt the terminology of [B] and [A$_2$].

Definition 2.1. *Let H be a hilbert space equipped with scalar product $<,>$, which associated norm is denoted by $\|\cdot\|$. A multivalued operator A from H is said to be monotone iff:*
$$\forall x_1, x_2 \in \text{dom}A \text{ (where dom } A := \{x \in H; Ax \neq \emptyset\})$$
$$< Ax_1 - Ax_2, x_1 - x_2 >\geq 0, \quad i.e. \quad \forall y_1 \in Ax_1, \quad \forall y_2 \in Ax_2$$
$$< y_1 - y_2, x_1 - x_2 >\geq 0.$$

Definition 2.2. *A monotone operator A is maximal monotone if it is maximal in the set of monotone operators, for the inclusion relation.*

As a consequence, given A a maximal monotone operator, the following implication holds:

$$\forall \xi \in \text{dom}A \quad < A\xi - y, \xi - x > \geq 0 \quad \Rightarrow y \in Ax.$$

Definition 2.3. *Let $\{A^n, A; n \in \mathbf{N}\}$ be a sequence of maximal monotone operators. We say that A^n graph-converges to A and we denote $A^n \overset{G}{\longrightarrow} A$ if:*

$$\forall (x, y) \in A, \quad \exists (x_n, y_n) \in A^n \quad \text{such that} \quad (x, y) = \lim_{n \to +\infty} (x_n, y_n).$$

Let us notice that we identify A with its graph i.e. $(x, y) \in A \Leftrightarrow y \in Ax$.

Proposition 2.4 [A$_1$], [A$_2$]. *Let $\{A^n, A; n \in \mathbf{N}\}$ be a family of maximal monotone operators such that $A^n \overset{G}{\longrightarrow} A$.*
Assuming that $(x_n, y_n) \in A^n$ for all $n \in \mathbf{N}$ and $(x, y) = \lim(x_n, y_n)$ then $(x, y) \in A$.

We define, for each $\epsilon > 0$, the resolvent of index ϵ of the maximal monotone operator A by: $J_\epsilon^A := (I + \epsilon A)^{-1}$, it is a contraction operator, and the corresponding Yosida approximation $A_\epsilon := \frac{1}{\epsilon}(I - J_\epsilon^A)$ which is again a single valued maximal monotone operator.

Proposition 2.5 [A₂]. *Let A and A^n be maximal monotone operators. Then the following statement are equivalent*

(i) $A^n \xrightarrow{G} A$

(ii) $J_\epsilon^{A^n}(x) \to J_\epsilon^A(x)$ for every $x \in H$ and every $\epsilon > 0$

(iii) $\exists \epsilon_o > 0$ such that $J_\epsilon^{A^n}(x) \to J_{\epsilon_o}^A(x)$ for all $x \in H$.

Proposition 2.6 [B]. *Let A be a maximal monotone operator, then*

(i) if $x \in \mathrm{dom}A$ $\|A_\epsilon x\| \xrightarrow{\epsilon \to 0} \|A^o x\|$

(ii) if $x \notin \mathrm{dom}A$ $\lim_{\epsilon \to 0} \|A_\epsilon x\| = +\infty$

where A^o is defined by: $A^o x := \mathrm{proj}_{Ax} 0$.

We can now state the main result of this section.

Theorem 2.7. *Let H be a finite dimensional space and $\Lambda = [0,1]$ the parametrized space. Let*

$$A : H \times \Lambda \longrightarrow H$$
$$(x, \lambda) \to A(x, \lambda)$$

be a parametrized family of multivalued operators. Let us denote $\{A_\lambda = A(\cdot, \lambda); \lambda \in \Lambda\}$ and $S_\lambda := A_\lambda^{-1}(0) = \{x \in H; 0 \in A_\lambda x\}$ the set of zeros of A_λ.

Suppose the following properties are satisfied:

(1) for every $\lambda \in \Lambda$, A_λ is a maximal monotone operator.

(2) the map $\lambda \to A_\lambda$ is graph-continuous.

(3) there exists a bounded open subset Ω of H such that

$$S_o \neq 0 \qquad and \qquad S_o \subset \Omega.$$

(4) $\forall \lambda \in \Lambda \qquad S_\lambda \cap \partial\Omega = \emptyset.$

Then the following conclusion holds:

for every $\lambda \in \Lambda \qquad S_\lambda \neq \emptyset \ and \ S_\lambda \subset \Omega.$

PROOF. Let $I = \{\lambda \in \Lambda; \emptyset \neq S_\lambda \subset \Omega\}$, we want to prove it to be both closed and open. Then from a connectedness argument we conclude that $I = \Lambda$.

a) I is closed. Indeed let us consider a sequence $\{\lambda_n; n \in \mathbf{N}\}$ such that $\lambda_n \to \lambda$ and $\lambda_n \in I$ for each $n \in \mathbf{N}$. One has $\lambda_n \in I \Leftrightarrow \exists x_n \in \Omega; 0 \in A_{\lambda_n} x_n$. Since Ω is bounded, we can extract a subsequence, which we still denote x_n, and some point $\overline{x}$ in $\overline{\Omega}$ such that $x_n \to \overline{x}$. hence $0 \in A_\lambda \overline{x}$ comes from Prop. 3.59 [A$_1$]. So $\overline{x} \in S_\lambda \cap \overline{\Omega}$, but $S_\lambda \cap \partial\Omega = \emptyset$, then $\overline{x} \in S_\lambda \cap \Omega$. Thanks to convexity of S_λ, we deduce $S_\lambda \subset \Omega$, that is $\lambda \in I$.

b) I is open. Let us argue by contradiction. We suppose that there exists a sequence $\{\lambda_n \notin I; n \in \mathbf{N}\}$ converging to some λ such that $\lambda \in I$. One has necessarily, for an infinite number of indexes n, either

b.i) $S_{\lambda_n} = \emptyset$
    or
b.ii) $S_{\lambda_n} \cap \Omega^C \neq \emptyset.$

For sake of simplicity of the notations we still denote this subsequence by λ_n. We now show that each case b.i) and b.ii) yields a contradiction.

In case b.i) one has $A_{\lambda_n}^{-1}(0) = \emptyset$. Let us consider for every $\epsilon > 0, x_{n.\epsilon}$ solution of the equation $0 \in (\epsilon I + A_{\lambda_n})x$, which exists since A_{λ_n} is maximal monotone. Let us show that for a fixed n, $\lim\limits_{\epsilon \to 0} \|x_{n.\epsilon}\| = +\infty$. After a simple computation we obtain that:

$$0 \in (\epsilon I + A_{\lambda_n})x_{n.\epsilon} \Leftrightarrow x_{n.\epsilon} = (A_{\lambda_n}^{-1})_\epsilon(0)$$

where $(A_{\lambda_n}^{-1})_\epsilon$ is the Yosida approximation of the operator $A_{\lambda_n}^{-1}$ of index $\epsilon > 0$. But $0 \notin \text{dom} A_{\lambda_n}^{-1}$, since $A_{\lambda_n}^{-1}(0) = \emptyset$. Then from Prop. 2.6 (iii), $\lim\limits_{\epsilon \to 0} \|(A_{\lambda_n}^{-1})_\epsilon(0)\| = +\infty$. Thus $\lim\limits_{\epsilon \to 0} \|x_{n.\epsilon}\| = +\infty$. For each integer n, we can choose $\epsilon = \epsilon_n$ such that:

$$0 < \epsilon_n \le \frac{1}{n}, \quad x_{n.\epsilon_n} \in (\overline{\Omega})^C \text{ and } \|x_{n.\epsilon_n}\| \ge n.$$

Take $x_n = x_{n.\epsilon_n}$, one has $\lim\limits_{n \to \infty} \|x_n\| = +\infty$.

On the other hand, $\lambda \in I$ implies that there exists some $\alpha \in \text{dom} A_\lambda$ such that $0 \in A_\lambda \alpha$. Since $A_{\lambda_n} \xrightarrow{G} A_\lambda$, there exists a sequence $\{(\alpha_n, \beta_n) \in A_{\lambda_n}; n \in \mathbf{N}\}$ converging to $(\alpha, 0)$. We notice that $(\alpha_n, \beta_n) \in A_{\lambda_n}$ can be rewritten as

$$(\alpha_n + \frac{\beta_n}{\epsilon_n}) \in (I + \frac{1}{\epsilon_n} A_{\lambda_n})(\alpha_n).$$

On the other hand $0 \in \epsilon_n x_n + A_{\lambda_n} x_n$, which is equivalent to

$$0 \in (I + \frac{1}{\epsilon_n} A_{\lambda_n})(x_n).$$

These two last relations and the contraction property of $(I + \frac{1}{\epsilon_n} A_{\lambda_n})^{-1}$ yield

$$\|x_n - \alpha_n\| \le \|\frac{\beta_n}{\epsilon_n} + \alpha_n\|.$$

Therefore $\|\epsilon_n x_n\| \le 2\epsilon_n \|\alpha_n\| + \|\beta_n\|$ and $\lim\limits_{n \to \infty} \|\epsilon_n x_n\| = 0$.

The case b.ii) is similar to b.i), take $\epsilon_n = 0$ for every $n \ge 0$, then one has existence of some $x_n \in S_{\lambda_n} \cap (\overline{\Omega})^C$ i.e.

$$x_n \in (\overline{\Omega})^C, 0 \in A_{\lambda_n} x_n \quad \text{and} \quad \epsilon_n x_n = 0.$$

Without restriction we can assume that the sequence $\{x_n; n \in \mathbf{N}\}$ is unbounded. Otherwise this would imply the existence of some $\overline{x} \in (\overline{\Omega})^C$ such that $A_\lambda \overline{x} \ni 0$ and contradict $\lambda \in I$. We summarize the above considerations in the following statement: in both cases b.i) and b.ii)

(H) $\begin{cases} \text{there exists} \quad \{\epsilon_n; n \in \mathbf{N}\} \quad \text{and} \quad \{x_n; n \in \mathbf{N}\} \quad \text{such that} \\ \lim \epsilon_n = 0, \qquad \lim \|x_n\| = +\infty, \qquad x_n \in \overline{\Omega}^C \\ \lim \|\epsilon_n x_n\| = 0 \quad \text{and} \quad 0 \in \epsilon_n x_n + A_{\lambda_n} x_n. \end{cases}$

Using the fact $\lambda \in I$, there exists some $\overline{x} \in \Omega$ with $0 \in A_\lambda \overline{x}$. Since $A_{\lambda_n} \xrightarrow{G} A_\lambda$, there exists some $(\alpha_n, \beta_n) \in A_{\lambda_n}$ with $(\overline{x}, 0) = \lim(\alpha_n, \beta_n)$ in $H \times H$. We notice that for n large enough $\alpha_n \in \Omega$, because Ω is open.

Let us consider for each $n \in \mathbf{N}$ a point θ_n defined by:

$$\theta_n = t_n \alpha_n + (1 - t_n)x_n \quad \text{and} \quad \theta_n \in \partial\Omega \quad \text{and} \quad t_n \in]0,1[.$$

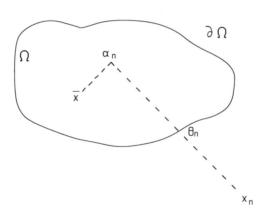

Figure 2

From the compactness of $\partial\Omega$, since H is a finite dimensional space, we can extract a converging subsequence of $\{\theta_n; n \in \mathbf{N}\}$. We still denote this subsequence by θ_n and its limit by $\overline{\theta} \in \partial\Omega$. Let us prove that $0 \in A_\lambda \overline{\theta}$. From Def. 2.2 this is equivalent to prove that:

$$\forall(\xi, \eta) \in A_\lambda, \quad < \eta, \overline{\theta} - \xi > \ \leq 0.$$

Let us give some $(\xi, \eta) \in A_\lambda$, since A_{λ_n} is graph-converging to A_λ, we assert the existence of $(\xi_n, \eta_n) \in A_{\lambda_n}$ for every $n \in \mathbf{N}$ such that $(\xi, \eta) = \lim(\xi_n, \eta_n)$. By definition of θ_n

$$< \eta_n, \theta_n - \xi_n > = t_n < \eta_n, \alpha_n - \xi_n > +(1 - t_n) < \eta_n, x_n - \xi_n >$$
$$\leq t_n < \beta_n, \alpha_n - \xi_n > +(1 - t_n) < \eta_n, x_n - \xi_n >$$

(since $\beta_n \in A_{\lambda_n} \alpha_n$ and $< \beta_n - \eta_n, \alpha_n - \xi_n > \geq 0$). On the other hand $0 \in \epsilon_n x_n + A_{\lambda_n} x_n$ and $\eta_n \in A_{\lambda_n} \xi_n$. Hence $< -\epsilon_n x_n - \eta_n, x_n - \xi_n > \geq 0$. Consequently

$$< \eta_n, \theta_n - \xi_n >$$
$$\leq t_n < \beta_n, \alpha_n - \xi_n > +(1 - t_n) < -\epsilon_n x_n, x_n - \xi_n >$$
$$\leq t_n \|\beta_n\| \cdot \|\alpha_n - \xi_n\| + (1 - t_n) \|\epsilon_n x_n\| \cdot \|\xi_n\|.$$

Using the following properties: $\lim \|\beta_n\| = \lim \|\epsilon_n x_n\| = 0$ and that the sequences $\{\xi_n; n \in \mathbf{N}\}$ and $\{\alpha_n; n \in \mathbf{N}\}$ are bounded, we derive

$$\limsup < \eta_n, \theta - \xi_n > \leq 0,$$

that is

$$< \eta, \overline{\theta} - \xi > \leq 0,$$

which yields the contradiction: $\overline{\theta} \in S_\lambda \cap \partial\Omega$. This completes the proof of Theorem 2.7.

Remark 2.8. Theorem 2.2 is closely related to Browder's degree theory for multivalued operators of monotone type. The basic results can be found in [Br₂], [Br₃] and [Br₄]. In fact, we have obtained a continuation property for solutions of equations governed by maximal monotone operators. Let us notice that we avoid using the degree theory, by relying on a direct continuation method, which not only provides existence results but also stability. We point out that in the previous section we investigated, with the same tools, stability results for minimization problems, which by opposition to degree methods, allow us to deal with more general situation (the Euler equation and hence the operators may be difficult to handle with !).

From this Theorem we easily derive the following proposition and corollaries:

Proposition 2.9. *Take assumptions (1), (2), (3) of Theorem 2.7 and instead of assumption (4) let us assume:*

$$(4') \qquad \forall \lambda \in [0,1[\quad S_\lambda \neq \emptyset \quad and \quad S_\lambda \cap \partial\Omega = \emptyset.$$

Then the following conclusions hold

$$\forall \lambda \in [0,1[\quad S_\lambda \neq \emptyset \quad and \quad S_\lambda \subset \Omega$$

$$(2.1) \qquad and \ for \quad \lambda = 1 \quad one \ has \quad S_1 \neq \emptyset.$$

PROOF. Take $\Lambda_t = [0,t]$ with $t < 1$. By Theorem 2.7, we can conclude that $\forall \lambda \in \Lambda_t \quad \emptyset \neq S_\lambda \subset \Omega$. Indeed all hypothesis of the previous Theorem are fulfilled by Λ_t. So

$$\forall t \in]0,1[, \forall \lambda \in \Lambda_t \quad \emptyset \neq S_\lambda \subset \Omega.$$

Therefore,

$$\text{for every } \lambda \in [0,1[\quad \text{one has} \quad \emptyset \neq S_\lambda \subset \Omega.$$

To verify the last statement of (2.1) it is sufficient to take $\lambda_n \to 1$ with $\lambda_n \in [0,1[$, then the end of the proof is similar to the first step of Theorem 2.7.

Corollary 2.10. *Let A be a maximal monotone operator which is coercive in the following sense:*

$$(2.2) \qquad for \ some \quad x_o \in H \qquad \lim_{|x| \to \infty} \frac{<A^o x, x - x_o>}{|x|} = +\infty.$$

$$(2.3) \qquad Then \ A \ is \ onto, \ i.e. \quad R(A) := \bigcup_{\xi \in H} A\xi = H.$$

PROOF. Let $y \in H$, we have to find $x \in H$ such that $y \in Ax$. It is sufficient to show that there exists $x \in H$ with $0 \in Ax$. Indeed if we replace A by $(A - y)$ we still have

$$\lim_{|x| \to \infty} \frac{< A^\circ x - y, x - x_o >}{|x|} = +\infty.$$

We now consider the convex homotopy $\{A_\lambda; \lambda \in [0,1]\}$ where

$$A_\lambda = \lambda A + (1 - \lambda)I \quad \text{for every} \quad \lambda \in [0,1]$$

and for which assumptions (3) and (4') of Proposition 2.9 can be condensed into

$$(2.4) \quad \begin{cases} \text{there exists a bounded set } \Omega \text{ in } H \\ \text{such that} \quad 0 \in \Omega \quad \text{and for every} \quad \lambda \in]0,1[\\ 0 \in (I + \frac{\lambda}{1-\lambda}A)x \Rightarrow x \notin \partial\Omega \end{cases}$$

Indeed, since A is maximal monotone, there exists x_λ such that

$$(I + \tfrac{\lambda}{1-\lambda}A)^{-1}(0) = \{x_\lambda\}.$$

We claim that the subset $\{x_\lambda; \lambda \in]0,1[\}$ is bounded. Otherwise there exists a sequence of elements x_{λ_ν} such that $\lim_{\nu \to \infty} |x_{\lambda_\nu}| = +\infty$. By coerciveness of A, we can find a non negative real number δ such that

$$< A^\circ x_{\lambda_\nu}, x_{\lambda_\nu} - x_o > \geq \delta|x_{\lambda_\nu}| \quad \text{for} \quad \nu \text{ sufficiently large.}$$

By definition of x_λ

$$0 \in x_{\lambda_\nu} + \frac{\lambda_\nu}{1 - \lambda_\nu} A x_{\lambda_\nu} \quad \text{for every} \quad \nu \in \mathbf{N}.$$

Hence

$$(2.5) \quad 0 \geq < x_{\lambda_\nu}, x_{\lambda_\nu} - x_o > + \frac{\lambda_\nu}{1 - \lambda_\nu} < A^\circ x_{\lambda_\nu}, x_{\lambda_\nu} - x_o >$$

$$0 \geq |x_{\lambda_\nu}|^2 - |x_{\lambda_\nu}|.|x_o| + \delta\frac{\lambda_\nu}{1 - \lambda_\nu}|x_{\lambda_\nu}|.$$

Therefore

(2.6) $$|x_{\lambda_\nu}| \leq |x_o| - \delta \frac{\lambda_\nu}{1 - \lambda_\nu} \leq |x_o|$$

which is impossible. Let us choose a real number $\rho > \sup_{\lambda \in]0,1[} |x_\lambda|$, then assumption (2.4) is satisfied with $\Omega = \text{int } B(0, \rho)$. Finally, from Proposition 2.9 one has $A_1^{-1}(0) = A^{-1}(0) \neq \emptyset$, it follows that there exists $x \in H$ such that $0 \in Ax$.

Corollary 2.11. *Let us consider X a finite dimensional space and an epi-continuous parametrized family $\{F_\lambda; \lambda \in \Lambda\}$ of proper convex l.s.c. functions from X into $\overline{\mathbf{R}}$, satisfying properties (iii) and (iv) of Theorem 1.6.*

Then the same conclusions as in Theorem 1.6 hold.

PROOF. It is sufficient to take the family of operators $A_\lambda = \partial F_\lambda$, the convex subdifferential of F_λ. Indeed assumptions (1), (3), (4) are immediately fulfilled. (2) is a consequence of the equivalence between Mosco-epi-convergence of convex proper lower semicontinuous functions and graph-convergence of their subdifferentials (cf. [A₁] Theorem 3.26 or [A₂] Theorem 1.2).

Definition 2.12. *(see [A-W₁], [A-W₂]). X and Y are now two reflexive Banach spaces.*

A sequence of saddle functions $\{K_n : X \times Y \longrightarrow \overline{\mathbf{R}}; n \in \mathbf{N}\}$ is said to be Mosco-epi/hypo-converging to K and we write $K = M - e/h - \lim K_n$, iff:

$$e_s/h_w - ls K_n \leq K \leq h_s/e_w - li K_n,$$

where

$$e_s/h_w - ls K_n(x, y) := \sup_{y_n \rightharpoonup y} \quad \inf_{x_n \to x} \quad \limsup K_n(x_n, y_n)$$

$$h_s/e_w - li K_n(x, y) := \inf_{x_n \rightharpoonup x} \quad \sup_{y_n \to y} \quad \liminf K_n(x_n, y_n).$$

Here s and $\to$ (respectively w and $\rightharpoonup$) denote the strong topology and the strong convergence (respectively the weak topology and the weak convergence).

Corollary 2.13. *Suppose X and Y are finite dimensional spaces. Consider $\{K_\lambda : X \times Y \to \overline{\mathbf{R}}; \lambda \in \Lambda\}$ an epi/hypo-continuous parametrized family of closed convex-concave saddle functions satisfying:*

There exists a bounded open subset Ω of $X \times Y$ for which;

$$\emptyset \neq S_0 \subset \Omega \quad and \quad S_\lambda \cap \partial\Omega = \emptyset \quad for\ every \quad \lambda \in \Lambda.$$

Then $S_\lambda \neq \emptyset$ and $S_\lambda \subset \Omega$ for every $\lambda \in \Lambda$ where S_λ is the set of saddle points of K_λ.

PROOF. Take in Theorem 2.7, for each $\lambda \in \Lambda$,

$$A_\lambda := \partial_x K_\lambda \times \partial_y(-K_\lambda)$$

i.e. $(\alpha, \beta) \in A_\lambda(x, y) \Leftrightarrow \begin{cases} \alpha \in \partial K_\lambda(\cdot, y)(x) \\ \beta \in \partial(-K_\lambda)(x, \cdot)(y). \end{cases}$

A_λ is a maximal monotone operator, see for instance Rockafellar [R_1], [R_2]. The family $\{A_\lambda; \lambda \in \Lambda\}$ satisfies assumptions (2), (3) and (4) of the above theorem, see [A-Az-W$_2$], Prop. 6.2.

The previous result still holds in the case where X and Y are general reflexive Banach spaces. However we have to strenghten the compactness assumption as follows:

For every converging sequence $\{\lambda_n; n \in \mathbf{N}\}$, every sequence of elements $(x_n, y_n) \in \overline{\Omega}$ such that

$$\liminf(\inf_X K_{\lambda_n}(x, y_n)) > -\infty$$

and

$$\limsup(\sup_Y K_{\lambda_n}(x_n, y)) < +\infty,$$

is strongly relatively compact.

References

[A$_1$] H.Attouch, *Variational convergence for functions and operators*, Applicable Math. Series, Pitman (1984).

[A₂] H.Attouch, *Famille d'opérateurs maximaux monotones et mesurabilité*, Ann. Mat. Pura e Appl. **120** (1979), 35-111.

[A-Az-W₁] H.Attouch, D.Azé, R.Wets, *Convergence of convex-concave saddle functions*, to appear in Ann. Inst. H. Poincaré.

[A-Az-W₂] H.Attouch, D.Azé, R.Wets, *On continuity properties of the partial Legendre-Fenchel transform: Convergence of sequences of augmented Lagrangian functions, Moreau-Yosida approximates and sub-differential operators*, Fermat days 85: Mathematics for optimization, J.B. Hiriart Urruty ed., North Holland Math. Studies **129** (1987).

[A-W₁] H.Attouch, R.Wets, *A convergence theory for saddle functions*, Trans. Am. Math. Soc. **280** (1983).

[A-W₂] H.Attouch, R.Wets, *A convergence theory for bivariate functions aimed at the convergence of saddle value*, Proceedings of S. Margherita Ligure on "Mathematical theories of Optimization", J.P. Cecconi and T.Zolezzi eds., lecture Notes in Math. **979**, Springer-Verlag (1981).

[A-W₃] H.Attouch, R.Wets, *Lipschitzian Stability of ε-Approximate Solutions In Convex Optimization*, I.I.A.S.A., A-2361, Laxenburg, Austria 1986, WP-00.

[A-W₄] H.Attouch, R.Wets, *Lipschitzian and Hölderian stability results for local minima of well-conditioned functions*, AVAMAC (1987).

[A-W₅] H.Attouch, R.Wets, *A quantitative approach via epigraphic distance to stability of strong local minimizers*, I.I.A.S.A., Working Paper, April 1987.

[A-E] J.P.Aubin, I.Ekeland, *Applied Nonlinear Analysis*, Wiley-Interscience, New York (1984).

[B] H.Brezis, *Opérateurs maximaux monotones et semi-groupes de contractions dans les espaces de Hilbert*, North-Holland (1973).

[Br₁] F.E.Browder, *Continuation of solutions of equations under homotopies of single-valued and multivalued mappings. Fixed Point Theory and its Applications*, S. Swaminathan ed., Acad. Press. (1976).

[Br₂] F.E.Browder, *Fixed point theory and nonlinear problems*, Bull. Am. Math. Soc. **9**, July 1983.

[Br₃] F.E.Browder, *Degree of mapping for nonlinear mappings of monotone type; Strongly nonlinear mappings*, Proc. Nat. Acad. Sci. USA **80** (1983), 2408-2409.

[Br4] F.E.Browder, *L'unicité du degré topologique pour les applications de type monotone*, C.R.A.S. Paris **296** (1983), 145-148.

[C] F.H.Clarke, *Optimization and Nonsmooth Analysis*, Wiley-Interscience, New York (1983).

[D.G] E.De Giorgi, *Convergence problems for functions and operators*. Proc. Int. Meet. On "Recent methods in nonlinear analysis", Roma 1978, Pitagora ed., Bologna (1979).

[D.G-F] E.De Giorgi, T.Franzoni, *Su un tipo di convergenza variazionale*, Atti Accad. Naz. Lincei **58** (1975), 842-850.

[E] I.Ekeland, *Non convex minimization problems*, Bull. Am. Math. Soc. **1** (1979), 443-474.

[E-T] I.Ekeland, R.Temam, *Analyse convexe et problèmes variationnels*, Dunod (1974).

[G] J.Guillerme, *Convergence of approximate saddle points*, Preprint 58 U.E.R. Science, Limoges (1987).

[Gu] J.Guddat, *Parametric Optimization: Pivoting and Predicator-Corrector Continuation*, A survey Didd. (A), Section Math. Der Humboldt-Universität zu Berlin (1987), 125-162.

[J-J-T] H.T.Jongen, P.Jonker, F.Twilt, *Critical sets in parametric optimization*, Math. Programming **35** (1986), 1-21.

[K-H] M.Kojima, A.Hirabayashi. *Continuous deformations of non linear programming*, Tokyo Inst. of Technology, Dept. of Information Science, Techn. Rep. **101** (1981).

[Kl] D.Klatte, *On the stability of local and global optimal solutions in parametric problems of nonlinear programming*, Part I and II. Seminarbericht **75**, Sektion Mathematik, Humboldt-Universität Berlin, November 1985.

[Kr1] E.Krauss, *A representation of maximal monotone operators by saddle functions*, Rev. Roum. de Math. Pures Appl. **30** (1985).

[Kr2] E.Krauss, *Maximal Monotone Operators and Saddle Functions* I, Zeitschrift für Analysis und ihre Anwendungen **5** (1986), 333-346.

[Ma] D.H.Martin, *Connected Level Sets, Minimizing Sets, and Uniqueness in Optimization*, I and II, CSIR Special Report WISK 276-277, Pretoria, South Africa (1977).

[Mo] J.J.Moreau, *Théorèmes "inf-sup"*, C.R.A.S. **258** (1964), 2720-2722.

[R1] R.T.Rockafellar, *Augmented Lagrangian multiplier functions and duality in non convex programming*, SIAM J. of Control

12 (1974), 268-285.

[R₂] R.T.Rockafellar, *Monotone operators associated with saddle functions and minimax problems*, in Nonlinear Functional Analysis, Amer. Math. Soc., Rhode Island (1970).

[Ro] S.M.Robinson, *Local epi-continuity and local optimization*, MRC Technical Summary Report **2798**, University of Wisconsin-Madison (1985).

[S] R.Schultz, *Estimates for Kuhn-Tucker points of perturbed convex programs*, Preprint 107, Humboldt University, East Berlin (1986).

[Sw] L.Schwartz, *Analyse. Topologie générale et analyse variationnelle* (1970).

[S-K] S.Shu-Zhong, C.Kung-Ching, *A local minimax Theorem without compactness*, Nonlinear and convex analysis, Lin and Simons eds., Marcel Dekker, New York (1987), 211-233.

[V] M.Volle, *Convergence en niveaux et en epigraphes*, CRAS **299** (1984).

[W] H.Wacker, *A continuation Method*, Academic Press. New York (1978), 1-35.

[Z-C-A] I.Zang, E.U.Choo, M.Avriel. *A Note on Functions Whose Local Minima are Global*, J. Opt. Th. Appl. **18** (1976), 555-559.

Avamac Université Perpignan

Avenue de Villeneuve

F-66025 PERPIGNAN Cedex

DISCRETIZATION OF EVOLUTION VARIATIONAL INEQUALITIES

Claudio Baiocchi[*]

Dedicated to Ennio De Giorgi on his sixtieth birthday

1. Introduction. In the framework of the usual Hilbert triplet $\{V, H, V'\}$ [1] let K, A be given with:

(1.1) K is a closed convex nonempty subset of V

and

[*] This work was partly supported by CNR through the IAN and partly supported by MPI through 40% and 60% funds.

[1] Say H, V are real Hilbert spaces (norms $|\ |$ and $\|\ \|$ respectively) with $V \subset H$, V dense in H. We identify H with its dual H'; the dual space V' is the completion of H with respect to the dual norm $\|h\|_* = \sup\{(h,v) \mid \|v\| \leq 1\}$ where $(\ ,\ )$ is the scalar product in H and can also be used for the pairing $V' - V$.

(1.2) $\begin{cases} A \text{ is a linear operator from } V \text{ in } V'; \\[2mm] \text{there exist } M, \alpha > 0 \text{ and } \lambda \text{ real with} \\[2mm] \|Av\|_* \le M\|v\| \quad \forall v \in V \\[2mm] (Av + \lambda v, v) \le \alpha \|v\|^2 \quad \forall v \in K - K \ ^{(2)}. \end{cases}$

We want to study (a weak form of) the following problem: we are given an initial value u_o in H and a function $f(t)$ with values in V'; we ask for $u(t)$ such that:

(1.3) $$u(t) \in K \quad \text{a.e. in} \quad t > 0$$

(1.4) $$(u'(t) + Au(t) - f(t), u(t) - v) \le 0 \quad \forall v \in K, \quad \text{a.e. in} \quad t > 0$$

(1.5) $$u(0) = u_o.$$

Problems of this kind have been deeply studied in [2] (for a slightly different approach see also [3]) in a very general framework: e.g. V could be a reflexive Banach space, A could be a (non linear) monotone operator. In particular, under very general assumptions on the data $u_o, f(t)$, an existence and uniqueness theorem holds for a weak formulation of (1.3), (1.4), (1.5); furthermore if u_o and f are "smooth and compatible" the solution u is smoother and (1.3), (1.4), (1.5) hold in the usual sense.

Let us recall that choosing $K = V$ in (1.1) the problem becomes simpler, because (1.4) can be written:

$$u'(t) + Au(t) = f(t) \quad \text{a.e. in} \quad t > 0$$

---

(2) $K_1 - K_2$ denotes the vector difference

$$\{k_1 - k_2 \mid k_1 \in K_1, k_2 \in K_2\}.$$

so that the problem is linear [3] ; in such a case the discussion of smoothness properties is simpler because, under suitable compatibility assumptions (e.g.: $u_o = 0$; $f(t)$ vanishes near $t = 0$), the smoothness of u is directly related to the smoothness of f by the simple formula:

$$f \in H^r(V') \Rightarrow u \in H^r(V) \cap H^{r+1}(V')$$

(say: if the r-th derivative of f is in L^2 as a function valued into V', then $u^{(r)}$ is L^2 valued into V and $u^{(r+1)}$ is L^2 valued into V').

On the contrary, in the inequalities framework, there exists a treshold: no matter how smooth and compatible $\{u_o, f\}$ are, already in the scalar case (say: $V = H = \mathbf{R}$) the function $u'(t)$ can have jump discontinuities; in particular one cannot directly apply to the inequalities the results which are known for the discretization of equations: in fact these results are usually obtained using sommability assumptions on $u''(t)$.

We will be concerned with a backward Euler scheme; given $k > 0$ we set:

$$(1.6) \qquad J_{k,n} = [nk, (n+1)k[$$

$$(1.7) \qquad u_{k,0} = u_o \; ; \; f_{k,n} = \frac{1}{k} \int_{J_{k,n}} f(t) dt \quad [4]$$

and we replace (1.3), (1.4) by the sequence of problems [5]:

$$(1.8) \qquad \begin{array}{c} u_{k,n+1} \in K \\ (u_{k,n+1} - u_{k,n} + kAu_{k,n+1} - kf_{k,n}, u_{k,n+1} - v) \leq 0 \\ \forall v \in K. \end{array}$$

---

[3] More generally: if K is a cone (closed, convex, vertex in 0) (1.4) splits into:

$$(u'(t)+Au(t)-f(t),v) \leq 0 \; \forall v \in K; \; (u'(t)+Au(t)-f(t),u(t))=0;$$

furthermore, if K is a subspace, (1.4) becomes:

$$(u'(t)+Au(t)-f(t),v)=0 \quad \forall v \in K.$$

[4] Of course different choices are possible.

[5] For $n=0,1,\dots$. Thanks to (1.2) there exists a $k_o>0$ such that, for $k \in]0,k_o]$, the operator $A+kI$ is coercive on $K-K$; by classical results on steady variational inequalities (see [4], [6]), for such values of k, problem (1.8) has a unique solution.

Defining $U_k(t)$ through:

(1.9)
$$\begin{cases} U_k(t) \text{ is the continuous function, linear on} \\ \\ \text{each } J_{k,n}, \text{ such that } U_k(nk) = u_{k,n+1}{}^{(6)} \end{cases}$$

we can conjecture that, in suitable topologies, U_k converges to a solution u of the continuous problem. We will prove such a conjecture,
and the result will be "constructive": independently from the existence results of [2] we will show that, under very general assumptions
on $\{u_o, f\}$, the family $\{U_k\}$ is a Cauchy family, for $k \to 0$, in suitable norms; and its limit is the unique solution, in a weak sense,
of (1.3), (1.4), (1.5). Furthermore, if $\{u_o, f\}$ satisfies compatibility ans smoothness assumptions, the family $\{U_k\}$ remains bounded
in stronger norms (which will imply that the solution is a "strong"
solution) and the distance between U_k and u is $O(h)$.

The order $O(h)$ is optimal, and our technique gives the regularity
results without assuming the validity of technical hypotheses like [7]:

(1.10) A is   symmetric

(1.11) $\forall h > 0$ it is $(I + hA)^{-1}(K) \subset K.$

On the other hand, under "intermediate" assumptions on $\{u_o, f\}$, we
will get intermediate regularity results for the solution u [8]; though
often the results seem "non optimal", this type of regularity seem
new: I don't know different proofs for them, also assuming the validity of (1.10), (1.11).

2. Problem and results: precise statements. In [1] we
showed that, for the discretization of evolution equations (say: K is

---

[6] $u_{k,n+1}$, and not $u_{k,n}$. The shift is needed because we want a function
valued into V, and in general the initial value $u_{k,0} = u_o$ is given in H.

[7] Assuming (1.10) and/or (1.11), some (different) regularity results hold
true; see [2], [3].

[8] Which is viewed as a function valued into intermediate spaces between
V and H; say, with the notations of [5], valued into $S(p,\theta,V;p,\theta-1,H)$ for suitable
$p \in [1, +\infty]$ and $\theta \in]0,1[$.

a subspace; see [3]) a good functional framework is obtained working in spaces of type "Intersection" and "Sum":

$$I = L^2(-\infty, +\infty; V) \cap H^{1/2}(-\infty, +\infty; H)$$

$$S = L^2(-\infty, +\infty : V') + H^{-1/2}(-\infty, +\infty; H) \quad (9)$$

respectively for the solution u and for the datum f (remark that $I = S'$).

I don't know if inequalities can be studied in a similar framework; however it will be still convenient to represent $f(t)$ in terms of a sum:

(2.1) $\qquad f = g + h \; ; \; g \in L^1(0, +\infty; H) \; ; h \in L^2(0, +\infty; V')$

For shorter notation we will set [10]:

(2.2)
$$\begin{cases} S(a,b) = L^1(a,b;H) + L^2(a,b;V'); \\[2mm] I(a,b) = L^2(a,b;V) \cap L^\infty(a,b;H); \end{cases}$$

we will assume u_o, f given with:

(2.3) $\qquad\qquad u_o \in H \quad ; \quad f \in S(0, +\infty)$

and we will search for u with:

(2.4) $\qquad u \in I(0, +\infty) \quad ; \quad u(t) \in K \quad$ a.e. in $\quad t > 0$ [11].

---

[9] Assume $u_o = 0$. By extending both $f(t)$ and $u(t)$ with values 0 for $t < 0$, we can study the problem on the whole **R**, instead of on $[0, +\infty[$. In fact [1] is devoted to problems with $t \in \mathbf{R}$; similar results for problems with $t \in [0, +\infty[$ are given in [8].

[10] In (2.2) we have $-\infty \leq a < b \leq +\infty$; remark that now I and S are no more reflexive; but the relation I=S' still holds, the pairing between $i(t) \in I$ and $s(t) \in S$ being given by $<i,s> = \int_a^b (s(t), i(t)) \, 0 \, dt$; any F linear continuous functional on S can be represented in the form $F(s) = <i,s>$, with $i \in I$ uniquely determined by F, and $\|F\|_{S'} = \|i\|_I$.

[11] So that (1.3) has been imposed, and we must only impose (1.4), (1.5).

In order to give a weak formulation of (1.4) let us start by assuming u "smooth", for instance:

$$(2.5) \qquad\qquad u' \in I(0,+\infty)$$

which is almost maximal, as u' cannot, in general, be a continuous function. Fix any function v with [12]

$$(2.6) \qquad v,v' \in I(0,+\infty) \quad ; \quad v(t) \in K \quad \text{for} \quad t > 0$$

and choose in (1.4) $v = v(t)$; it follows (see [12]):

$$(2.7) \quad \left(\frac{1}{2}|u(t) - v(t)|^2\right)' + \left(v'(t) + Au(t) - f(t), u(t) - v(t)\right) \leq 0$$

which still has a meaning also dropping the assumption (2.5); of course, if u satisfies only (2.4), derivative and sign in (2.7) must be interpreted in the distributions sense, or (equivalently) by imposing that the function:

$$F(T) = \frac{1}{2}|u(T) - v(T)|^2 +$$

(2.8)

$$+ \int_o^T (v'(t) + Au(t) - f(t), u(t) - v(t))dt$$

is non increasing.

Remark 2.1. For f given with (2.1), assumption (2.4) on u is needed in order to ensure the sommability of $(f(t),u(t))$. We could replace such a framework by:

$$f \in L^2(0,+\infty; V') \quad ; \quad u \in L^2(0,+\infty; V)$$

but the results would be similar: the unique solution u would belong to the space $I(0,+\infty)$; so that we prefer to start directly with (2.4), allowing a greater generality for f.

---

[12] Of course we could work with weaker assumptions both on u and v; the essential need is the validity of $(|d(t)|^2)'=2(d'(t),d(t))$ for the difference $d(t)=u(t)-v(t)$.

In order to impose (1.5) we could ask a little bit more for u:

(2.9) $\qquad t \to u(t)$ is continuous from $[0, +\infty[$ into H

and impose (1.5) in the usual sense; remark however that from (2.4), (2.9) it follows that, for $h > 0$, the quantity $\frac{1}{h} \int_o^h u(t) dt$ belongs to K, and converges in H to $u(0)$ as $h \to 0$; a necessary condition is then given by:

(2.10) $\qquad u_o \in \mathcal{K}$, where $\mathcal{K}$ is the closure of K in H

A different approach is obtained by englobing (1.5) into the integral form of (2.7); say we could ask that, for any v with (2.6), the function $F(t)$ considered in (2.8) is non increasing and bounded by $\frac{1}{2}|u_o - v(0)|^2$. We will impose the condition (which is stronger if (2.10) fails):

$$(2.11) \quad \begin{cases} \text{for any } v \text{ given with (2.6) the function} \\[2mm] T \to 1/2|u(T) - v(T)|^2 + \\[2mm] + \int_o^T (v'(t) + Au(t) - f(t), u(t) - v(t)) dt \\[2mm] \text{is non increasing and bounded by } \frac{1}{2}|\mathcal{P}u_o - v(0)|^2 \end{cases}$$

where $\mathcal{P}$ is the projection of H on $\mathcal{K}$ (with respect to the norm of H).

We will see that if u_o is given with (2.10) (so that $\mathcal{P}u_o \equiv u_o$) each solution of (2.11) satisfies (2.9) and (1.5).

Remark 2.2. If (2.10) fails, dropping in (2.11) the operator $\mathcal{P}$ one gets a problem without uniqueness (easy counterexamples already with $V = H = \mathbf{R}$); for such a problem an existence result is true, because any solution with $\mathcal{P}u_o$ solves the problem with u_o.

We will deal with the following formulation:

Problem 2.1. *Assume (1.1), (1.2) and:*

(2.12) $\qquad 0 \in K$; *(1.2) holds true with* $\lambda = 0$.

Given $\{u_o, f\}$ with (2.3), we ask for u with (2.4) such that, for any v with (2.6) it is:

$$(2.13) \quad \int_o^{+\infty} (v'(t) + Au(t) - f(t), u(t) - v(t))dt \leq \frac{1}{2}|\mathcal{P}u_o - v(0)|^2$$

Remark 2.3. Formula (2.13) is an obvious consequence of (2.11); remark that in fact the "natural" initial condition imposed in (2.13) is:

$$(2.14) \quad \begin{aligned} &u(0) = \mathcal{P}u_o \quad ; \quad \text{say} : \\ &u_o \in \mathcal{K} \quad \text{and} \quad (u(0) - u_o, u(0) - v) \leq 0 \quad \forall v \in \mathcal{K} \end{aligned}$$

which looks better adapted to (1.4) than (1.5). Of course, if (2.10) holds true, we have $\mathcal{P}u_o = u_o$ and we are in fact imposing (1.5).

Remark 2.4. Assumption (2.12) is needed because of the sommability assumption (2.4): if $0 \notin K$ the set of u with (2.4) is empty. On the other hand, if $\lambda > 0$, sommability assumptions on $u(t)$, $f(t)$ must be replaced by similar assumptions on $\exp(-\lambda t)u(t)$ and $\exp(-\lambda t)f(t)$ (and we must multiply (2.7) by $\exp(-2\lambda t)$ *before* the integration leading to (2.10)). A different approach in the case $\lambda > 0$ is obtained by asking for *local* solutions; see Rem. 2.5 later on.

In the following, given $\{u_o, f\}$ with (2.3), we will work with $u_{k,o}$ and $f_{k,n}$ defined through (1.7); we will set (compare to (1.9) and [6]):

$$(2.15) \quad \begin{cases} \tilde{u}_k(t) \text{ is the continuous function, linear on each} \\ J_{k,n}, \text{ with } \tilde{u}_k(nk) = u_{k,n} \; ; \; U_k(t) = u_k(t+k) \end{cases}$$

and we will prove the following result:

Theorem 2.1. *The operator T_k from $\{u_o, f\}$ to U_k is bounded and Lipschitz continuous from $H \times S(0, +\infty)$ into $I(0, +\infty)$, uniformly in k. For any $\{u_o, f\}$ with (2.3) the family $\{U_k\}$ is a Cauchy* *family in $I(0, +\infty)$; its limit u is the unique solution of Pb. 2.1 [13], and it satisfies (2.9), (2.11), (2.7), (2.14).*

Furthermore, assuming:

(2.16)

$$g \in BV(0, +\infty; H)^{(14)}; \quad h \in H^1(0, +\infty; V') ;$$

$$\mathcal{P}u_o \in K ; \quad A\mathcal{P}u_o - h(0) \in H$$

the family $\{U'_k\}$ is bounded in $(I(0, +\infty); u$ satisfies (2.5), (1.4); and $\|U_k - u\|_{I(0,+\infty)} = O(k)$.

As we already pointed out (see [3]) in the framework of equations some further properties hold true, e.g.:

(2.17) $$(u'(t) + Au(t) - f(t), v) = 0 \quad \forall v \in K ;$$

(2.18) $$\left(\frac{1}{2}|u(t)|^2\right)' + \left(Au(t) - f(t), u(t)\right) = 0$$

(in particular the function $t \to |u(t)|^2$ is absolutely continuous) and also:

(2.19) $$\forall s > 0 \quad \text{it is} \quad \|u(t+s) - u(t)\|_{L^2(0,+\infty;H)} \leq C \cdot \sqrt{s}.$$

If K is a cone we can expect the validity of (2.18) and:

(2.20) $$(u'(t) + Au(t) - f(t), v) \geq 0 \quad \forall v \in K.$$

From our results on problem (1.8) it will follow:

Theorem 2.2. *If K is a cone, the solution u of Pb. 2.1 satisfies (2.18), (2.19), (2.20); if K is a subspace u satisfies (2.17).*

---

[13] In particular, the operator T from $\{u_o, f\}$ to u is bounded and Lipschitz continuous from $H \times S(0, +\infty)$ in $I(0, +\infty)$.

[14] We will set $V(a, b; g) = \sup \sum_{m=0}^{n-1} |g(t_{m+1}) - g(t_m)|$ where the supremum is taken over all choices of $\{t_m\}$ with $a < t_o < t_1 ... < t_n < b$. If the supremum is finite we will write $g \in BV(a, b; H)$. Remark that any g in BV is bounded; this property, as well as other interesting properties of BV can be found in the Appendix of [3].

Remark 2.5. If (2.12) fails we could replace (2.3), (2.4) by the relations:

$$(2.21) \qquad u_o \in H \quad ; \quad f \in S_{\text{loc}}([0,+\infty[)$$

$$(2.22) \qquad u \in I_{\text{loc}}([0,+\infty[) \quad ; \quad u(t) \in K \text{ a.e. in } t > 0$$

(with the obvious meaning for S_{loc}, I_{loc}); we then replace (2.13) (or better: (2.11)) by:

$$(2.23) \begin{cases} \text{for any } v \text{ given with (2.6) the function :} \\[4pt] T \to \tfrac{1}{2}|\exp(-\lambda T)[u(T) - v(T)]|^2+ \\[4pt] \quad + \int_o^T \big(\exp(-\lambda t)[v'(t) + Au(t) + \lambda u(t) - \lambda v(t) - f(t)], \\[4pt] \quad \exp(-\lambda t)[u(t) - v(t)]\big)\,dt \\[4pt] \text{is non increasing , and bounded by } \tfrac{1}{2}|\mathcal{P}u_o - v(0)|^2. \end{cases}$$

Also for such problem we will get results similar to the ones given in Thm. 2.1 and Thm. 2.2.

Remark 2.6. From the theoretical point of view it could be interesting to dipose of alternate ways of writing the "compatibility condition": $A\mathcal{P}u_o - h(0) \in H$. In the framework of Rem. 2.5 one can work with the unknown function $u_*(t) = u(t) - \mathcal{P}u_o$, which vanishes for $t = 0$ (see (2.14)) and solves a similar problem where K is replaced by $K_* = K - \{\mathcal{P}u_o\}$ (vector difference, see [2]) and $\{u_o, f\}$ is replaced by $\{0, f_*\}$, where $f_*(t) = f(t) - A\mathcal{P}u_o$. Such f_* can be decomposed as $g_* + h_*$, with $g_*(t) = g(t) + h(0) - A\mathcal{P}u_o$, $h_*(t) = h(t) - h(0)$; and now one can work as suggested in [9]: the solution u_* is viewed as the restriction to $(0,+\infty)$ of the solution of a problem posed on the whole of $\mathbf{R}$ (or simply in $(-1,+\infty)$) with f_* vanishing for $t < 0$. After a translation (say: $t \to t + 1$, in order to work again with t in $(0,+\infty)$) and suppressing the index $*$, we are working with:

$$(2.24) \qquad u_o = 0 \quad ; \quad f(t) \equiv 0 \text{ for } t \in (0,1)$$

which will imply, for the discrete solution of (1.8), that the "first" $u_{k,n}$ vanish.

Remark 2.7. Let us recall that, concerning *equations*, in the framework of (2.24) the regularity of u is strictly related to the regularity of f; e.g. we have, for any $\theta > 0$:

$$f \in H^\theta(0,+\infty;V') \Rightarrow u \in H^{\theta+\frac{1-s}{2}}(0,+\infty;V_s)$$

(2.25)

$$\text{for} \quad 0 \leq s \leq 1 \,^{(15)};$$

a similar result for inequalities cannot hold if $\theta + \frac{1-s}{2} \geq 3/2$ (because of discontinuities in $u'(t)$); and is an open problem (also under the additional hypotheses (1.10), (1.11)) if $\theta + \frac{1-s}{2} < 3/2$. Our results, which do not assume (1.10), (1.11), give a first contribution in such a direction; for instance we will show that, if (2.16) holds true, the solution u satisfies $u \in H^{(4/3)-\epsilon}(0,+\infty;V_{1/3}) \quad \forall \epsilon > 0$.

3. Stability estimates. Let us firstly consider the problem (1.8) with $n = 0$; $k > 0$ being fixed, we will write u_1 instead of $u_{k,1}$; similarly for f_o (instead of $f_{k,0}$), decomposed into $f_o = g_o + h_o$ where, with notations (2.1):

$$(3.1) \qquad g_o = \frac{1}{k}\int_o^k g(t)dt \quad ; \quad h_o = \frac{1}{k}\int_o^k h(t)dt.$$

We will derive general estimates for:

$$u_1 \in K$$

$$(3.2) \qquad (u_1 + kAu_1, u_1 - v) \leq (G_o, u_1 - v) + \sqrt{k}(H_o, u_1 - v)$$

$$\forall v \in K$$

$^{(15)}$ In (2.25) we used the short notation V_s for the space that in [5] is denoted by $S\left(2,\frac{s+1}{2},V;2,\frac{s-1}{2},V'\right)$. The assumption on f implies $u \in H^\theta(0,+\infty;V)$ and $u \in H^{\theta+1}(0,+\infty;V')$, so that u is in each intermediate space, in particular in

$$S\left(2,\frac{s+1}{2},H^\theta(0,+\infty;V) ; 2,\frac{s-1}{2},H^{\theta+1}(0,+\infty;V')\right)$$

which coincides with $H^{\theta+\frac{1-s}{2}}(0,+\infty;V_s)$.

where G_o, H_o, for problem (1.8), are given by:

$$(3.3) \qquad G_o = u_{k,o} + kg_o \quad ; \quad H_o + \sqrt{k}h_o$$

so that, under (1.7), it will be:

$$(3.4) \quad G_o \to u_o \quad \text{in} \quad H \; ; \; H_o \to 0 \quad \text{in} \quad V' \text{ (strong convergences)}$$

On choosing $v = 0$ in (3.2) one gets easily [16]:

$$(3.5) \qquad |u_1|^2 + k\|u_1\|^2 \le C\{|G_o|^2 + \|H_o\|_*^2\}$$

where, as in the following, we denote by C a generic constant depending only fron constants α and M of (1.2); if (2.12) fails, C could eventually depend also from λ and from the element used in the translation (see [16]).

Let us show that:

$$(3.6) \qquad |u_1 - \mathcal{P}u_o|^2 + k\|u_1\|^2 \to 0 \quad \text{as} \quad k \to 0.$$

In fact, by definition of $\mathcal{P}$ (and because of $u_1 \in K \subset \mathcal{K}$) it is $(\mathcal{P}u_o - u_o, \; u_1 - \mathcal{P}u_o) \ge 0$; so that $|u_1 - \mathcal{P}u_o|^2 \le (u_1 - u_o, \; u_1 - \mathcal{P}u_o)$. From (1.2), (2.11), (3.2) it follows that, for any $v \in K$:

$$|u_1 - \mathcal{P}u_o|^2 + \alpha k\|u_1 - v\|^2 \le$$

$$\le (u_1 - u_o, u_1 - \mathcal{P}u_o) + k(A(u_1 - v), u_1 - v)$$

$$\le (u_1 - u_o, v - \mathcal{P}u_o) + (u_1 - u_o + kAu_1, u_1 - v) -$$

$$- k(Av, u_1 - v) \le |u_1 - u_o| \, |v - \mathcal{P}u_o| + (G_o - u_o, u_1 - v) =$$

$$= \sqrt{k}(H_o - \sqrt{k}Av, u_1 - v).$$

---

[16] We are assuming (2.12); however, concerning next Thm. 3.1, only minor changes are needed: the assumption $0 \in K$ becomes true after a translation, and if $k \in]0, k_o]$ (see [5]) all remains valid also if (1.2) holds with $\lambda > 0$.

In particular we have:

$$
(3.7) \quad
\begin{cases}
\forall v \in K \quad \text{one has} \quad |u_1 - \mathcal{P}u_o|^2 + k\|u_1 - v\|^2 \leq \\[2mm]
\leq C\Big\{ |u_1 - u_o|\, |v - \mathcal{P}u_o| + |(G_o - u_o, u_1 - v) + \\[2mm]
+ \sqrt{k}(H_o - \sqrt{k}Av, u_1 - v)| \Big\}
\end{cases}
$$

Let now $v(k)$ be a family of elements of K such that as $k \to 0$, one has $k\|v(k)\|^2 \to 0$ and $v(k) \to \mathcal{P}u_o$ strongly in H (such a family must exist because $\mathcal{P}u_o \in K$); choosing $v = v(k)$ in (3.7), we get (3.6) by means of (3.4), (3.5).

Theorem 3.1. *For any* $i > 0$ *there exists* $C = C_i$ *such that*

$$
(3.8) \qquad |u_{k,i} - \mathcal{P}u_o|^2 + k\|u_{k,i}\|^2 \leq C\{|u_o|^2 + \|f\|_{S(0,ik)}^2\};
$$

furthermore one has

$$
(3.9) \qquad \forall i > 0 \quad |u_{k,i} - \mathcal{P}u_o|^2 + k\|u_{k,i}\|^2 \to 0 \quad as \quad k \to 0
$$

(but the limit is not uniform in i*).*

PROOF. By induction on i. The case $i = 1$ follows from (3.5), (3.6) (remark that the right hand side of (3.5) can be estimated by $C\{|u_o|^2 + \|g\|_{L^1(0,k;H)}^2 + \|h\|_{L^2(0,k;V')}^2\}$ for any decomposition $f = g + h$; see (3.3)).

In general the property for $i+1$ follows from the property for i, by reproducing the proof of (3.5), (3.6).

Theorem 3.2. *Under assumption (2.16) one has*

$$
|u_1 - \mathcal{P}u_o|^2 + k\|u_1 - \mathcal{P}u_o\|^2 = O(k^2)
$$

PROOF. By using (3.3), choosing $v = \mathcal{P}u_o$ in (3.7) one gets:

(3.10)
$$\begin{cases} |u_1 - \mathcal{P}u_o|^2 + k\|u_1 - \mathcal{P}u_o\|^2 \leq \\[2mm] \leq C\Big(k|u_1 - \mathcal{P}u_o|\int_o^k |g(t)|dt+ \\[2mm] +k|(h(0) - A\mathcal{P}u_o, u_1 - \mathcal{P}u_o)|+ \\[2mm] +\|u_1 - \mathcal{P}u_o\|\int_o^k \|h(t) - h(0)\|_* dt\Big) \end{cases}$$

It is then sufficient to estimate the integrals in (3.10) by $k\|g\|_{L^\infty(0,k;H)}$ and $k^{3/2}\|h'\|_{L^2(0,k;V')}$ respectively.

Remark 3.1. By choosing $v = u_{k,1}$ into (1.8) with $n = 1$ (say in the problem defining $u_{k,2}$) similar arguments show the validity of:
(3.11)
$$\begin{cases} |u_{k,2} - u_{k,1}|^2 + k\|u_{k,2} - u_{k,1}\|^2 \\[2mm] \leq (f_{k,1} - kAu_{k,1}, u_{k,2} - u_{k,1}) \\[2mm] \leq Ck^2\Big\{ \sup_{0<t<2k} |g(t)|^2 + |A\mathcal{P}u_o - h(0)|^2 + \int_o^{2k} \|h'(t)\|_*^2 dt\Big\} \end{cases}$$

Remark 3.2. Of course, if $\{u_o, f\}$ satisfies "intermediate" assumptions between (2.3) and (2.16), one must expect an intermediate order of zero for $|u_{k,1}-\mathcal{P}u_o|^2$, $k\|u_{k,1}-\mathcal{P}u_o\|^2$, $|u_{k,2}-u_{k,1}|^2$, $k\|u_{k,2}-u_{k,1}\|^2$. However, in general, it is not obvious what are "intermediate" hypotheses; the only obvious case corresponds to $\mathcal{P}u_o \in \mathcal{K}$, $A\mathcal{P}u_o - h(0) \in H$: one can then work assuming (2.24), and the "first" $u_{k,n}$ will vanish for k small enough.

In order to prove Thm. 2.1, the estimates (of "elliptic" type) we obtained until now are not sufficient; we need some "parabolic" estimates, which will follow from a Gronwall-like result; we will deal both with a continuous and a discrete formulation of such a Lemma (Thm. 3.4 and 3.3 respectively).

Theorem 3.3. *Let the sequences* $\{w_n\}$, $\{p_n\}$, $\{q_n\}$ *be given* *with:*

(3.12) $w_o \in H$; for $n \geq 0$ $w_{n+1} \in K - K, p_n \in H, q_n \in V'$

such that, for $n \geq 0$, one has:

(3.13) $\left(w_{n+1} - w_n + kAw_{n+1}, w_{n+1}\right) \leq k(p_n + q_n, w_{n+1})^{(17)}$

Setting, for $N > 0$:

(3.14)
$$
\begin{cases}
X_N = \max\left[|w_n| \ ; \ 1 \leq n \leq N\right]; \ Y_N = \sqrt{\alpha k \sum_{n=1}^{N} \|w_n\|^2} \\
P_N = k \sum_{n=o}^{N-1} |p_n| \ ; \ Q_N = \sqrt{|w_o|^2 + \frac{1}{\alpha} k \sum_{n=o}^{N-1} \|q_n\|_*^2}
\end{cases}
$$

one has, for any $N > 0$:

(3.15) $\max\left[X_N, Y_N\right] \leq P_N + \sqrt{P_N^2 + Q_N^2}$

PROOF. From (3.13) and the coerciveness it follows:

$$2(w_{n+1} - w_n, w_{n+1}) + 2k\alpha\|w_{n+1}\|^2$$

$$\leq 2k|p_n| \, |w_{n+1}| + 2k\|q_n\|_* \|w_{n+1}\|$$

$$\leq 2k|p_n| \, |w_{n+1}| + \frac{1}{\alpha}k\|q_n\|_*^2 + \alpha k\|w_{n+1}\|^2.$$

By simplifying, and by applying the inequality:

(3.16) $(w_{n+1} - w_n, w_{n+1}) \geq |w_{n+1}|^2 - |w_n|^2$

we get a family of inequalities depending on n; summing up such inequalities for $n = 0, 1, \ldots, m - 1$, most of the terms cancel and we get (see (3.14)):

$$|w_m|^2 + Y_m^2 \leq 2k \sum_{n=o}^{m-1} |w_{n+1}| \, |p_n| + Q_m^2 \leq 2X_m P_m + Q_m^2$$

---

(17) Of course the "true" right hand side is the sum $s_n = p_n + q_n$, and not "p_n and q_n".

From such inequalities we get on one hand:

(3.17) $Y_N^2 \leq 2X_N P_N + Q_N^2 \quad (N \geq 1)$

and on the other hand, because of the monotonicity of all terms, we get also:

(3.18) $|w_m|^2 \leq 2X_N P_N + Q_N^2 \quad (1 \leq m \leq N).$

Taking the maximum for m in $\{1,\ldots,N\}$, (3.18) gives $X_N^2 \leq 2X_N P_N + Q_N^2$, say "half" of (3.15); putting such estimate for X_N into (3.17) we get the remaining part of (3.15).

Remark 3.3. The right hand side of (3.15) also bounds the quantity $\sum_{n=0}^{N-1} |w_{n+1} - w_n|^2$: it is sufficient to replace the estimate (3.16) by the identity $2(x - y, x) = |x - y|^2 + |x|^2 - |y|^2$.

Remark 3.4. The family: $\{n \rightarrow w_{n+1}\}$ satisfies a similar estimate, with initial datum w_1 and right hand side (see [17]) $s_n = p_{n+1} + q_{n+1}$.

In the continuous version of Thm. 3.3 we need to treat a further term, but arguments are quite the same:

Theorem 3.4. *Let $w(t)$, $s(t)$, $r(t)$ be functions with values in V, V', $\mathbf{R}$ such that:*

(3.19)
$$w, w' \in I_{\mathrm{loc}}([0,+\infty[); \quad w(t) \in K - K \quad \text{for} \quad t > 0 \ ;$$
$$r(t) \in L^1_{\mathrm{loc}}([0,+\infty[)$$

(3.20)
$$s(t) = p(t) + q(t) \ ; \quad p(t) \in L^1_{\mathrm{loc}}([0,+\infty[; H);$$
$$q(t) \in L^2_{\mathrm{loc}}([0,+\infty[; V')$$

(3.21) $(w'(t) + Aw(t), w(t)) \leq (s(t), w(t)) + r(t).$

Setting, for $T > 0$ and for any decomposition (3.20):

(3.22) $X(T) = \|w\|_{L^\infty(0,T;H)} \ ; \quad Y(T) = \sqrt{\alpha}\|w\|_{L^2(0,T;V)}$

(3.23)
$$R(T) = \sup_{0 < \theta < T} \int_o^\theta r(t)dt$$

(3.24)
$$P(T) = \|p\|_{L^1(0,T;H)}$$
$$Q(T) = \sqrt{|w(0)|^2 + \frac{1}{\alpha}\|q\|_{L^2(0,T;V')}^2 + 2R(T)}$$

one has:

(3.25)
$$\max[X(T), Y(T)] \le P(T) + \sqrt{P(T)^2 + Q(T)^2}$$

(3.26)
$$\|w\|_{I(0,T)} \le C\left\{|w(0)| + \|s\|_{S(0,T)} + \sqrt{R(T)}\right\}.$$

PROOF. By integration of (3.21) on $(0,T)$, because of coerciveness, one gets:

$$|w(T)|^2 + 2\alpha \int_o^T \|w(t)\|^2 dt$$

$$\le |w(0)|^2 + 2\int_o^T |p(t)|\,|w(t)|dt + 2\int_o^T \|q(t)\|_*\|w(t)\|dt + 2R(T)$$

from which $|w(T)|^2 + Y(T)^2 \le 2X(T)P(T) + Q(T)^2$. Now one can reproduce the arguments used in Thm. 3.3, thus getting (3.25); taking the infimum of the right hand side of (3.25) (with respect to the decompositions $s = p + q$ as in (3.20)) one gets (3.26) with a C which depends only from α.

Remark 3.5. Let us assume $\lambda > 0$ in (1.2). Let us define $\tilde{w}(t) = \exp(-\lambda t)w(t)$, and similarly for $\tilde{s}(t), \tilde{p}(t), \tilde{q}(t)$; let $\tilde{r}(t)$ be defined by $\exp(-2\lambda t)r(t)$. Such new quantities will satisfy a relation similar to (3.21), with A replaced by $\tilde{A} = A + \lambda I$ (and for $\tilde{A}$ coerciveness holds with $\lambda = 0$). Though the property $\tilde{w}(t) \in K - K$ fails, the coerciveness can still be applied; and (3.26) holds true for $\tilde{w}, \tilde{s}, \tilde{r}$; that is, in terms of w, s, r we have:

$$
(3.27) \quad
\begin{cases}
\|\exp(-\lambda t)w(t)\|_{I(0,T)} \leq \\[2mm]
\leq C\Big(|w(0)| + \|\exp(-\lambda t)s(t)\|_{S(0,T)} + \\[2mm]
\qquad + \sqrt{\sup_{0<\theta<T} \int_o^\theta \exp(-2\lambda t)r(t)dt}\ \Big).
\end{cases}
$$

Remark 3.6. If we replace $r(t)$ by a sum $\sum_{n=1}^{l} r_n(t)$, it will be sufficient to replace $R(T)$ by the sum of the corresponding $R_n(T)$. As a typical example we could perturbe the right hand side of (3.21) by an addendum like $|(m(t), w(t))|$ with $m \in S_{\text{loc}}([0, +\infty[)$. The corresponding $R(T)$ will have the form $\int_o^T |(m(t), w(t))|dt \leq \|m\|_{S(0,T)}\|w\|_{I(0,T)} \leq \epsilon\|w\|_{I(0,T)}^2 + \frac{1}{\epsilon}\|m\|_{S(0,T)}^2$; so that, in order to handle such a perturbation, it will be sufficient to increase the constant C and to add $\|m\|_{S(0,T)}$ in the right hand side of (3.26).

Remark 3.7. A somewhat different type of perturbation will be a term of the form $r(t) = m(t) - m(t + i)$, where m is a non negative function of $L_{\text{loc}}^1([0, +\infty[)$ and $i > 0$. The corresponding $R(T)$ will be, for each T, bounded by $\int_o^i m(t)dt$: in fact, for $T \leq i$, $\int_o^T r(t)dt \leq \int_o^T m(t)dt \leq \int_o^i m(t)dt$; and, for $T > i$, $\int_o^T r(t)dt = \int_o^T m(t)dt - \int_i^{i+T} m(t)dt \leq \int_o^i m(t)dt$.

Now we come back to Thm. 3.3, by assuming that the datum $p_n + q_n$ (see [17]) has the form:

$$
(3.28) \quad s_n = p_n + q_n \ ; \ p_n = \frac{1}{k}\int_{J_{k,n}} p(t)dt \ ; \ q_n = \frac{1}{k}\int_{J_{k,n}} q(t)dt
$$

where $s(t) = p(t) + q(t)$ is like in (3.20). Then we have $P_N \leq \|p\|_{L^1(0,Nk;H)}$, $Q_N \leq |w_o| + \frac{1}{\sqrt{\alpha}}\|q\|_{L^2(0,Nk;V')}$ so that, taking the infimum with respect to the decompositions (3.20), (3.15) becomes:

$$
(3.29) \quad \max[X_N, Y_N] \leq C\Big\{|w_o| + \|s\|_{S(0,Nk)}\Big\}.
$$

Remark 3.8. If in (1.2) it is $\lambda > 0$, it could be convenient to replace (3.28) with:

$$(3.30) \quad \begin{cases} p_n = \int_{J_{k,n}} \exp(-\lambda t)p(t)dt \mid \int_{J_{k,n}} \exp(-\lambda t)dt \\ q_n = \int_{J_{k,n}} \exp(-\lambda t)q(t)dt \mid \int_{J_{k,n}} \exp(-\lambda t)dt \end{cases}$$

and, for $k \in]0,k_o]$ (see [5]), to define $\tilde{w}_n$ by means of $\tilde{w}_n = (1 - \lambda k)^n w_n$; with similar notations for $\tilde{p}_n, \tilde{q}_n$, the new problems involves $\tilde{A} = A + \lambda I$; working as in Rem. 3.5 one ends up with [18]:

$$(3.31) \quad \max[\tilde{X}_N, \tilde{Y}_N] \leq C\{|w_o| + \|\exp(-\lambda t)s(t)\|_{S(0,Nk)}\}.$$

Coming back to (3.29) we want now express the left hand member in terms of functions (instead of sequences); to this end we define:

$$(3.32) \quad \begin{cases} w_k(t) \text{ is the step function which on} \\ J_{k,n} \text{ takes the value } w_n. \end{cases}$$

In our previous hypotheses, and using such a notation, let us prove the following result:

Theorem 3.5. *There exists $C = C(\alpha)$ such that:*

$$(3.33) \quad \|w_k(t+k)\|_{I(0,T)} \leq C(|w_o| + \|s(t)\|_{S(0,T+k)})$$

$$\|w_k(t+k)\|_{I(0,T)} \leq$$

$$(3.34)$$

$$\leq C\left(|w_1| + \sqrt{k}\|w_1\| + \|s(t)\|_{S(0,T+k)}\right).$$

PROOF. For T of the form $N \cdot k$, (3.29) gives a more precise relation than (3.33) (we can replace $\|s\|_{S(0,T+k)}$ by $\|s\|_{S(0,T)}$). For

---

[18] Thanks to $1/(1-\lambda k)^N \leq \exp(-\lambda k^N/(1-\lambda k))$; in particular C can depend from k_o, but is independent from N and from $k \in]0,k_o]$.

general T we choose $N = 1 + \text{int}(T/k)$; in (3.33) the left hand side is bounded by $\|w_k(t+k)\|_{I(0,Nk)}$, which in turn is bounded, as we already saw, by $C\{|w_o| + \|s\|_{S(0,Nk)}\}$; by definition of N, the last quantity is bounded by the right hand side of (3.33).

The validity of (3.34) for $T \leq k$ is obvious; so that we only need to prove that, for $T > k$, the right hand side of (3.34) bounds the norm of $w_k(t+k)$ in the space $I(k,T)$ or, which is the same, the norm of $w_k(t+2k)$ in $I(0,T-k)$. This estimate follows from (3.33) in the framework of Rem. 3.4.

We are now ready to prove stability estimates for problem (1.8) with data given by (1.7); with notations similar to (3.32) we define:

$$(3.35) \qquad \begin{cases} u_k(t) \ (\text{resp. } f_k(t)) \ \text{ is the step function which }, \\ \\ \text{on } J_{k,n}, \text{takes the value } u_{k,n} \ (\text{resp. } f_{k,n}) \end{cases}$$

Corollary 3.1. *Under assumption (2.11) there exists* $C = C(\alpha)$ *such that, for* $T > 0$:

$$(3.36) \qquad \|u_k(t+k)\|_{I(0,T)} \leq c\{|w_o| + \|f_k(t)\|_{S(0,T+k)}\}$$

$$(3.37) \qquad \begin{aligned} &\|u_k(t+k)\|_{I(0,T)} \leq \\ &\leq C\{|u_{k,1}| + \sqrt{k}\|u_{k,1}\| + \|f_k(t)\|_{S(0,T+k)}\} \ . \end{aligned}$$

PROOF. Let us choose $v = 0$ in (1.8); because of $u_{k,n} = u_{k,n} - 0 \in K - K$ for $n > 0$, we have (3.12), (3.13) with $w_n = u_{k,n}$, $p_n + q_n = f_{k,n}$; from (3.33), (3.34) we get (3.36), (3.37).

Theorem 3.6. *With the choices (1.7), the method (1.8) is stable, in the sense that the function* $U_k(t)$ *defined in (2.15) satisfies:*

$$(3.38) \qquad \|U_k(t)\|_{I(0,T)} \leq C\{|u_o| + \|f(t)\|_{S(0,T+2k)}\}$$

PROOF. As usually, we firstly work with T of the form $N \cdot k$; for such T we have trivially:

$$\|f_k(t)\|_{S(0,T)} \leq \|f(t)\|_{S(0,T+k)} \; ; \; \|U_k(t)\|_{I(0,T)} \leq \|u_k(t)\|_{I(0,T+k)}$$

such relations, inserted into (3.36), give (3.38) for the special values $T = N \cdot k$; the general case of $T > 0$ follows as usually by taking a greater interval for the norm of f in the space S. Remark that (3.38) gives directly a bound for U_k which does not depend on k; however, working with (3.37) instead of (3.36) we end up with an estimate "more precise" for $k \to 0$: because of (3.9) the term $|u_o|$ would be replaced by $|\mathcal{P}u_o|$.

Remark 3.9. If (1.2) holds with $\lambda > 0$, one can modify (1.7) as suggested in Rem. 3.5.

Theorem 3.7. *Setting:*

(3.39)
$$\begin{cases} E(k,T) = \|f_k(t) - f(t)\|_{S(0,T+3k)} + \\ + \|f(t+k) - f(t)\|_{S(0,T+3k)} + |u_{k,2} - u_{k,1}| + \\ + \sqrt{k} \, \|u_{k,2} - u_{k,1}\| + |u_{k,1} - \mathcal{P}u_{k,0}| \end{cases}$$

we have, for $T > 0$:

(3.40)
$$\|kU_k'(t)\|_{I(0,T)} \leq C \, E(k,T).$$

PROOF. $kU_k'(t)$ is the step function taking the value $w_n = u_{k,n+2} - u_{k,n+1}$ on the interval $J_{k,n}$.
The choices $v = u_{k,n+2}$ in (1.8), and $v = u_{k,n+1}$ in (1.8) with n replaced by $n+1$ give rise to two inequalities which, summed together, imply (3.12), (3.13) with $p_n + q_n = f_{k,n+2} - f_{k,n+1}$. From (3.33) with $s(t) = f_k(t+k)$ and $w_o = u_{k,2} - u_{k,1}$, one gets easily (3.40).

Remark 3.10. Working with $u_{k,n+p+1}$ instead of $u_{k,n+2}$, we could estimate $u_k(t + pk + k) - u_k(t + k)$ by $u_{k,p+1} - u_{k,1}$ and $f(t + pk) - f(t)$.

From now on we will work assuming (1.7); however the essential property we will use is given by:

(3.41) $E(k,T) = o(1)$; and $E(k,T) = O(k)$ if (2.6) holds true.

Remark 3.11. Let us control that (3.41) is valid if we choose the discretization (1.7): the part of $E(k,T)$ referring to initial values has been checked in (3.11) and Thm. 3.1, and the remaining part is well known. Let us give a sketch for the latter. Fix any decomposition $f = g + h$ with (2.1); with the obvious meaning for $g_k(t)$, $h_k(t)$ we have (as in Thm. 3.5 we work initially with T of the form $N \cdot k$):

$$\|h(t+k) - h(t)\|_{L^2(0,T;V')} \le k\|h'(t)\|_{L^2(0,T+k;V')};$$

$$\|h_k(t) - h(t)\|_{L^2(0,T;V')} \le k\|h'(t)\|_{L^2(0,T+k;V')};$$

$$\|g(t+k) - g(t)\|_{L^1(0,T;H)} = \sum_{n=0}^{N-1} \int_{J_{k,n}} |g(t+k) - g(t)|dt =$$

$$= \int_o^k \sum_{n=0}^{N-1} |g(t+(n+1)k) - g(t+nk)|dt \le$$

$$\le k \sum_{n=o}^{N-1} V(J_{k,n};H) \le kV(0,T;H);$$

$$\|g_k(t) - g(t)\|_{L^1(0,T;H)} = \sum_{n=0}^{N-1} \int_{J_{k,n}} \left| \frac{1}{k} \int_{J_{k,n}} g(s)ds - g(t) \right| dt \le$$

$$\le \frac{1}{k} \sum_{n=0}^{N-1} \int_{J_{k,n}} \int_{J_{k,n}} |g(s) - g(t)|dt \, ds$$

$$\le k \sum_{n=0}^{N-1} V(J_{k,n};H) \le kV(0,T;H).$$

As usual, we pass to the case of general T by enlarging the interval; by taking the infimum over the decompositions (2.1) one gets the estimate $O(k)$ under the assumption (2.16); by density one gets the estimate $o(1)$ for general f.

From (3.38), (3.40), (3.41) one obviously gets (see [10]):

Theorem 3.8. *The family $U_k(t)$ admits weakly* adherent points in $I(0, +\infty)$; under the assumption (2.16) such points have first derivative in $I(0, +\infty)$.*

In the framework of *linear* problems theorem 3.8 would already be sufficient in order to pass to the limit for $k \to 0$ [19]; here we will need the more precise estimate:

$$(3.42) \qquad \|U_k(t) - U_h(t)\|_{I(0,T)} \leq C\{E(k,T) + E(h,T)\}$$

that will be proved in Section 4 and which will easily imply Thm. 2.1. Let us also remark that, once (3.42) has been proved, from (3.41) it follows that $\{U_k\}$ is a Cauchy family, and its limit U verifies the relation:

$$(3.43) \qquad \|U_k(t) - U(t)\|_{I(0,T)} \leq C \, E(k,T).$$

Remark 3.12. The solution u being, in general, "not too smooth", from a practical point of view there is no scope to ask for "smooth" approximations. However, from a theoretical point of view, it could be interesting to remark that this is possible, with no influence on the error estimates. In fact let $\chi_o(t)$ be the characteristic function of the interval $[-1/2, +1/2[$; and let $\chi_m(t)$ be the function defined by induction with the formula $\chi_{n+1} = \chi_n \star \chi_o$ (convolution product). Setting $\varphi_{k,m}(t) = \sum_{n=0}^{\infty} \chi_m(t/k - n)u_{k,n+1}$ the function $\varphi_{k,m}$ give, for increasing values of m, smoother and smoother approximations of u (by a translation, u_k and U_k coincide with $\varphi_{k,0}$ and $\varphi_{k,1}$

---

[19] Though it is not yet easy to get strong convergence nor optimal error estimate. It is here that, usually, in the framework of equations, one uses sommability assumptions on $u''(t)$ that for inequalities cannot hold true.

respectively). Because of $\|U_k - \varphi_{k,m}\|_{I(0,T)} \leq C_m \|kU_k'\|_{I(0,T)}$ once (3.43) has been checked we will have:

$$(3.44) \qquad \|U(t) - \varphi_{k,m}(t)\|_{I(0,+\infty)} \leq C_m \, E(k,+\infty).$$

Remark 3.13. By applying Rem. 3.3 to Thm. 3.3 one gets:

$$(3.45) \qquad k\|U_k'(t)\|_{L^2(0,+\infty;H)}^2 \leq C$$

so that, in terms of functions $\varphi_{k,m}$ introduced in Rem. 3.12, one has:

$$(3.46) \qquad k^{r+1/2}\|\varphi_{k,m}\|_{H^{r+1}(0,+\infty;H)} \leq C_m \quad (0 \leq r \leq m-1).$$

Always from Rem. 3.3, in connection with Thm. 3.7, one derives $k\|U_k'(t+k) - U_k'(t)\|_{I(0,+\infty)}^2 \leq C \, E(k,+\infty)$; or, in an equivalent form:

$$(3.47) \qquad k^{3/2}\|\varphi_{k,2}\|_{H^2(0,+\infty;H)} \leq C \, E(k,+\infty)$$

and in general it follows:

$$(3.48) \quad k^{r+3/2}\|\varphi_{k,m}\|_{H^{r+2}(0,+\infty;H)} \leq C_m E(k,+\infty) \quad (0 \leq r \leq m-2).$$

We are dealing with estimates which are typical of the interpolation theory; we will see in Section 5 that they can fournish further smoothness properties of the limit function u.

4. The limit problem. Let us firstly remark that, with notations (3.25), (2.15), one can rewrite (1.8) in the form:

$$(4.1) \qquad \left(\tilde{u}_k'(t) + Au_k(t+k) - f_k(t), u_k(t+k) - v\right) \leq 0 \quad \forall v \in K$$

and that, setting:

$$(4.2) \qquad \ell_k(t) = n + 1 - t/k \quad \text{for} \quad t \in J_{k,n}$$

The functions $u_k(t), \hat{u}_k(t)$ are related through:

$$(4.3) \qquad \tilde{u}_k(t) = \ell_k(t)u_k(t) + [1 - \ell_k(t)]u_k(t+k)$$

In particular, from the obvious relation:

(4.4) $$0 \leq \ell_k(t) \leq 1$$

one gets:

(4.5) $$\|\tilde{u}_k(t) - u_k(t+k)\| \leq \|u_k(t) - u_k(t+k)\| = \|k\tilde{u}'_k(t)\|;$$

an analogous estimate holds for the H-norms; and furthermore, for each interval (a,b):

(4.6)
$$\|\tilde{u}_k(t)\|_{I(a,b)} \leq \|u_k(t)\|_{I(a,b)} +$$
$$+ \|u_k(t+k)\|_{I(a,b)} \leq 2\|u_k(t)\|_{I(a,b+k)}.$$

We will now translate (4.1) into a relation in terms of U_k (given by (2.15)) with a right hand side involving terms of type (4.5) (say "small" terms, because of Thm. 3.7); we will then apply Thm. 3.4 to the difference $U_k - U_h$ in order to get (3.42); as an intermediate result we will get a less precise bound involving $\sqrt{E(k,T)}$ instead of $E(k,T)$.

Lemma 4.1. *For any v in K one has:*

(4.7)
$$\begin{cases} (U'_k(t) + AU_k(t) - f_k(t+k), U_k(t) - v) \leq \\ \leq M\|kU'_k(t)\| \, \|U_k(t) - v\| + \\ + \ell_k(t)(U'_k(t) + Au_k(t+k), -kU'_k(t)). \end{cases}$$

PROOF. The term $(AU_k(t), U_k(t) - v)$ can be bounded by:

$$(Au_k(t+2k), U_k(t) - v) + (A\tilde{u}_k(t+k) -$$

$$- Au_k(t+2k), U_k(t) - v) \leq (Au_k(t+2k), U_k(t) - v) +$$

$$+ M\|\tilde{u}_k(t+k) - u_k(t+2k)\| \, \|U_k(t) - v\|.$$

PROOF. The term $(AU_k(t), U_k(t) - v)$ can be bounded by:

$$(Au_k(t + 2k), U_k(t) - v) + (A\tilde{u}_k(t + k) -$$

$$- Au_k(t + 2k), U_k(t) - v) \leq (Au_k(t + 2k), U_k(t) - v) +$$

$$+ M\|\tilde{u}_k(t + k) - u_k(t + 2k)\| \, \|U_k(t) - v\|$$

The last term can be bounded (see (4.5)) by means of $M\|U_k'(t)\|$ $\|U_k(t) - v\|$; the remaining part of the left hand side in (4.7) is now given by: $(U_k'(t) + Au_k(t + 2k) - f_k(t + k), U_k(t) - v)$. We decompose $U_k(t) - v$ in the form $[U_k(t) - u_k(t+2k)] + [u_k(t+2k) - v] = -\ell_k(t)kU_k'(t) + [u_k(t+2k) - v]$ (see (4.3)); the part corresponding to the last term is ≤ 0 because of (4.1), and (4.7) is proved.

Corollary 4.4. *For any $v \in K$, a.e. in $(0, +\infty)$, one has:*

$$(4.8) \quad \begin{cases} (U_k'(t) + AU_k(t) - f(t), U_k(t) - v) \leq \\ \\ \leq M\|kU_k'(t)\| \, \|U_k(t) - v\| + M\|u_k(t + 2k)\| \, \|U_k(t) - v\| \\ \\ + \big|(f_k(t + k), -kU_k'(t))\big| + \big|(f_k(t + k) - f(t), U_k(t) - v)\big|. \end{cases}$$

PROOF. On the left hand side we replaced the term $f_k(t + k)$ appearing in (4.7) by $f(t)$; this is taken into account from the last term on the right hand side of (4.8). In particular, starting from (4.7), we just need to estimate the last term by $|(Au_k(t + 2k) - f_k(t + k), -kU_k'(t))|$ (see (4.4): the part $\ell_k(t)(U_k'(t), -kU_k'(t))$ is ≤ 0 and in the remaining part we estimate $\ell_k(t)$ by 1).

Corollary 4.2. *For any choice of $h, k > 0$ one has:*

$$(4.9) \quad \|U_k(t) - U_h(t)\|_{I(0,T)} \leq C\left\{\sqrt{E(k,T)} + \sqrt{E(h,T)}\right\} \quad \forall T > 0.$$

PROOF. Let us choose, a.e. in $(0, +\infty)$, $v = U_h(t)$ into (4.8); then reverse the roles of h and k. Adding together such two inequalities one can treat the resulting inequality by means of Thm.

3.4 with $w(t) = U_k(t) - U_h(t)$. The initial value $w(0)$ splits into $u_{k,1} - u_{h,1} = [u_{k,1} - \mathcal{P}u_o] - [u_{h,1} - \mathcal{P}u_o]$ so that its norm in H is bounded by $E(k,T) + E(h,T)$. The remaining part is bounded by the term $M\|u_k(t+2k)\|_{L^2(0,T;V)}\|kU_k'(t)\|_{L^2(0,T;V)}$ plus a similar term in h; because of Thm. 3.6, 3.7, all terms are bounded by $E(k,T) + E(h,T)$.

The estimate (4.9) is less strict than (3.42), but it is sufficient to pass to the limit; in fact:

Theorem 4.1. *The family $\{U_k(t)\}$ is a Cauchy family in $I(0,+\infty)$ as $k \to 0$; its limit U solves Pb. 2.1 and satisfies (2.8), (2.10), (2.13); any other solution u of Pb. 2.1 must coincide with U.*

PROOF. $\{U_k\}$ is a Cauchy family because of (4.9), (3.41); its limit U solves Pb. 2.1 and satisfies obviously $U \in C^o([0,+\infty[;H)$; and $U(0) = \lim_{k \to 0} U_k(0) = \lim_{k \to 0+} u_{k,1} = \mathcal{P}u_o$ (see Section 3).

Fixed any $v(t)$ with (2.6), one can choose $v = v(t)$ in (4.8); by applying to $U_k(t) - v(t)$ the formula in [12] one gets an inequality which, in the distribution sense, passes to the limit giving (2.10).

Let u be another solution of Pb. 2.1. Choose in (4.8) $v \equiv u$ and integrate on $(0,+\infty)$; then add this inequality to (2.12) written with $v \equiv U_k$. One gets, with obvious notations:

$$\int_o^{+\infty} (A(U_k - u), U_k - u)dt \le$$

$$\le C\Big(\int_o^{+\infty} (f_k(t+k) - f(t), U_k(t) - u(t))dt +$$

$$+ \|kU_k'\|_{L^2(0,+\infty;V)}\|U_k - u\|_{L^2(0,+\infty;V)}$$

$$+ \|kU_k'\|_{I(0,+\infty)}\|f_k - f\|_{S(0,+\infty)} \Big)$$

so that, by coerciveness, $U_k \to u$ in $L^2(0,+\infty;V)$.

Remark 4.1. Let us assume only $f \in L^2(0, +\infty; V')$ (see Rem. 2.2); (2.12) has a meaning for u only in $L^2(0, +\infty; V)$; the uniqueness we just proved still holds in the framework of Rem. 2.2.

Remark 4.2. If (2.11) fails Thm. 4.1 has an obvious extension in the framework of Rem. 2.5; by modification of $f_{k,n}$ as suggested in Rem. 3.8 one can also treat the problem in weighted spaces.

Theorem 4.4. *The operator* $T : \{u_o, f\} \to u$ *(u solution of Pb. 2.1) is Lipschitz continuous from* $H \times S(0, +\infty)$ *into* $I(0, +\infty)$; *furthermore, under assumption (2.16), one has* $u' \in I(0, +\infty)$ *and the map* $\{u_o, f\} \to u'$ *is bounded from the space (2.16) into* $I(0, +\infty)$.

PROOF. The last claim follows from (3.40). The first statement follows from the fact that the map $\{u_o, f\} \to U_k$ is Lipschitz continuous uniformly in k: in fact, starting from a second couple $\{u_o^*, f^*\}$ of data, with obvious notations one can choose $v = u_{k,n+1}^*$ into (1.8); and $v = u_{k,n+1}$ into (1.8) related to $u_{k,n+1}^*$. To the difference $w_n = u_{k,n} - u_{k,n}^*$ one can apply Thm. 3.5, so getting:

$$\|u_k(t+k) - u_k^*(t+k)\|_{I(0,+\infty)} \leq$$

$$\leq C\{|u_{k,1} - u_{k,1}^*| + \|f_k(t) - f_k^*(t)\|_{S(0,+\infty)}\};$$

the estimate on the "step extension" $u_k - u_k^*$ holds true also for the "linear extension" $U_k - U_k^*$, because of (4.6).

Let us now come back to (4.7), in order to give a stronger estimate of the last term.

Lemma 4.2. Setting:

$$(4.10) \qquad m_k(t) = k\ell_k(t)|\tilde{u}_k'(t)|^2/2$$

one has, for any $v \in K$, a.e. in $t > 0$:

$$(4.11) \quad \begin{cases} (U_k'(t) + AU_k(t) - f_k(t+k), U_k(t) - v) \leq \\ \\ \leq M\|kU_k'(t)\| \, \|U_k(t) - v\| + M\|kU_k'(t)\|^2 \\ \\ +|(f_k(t+k) - f_k(t), -kU_k'(t))| + m_k(t) - m_k(t+k). \end{cases}$$

PROOF. Let us estimate the factor of $\ell_k(t)$ in the last term of (4.7); because of $\ell_k(t) \geq 0$ we can then insert such estimate in (4.7). On the interval $J_{k,n}$ such a factor takes the value:

$$\left(u_{k,n+2} - u_{k,n+1} + kAu_{k,n+2} - kf_{k,n+1}, u_{k,n+1} - u_{k,n+2}\right) =$$

$$= \left(\left[u_{k,n+2} - u_{k,n+1}\right] - \left[u_{k,n+1} - u_{k,n}\right], u_{k,n+1} - u_{k,n+2}\right) +$$

$$+ k\left(Au_{k,n+2} - Au_{k,n+1}, u_{k,n+1} - u_{k,n+2}\right)$$

$$+ k\left(f_{k,n} - f_{k,n+1}, u_{k,n+1} - u_{k,n+2}\right)$$

$$+ \left(u_{k,n+1} - u_{k,n} + kAu_{k,n+1} - kf_{k,n}, u_{k,n+1} - u_{k,n+2}\right).$$

The last term is ≤ 0 because of (1.8) with $v = u_{k,n+2}$; the second term is ≤ 0 if $\lambda = 0$ in (1.2), but in general, being on $J_{k,n}$, it can be estimated by $k|(A[u_k(t + 2k) - u_k(t + k)], -kU_k'(t))|$.
The third term can be written as $k(f_k(t) - f_k(t+k), -kU_k'(t))$; finally the first term, because of Rem. 3.3, can be bounded by:

$$\{|u_{k,n+1} - u_{k,n}|^2 - |u_{k,n+2} - u_{k,n+1}|^2\}/2 =$$

$$= k\{|\tilde{u}_k'(t)|^2 - |\tilde{u}_k'(t + k)|^2\}/2$$

Such a quantity, multiplied by $\ell_k(t)$, coincides (because of the periodicity of ℓ_k) with $m_k(t) - m_k(t + k)$. In the remaining terms we estimate the factor $\ell_k(t)$ by 1; (4.11) then follows from (4.7).

Theorem 4.3. *The estimate (3.42) holds true; in particular one has:*

(4.12) $\|U_k(t) - u(t)\|_{I(0,T)} \leq C\, E(k,T) \quad \forall T > 0$

PROOF. Let us firstly work with $V_k(t) = U_k(t + k)$, $V_h(t) = U_h(t + h)$. Writing (4.11) in terms of $V_k(t)$, we can choose in it $v \equiv V_h(t)$; by inverting the roles of h and k, and by adding such two inequalities, we get a problem in $V_k - V_h$ which can be handled by means of Thm. 3.4 and Rem. 3.7: $m_k(t)$ being given by (4.10) we have:

$$\|V_k(t) - V_h(t)\|_{I(0,T)} \leq C\{\tilde{E}(k,T) + \tilde{E}(h,T)\}$$

where $\tilde{E}(k,T) = E(k,T) + \sqrt{\int_o^k m_k(t+k)dt} = E(k,T) + \sqrt{(|u_{k,2} - u_{k,1}|^2)/4} = E(k,T) + |u_{k,2} - u_{k,1}|/2$ and the last term is bounded by $E(k,T)$[20] In order to conclude we just need an estimate for $U_k(t) - V_k(t)$ (and the similar estimate with k replaced by h); because of

$$\|U_k(t) - V_k(t)\|_{I(0,T)} = \|U_k(t) - U_k(t+k)\|_{I(0,T)} \leq$$

$$\leq \|kU_k'(t)\|_{I(0,T+k)}$$

also such a term is bounded by $E(k,T)$; so that (3.42) is proved and its limit for $k \to 0$ gives (4.12).

Remark 4.3. The ideas formally developped in Rem. 3.12, 3.13 are now completely justified; formulae (3.44),...,(3.47) hold true with $U = u$ the unique solution of Pb.2.1.

Remark 4.4. Thm. 2.1 has been entirely proved: formula (1.4), under assumption (2.16), follows by applying in (2.7) the relation [12] from right to left.

5. Further remark and open problems. Let us prove Thm. 2.2. If K is a cone we can choose in (4.8) $v = 0$; then $v = 2U_k(t)$. By rewriting $(U_k'(t), U_k(t))$ in the form $(1/2|U_k(t)|^2)'$, we get two inequalities which pass to the limit (in the distribution sense) giving

---

[20] If we had worked directly with U_k, U_h (instead of V_k, V_h) the corresponding term would have been $\int_o^k m_k(t)dt = (|u_{k,1} - u_{k,o}|^2)/4 \to 0$. Of course, for suitably choosen values of $u_{k,0}$, we could work directly with U_k, U_h.

(2.17). Choosing in (4.8) $v = U_k(t) + w$, w in K, we get (2.18); from (2.18), if K is a subspace, follows obviously (2.20). In order to get $(2.19)^{(21)}$ let us choose in (1.8) $v = u_{k,n+1} + u_{k,m}$ with a fixed $m > 0$; summing up such inequalities for $n = m$, $m + 1, ..., m + p - 1$, we get $\|u_k(t+s) - u_k(t)\|_{L^2(0,+\infty;H)} \leq Cs$ if p, k are related by $s = pk$; from such a relation we get (2.19).

The following considerations are to be viewed as a hint for possible developments which, at a first glance, seem to be of some interest.

(i) We never assumed the validity of (1.10), (1.11); under such hypotheses one could derive both further smoothness results for the solution (results already known; see [2]) and new stability estimates for the scheme (1.8).

(ii) The map T from $\{u_o, f\}$ to the solution u of Pb. 2.1 is Lipschitz continuous from $H \times S(0, +\infty)$ into $I(0, +\infty)$; and it is bounded from the *set* (2.16) [22] into the space $\{u \in I(0, +\infty);\ u' \in I(0, +\infty)\}$. In particular the general hypotheses for applying the theory of non linear interpolation hold true (see [7]). However the problem if explicit the intermediate *set* between (2.3) and (2.16) seems to be delicate. The same type of problem comes out if we fix a $\theta \in]0, 1[$ and we look for the couples $\{u_o, f\}$ such that

$$(5.1) \qquad\qquad E(k, T) = O(k^\theta)$$

(the extremal cases $\theta = 0$ and $\theta = 1$ are solved by (3.41)). It is a problem connecting non linear interpolation and singular perturbation of steady variational inequalities (this being the type of problem solved by $u_{k,1}$: see (1.8) with $n = 1$, where k must tend to 0).

(iii) Formulae (5.1), (4.12) give a precise estimate for the error order. On the other hand they can also give smoothness properties of the solution u: if we represent u as the sum of two terms:

$$(5.2) \qquad\qquad u = a_k + b_k$$

---

[21] Such a regularity result, as well as the ones we will show later on, seem to be new; all of them, as the previous results, are independent from the validity of (1.10), (1.11).

[22] Remark that we must deal with a set, and not a subspace, because of the (convex) restriction $u_o \in K$.

where $a_k = u - U_k$, $b_k = U_k$, from (4.12), (5.1) on one hand, and (3.45) from the other hand, we get that u belongs to an intermediate space between $I(0, +\infty)$ and $H^1(0, +\infty; H)$.[23] More generally, let us start from (3.44) (with $U = u$), (5.1) and (3.46); fixed any s with $1/(1 + 2\theta) < s < 1$, let us choose r such that $\frac{r+1/2}{r+1/2+\theta} > s$ (say $r > \frac{\theta s}{1-s} - 1/2$); with a notation already used in (2.25) (see [15]) we get $u \in H^{\frac{\theta(r+1)}{r+1/2+\theta} - \epsilon}(0, +\infty : V_s)$ for any $\epsilon > 0$ [24]. If (5.1) holds with $\theta \geq 1/2$ it will be convenient to choose $r \to +\infty$ (and one can choose s close to 1); otherwise it will be convenient to choose r "small" in function of s. Such two cases correspond to:

(5.3)
$$\begin{cases} \text{If (5.1) holds true for a } \theta \geq 1/2, \text{ one has} \\[2mm] u \in H^{\theta - \epsilon}(0, +\infty; V_{1-\epsilon}) \text{ for any } \epsilon > 0 \end{cases}$$

(5.4)
$$\begin{cases} \text{If (5.1) holds true for a } \theta > 1/2, \text{ one has} \\[2mm] u \in H^{\theta s + \frac{1-s}{2} - \epsilon}(0, +\infty; V_s) \text{ for } \epsilon > 0, \frac{1}{1+2\theta} < s < 1. \end{cases}$$

(iv) The smoothness obtained in (5.3) is surely non optimal: if (5.1) is obtained by interpolation between the cases $\theta = 0$ and $\theta = 1$, we must expect for u a smoothness "intermediate" between $u \in I(0, +\infty)$ and $u, u' \in I(0, +\infty)$; in particular we must expect the estimate:

(5.5)
$$\|u(t + \tau) - u(t)\|_{I(0, +\infty)} \leq C \tau^\theta$$

---

[23] It is the space $\underline{S}(\infty, -\theta, I(0, \infty); \infty, 1/2, H^1(0, \infty; H))$ with the notations of [5]. In order to avoid the (unessential) difficulties related to the measurability with respect to k one can work just with a sequence $\{k_n\}$ of k, with $k_n \to 0$, say with the space that in [5] is denoted by $\underline{s}(+\infty, -\theta, I(0, +\infty); +\infty, 1/2, H^1(0, +\infty; H))$.

[24] Of course, in order to drop the ϵ, one could use instead of $V_s = S\left(2, \frac{s+1}{2}, V; 2, \frac{s-1}{2}, V'\right)$, some intermediate spaces between V and V' (or V and H) obtained by using different sommability exponent. In such a framework one could also enlarge (5.1) into a more general assumption of sommability instead of a boundedness.

(v) Let us show the validity of (5.5) in a particular case: we will assume (2.24), and we further assume that, for a suitable $\theta \in$ $]0,1]$, it is:[25]

(5.6) $f = g + h \; ; \; g \in W^{\theta,1}(0,+\infty;H) \; ; \; h \in H^{\theta}(0,+\infty;V').$

Under hypotheses (5.6) the validity of (5.1) is quite trivial; and (5.5) follows directly from Rem. 3.10.

(vi) We end up with a new smoothness result which is in same sense stronger than the ones until now obtained and, for some values of θ, cannot be obtained "by interpolation between two different results". Under the assumption (5.6) (and more generally under the assumption (5.1)) let us come back to (5.2) with $b_k = \varphi_{k,m}$ (notations of (3.48), where now it is $U \equiv u$); working as we did for (5.4) (also now it is convenient to choose r small) fix any s with $\frac{3-2\theta}{3} < s < 1$, and any r with $\frac{r+3/2-\theta}{r+3/2} > s$ (say $r < \frac{\theta}{1-s} - 3/2$); we get $u \in H^{\theta \frac{r+2}{r+3/2} - \epsilon}(0,+\infty;V_s)$ for any $\epsilon > 0$; and we can rewrite such a relation in the form:

(5.7)
$$\begin{cases} \text{under the assumtion (5.6), it is} \\ u \in H^{\theta + \frac{1-s}{2} - \epsilon}(0,+\infty;V_s), \; \frac{3-2\theta}{3} < s < 1, \; \epsilon > 0 \end{cases}$$

and in particular, for $\theta = 1$ (the assumption (2.16) is sufficient to give (5.1) with $\theta = 1$):

(5.8)
$$\begin{cases} \text{under the assumption (2.16), for any } \epsilon > 0 \\ \text{it is } u \in H^{\frac{3-s}{2} - \epsilon}(0,+\infty;V_s) \text{ for } \frac{1}{3} < s < 1; \end{cases}$$

(in particular it is $u \in H^{\sigma}(0,+\infty;V_{1/3})$ for any $\sigma < 4/3$). Remark that, but the ϵ, and in a restricted range for s, relations (5.7), (5.8) are optimal: compare with (2.25) in the framework of equations.

---

[25] With the choices (1.7), for small k, the terms in $u_{k,2}, u_{k,1}, u_{k,0}$ disappear from E(k,T). In connection with Rem. 2.6 let us also remark that, for $\theta < 1/2$, the extension of g,h (values 0 for $t<0$) does not destroy the smoothness (5.6); the same holds for $\theta > 1/2$ if h(0)=0; as usual, the case $\theta = 1/2$ is more delicate.

References

[1] C. Baiocchi, F.Brezzi, *Optimal Error Estimates for Linear Parabolic Problems under Minimal Regularity Assumption*, Calcolo **20** (1983), 143-176.

[2] H.Brezis, *Equations et inéquations non linéaires dans les espaces vectoriels en dualité*, Ann.Inst.Fourier **18** (1968), 115-175.

[3] H.Brezis, *Opérateurs maximaux monotones et sémi-groupes de contractions dans les espaces de Hilbert*, North-Holland, Amsterdam (1973).

[4] G.Fichera, *Boundary Value Problems in Elasticity with Unilateral Constraints*, Handbuch der Physik **6**, Ab.2, Springer Verlag (1972), 347-389.

[5] J.L.Lions, J.Peetre, *Sur une classe d'espaces d'interpolation*, Pubblicazioni I.H.E.S. **19** (1964), 1-68.

[6] J.L.Lions, G.Stampacchia, *Variational Inequalities*, Comm. P.A.M. **20** (1967), 493-519.

[7] L.Tartar, *Théorèmes d'interpolation non linéaire et applications*, C.R.A.S. **270** (1970), A 1729-A 1731.

[8] F. Tomarelli, *Regularity Theorems and Optimal Error Estimates for Linear Parabolic Cauchy Problems*, Numerische Mathematik **45** (1984), 23-50.

Dipartimento di Matematica

Università di Pavia

Strada Nuova, 65

I-27100 PAVIA

HOMOGENIZATION FOR
NON LINEAR ELLIPTIC EQUATIONS
WITH RANDOM HIGHLY OSCILLATORY COEFFICIENTS

ALAIN BENSOUSSAN

Dedicated to Ennio De Giorgi on his sixtieth birthday

Introduction. We consider in this article non linear elliptic equations of the form

$$-\frac{\partial}{\partial x_i}(a_{ij}^\epsilon(x;\omega)\frac{\partial u^\epsilon}{\partial x_j}) = H^\epsilon(x, Du^\epsilon, u^\epsilon, \omega), \quad x \in O$$

$$u_{|\partial O}^\epsilon = 0$$

where $a_{ij}^\epsilon(x;\omega)$ is a sequence of stochastic processes, the matrix a_{ij}^ϵ being uniformly bounded and coercive. The non linear operator H^ϵ has quadratic growth in Du^ϵ.

As far as the linear part is concerned, our approach is to follow ideas of generalized homogenization, introduced by L.Tartar [17], [18], [19], see also F.Murat [13], F.Murat,L.Tartar [14]. We provide sufficient conditions to infer homogenization. Additional conditions are necessary, unlike in the deterministic case where boundedness and coercivity of the matrix a_{ij}^ϵ suffice to derive homogenization results (at least for subsequences).

We recover the classical periodic case considering

$$a_{ij}^{\epsilon}(x;\omega) = \tilde{a}_{ij}(\frac{x}{\epsilon} + \omega)$$

where $\tilde{a}_{ij}(\omega)$ is a periodic function in all components, with period 1.
The main example of the theory concerns the case

$$a_{ij}^{\epsilon}(x;\omega) = a_{ij}(\frac{x}{\epsilon},\omega)$$

where $a_{ij}(y;\omega)$ is a *stationary process*. This situation has been considered in Jurinski [10] Kozlov [11] Papanicolau-Varadhan [15], Bensoussan-Blankenship [1], and the results are reviewed. We then deal with the nonlinear problem itself in a way simular to that of the article Bensoussan-Blankenship [1].

We cannot treat however the case of non divergence form operators. The extension to the random case of the results of Bensoussan-Boccardo-Murat [2] for the deterministic case remain an open problem.

Non linear stochastic homogenization has been also considered by G.Dal Maso and L.Modica [6], [7] in a different context, namely that of random integral functionals to be minimized. The theory of Γ-convergence (De Giorgi [8]) is the main tool of this approach. The random case has been also initiated by De Giorgi (see L.Modica [12]) in dimension 1.

A complete treatment can be found in G.Facchinetti, L.Russo [9].

The author would like to thank very much L.Boccardo and F.Murat for their help and suggestions. This paper owes a lot to both of them.

1.General homogenization theory.

1.1.Assumptions. Notations. Let (Ω, A, P) be a probability space with A countably generated. We shall consider in the sequel a family of matrices $A^{\epsilon}(x;\omega), x \in \mathbf{R}^n$. We shall say that $A^{\epsilon} \in L(\mathbf{R}^n; \mathbf{R}^n)$ belongs to $M(\alpha, \beta, O)$ where O is a bounded domain of $\mathbf{R}^n$, and $\alpha, \beta > 0$, whenever

(1.1) A^ϵ is measurable and

$$A^\epsilon(x;\omega)\xi.\xi \geq \alpha|\xi|^2, \ (A^\epsilon)^{-1}(x;\omega)\xi.\xi \geq \beta|\xi|^2,$$
$$\alpha,\beta > 0 \ , \ \forall \xi \in \mathbf{R}^n \ ; \text{ a.e. } x \in O, \text{ a.s.}$$

Generalizing a concept of L.Tartar [17] (cf. also F.Murat, L.Tartar [14]) we shall say that:

(1.2) A^ϵ converges to A^o in O, whenever

$A^o \in M(\alpha,\beta,O)$, and $\forall f \in L^2(\Omega, A, P, H^{-1}(O))$

defining u^ϵ and u^o by

$$-\text{div } [A^\epsilon \text{ grad } u^\epsilon] = -\text{div } [A^o \text{ grad } u^o] = f$$

one has

a) $u^\epsilon \to u^o$ in $L^2(\Omega, A, P, H^{-1}(O))$ weakly

b) $A^\epsilon \text{ grad } u^\epsilon \to A^o \text{ grad } u^o$ in $L^2(\Omega, P; (L^2(O))^n)$ weakly.

In the sequel we shall denote by $a_{ij}^\epsilon(x;\omega)$ the coefficients of the matrix A^ϵ. We shall also denote by $<,>$ the duality product between $H_0^1(O)$ and $H^{-1}(O)$.

1.2. A sufficient condition for H convergence. We shall consider the following property: there exists a sequence $w_\ell^\epsilon(x;\omega)$ such that

a) $w_\ell^\epsilon(x;\omega) \in L^2(\Omega, A, P; H^{-1}(O)), \ \ell = 1...n$

$w_\ell^\epsilon - x_\ell \to 0$ in $L^2(\Omega, A, P; H^{-1}(O))$ weakly

and $L^2(\Omega, A, P; L^2(O))$ strongly

(1.3)   b) $a_{ij}^\epsilon \dfrac{\partial x_\ell^\epsilon}{\partial x_j} - q_{\ell i} \to 0$, in $L^2(\Omega, A, P; H^{-1}(O))$ strongly,

where $q_{\ell i} \in L^2(\Omega, A, P; L^2(O))$, for any i, ℓ

c) $\dfrac{\partial}{\partial x_i}\left[a_{ij}^\epsilon \dfrac{\partial w_\ell^\epsilon}{\partial x_j}\right] \to \dfrac{\partial}{\partial x_i} q_{\ell i}$ in

$L^2(\Omega, A, P; H^{-1}(O))$ strongly, for any ℓ.

Remark 1.1. Since $a_{ij}^\epsilon \frac{\partial w_\ell^\epsilon}{\partial x_j}$ is bounded in $L^2(\Omega, A, P; L^2(O))$, it converges to $q_{\ell i}$ in $L^2(\Omega, A, P; L^2(O))$ weakly.

We then assert the following

Theorem 1.1. *We assume that the matrix* $A^\epsilon(x; \omega) = a_{ij}^\epsilon(x; \omega)$ *belongs to* $M(\alpha, \beta; O)$, *and that (1.3) is satisfied. Then the matrix* $A^o(x; \omega) = q_{ij}(x; \omega)$ *belongs to* $M(\alpha, \beta; O)$ *and* A^ϵ H *converges to* A^o.

PROOF. 1. *The matrix* A^o *belongs to* $M(\alpha, \beta; O)$.
Let

$$\phi \in C^\infty(\bar{O}), \theta \in C_0^\infty(O) \quad \text{and} \quad s \in L^\infty(\Omega, A, P), \ \theta \geq 0, \ \zeta \geq 0.$$

Define

$$z^\epsilon = \phi + \frac{\partial \phi}{\partial x_\ell}(w_\ell^\epsilon - x_\ell) \in L^2(\Omega, A, P; H^1(O))$$

and

$$z^\epsilon \to \phi \quad \text{in} \quad L^2(\Omega, A, P; H^1(O)) \quad \text{weakly.}$$

We set

$$
\begin{aligned}
X^\epsilon &= E \int_O \zeta \theta a_{ji}^\epsilon \frac{\partial z^\epsilon}{\partial x_j} \frac{\partial z^\epsilon}{\partial x_i} dx \\
&= -E < \frac{\partial}{\partial x_i}(a_{ji}^\epsilon \frac{\partial z^\epsilon}{\partial x_j}), \zeta \theta z^\epsilon > -E \int_O \zeta \frac{\partial \theta}{\partial x_i} a_{ji}^\epsilon \frac{\partial z^\epsilon}{\partial x_j} z^\epsilon dx \\
&= -E < \zeta \theta z^\epsilon \frac{\partial \phi}{\partial x_\ell}, \frac{\partial}{\partial x_i}(a_{ji}^\epsilon \frac{\partial w_\ell^\epsilon}{\partial x_j}) > \\
&\quad - E \int_O \zeta \theta z^\epsilon a_{ji}^\epsilon \frac{\partial w_\ell^\epsilon}{\partial x_j} \frac{\partial^2 \phi}{\partial x_i \partial x_\ell} dx \\
&\quad + E \int_O \zeta \theta \frac{\partial^2 \phi}{\partial x_\ell \partial x_j}(w_\ell^\epsilon - x_\ell) a_{ji}^\epsilon \frac{\partial z^\epsilon}{\partial x_i} dx \\
&\quad - E \int_O \zeta \frac{\partial \theta}{\partial x_i} a_{ji}^\epsilon \frac{\partial \theta}{\partial x_\ell} \frac{\partial w_\ell^\epsilon}{\partial x_j} z^\epsilon dx
\end{aligned}
$$

Therefore from (1.3), one can assert that

$$X^\epsilon \to - E < \zeta\theta\phi\frac{\partial\phi}{\partial x_\ell}, \frac{\partial}{\partial x_i}q_{\ell i} > -E\int_O \zeta\theta\phi q_{\ell i}\frac{\partial^2\phi}{\partial x_i\partial x_\ell}dx$$

$$- E\int_O \zeta\frac{\partial\theta}{\partial x_i}q_{i\ell}\frac{\partial\phi}{\partial x_\ell}\phi dx = E\int_O \zeta\theta q_{\ell i}\frac{\partial\phi}{\partial x_i}\frac{\partial\phi}{\partial x_\ell}dx.$$

On the other hand,

$$X^\epsilon \geq \alpha E\int_O \zeta\theta|Dz^\epsilon|^2 dx$$

and by Fatou's Lemma

$$\alpha E\int_O \zeta\theta|D\phi|^2 dx \leq E\int_O \zeta\theta q_{\ell i}\frac{\partial\phi}{\partial x_i}\frac{\partial\phi}{\partial x_\ell}dx.$$

Pick $\phi = \xi \cdot x$, where $\xi \in \mathbf{R}^n$, we deduce

$$\alpha|\xi|^2 E\int_O \zeta\theta dx \leq E\int_O \zeta\theta q_{\ell i}\xi_i\xi_\ell dx.$$

Since θ, ζ are arbitrary it follows that

$$q_{\ell i}\xi_i\xi_\ell \geq \alpha|\xi|^2 \quad \text{a.e. in} \quad O, \quad \text{a.s.}$$

Moreover, noting that

$$[(A^\epsilon)^*]^{-1} \geq \beta I$$

we have

$$X^\epsilon = E\int_O \zeta\theta(A^\epsilon)^* Dz^\epsilon Dz^\epsilon dx$$

$$= E\int_O \zeta\theta[(A^\epsilon)^*]^{-1}(A^\epsilon)^* Dz^\epsilon(A^\epsilon)^* Dz^\epsilon dx$$

$$\geq \beta E\int_O \zeta\theta|(A^\epsilon)^* Dz^\epsilon|^2 dx.$$

But

(1.4) $\quad (A^\epsilon)^* Dz^\epsilon \to (A^o)^* D\phi$ in $L^2(\Omega, A, P; (L^2(O))^n)$ weakly.

Using again Fatou's Lemma we deduce

$$\beta E \int_O \zeta \theta |(A^o)^* D\phi|^2 dx \le E \int_O \zeta \theta q_{\ell i} \frac{\partial \phi}{\partial x_i} \frac{\partial \phi}{\partial x_\ell} dx.$$

Picking again $\phi = \xi_\ell x_\ell$ yields

$$\beta |(A^o)^* \xi|^2 \le (A^o)^* \xi \cdot \xi \quad \text{a.e.} \quad x \in O, \quad \text{a.s.}$$

hence $(A^o)^{*-1} \eta \cdot \eta \ge \beta |\eta|^2 \; \forall \eta$, which proves that $A^o \in M(\alpha, \beta; O)$.

2. A^ϵ converges to A^o.

Let $f \in L^2(\Omega, A, P; H^{-1}(O))$ and u^ϵ the solution of

$$(1.5) \qquad -\text{div}\, [A^\epsilon \, \text{grad}\, u^\epsilon] = f, \; u^\epsilon \in L^2(\Omega, A, P; H_0^1(O)).$$

Clearly

$$(1.6) \qquad \|u^\epsilon\|_{L^2(\Omega, A, P; H_0^1(0))} \le C$$

and we can extract a subsequence, still denoted u^ϵ such that

$$(1.7) \qquad u^\epsilon \to u^o \; \text{in} \; L^2(\Omega, A, P; H_0^1(O)) \quad \text{weakly.}$$

Now let $v \in L^2(\Omega, A, P; H_0^1(O))$ we can write (1.5) as

$$(1.8) \qquad -E < u^\epsilon, \frac{\partial}{\partial x_j} (a_{ij}^\epsilon \frac{\partial v}{\partial x_i}) >= E < f, v > .$$

Let $\phi \in C_0^\infty(O)$ and $\zeta \in L^\infty(\Omega, A, P)$, we take

$$v = \zeta(\phi + \frac{\partial \phi}{\partial x_\ell} \frac{\partial w_\ell^\epsilon}{\partial x_i} + \frac{\partial^2 \phi}{\partial x_i \partial x_\ell} (w_\ell^\epsilon - x_\ell))$$

and (1.8) yields
(1.9)

$$- E\zeta < u^\epsilon, \frac{\partial \phi}{\partial x_\ell} \frac{\partial}{\partial x_j} (a_{ij}^\epsilon \frac{\partial w_\ell^\epsilon}{\partial x_j}) + a_{ij}^\epsilon \frac{\partial w_\ell^\epsilon}{\partial x_i} \frac{\partial^2 \phi}{\partial x_\ell \partial x_j}$$

$$+ \frac{\partial}{\partial x_j} (a_{ij}^\epsilon \frac{\partial^2 \phi}{\partial x_i \partial x_\ell} (w_\ell^\epsilon - x_\ell)) >= E\zeta < f, \phi + \frac{\partial \phi}{\partial x_\ell} (w_\ell^\epsilon - x_\ell) > .$$

We can pass to the limit in the above expression. Indeed

$$(1.10) \quad E\zeta < u^\epsilon, \frac{\partial \phi}{\partial x_\ell} \frac{\partial}{\partial x_j} (a_{ij}^\epsilon \frac{\partial w_\ell^\epsilon}{\partial x_j}) > \to E\zeta < u^o, \frac{\partial \phi}{\partial x_\ell} \frac{\partial}{\partial x_i} q_{\ell i} >$$

(1.11)

$$E\zeta < u^\epsilon, a_{ij}^\epsilon \frac{\partial w_\ell^\epsilon}{\partial x_i} \frac{\partial^2 \phi}{\partial x_\ell \partial x_j} > = E \int_O \zeta u^\epsilon a_{ij}^\epsilon \frac{\partial w_\ell^\epsilon}{\partial x_i} \frac{\partial^2 \phi}{\partial x_\ell \partial x_j} dx$$

$$= E \int_O \zeta u^o q_{\ell i} \frac{\partial^2 \phi}{\partial x_\ell \partial x_j} dx + E \int_O \zeta u^\epsilon (a_{ij}^\epsilon \frac{\partial w_\ell^\epsilon}{\partial x_i} - q_{\ell j}) \frac{\partial^2 \phi}{\partial x_\ell \partial x_j} dx$$

$$+ E \int_O \zeta (u^\epsilon - u^o) q_{\ell i} \frac{\partial^2 \phi}{\partial x_\ell \partial x_j} dx \to E \int_O \zeta u^o q_{\ell i} \frac{\partial^2 \phi}{\partial x_\ell \partial x_j} dx$$

and

$$E\zeta < u^\epsilon, \frac{\partial}{\partial x_j} (a_{ij}^\epsilon \frac{\partial^2 \phi}{\partial x_i \partial x_\ell} (w_\ell^\epsilon - x_\ell)) >$$

$$(1.12)$$

$$= -E\zeta \int_O a_{ij}^\epsilon \frac{\partial u^\epsilon}{\partial x_j} \frac{\partial^2 \phi}{\partial x_i \partial x_\ell} (w_\ell^\epsilon - x_\ell) dx \to 0.$$

We deduce the relation

$$-E\zeta < u^o, \frac{\partial \phi}{\partial x_\ell} \frac{\partial}{\partial x_i} q_{\ell i} > -E \int_O \zeta u^o q_{\ell j} \frac{\partial^2 \phi}{\partial x_\ell \partial x_j} dx$$

$$= E\zeta < f, \phi >$$

or

$$-E\zeta < u^o, \frac{\partial}{\partial x_i} (q_{\ell i} \frac{\partial \phi}{\partial x_\ell}) > = E\zeta < f, \phi >$$

i.e

$$-E < \frac{\partial}{\partial x_i} (q_{ij} \frac{\partial u^o}{\partial x_j}), \phi > \zeta = E < f, \phi > \zeta.$$

Since $L^2(\Omega, A, P)$ is separable, this suffices to imply

$$-\text{div} [A^o \text{ grad } u^o] = f.$$

Let us then prove the part b) of (1.2).
    Define

$$\xi_i^\epsilon = a_{ij}^\epsilon \frac{\partial u^\epsilon}{\partial x_j}$$

which is bounded in $L^2(\Omega, A, P; L^2(O))$. We can extract a subsequence (still denote ξ_i^ϵ) such that

$$\xi_i^\epsilon \to \xi_i^o \quad \text{in} \quad L^2(\Omega, A, P; L^2(O)) \quad \text{weakly}.$$

Since

(1.13)
$$-\frac{\partial}{\partial x_i}\xi_i^o = f$$

we deduce

(1.14)
$$-\frac{\partial}{\partial x_i}\xi_i^o = f.$$

Now write, using again

$$\phi \in C_0^\infty(O), \zeta \in L^\infty(\Omega, A, P),$$

$$E \int_O \xi_i^\epsilon \frac{\partial w_\ell^\epsilon}{\partial x_i}\phi\zeta dx$$

(1.15)
$$= -E\zeta < \frac{\partial \xi_i^\epsilon}{\partial x_i}, w_\ell^\epsilon\phi > -E \int_O \zeta\xi_i^\epsilon w_\ell^\epsilon \frac{\partial\phi}{\partial x_i}dx$$

$$= -E\zeta < \frac{\partial \xi_i^o}{\partial x_i}, w_\ell^\epsilon\phi > -E \int_O \zeta\xi_i^\epsilon w_\ell^\epsilon \frac{\partial\phi}{\partial x_i}dx$$

$$\to -E\zeta < \frac{\partial \xi_i^o}{\partial x_i}, x_\ell\phi > -E \int_O \zeta\xi_i^o x_\ell \frac{\partial\phi}{\partial x_i}dx$$

$$= E \int_O \zeta\phi\xi_\ell^o dx.$$

On the other hand, one has

$$E \int_O \zeta\phi\xi_i^\epsilon \frac{\partial w_\ell^\epsilon}{\partial x_i}dx = E \int_O a_{ji}^\epsilon \frac{\partial w_\ell^\epsilon}{\partial x_j}\frac{\partial u^\epsilon}{\partial x_i}\phi\zeta dx$$

$$= -E\zeta < \frac{\partial}{\partial x_i}(a_{ji}^\epsilon \frac{\partial w_\ell^\epsilon}{\partial x_j}), u^\epsilon\phi > -E \int_O \zeta a_{ji}^\epsilon \frac{\partial w_\ell^\epsilon}{\partial x_j}u^\epsilon \frac{\partial\phi}{\partial x_i}dx$$

$$\to -E\zeta < \frac{\partial}{\partial x_i}q_{\ell i}, u^o\phi > -E \int_O \zeta q_{\ell i}u^o \frac{\partial\phi}{\partial x_i}dx$$

$$= E \int_O \zeta\phi q_{\ell i}\frac{\partial u^o}{\partial x_i}dx$$

therefore

$$\xi_\ell^o = q_{\ell i} \frac{\partial u^o}{\partial x_i}.$$

Since the limit u^o is uniquely defined, we can assert that for the whole sequence

$$u^\epsilon \to u^o \text{ in } L^2(\Omega, A, P; H_0^1(O)) \text{ weakly}$$

$$a_{ij}^\epsilon \frac{\partial u^\epsilon}{\partial x_j} \to q_{ij} \frac{\partial u^o}{\partial x_j} \text{ im } L^2(\Omega, A, P; (L^2(O))^n) \text{ weakly.}$$

The proof of the desired results has been completed.

Let us also prove a uniqueness result for the H-convergence limit.

Proposition 1.1. *The H-convergence limit is necessarily unique.*

PROOF. Suppose that

(1.16) $A^\epsilon\ H - $ converges to A^o and $\tilde{A}^o$ in O.

Let $z \in L^2(\Omega, A, P; H_0^1(O))$, set

$$f = -\text{div } [A^o \text{ grad } z] \in L^2(\Omega, A, P; H^{-1}(O)).$$

Consider u^ϵ defined by

$$-\text{div } [A^\epsilon \text{ grad } u^\epsilon] = f$$

then from (1.2) a), b) we can assert that

$$u^\epsilon \to z \text{ in } L^2(\Omega, A, P; H_0^1(O)) \quad \text{weakly}$$

$$A^\epsilon \text{ grad } u^\epsilon \to A^o \text{ grad } z \text{ in } L^2(\Omega, A, P; (H_0^1(O))^n) \text{ weakly.}$$

Since A^ϵ also H converges to $\tilde{A}^o$ in O, we have necessarily

$$A^o \text{ grad } z = \tilde{A}^o \text{ grad } z \quad \text{a.e. in } O, \quad \text{a.s.}$$

Pick Δ with $\bar{\Delta} = O$, and let $\phi \in C_0^\infty(O), \phi = 1$ on Δ. We can take

$$z = \lambda \cdot x\phi$$

hence

$$A^o\lambda = \tilde{A}^o\lambda \quad \text{on} \quad \Delta \quad \text{a.s.}$$

which implies

$$A^o = \tilde{A}^o \quad \text{a.e. in} \quad \Delta, \quad \text{a.s.}$$

But $O = \uparrow \Delta_n$, $\bar{\Delta}_n \subset O$, therefore

$$A^o = \tilde{A}^o \quad \text{a.e. in} \quad O, \quad \text{a.s.}$$

and the desired result follows.

Remark 1.1. Suppose that $A^\epsilon \in M(\alpha,\beta;O)$ $\forall O$ and there exist $w_\ell^\epsilon(x,\omega)$ satisfying (1.3) $\forall O$, then there exists a unique matrix $A^o \in M(\alpha,\beta;O)$ $\forall O$ and $A^\epsilon H$ converges to A^o in O, $\forall O$.

1.3. Correctors. We shall consider in this paragraph the analogue of condition (1.3), for the transpose matrix $(A^\epsilon)^*$. More precisely we assume that there exists a sequence $v_\ell^\epsilon(x,\omega)$ such that

$$
\begin{aligned}
&a) \quad v_\ell^\epsilon \in L^2(\Omega,A,P;H^1(O)), \quad \ell = 1...n \\
&\quad\quad v_\ell^\epsilon - x_\ell \to 0 \quad \text{in} \quad L^2(\Omega,A,P;H^1(O)) \quad \text{weakly} \\
&\quad\quad \text{and} \quad L^2(\Omega,A,P;L^2(O)) \quad \text{strongly}
\end{aligned}
$$

(1.17)
$$
\begin{aligned}
&b) \quad a_{ij}^\epsilon \frac{\partial v_\ell^\epsilon}{\partial x_j} - q_{i\ell} \to 0 \quad \text{in} \quad L^2(\Omega,A,P;H^{-1}(O)) \quad \text{strongly} \\
&\quad\quad \text{where} \quad q_{i\ell} \in L^2(\Omega,A,P;L^2(O)) \quad \forall i,\ell = 1...n \\
&c) \quad \frac{\partial}{\partial x_i}[a_{ij}^\epsilon \frac{\partial v_\ell^\epsilon}{\partial x_j}] \to \frac{\partial}{\partial x_i}q_{i\ell} \quad \text{in} \quad L^2(\Omega,A,P;H^{-1}(O)) \\
&\quad\quad \text{strongly, for any} \quad \ell.
\end{aligned}
$$

The condition (1.17) is clearly the analogue of (1.3) for the matrix $(A^\epsilon)^*$. From Theorem 1.1 we immediately deduce that

(1.18) $(A^\epsilon)^* H$ converges to $(A^o)^*$ in O

where we have set $A^o = q_{i\ell}$. Moreover $A^o \in M(\alpha, \beta; O)$, since $(A^o)^*$ belongs to $M(\alpha, \beta; O)$. Let $f \in L^2(\Omega, P,; H^{-1}(O))$ and consider

(1.19)
$$- \operatorname{div} [A^\epsilon \operatorname{grad} u^\epsilon] = f$$
$$u^\epsilon \in L^2(\Omega, A, P; H_0^1(O))$$

we have the following

Theorem 1.2. *We assume that $A^\epsilon \in M(\alpha, \beta; O)$ and that there exist v_ℓ^ϵ satisfying the condition (1.17). Then one has*

(1.20)
$$u^\epsilon \to u^o \ in \ L^2(\Omega, A, P; H_0^1(O)) \ weakly \ and$$
$$L^2(\Omega, A, P; L^2(O)) \ strongly$$

(1.21)
$$\frac{\partial u^\epsilon}{\partial x_i} - \frac{\partial u^o}{\partial x_\ell} \frac{\partial v_\ell^\epsilon}{\partial x_i} \to 0 \ in \ L^1(\Omega, A, P; L^1(O))$$
$$a_{ij}^\epsilon \frac{\partial u^\epsilon}{\partial x_j} \to q_{ij} \frac{\partial u^o}{\partial x_j} \ in \ L^2(\Omega, A, P; L^2(O)) \ weakly$$

where u^o is the solution of

(1.22)
$$-\operatorname{div} [A^o \operatorname{grad} u^o] = f, \quad u^o \in L^2(\Omega, A, P; H_0^1(O)).$$

PROOF. Since u^ϵ remains in a bounded subset of $L^2(\Omega, A, P; H_0^1(O))$ and

$$\xi_i^\epsilon = a_{ij}^\epsilon \frac{\partial u^\epsilon}{\partial x_j}$$

remains in a bounded subset of $L^2(\Omega, A, P; L^2(O))$, we can extract subsequences still denoted by u^ϵ and ξ_i^ϵ such that

$$u^\epsilon \to u^o \ in \ L^2(\Omega, A, P; H_0^1(O)) \ weakly$$

$$\xi_i^\epsilon \to \xi_i \ in \ L^2(\Omega, A, P; L^2(O)) \ weakly.$$

We shall consider a sequence u_k such that

$$(1.23)$$
u_k is infinitely differentiable with respect to x
and is bounded as well as its derivatives, a.e.a.s.
u_k has compact support in O
$u_k \to u^o$ in $L^2(\Omega, A, P; H_0^1(O))$ strongly.

To construct such a sequence pick e_n to be an orthonormal basis of $L^2(\Omega, A, P)$ (recall that it is separable), and let

$$\mu_n(x) = Eu^o(x)e_n$$

which belongs to $H_0^1(O)$. Let $\mu_n^P(x) \in C_0^\infty(O)$ with

$$\mu_n^P \to \mu_o \text{ in } H_o^1(O) \; \forall n \text{ fixed, as } P \to \infty.$$

Consider the sequence

$$u^{N,M,P}(x,\omega) = \sum_{n=1}^N \mu_n^P(x)e_n 1_{|e_n|<M}$$

it satisfyies the required property (1.23).
     We define
$$u_k^\epsilon = u_k + \frac{\partial u_k}{\partial x_\ell}(v_\ell^\epsilon - x_\ell)$$

which belongs to $L^2(\Omega, A, P; H_0^1(O))$. Note that the gradient is given by

$$\frac{\partial u_k^\epsilon}{\partial x_i} = \frac{\partial u_k}{\partial x_\ell}\frac{\partial v_k^\epsilon}{\partial x_i} + \frac{\partial^2 u_k}{\partial x_i \partial x_\ell}(v_\ell^\epsilon - x_\ell).$$

We compute

$$E\int_O a_{ij}^\epsilon \frac{\partial}{\partial x_j}(u^\epsilon - u_k^\epsilon)\frac{\partial}{\partial x_i}(u^\epsilon - u_k^\epsilon)dx = E<f, u^\epsilon - u_k^\epsilon>$$

$$- E\int_O a_{ij}^\epsilon \frac{\partial u^\epsilon}{\partial x_i}\frac{\partial^2 u_k}{\partial x_j \partial x_\ell}(v_\ell^\epsilon - x_\ell)dx+$$

$$+ E\int_O a_{ij}^\epsilon \frac{\partial v_\ell^\epsilon}{\partial x_j}\frac{\partial^2 u_k}{\partial x_i \partial x_\ell}u^\epsilon dx$$

$$+ E < \frac{\partial}{\partial x_i}(a_{ij}^\epsilon \frac{\partial v_\ell^\epsilon}{\partial x_j}), \frac{\partial u_k}{\partial x_\ell} u^\epsilon >$$

$$(1.24) - E < \frac{\partial}{\partial x_i}(a_{ij}^\epsilon \frac{\partial v_\ell^\epsilon}{\partial x_j}), \frac{\partial u_k}{\partial x_\ell} \frac{\partial u_k}{\partial x_h} v_h^\epsilon >$$

$$- E \int_O a_{ij}^\epsilon \frac{\partial v_\ell^\epsilon}{\partial x_j} \frac{\partial}{\partial x_i}(\frac{\partial u_k}{\partial x_\ell} \frac{\partial u_k}{\partial x_h}) v_h^\epsilon dx$$

$$+ E \int_O [a_{ij}^\epsilon \frac{\partial u_k}{\partial x_j \partial x_\ell}(v_\ell^\epsilon - x_\ell) + a_{ij}^\epsilon \frac{\partial^2 u_h}{\partial x_i \partial x_h}(v_h^\epsilon - x_h)$$

$$+ a_{ij}^\epsilon \frac{\partial^2 u_k}{\partial x_j \partial x_\ell} \frac{\partial^2 u_k}{\partial x_i \partial x_h}(v_\ell^\epsilon - x_\ell)(v_h^\epsilon - x_h)] dx.$$

We have

$$u_k^\epsilon \to u_k \text{ in } L^2(\Omega, A, P; L^2(O)) \text{ strongly, and}$$
$$L^2(\Omega, A, P; H_0^1(O)) \text{ weakly.}$$

We thus deduce from (1.24)

$$(1.25) \quad \begin{aligned} \alpha \limsup_{\epsilon \to 0} E \int_O |Du^\epsilon - Du_k^\epsilon|^2 dx &\leq E < f, u^o - u_k > \\ &+ E \int_O q_{i\ell} \frac{\partial^2 u_k}{\partial x_i \partial x_\ell} u^o dx + E < \frac{\partial}{\partial x_i} q_{i\ell}, \frac{\partial u_k}{\partial x_\ell} u^o > \\ &- E < \frac{\partial}{\partial x_i} q_{i\ell}, \frac{\partial u_k}{\partial x_\ell} \frac{\partial u_k}{\partial x_h} x_h > - E \int_O q_{i\ell} \frac{\partial}{\partial x_i}(\frac{\partial u_k}{\partial x_\ell} \frac{\partial u_k}{\partial x_h}) x_h dx \\ &= E < f, u^o - u_k > - E \int_O q_{i\ell}(\frac{\partial u^o}{\partial x_i} - \frac{\partial u_k}{\partial x_i}) \frac{\partial u_k}{\partial x_\ell} dx. \end{aligned}$$

From Poincaré inequality we also have

$$(1.26) \quad \begin{aligned} C \limsup_{\epsilon \to 0} E \int_O (u^\epsilon - u_k^\epsilon)^2 dx &\leq E < f, u^o - u_k > \\ &- E \int_O q_{i\ell}(\frac{\partial u^o}{\partial x_i} - \frac{\partial u_k}{\partial x_i}) \frac{\partial u_k}{\partial x_\ell} dx \end{aligned}$$

hence

$$\begin{aligned} C \limsup_{\epsilon \to 0} E \int_O (u^\epsilon - u^o)^2 dx &\leq C_1 E \int_O (u^o - u_k)^2 dx \\ &+ E < f, u^o - u_k > - E \int_O q_{i\ell}(\frac{\partial u^o}{\partial x_i} - \frac{\partial u_k}{\partial x_i}) \frac{\partial u_k}{\partial x_\ell} dx. \end{aligned}$$

Letting k tend to $+\infty$, we deduce from (1.23) that

$$(1.27) \qquad u^\epsilon \to u^o \text{ in } L^2(\Omega, A, P; L^2(O)) \text{ strongly.}$$

Next we have

$$\sum_i E \int_O \left| \left(\frac{\partial u^\epsilon}{\partial x_i} - \frac{\partial u^o}{\partial x_\ell} \right) \frac{\partial v_\ell^\epsilon}{\partial x_i} \right| dx \le \sum_i E \int_O \left| \left(\frac{\partial u_k}{\partial x_\ell} - \frac{\partial u^o}{\partial x_\ell} \right) \frac{\partial v_\ell^\epsilon}{\partial x_i} \right| dx$$

$$+ \sum_i E \int_O \left| \frac{\partial u^\epsilon}{\partial x_i} - \frac{\partial u_k^\epsilon}{\partial x_i} \right| dx + \sum_i E \int_O \left| \frac{\partial^2 u_k}{\partial x_i \partial x_\ell} (v_\ell^\epsilon - x_\ell) \right| dx$$

$$\le C(E \int_O |Du_k - Du^o|^2 dx)^{1/2} + C(E \int_O |Du^\epsilon - Du_k^\epsilon|^2 dx)^{1/2} +$$

$$+ C_k \sum_\ell (E \int_O (v_\ell^\epsilon - x_\ell)^2 dx)^{1/2}.$$

From this inequality, making use of (1.26) we easily deduce the result (1.21).

Let $\zeta \in L^\infty(\Omega, A, P), \phi \in C_0^\infty(O)$, then

$$|E \int_O (\xi_i^\epsilon - a_{ij}^\epsilon \frac{\partial v_\ell^\epsilon}{\partial x_j} \frac{\partial u^o}{\partial x_\ell}) \phi \zeta dx| \le C \sum_j E \int_O |\frac{\partial u^\epsilon}{\partial x_j} - \frac{\partial v_\ell^\epsilon}{\partial x_j} \frac{\partial u^o}{\partial x_\ell}| dx$$

$$\to 0 \text{ as } \epsilon \to 0.$$

From (1.17) b) we also have

$$E \int a_{ij}^\epsilon \frac{\partial v_\ell^\epsilon}{\partial x_j} \frac{\partial u^o}{\partial x_\ell} \phi \zeta dx \to E \int q_{i\ell} \frac{\partial u^o}{\partial x_\ell} \phi \zeta dx.$$

Therefore it follows that

$$E \int_O \xi_i^\epsilon \phi \zeta dx \to E \int_O q_{i\ell} \frac{\partial u^o}{\partial x_\ell} \phi \zeta dx,$$

hence

$$E \int_O \xi_i \phi \zeta dx = E \int_O q_{i\ell} \frac{\partial u^o}{\partial x_\ell} \phi \zeta dx$$

and this completes the 2nd part of (1.21).

From (1.19) it is then clear that u^o is the unique solution of (1.22).

From the Theorem 1.2 we deduce in particular that

(1.28) $A^\epsilon H$ converges to A^o in O.

But we also get strong convergences results. Therefore (1.3) is more adapted to derive just the result (1.28).

To practice, they will occur simultaneously. Note that (1.28) does not follow from (1.18) in general.

1.4. The periodic case. Let us compare the results of Theorem 1.1 and 1.2 with the results of classical homogenization. We shall consider a matrix

$$\tilde{A}(\omega) = \tilde{a}_{ij}(\omega)$$

where $\tilde{a}_{ij}$ is periodic in ω. We shall take Ω to be the n dimensional torus with size 1. A is the σ-algebra of Lebesgue measurable sets and P is the Lebesgue measure on Ω. We take

$$a_{ij}^\epsilon(x;\omega) = \tilde{a}_{ij}(\frac{x}{\epsilon} + \omega)$$

which belong to $M(\alpha, \beta; O)$, provided

$$\tilde{a}_{ij}(\omega)\xi_i\xi_j > \alpha|\xi|^2$$
$$(\tilde{a}^{-1})_{ij}(\omega)\xi_i\xi_j \geq \beta|\xi|^2 , \ \forall\xi.$$

Let us check that the assumption (1.17) is satisfied. We look for $v_\ell^\epsilon(x;\omega)$ of the form

$$v_\ell^\epsilon(x;\omega) = x_\ell + \epsilon\tilde{\chi}^\ell(\frac{x}{\epsilon} + \omega)$$

where $\tilde{\chi}^\ell(\omega)$ is periodic and satisfies

(1.29) $$-\frac{\partial}{\partial\omega_i}[\tilde{a}_{ij}\frac{\partial\tilde{\chi}^\ell}{\partial\omega_j}] = \frac{\partial}{\partial\omega_i}\tilde{a}_{i\ell}.$$

There exists a unique solution $\tilde{\chi}^\ell$ of (1.29) in $H^1(\Omega)$, periodic such that

$$\int_\Omega \tilde{\chi}^\ell d\omega = 0.$$

Note that

$$\frac{\partial v_\ell^\epsilon}{\partial x_j} = \delta_{\ell j} + \frac{\partial \tilde{\chi}^\ell}{\partial w_j}(\frac{x}{\epsilon} + w)$$

and

$$\|v_\ell^\epsilon - x_\ell\|^2_{L^2(\Omega,A,P;L^2(O))} = \epsilon^2 \int_\Omega \int_O (\tilde{\chi}^\ell(\frac{x}{\epsilon} + w))^2 dx\ dw$$

$$= \epsilon^2 \text{ Meas } O \cdot \int_\Omega (\tilde{\chi}^\ell(w))^2 dw \to 0 \text{ as } \epsilon \to 0.$$

$$\|\frac{\partial v_\ell^\epsilon}{\partial x_j} - \delta_{\ell j}\|^2_{L^2(\Omega,A,P;L^2(O))} = \int_\Omega \int_O (\frac{\partial \tilde{\chi}^\ell}{\partial w_j}(\frac{x}{\epsilon} + w))^2 dx\ dw$$

$$= \text{ Meas } O \cdot \int_\Omega (\frac{\partial \tilde{\chi}^\ell}{\partial w_j}(w))^2 dw, \text{ bounded.}$$

Next

(1.30) $$a_{ij}^\epsilon \frac{\partial v_\ell^\epsilon}{\partial x_j} = (\tilde{a}_{i\ell} + \tilde{a}_{ij}\frac{\partial \tilde{\chi}^\ell}{\partial w_j})(\frac{x}{\epsilon} + w)$$

and

$$\frac{\partial}{\partial x_i}(a_{ij}^\epsilon \frac{\partial v_\ell^\epsilon}{\partial x_j}) = \frac{1}{\epsilon}\frac{\partial}{\partial w_i}(\tilde{a}_{ij}\frac{\partial \tilde{\chi}^\ell}{\partial w_j})(\frac{x}{\epsilon} + w) = 0$$

according to (1.29).

The results (1.17) b) and c) with

$$q_{i\ell} = \int_\Omega (\tilde{a}_{i\ell} + \tilde{a}_{ij}\frac{\partial \tilde{\chi}^\ell}{\partial w_j})(w)dw$$

will follow from the

Lemma 1.1. *Let* $\tilde{\phi}(w)$ *be periodic and* $L^2(\Omega)$*, then one has*

(1.31) $$\tilde{\phi}(\frac{x}{\epsilon} + w) \to \mathcal{M}(\tilde{\phi}) = \int \tilde{\phi}(w)dw$$

in $L^2(\Omega; H^{-1}(O))$ *strongly.*

PROOF. We may assume that $\mathcal{M}(\tilde{\phi}) = 0$. Therefore there exists $\tilde{\psi} \in H^1(\Omega)$, periodic such that

$$-\Delta_\omega \tilde{\psi}(\omega) = \tilde{\phi}(\omega), \quad \int \tilde{\psi} d\omega = 0.$$

Define

$$z^\epsilon(x; \omega) \text{ to be the solution of}$$

$$-\Delta_x z^\epsilon = \tilde{\phi}(\frac{x}{\epsilon} + \omega)$$

$$z^\epsilon|_{\partial O = 0}$$

then we must prove that

$$(1.32) \qquad \int_\Omega \int_O |D_x z^\epsilon|^2 dx \to 0 \quad \text{as} \quad \epsilon \to 0.$$

But the left hand side of (1.32) is equal to

$$\int_\Omega \int_O z^\epsilon(x)\tilde{\phi}(\frac{x}{\epsilon} + \omega)dx d\omega = -\int_\Omega \int_O z^\epsilon(x)\Delta_\omega \tilde{\phi}(\frac{x}{\epsilon} + \omega)dx \ d\omega$$

$$= -E \int_\Omega \int_O z^\epsilon(x) \ \text{div}_x D_\omega \tilde{\psi}(\frac{x}{\epsilon} + \omega)dx \ d\omega$$

$$= E \int_\Omega \int_O D_x z^\epsilon D_\omega \tilde{\psi}(\frac{x}{\epsilon} + \omega)dx \ d\omega$$

which tends to 0 as $\epsilon \to 0$, since

$$\int_\Omega \int_O |D_x z^\epsilon|^2 dx \ d\omega \leq \int_\Omega \int_O |\tilde{\phi}(\frac{x}{\epsilon} + \omega)|^2 dx \ d\omega$$

$$\leq C \ \text{Meas} \ O \int_\Omega (\tilde{\phi}(\omega))^2 d\omega$$

and

$$\int_\Omega \int_O |D_\omega \tilde{\psi}(\frac{x}{\epsilon} + \omega)|^2 dx \ d\omega = \ \text{Meas} \ O \int_\Omega |D_\omega \tilde{\psi}(\omega)|^2 d\omega.$$

Remark 1.2. If we consider a sequence $A^\epsilon \in M(\alpha, \beta; O)$ we cannot expect a result like that of Theorem 1.2, even for subsequences, without additional assumptions. This is true when $A^\epsilon(x)$ is deterministic (cf. L.Tartar [17]). A very simple counterexample is the following

$$A^\epsilon(x;\omega) = a^\epsilon(\omega)I,$$

for which

$$u^\epsilon(x;\omega) = \frac{1}{a^\epsilon(\omega)} z(x)$$

where

$$-\Delta z = f \quad z \in H_0^1(O)$$

A strong convergence of u^ϵ in L^2 is impossible, even for subsequences if Ω is not countable. For a more significant counter example, see L.Tartar [18].

2. Example of *H*-convergence: stationary processes.

2.1. Notation.Assumptions. Let $F = L^2(\Omega, A, P; \mathbf{C})$ where $\mathbf{C}$ is the set of complex numbers. If $\tilde{f}, \tilde{g} \in F$, then the scalar product is given by $E\tilde{f}\bar{\tilde{g}}$. We assume that there exists

(2.1)
$$T_y, \ y \in \mathbf{R}^n, \ strongly \ continuous \ unitary \ group \ on \ F$$
$$T_y \ is \ ergodic, \ i.e. \ if \ \tilde{f} \in F \ satisfies \ T_y\tilde{f} = \tilde{f} \ \forall y$$
$$then \ \tilde{f} \ is \ a \ constant$$
$$T_y 1 = 1, \quad T_y\tilde{f} \geq 0 \ if \ \tilde{f} \geq 0.$$

The group T_y has the spectral resolution

(2.2)
$$T_y = \int_{\mathbf{R}^n} e^{i\lambda \cdot y} \cup (d\lambda)$$

where $\cup$ is a projection valued measure. It satisfies

(2.3)
$$E \cup (\Delta)\tilde{f} \ \overline{\cup(\Delta')\tilde{g}} = 0 \ \forall \tilde{f}, \tilde{g} \in F, \ if \ \Delta \cap \Delta' = \emptyset$$
$$E \cup (\Delta)\tilde{f} \ \overline{\cup(\Delta)\tilde{g}} = E \cup (\Delta)\tilde{f}\bar{\tilde{g}} \ \forall \tilde{f}, \tilde{g} \in F$$

and by ergodicity

(2.4)
$$\cup(\{0\})\tilde{f} = E\tilde{f}$$

We next define

$$(2.5) \qquad D_i \tilde{f}(\omega) = \frac{\partial}{\partial y_i}(T_y \tilde{f})(\omega)|_{y=0}$$

which are closed, densely defined linear operators with domains $\mathcal{D}(D_i)$ in F. Note that

$$(2.6) \qquad D_j \tilde{f} = i \int_{\mathbf{R}^n} \lambda_j \cup (d\lambda) \tilde{f} \quad \forall \tilde{f} \in \mathcal{D}(D_i).$$

The space $F^1 = \cap_j (D_j)$ is dense in F. We equip F^1 with the Hilbert scalar product

$$(2.7) \qquad ((\tilde{f}, \tilde{g}))_{F^1} = E\tilde{f}\bar{\tilde{g}} + \sum_{j=1}^{d} ED_j\tilde{f} \; \overline{D_j\tilde{g}}.$$

We identify F with its dual and call F^{-1} the dual of F^1. We have the inclusions

$$F^1 \subset F \subset F^{-1}$$

each space being dense in the next one with continuous injection.

The family T_y is also a strongly continuous unitary group in F^1 and F^{-1}. We also assume that T_y is extended to complex random variables not necessarily in F. It is a linear group verifying the property

$$\begin{aligned} & E\phi(T_y\tilde{\eta}_1, \ldots, T_y\tilde{\eta}_k) = E\phi(\tilde{\eta}_1, \ldots, \tilde{\eta}_k) \\ (2.8) \quad & \forall \tilde{\eta}_1, \ldots \tilde{\eta}_k \text{ complex random variables, and} \\ & \phi \text{ Borel bounded function of } \mathbf{C}^k. \end{aligned}$$

$$\begin{aligned} (2.9) \quad & y \cdot \omega \to T_y\tilde{\eta}(\omega) \text{ is measurable} \\ & T_y\tilde{\eta} \geq 0 \text{ if } \tilde{\eta} \geq 0. \end{aligned}$$

A *stationary process* is a stochastic process which can be represented in the form

$$(2.10) \qquad \eta(y; \omega) = T_y\tilde{\eta}(\omega).$$

The space of square integrable stationary processes can be identified with F. They are necessarily continuous function of y with values in F.

Example: *periodic functions.* Let $\Omega =$ unit n−dimensional torus, $A = \sigma$-algebra of Lebesgue measurable sets and $P =$Lebesgue measure on Ω. The space F is the space of measurable periodic functions (period 1 in each component) such that

$$\int_\Omega (\tilde{f}(\omega))^2 d\omega < \infty$$

We define $T_y \tilde{f}(\omega) = \tilde{f}(\omega + y)$. The operator T_y satisfies all the required properties.

2.2. A fundamental technical result. We have the

Lemma 2.1. *Let* $\tilde{\phi} \in F$ *and* $\phi(y; \omega) = T_y \tilde{\phi}(\omega)$. *If* $E\tilde{\phi} = 0$, *there exists* $\psi_\ell(y; \omega) \in C^1(\mathbf{R}^n; F)$, $\ell = 1...n$, *such that*

(2.11)
$$\psi_\ell(0, \omega) = 0$$
$$\sum_\ell \frac{\partial \psi_\ell}{\partial y_\ell}(y; \omega) = \phi(y; \omega)$$
$$\sum_\ell E|\psi_\ell(y)|^2 \le |y|^2 E|\tilde{\phi}|^2$$

(2.12)
$$\frac{\sum_\ell E|\psi_\ell(y)|^2}{|y|^2} \to 0 \text{ as } |y| \to \infty.$$

Example of periodic functions. Let $\tilde{\phi}$ be periodic with period 1 in all components, and

$$\phi(y; \omega) = \tilde{\phi}(y + \omega), \quad \int_\Omega \tilde{\phi}(\omega) d\omega = 0$$

then we solve the problem

(2.13)
$$- \Delta \tilde{\psi}(\omega) = \tilde{\phi}$$
$$\tilde{\psi} \text{ periodic}, \quad \int \tilde{\psi}(\omega) d\omega = 0$$

which defines $\tilde{\psi}$ uniquely. Then set

(2.14)
$$\psi_\ell(y; \omega) = -\frac{\partial \tilde{\psi}}{\partial \omega_\ell}(y + \omega) + \frac{\partial \tilde{\psi}}{\partial \omega_\ell}(\omega)$$

is satisfies the properties of the Lemma.
Indeed by Fourier transform one has

$$\tilde{\phi}(\omega) = \sum_{\substack{k \in Z^n \\ k \neq 0}} \phi_k e^{2nik \cdot \omega}$$

with

$$\sum_k |\phi_k|^2 = \int |\tilde{\phi}(\omega)|^2 d\omega.$$

We have

$$\tilde{\psi}(\omega) = \sum_{\substack{k \in Z^n \\ k \neq 0}} \frac{\phi_k}{4n^2 |k|^2} e^{2nik \cdot \omega}$$

and

$$\psi_\ell(y; \omega) = -\sum_{\substack{k \in Z^n \\ k \neq 0}} \frac{i\phi_k k_\ell}{2n|k|^2} e^{2nik \cdot \omega} \left(e^{2nik \cdot y} - 1\right)$$

and

$$\sum_\ell E|\psi_\ell(y)|^2 = \sum_{\substack{k \in Z^n \\ k \neq 0}} \frac{|\phi_k|^2}{4n^2 |k|^2} |e^{2nik \cdot y} - 1|^2.$$

Proof of Lemma 2.1. We define

$$\psi_\ell(y; \omega) = \int_{\mathbf{R}^n} (e^{i\lambda y} - 1) \frac{(-i\lambda_\ell)}{|\lambda|^2} \cup (d\lambda) \tilde{\phi}(\omega)$$

then (2.11) is satisfied. Moreover

$$\sum_\ell E|\psi_\ell(y)|^2 = E \int_{\mathbf{R}^n} \frac{|e^{i\lambda y} - 1|^2}{\lambda^2} \cup (d\lambda)\tilde\phi(\omega)\overline{\tilde\phi(\omega)}$$

$$\le |y|^2 E|\tilde\phi|^2$$

and

$$\sum_\ell \frac{E|\psi_\ell(y)|^2}{|y|^2} = E \int_{\mathbf{R}^n} \frac{|e^{i\lambda y} - 1|^2}{|\lambda y|^2} \cup (d\lambda)\tilde\phi(\omega)\overline{\tilde\phi(\omega)}$$

$$\to E(\cup\{0\}\tilde\phi\overline{\tilde\phi}) = |E\tilde\phi|^2 = 0$$

as $|y| \to \infty$ which is the desired result.

2.3. The cell problem. Consider stationary processes

$$a_{ij}(y;\omega) = T_y \tilde a_{ij}(\omega)$$

such that

(2.15) $$\qquad \alpha|\xi|^2 \le \sum_{i,j} \tilde a_{ij}(\omega)\xi_i\xi_j \le \beta|\xi|^2 \quad a.s. \quad \forall \xi \in \mathbf{R}^n,$$

we define the cell problem as follows. Let

(2.16)
$$g_j(y;\omega) = T_y \tilde g_j(\omega)$$
square integrable stationary processes, $j = 1...n$

then we look for

$$\chi(y;\omega) \in C^1(\mathbf{R}^n; F), \ \chi(\sigma;\omega) = 0, \ E\chi(y) = 0 \ \forall y,$$

$$\frac{\partial \chi}{\partial y_j} \text{ is a square integrable stationary process, } \forall j$$

(2.17)
$$-\frac{\partial}{\partial y_i}(a_{ij}(y,\omega)\frac{\partial \chi}{\partial y_j}) = \frac{\partial g_j}{\partial y_j}(y,\omega)$$
equality in $C^o(R^n, F^{-1})$.

Then (2.17) has one and only one solution (G.Papanicolau-S.R.S. Varadhan[15], see also A.Bensoussan-G.Blankenship [1]).

Example: periodic case. Let $\tilde{g}_j$ be periodic L^2 functions, we look for $\tilde{\chi}(\omega)$ periodic such that

$$-\frac{\partial}{\partial \omega_i}(\tilde{a}_{ij}\frac{\partial}{\partial \omega_j}\tilde{\chi}) = \frac{\partial \tilde{g}_j}{\partial \omega_j}$$

which has a unique solution in F^1. The function

$$\chi(y;\omega) = \tilde{\chi}(y+\omega) - \tilde{\chi}(\omega)$$

is the solution of (2.17).

2.4. Application. Consider now the matrix

$$A^\epsilon(x;\omega) = a^\epsilon_{ij}(x;\omega) = a_{ij}(\frac{x}{\epsilon};\omega),$$

where $a_{ij}(y;\omega)$ are stationary processes.

Then we shall show that there exist v^ϵ_ℓ satisfying the condition (1.17). We look for v^ϵ_ℓ as follows

$$(2.18) \qquad v^\epsilon_\ell(x;\omega) = x_\ell + \epsilon\chi^\ell(\frac{x}{\epsilon},\omega)$$

where $\chi^\ell(y;\omega)$ if the solution of

$$(2.19) \qquad -\frac{\partial}{\partial y_i}(a_{ij}(y;\omega)\frac{\partial \chi^\ell}{\partial y_i}) = \frac{\partial a_{j\ell}}{\partial y_j}(y;\omega)$$

corresponding to (2.17) with $g_j(y,\omega) = a_{j\ell}(y,\omega)$. Let us check that the condition (1.17) hold.

Define

$$\tilde{\chi}^\ell_j(\omega) \in F$$

such that

$$\frac{\partial \chi^\ell}{\partial y_i}(y;\omega) = T_j\tilde{\chi}^\ell_j(\omega)$$

then the function $\chi^\ell(y;\omega)$ is explicity given by the formula

$$(2.20) \qquad \chi^\ell(y,\omega) = \int_{\mathbf{R}^n}(e^{i\lambda y}-1)\frac{1}{|\lambda|^2}\sum_j(-i\lambda_j)\cup(d\lambda)\tilde{\chi}^\ell_j(\omega)$$

hence

$$E(\chi^\ell(y))^2 = \int_{\mathbf{R}^n} \frac{|e^{i\lambda y} - 1|^2}{|\lambda|^4} \sum_{j,k} \lambda_j \lambda_k E \cup (d\lambda) \tilde{\chi}_j^\ell \overline{\tilde{\chi}_k^\ell}.$$

Therefore

(2.21)
$$E(\tilde{\chi}^\ell(y))^2 \le |y|^2 \sum_j E(\tilde{\chi}_j^\ell)^2$$

and

(2.22)
$$\frac{E(\tilde{\chi}^\ell(y))^2}{|y|^2} \to 0 \quad \text{as} \quad |y| \to \infty.$$

Therefore

(2.23)
$$\epsilon^2 E \int_O (\chi^\ell(\frac{x}{\epsilon}))^2 dx \to 0 \quad \text{as} \quad \epsilon \to 0.$$

Next we have

(2.24)
$$\epsilon \frac{\partial \chi^\ell}{\partial x_j}(\frac{x}{\epsilon}, \omega) = \frac{\partial \chi^\ell}{\partial y_j}(\frac{x}{\epsilon}, \omega)$$

$$a_{ij}^\epsilon \frac{\partial v_\ell^\epsilon}{\partial x_j} = (a_{i\ell} + a_{ij} \frac{\partial \chi^\ell}{\partial y_j})(\frac{x}{\epsilon}, \omega)$$

Set

(2.25)
$$q_{i\ell} = E(\tilde{a}_{ij} + \tilde{a}_{ij} \tilde{\chi}_j^\ell)$$

and note that $E\tilde{\chi}_j^\ell = 0 \; \forall j$. Then the properties (1.17) a) and b) will follow from the

Lemma 2.2. *Let $\tilde{\phi} \in F$ and $\phi(y; \omega) = T_y \tilde{\phi}(\omega)$. If $E\tilde{\phi} = 0$, then for any z^ϵ in $H_0^1(O; F)$ such that*

(2.26)
$$\|z^\epsilon\|_{H^1(O;F)} \le C$$

one has

(2.27)
$$E \int_O \phi(\frac{x}{\epsilon}, \omega) z^\epsilon(x, \omega) dx \to 0, \quad \text{as} \quad \epsilon \to 0.$$

PROOF. It is a consequence of Lemma 2.1. Indeed from Lemma 2.1, we have

$$E \int_O \phi(\frac{x}{\epsilon}, \omega) z^\epsilon(x, \omega) dx = \epsilon E \int_O \frac{\partial \psi_\ell}{\partial x_\ell}(\frac{x}{\epsilon}, \omega) z^\epsilon(x, \omega) dx$$

$$= -\epsilon E \int_O \psi_\ell(\frac{x}{\epsilon}, \omega) \frac{\partial z^\epsilon}{\partial x_\ell}(x, \omega) dx$$

$\to 0$ as $\epsilon \to 0$, by virtue of (2.26) and (2.11),(2.12).

Finally we have

$$\frac{\partial}{\partial x_i}(a_{ij}^\epsilon \frac{\partial v_\ell^\epsilon}{\partial x_j}) = \frac{1}{\epsilon} \frac{\partial}{\partial y_i}(a_{i\ell} + a_{ij} \frac{\partial \chi^\ell}{\partial y_i})(\frac{x}{\epsilon}, \omega) = 0$$

by virtue of (2.19).

The verification of (1.17) will follow. Indeed to show that

$$\phi(\frac{x}{\epsilon}, \omega) \to 0, \quad \text{in } L^2(\Omega, A, P; H^{-1}(O)) \text{ strongly}$$

one has to check that

$$E \int_O |Dz_\epsilon|^2 dx \to 0 \quad \text{where } z_\epsilon \text{ is the solution of}$$

$$-\Delta z_\epsilon = \phi(\frac{x}{\epsilon}, \omega) \qquad z_{\epsilon|\partial O} = 0$$

but

$$E \int_O |Dz_\epsilon|^2 dx = E \int_O z_\epsilon \phi(\frac{x}{\epsilon}, \omega) dx \to 0$$

by virtue of (2.27).

3. The non linear problem. Setting

3.1. Notation. We shall consider the following nonlinear problem

(3.1) $$-\frac{\partial}{\partial x_i}(a_{ij}^\epsilon(x, \omega) \frac{\partial u^\epsilon}{\partial x_j}) = H^\epsilon(x, Du^\epsilon, u^\epsilon, \omega)$$

$$u^\epsilon|_{\partial O} = 0.$$

Let us make precise the operator H^ϵ. We note $H^\epsilon(x, p, z, \omega)$ the function of the arguments $x \in \mathbf{R}^n, p \in \mathbf{R}^n, z \in \mathbf{R}, \omega \in \Omega$. We assume that H^ϵ is a measurable function satisfying

$$(3.2) \qquad \begin{aligned} &H^\epsilon(x, p, z + b, \omega) \le -\beta b + H^\epsilon(x, p, z, \omega) \\ &\forall b > 0, \beta > 0 \end{aligned}$$

$$(3.3) \qquad \begin{aligned} &H^\epsilon(x, p, z + b, \omega) \ge -\beta b + H^\epsilon(x, p, z, \omega) \\ &\forall b < 0, \beta > 0 \end{aligned}$$

$$(3.4) \qquad |H^\epsilon(x, 0, 0, \omega)| \le M$$

$$(3.5) \qquad |H^\epsilon(x, p, z_1, \omega) - H^\epsilon(x, p, z_2, \omega)| \le \bar{H}|z_1 - z_2|$$

$$(3.6) \quad |H^\epsilon(x, p, z, \omega) - H^\epsilon(x, q, z, \omega)| \le \bar{H}|p - q|(1 + |p| + |q| + |z|^{1/2})$$

$$(3.7) \qquad \begin{aligned} &|H^\epsilon(x, p, z, \omega) - H^\epsilon(x', p, z, \omega)| \le \bar{H}|x - x'|^\theta(1 + |p|^2 + |z|) \\ &0 < \theta \le 1 \end{aligned}$$

$$(3.8) \qquad \begin{aligned} &E \int_O [H^\epsilon(x, \sum_\ell \frac{\partial \psi}{\partial x_\ell} Dv_\ell^\epsilon, \psi) - \mathcal{H}(x, D\psi, \psi)]z^\epsilon dx \to 0 \\ &\forall z^\epsilon \text{ bounded in } L^2(\Omega, A, P; H_0^1(O)) \cap L^\infty(O \times \Omega) \text{ and} \\ &\forall \psi \in L^2(\Omega, A, P; H_0^1(O)) \text{ such that } \psi, D\psi \in L^\infty(O \times \Omega) \end{aligned}$$

$$(3.9) \qquad \begin{aligned} &\mathcal{H}(x, p, z, \omega) \text{ is measurable and satisfies} \\ &\mathcal{H}(x, p, z + b, \omega) \le -\beta b + \mathcal{H}(x, p, z, \omega) \; \forall b > 0 \\ &\mathcal{H}(x, p, z + b, \omega) \ge -\beta b + \mathcal{H}(x, p, z, \omega) \; \forall b < 0 \\ &|\mathcal{H}(x, 0, 0, \omega)| \le M \\ &|\mathcal{H}(x, p, z_1, \omega) - \mathcal{H}(x, p, z_2, \omega)| \le \bar{\mathcal{H}}|z_1 - z_2| \\ &|\mathcal{H}(x, p, z, \omega) - \mathcal{H}(x, q, z, \omega)| \le \bar{\mathcal{H}}|p - q|(1 + |p| + |q| + |z|^{1/2}) \end{aligned}$$

where $\bar{H}, \bar{\mathcal{H}}$ are constants. Define

$$A = -\frac{\partial}{\partial x_i}\left(q_{ij}\frac{\partial}{\partial x_j}\right)$$

and assume that

(3.10) $$q_{ij}, Dq_{ij} \in L^\infty(O \times \Omega).$$

Consider the problem

(3.11)
$$\begin{aligned}
&\mathcal{A}u = \mathcal{H}(x, Du, u, \omega) \\
&u \in L^2(\Omega, A, P; H_0^1(O)) \cap L^\infty(O \times \Omega).
\end{aligned}$$

In fact for any fixed ω equation (3.11) is a standard quasi linear elliptic equation with quadratic growth 1st order terms. By virtue of the assumptions (3.9) and (3.10)

$$\|u\|_{W^{2,p}} \leq C \quad \text{a.s.}$$

with a constant independent of ω. In particular we can assert that

(3.12) $$Du \in L^\infty(O \times \Omega).$$

We write

$$v^\epsilon = (v_1^\epsilon, ..., v_n^\epsilon)$$

and Dv^ϵ the matrix

$$(Dv^\epsilon)_{ij} = \frac{\partial v_i^\epsilon}{\partial x_j}.$$

We shall assume the property

(3.13)
$$E\int_O (a_{ij}^\epsilon \frac{\partial v_\ell^\epsilon}{\partial x_j}\frac{\partial v_\ell^\epsilon}{\partial x_i} - q_{\ell\ell})z^\epsilon dx \to 0$$
$$\forall z^\epsilon \in L^2(\Omega, A, P; H_0^1(O)) \cap L^\infty,$$
with norm bounded with respect to ϵ.

The main result is the following

Theorem 3.1. *We assume (1.1), (1.17), (3.2) to (3.10), and (3.13). Then the solution u^ϵ of (3.1) satisfies*

(3.14)
$$u^\epsilon \to u \text{ in } L^2(\Omega, A, P; H_0^1(O)) \text{ weakly and}$$
$$L^2(\Omega, A, P; L^2(O)) \text{ strongly}$$

(3.15) $\quad Du^\epsilon - \dfrac{\partial u}{\partial x_\ell} Dv_\ell^\epsilon \to 0 \text{ in } L^2(\Omega, A, P; L^2(O)) \text{ strongly},$

where u is the solution of (3.11).

3.2. A Priori estimates. In this paragraph, we shall derive a priori estimates on the solution u^ϵ of (3.1). This will permit to study the limit in the next section.

3.2.1. An L^∞ estimate

Lemma 3.1. *One has the estimate*

(3.16)
$$|u^\epsilon(x,\omega)| \leq \frac{M}{\beta}.$$

PROOF. We proceed formally considering that a.s. (3.1) holds $\forall x$. At a point of maximum x_o, which is not on the boundary one has from (3.1)

$$H^\epsilon(x_o, 0, u^\epsilon(x_o), \omega) \geq 0$$

and from (3.2), it follows that, if $u^\epsilon(x_o) > 0$,

$$0 \leq -\beta u^\epsilon(x_o) + M$$

hence $u^\epsilon(x_o) \leq M/\beta$. Similarly one has $u^\epsilon(x_o) \geq -M/\beta$, hence (3.16) follows.

3.2.2. H_0^1 estimate

Lemma 3.2. *One has the estimate*

$$\|u^\epsilon\|_{L^2(\Omega,A,P;H_0^1(O))} \leq C.$$

PROOF. Set

$$H^\epsilon = H^\epsilon(x, Du^\epsilon, u^\epsilon, \omega).$$

For $s > 0$, note that

$$\exp\{s(u^\epsilon)^2\}u^\epsilon \in L^2(\Omega, A, P; H_0^1(O)) \cap L^\infty(O \times \Omega).$$

Multiply (3.1) by $\exp\{s(u^\epsilon)^2\}u^\epsilon$, integrate over O and take the mathematical expectation. We deduce

$$E \int_O a_{ij}^\epsilon \frac{\partial u^\epsilon}{\partial x_j} \frac{\partial u^\epsilon}{\partial x_i} (1 + 2s(u^\epsilon)^2) e^{s(u^\epsilon)^2} dx = E \int_O H^\epsilon u^\epsilon e^{s(u^\epsilon)^2} dx$$

and from the assumptions (3.4),(3.5),(3.6) we deduce

$$\alpha_o E \int_O (2s(u^\epsilon)^2 + 1) e^{s(u^\epsilon)^2} |Du^\epsilon|^2 dx \leq$$

$$\leq CE \int_O e^{s(u^\epsilon)^2} |u^\epsilon|(1 + |Du^\epsilon|^2 + |u^\epsilon|) dx$$

and picking s sufficiently large, we deduce

$$E \int_O e^{s(u^\epsilon)^2} |Du^\epsilon|^2 dx \leq C$$

which implies the desired result.

4. Convergence

4.1. A sequence of problems related to the limit. The solution u of the limit problem (3.11) satisfies

$$\|u\|_{L^\infty} \leq \frac{M}{\beta}$$

and
$$\|Du\|_{L^\infty} \leq C.$$

We shall consider in the sequel the following sequence

(4.1)
$$\mathcal{A}u^k + Nu^k = \mathcal{H}(x, Du^k, u^k) + Nu^{k-1}$$
$$u^k \in L^2(\Omega, A, P; H_0^1(O)), \ u^k \in L^\infty(O \times \Omega)$$

which we initialize with $u^o = 0$. In (4.1), N is a constant which will be fixed later on.

Lemma 4.1. *The sequence u^k satisfies*

(4.2)
$$\|u^{k+1} - u^k\|_{L^\infty} \leq \frac{N}{N + \beta}\|u^k - u^{k-1}\|_{L^\infty}$$

(4.3)
$$\|u^k\|_{L^\infty} \leq \frac{M}{\beta}.$$

PROOF. We prove (4.2) by maximum principle arguments, assuming for any ω regularity. Let x_o be a point of maximum of $u^{k+1} - u^k$ and suppose that $(u^{k+1} - u^k)(x_o) > 0$. Note that from (3.6)

$$\mathcal{H}(x_o, Du^{k+1}(x_o), u^k(x_o), \omega) = \mathcal{H}(x_o, Du^k(x_o), u^k(x_o), \omega)$$

and
$$\mathcal{A}(u^{k+1} - u^k)(x_o) > 0.$$

Moreover from (3.9)

$$\mathcal{H}(x_o, Du^{k+1}(x_o), u^{k+1}(x_o)) - \mathcal{H}(x_o, Du^{k+1}(x_o), u^k(x_o))$$
$$\leq -\beta(u^{k+1} - u^k)(x_o).$$

Since

$$\mathcal{A}(u^{k+1} - u^k)(x_o) + N(u^{k+1} - u^k)(x_o) =$$
$$\mathcal{H}(x_o, Du^{k+1}(x_o), u^{k+1}(x_o)) - \mathcal{H}(x_o, Du^k(x_o), u^k(x_o))$$
$$+ N(u^k - u^{k-1})(x_o)).$$

Therefore

$$N(u^{k+1} - u^k)(x_o) \le -\beta(u^{k+1} - u^k)(x_o) + N\|u^k - u^{k-1}\|$$

hence

$$(u^{k+1} - u^k)(x_o) \le \frac{N}{N+\beta}\|u^k - u^{k-1}\|$$

and the result (4.2) is proved.

We deduce from (4.2) that

$$\|u^{k+1} - u^k\|_{L^\infty} \le \left(\frac{N}{N+\beta}\right)^k \|u^1\|$$

hence u^k converges in $L^\infty(O \times \Omega)$. Note that

$$\|u^1\| \le \frac{M}{N+\beta}.$$

From (4.1) it follows that

$$|\mathcal{A}u^k| \le M + |\mathcal{H}(x, Du^k, u^k)|$$
$$\le C_o(1 + |Du^k|^2).$$

From this estimate, it also follows that

(4.4) $$\|u^k(\cdot, \omega)\|_{W^{2,p}} \le C_p, \quad \text{a.s.}$$

where the constant is deterministic.
We have

$$u^k \to u \quad \text{in} \quad L^\infty(O \times \Omega)$$

hence in particular

$$\text{a.s.} \quad u^k(\cdot, \omega) \to u(\cdot, \omega).$$

From (4.4)

$$\text{a.s.} \quad u^k(\cdot, \omega) \to u(\cdot, \omega) \quad \text{in} \quad W^{2,p} \quad \text{weakly}$$
$$\text{and} \quad W^{1,\infty} \quad \text{strongly}.$$

Clearly u is the solution of (3.10).

4.2. A fundamental relation. We introduce the sequence $u^{\epsilon,k}$ defined by

(4.5)
$$-\frac{\partial}{\partial x_i}(a_{ij}^{\epsilon}(x;\omega)\frac{\partial u^{\epsilon,k}}{\partial x_j}) + Nu^{\epsilon,k} = H^{\epsilon}(x, Du^{\epsilon,k}, u^{\epsilon,k}, \omega) + Nu^{\epsilon,k-1}$$

$$u^{\epsilon,k}|_{\partial O} = 0.$$

As in Lemma 4.1 we deduce the relations

$$\|u^{\epsilon,k+1} - u^{\epsilon,k}\|_{L^{\infty}} \leq \frac{N}{N+\beta}\|u^{\epsilon,k} - u^{\epsilon,k-1}\|_{L^{\infty}}$$

(4.6) $$\|u^{\epsilon,k}\|_{L^{\infty}} \leq \frac{M}{\beta}$$

$$\|u^{\epsilon,k+1} - u^{\epsilon,k}\|_{L^{\infty}} \leq \frac{M}{N+\beta}.$$

We shall consider next a sequence u_{δ}^k of smooth functions of x, which is bounded as well as its derivatives a.s., and such that

(4.7)
$$u_{\delta}^k \to u^k \text{ in } H_0^1(O) \quad a.s. \text{ as } \delta \to 0$$

$$\|u_{\delta}^k(\cdot,\omega)\|_{W^{1,\infty}} \leq \bar{u}$$

where $\bar{u}$ does not depend on k, δ, ω. We define

(4.8) $$u_{\delta}^{\epsilon,k} = u_{\delta}^k + \frac{\partial u_{\delta}^k}{\partial x_{\ell}}\frac{(v_{\ell}^{\epsilon} - x_{\ell})}{[1 + (v_{\ell}^{\epsilon} - x_{\ell})^2]^{1/2}}$$

hence

(4.9) $$\|u_{\delta}^{\epsilon,k}\|_{L^{\infty}} \leq C_o\bar{u}.$$

Next we have

(4.10)
$$\frac{\partial u_{\delta}^{\epsilon,k}}{\partial x_j} = \frac{\partial u_{\delta}^k}{\partial x_{\ell}}\frac{\partial v_{\ell}^{\epsilon}}{\partial x_j} + \frac{\partial^2 u_{\delta}^k}{\partial x_{\ell}\partial x_j}\frac{(v_{\ell}^{\epsilon} - x_{\ell})}{[1 + (v_{\ell}^{\epsilon} - x_{\ell})^2]^{1/2}}$$

$$+ \frac{\partial u_{\delta}^k}{\partial x_j}[1 - \frac{1}{(1 + (v_j^{\epsilon} - x_j)^2)^{3/2}}]$$

$$- \frac{\partial u_{\delta}^k}{\partial x_{\ell}}\frac{\partial v_{\ell}^{\epsilon}}{\partial x_j}[1 - \frac{1}{(1 + (v_{\ell}^{\epsilon} - x_{\ell})^2)^{3/2}}]$$

hence we deduce the estimate (pointwise in x, ω)

$$(4.11) \qquad |Du_\delta^{\epsilon,k}| \leq C_1 \bar{u}(1 + \|Dv^\epsilon\|) + H_\delta^k |v^\epsilon - x|$$

where H_δ^k is a constant such that

$$(4.12) \qquad \|D^2 u_\delta^k\| \leq H_\delta^k.$$

We shall use the notation

$$F_\delta^{\epsilon,k} = \exp\left[s(u^{\epsilon,k} - u_\delta^{\epsilon,k})^2\right]$$

where s is to be chosen later on.

Lemma 4.2. *The following relation holds*

$$
\begin{aligned}
&E \int_O a_{ij}^\epsilon \frac{\partial}{\partial x_j}(u^{\epsilon,k} - u_\delta^{\epsilon,k}) \frac{\partial}{\partial x_i}(u^{\epsilon,k} - u_\delta^{\epsilon,k}) \cdot \\
&\qquad \cdot [1 + 2s(u^{\epsilon,k} - u_\delta^{\epsilon,k})^2] F_\delta^{\epsilon,k} dx \\
&+ NE \int_O (u^{\epsilon,k} - u_\delta^{\epsilon,k})^2 F_\delta^{\epsilon,k} dx \\
&+ E < -\frac{\partial}{\partial x_i}(a_{ij}^\epsilon \frac{\partial v_\ell^\epsilon}{\partial x_j}), \frac{\partial u_\delta^k}{\partial x_\ell}(u^{\epsilon,k} - u_\delta^{\epsilon,k}) F_\delta^{\epsilon,k} > \\
&- E \int_O a_{ij}^\epsilon \frac{\partial^2 u_\delta^k}{\partial x_\ell \partial x_i} \frac{\partial v_\ell^\epsilon}{\partial x_j}(u^{\epsilon,k} - u_\delta^{\epsilon,k}) F_\delta^{\epsilon,k} dx \\
(4.13)\quad &+ E \int_O \{a_{ij}^\epsilon \frac{\partial^2 u_\delta^k}{\partial x_\ell \partial x_j} \frac{(v_\ell^\epsilon - x_\ell)}{(1 + (v_\ell^\epsilon - x_\ell)^2)^{1/2}} + \\
&\qquad + (a_{ij}^\epsilon - a_{i\ell}^\epsilon \frac{\partial v_j^\epsilon}{\partial x_\ell}) \frac{\partial u_\delta^k}{\partial x_j}[1 - \frac{1}{(1 + (v_j^\epsilon - x_j)^2)^{3/2}}]\} \cdot \\
&\qquad \cdot \frac{\partial}{\partial x_i}(u^{\epsilon,k} - u_\delta^{\epsilon,k})(1 + 2s(u^{\epsilon,k} - u_\delta^{\epsilon,k})^2) F_\delta^{\epsilon,k} dx \\
&+ NE \int_O u_\delta^k(u^{\epsilon,k} - u_\delta^{\epsilon,k}) F_\delta^{\epsilon,k} dx \\
&+ NE \int_O \frac{\partial u_\delta^k}{\partial x_\ell} \frac{(v_\ell^\epsilon - x_\ell)}{(1 + (v_\ell^\epsilon - x_\ell)^2)^{1/2}}(u^{\epsilon,k} - u_\delta^{\epsilon,k}) F_\delta^{\epsilon,k} dx =
\end{aligned}
$$

$$= E \int_O H^\epsilon (u^{\epsilon,k} - u^{\epsilon,k}_\delta) dx$$

$$+ NE \int_O (u^{\epsilon,k-1} - u^{\epsilon,k-1}_\delta)(u^{\epsilon,k} - u^{\epsilon,k}_\delta) F^{\epsilon,k}_\delta dx$$

$$+ NE \int_O u^{k-1}_\delta (u^{\epsilon,k} - u^{\epsilon,k}_\delta) F^{\epsilon,k}_\delta dx$$

$$+ NE \int_O \frac{\partial u^{k-1}_\delta}{\partial x_\ell} \frac{(v^\epsilon_\ell - x_\ell)}{(1 + (v^\epsilon_\ell - x_\ell)^2)^{1/2}} (u^{\epsilon,k} - u^{\epsilon,k}_\delta) F^{\epsilon,k}_\delta dx.$$

PROOF. We multiply (4.5) by $(u^{\epsilon,k} - u^{\epsilon,k}_\delta) F^{\epsilon,k}_\delta$ which belongs to $L^2(\Omega, A, P; H^1_0) \cap L^\infty$ and perform some integrations by parts.

4.3. Estimating the Hamiltonian. We first have

(4.14)

$$E \int_O (H^\epsilon(x, Du^{\epsilon,k}, u^{\epsilon,k}) - H^\epsilon(x, Du^{\epsilon,k}, u^{\epsilon,k}_\delta)).$$

$$\cdot (u^{\epsilon,k} - u^{\epsilon,k}_\delta) F^{\epsilon,k}_\delta dx \le -\beta E \int_O (u^{\epsilon,k} - u^{\epsilon,k}_\delta)^2 F^{\epsilon,k}_\delta dx$$

(4.15)

$$\left| E \int_O (H^\epsilon(x, Du^{\epsilon,k}, u^{\epsilon,k}_\delta) - H^\epsilon(x, Du^{\epsilon,k}, u^k_\delta))(u^{\epsilon,k} - u^{\epsilon,k}_\delta) F^{\epsilon,k}_\delta dx \right|$$

$$\le \bar{H} E \int_O \left| \frac{\partial u^k_\delta}{\partial x_\ell} \frac{(v^\epsilon_\ell - x_\ell)}{[1 + (v^\epsilon_\ell - x_\ell)^2]^{1/2}} \right| \cdot |u^{\epsilon,k} - u^{\epsilon,k}_\delta| F^{\epsilon,k}_\delta dx.$$

Then also using (3.6) we have

$$\left| E \int_O (H^\epsilon(x, Du^{\epsilon,k}, u^k_\delta) - H^\epsilon(x, Du^{\epsilon,k}_\delta, u^k_\delta))(u^{\epsilon,k} - u^{\epsilon,k}_\delta) F^{\epsilon,k}_\delta dx \right|$$

$$\le \bar{H} E \int_O |Du^{\epsilon,k} - Du^{\epsilon,k}_\delta|(1 + |Du^{\epsilon,k}| + |Du^{\epsilon,k}_\delta| +$$

$$+ |u_\delta^k|^{1/2}) |u^{\epsilon,k} - u_\delta^{\epsilon,k}| F_\delta^{\epsilon,k} dx$$

$$\leq \bar{H} E \int_O |Du^{\epsilon,k} - Du_\delta^{\epsilon,k}|^2 |u^{\epsilon,k} - u_\delta^{\epsilon,k}| F_\delta^{\epsilon,k}$$

$$+ \bar{H} E \int_O |Du^{\epsilon,k} - Du_\delta^{\epsilon,k}| (1 + 2|Du_\delta^{\epsilon,k}| +$$

$$+ |u_\delta^k|^{1/2}) |u^{\epsilon,k} - u_\delta^{\epsilon,k}| F_\delta^{\epsilon,k} dx$$

$$(4.16) \quad \leq \bar{H} \frac{\rho}{2} E \int_O |Du^{\epsilon,k} - Du_\delta^{\epsilon,k}|^2 F_\delta^{\epsilon,k} dx$$

$$+ \frac{\bar{H}}{2\rho} E \int_O |Du^{\epsilon,k} - Du_\delta^{\epsilon,k}|^2 (u^{\epsilon,k} - u_\delta^{\epsilon,k})^2 F_\delta^{\epsilon,k} dx$$

$$+ \bar{H} \frac{\sigma}{2} E \int_O |Du^{\epsilon,k} - Du_\delta^{\epsilon,k}|^2 F_\delta^{\epsilon,k} dx$$

$$+ \frac{\bar{H}}{\sigma} E \int_O (1 + 4|Du_\delta^{\epsilon,k}|^2 + |u_\delta^k|)(u^{\epsilon,k} - u_\delta^{\epsilon,k})^2 F_\delta^{\epsilon,k} dx.$$

We also use the fact (cf.(4.11))

$$E \int_O |Du_\delta^{\epsilon,k}|^2 (u^{\epsilon,k} - u_\delta^{\epsilon,k})^2 F_\delta^{\epsilon,k} dx$$

$$\leq 2 C_1^2 \bar{u}^2 E \int_O (u^{\epsilon,k} - u_\delta^{\epsilon,k})^2 F_\delta^{\epsilon,k} dx$$

$$(4.17) \qquad + 2 C_1^2 \bar{u}^2 E \int_O \|Dv^\epsilon\|^2 (u^{\epsilon,k} - u_\delta^{\epsilon,k})^2 F_\delta^{\epsilon,k} dx$$

$$+ C (H_\delta^k)^2 E \int_O |v^\epsilon - x|^2 dx.$$

We shall particularly consider the term

$$E \int_O \|Dv^\epsilon\|^2 (u^{\epsilon,k} - u_\delta^{\epsilon,k})^2 F_\delta^{\epsilon,k} dx$$

$$= \sum_\ell E \int_O |Dv_\ell^\epsilon|^2 (u^{\epsilon,k} - u_\delta^{\epsilon,k})^2 F_\delta^{\epsilon,k} dx$$

$$(4.18) \qquad \leq \frac{1}{\alpha} \sum_\ell E \int_O a_{ij}^\epsilon \frac{\partial v_\ell^\epsilon}{\partial x_j} \frac{\partial v_\ell^\epsilon}{\partial x_i} (u^{\epsilon,k} - u_\delta^{\epsilon,k})^2 F_\delta^{\epsilon,k} dx$$

$$= \frac{1}{\alpha} \sum_\ell E \int_O q_{\ell\ell} (u^{\epsilon,k} - u_\delta^{\epsilon,k})^2 F_\delta^{\epsilon,k} dx +$$

$$+\frac{1}{\alpha}\sum_{\ell}E\int_{O}[a_{ij}^{\epsilon}\frac{\partial v_{\ell}^{\epsilon}}{\partial x_j}\frac{\partial v_{\ell}^{\epsilon}}{\partial x_i}-q_{\ell\ell}](u^{\epsilon,k}-u_{\delta}^{\epsilon,k})^2F_{\delta}^{\epsilon,k}dx.$$

Next we have

$$\left|E\int_{O}(H^{\epsilon}(x,Du_{\delta}^{\epsilon,k},u_{\delta}^{\epsilon})-H^{\epsilon}(x,\frac{\partial u_{\delta}^{k}}{\partial x_{\ell}}Dv_{\ell}^{\epsilon},u_{\delta}^{k}))(u^{\epsilon,k}-u_{\delta}^{\epsilon,k})F_{\delta}^{\epsilon,k}dx\right|$$

$$\leq\bar{H}E\int_{O}\{\sum_{j}|\frac{\partial^2 u_{\delta}^{k}}{\partial x_{\ell}\partial x_j}\frac{(v_{\ell}^{\epsilon}-x_{\ell})}{[1+(v_{\ell}^{\epsilon}-x_{\ell})^2]^{1/2}}+$$

$$+\frac{\partial u_{\delta}^{k}}{\partial x_j}(1-\frac{1}{[1+(v_j^{\epsilon}-x_j)^2]^{1/2}})+$$

$$-\frac{\partial u_{\delta}^{k}}{\partial x_{\ell}}\frac{\partial v_{\ell}^{\epsilon}}{\partial x_j}(1-\frac{1}{[1+(v_{\ell}^{\epsilon}-x_{\ell})^2]^{1/2}})^2\}|u^{\epsilon,k}-u_{\delta}^{\epsilon,k}|F_{\delta}^{\epsilon,k}dx$$

$$\leq\bar{H}E\int_{O}\{\sum_{j}|\frac{\partial^2 u_{\delta}^{k}}{\partial x_{\ell}\partial x_j}\frac{(v_{\ell}^{\epsilon}-x_{\ell})}{[1+(v_{\ell}^{\epsilon}-x_{\ell})^2]^{1/2}})+$$

$$+\frac{\partial u_{\delta}^{k}}{\partial x_j}(1-\frac{1}{[1+(v_j^{\epsilon}-x_j)^2]^{1/2}})+$$

$$-\frac{\partial u_{\delta}^{k}}{\partial x_{\ell}}\frac{\partial v_{\ell}^{\epsilon}}{\partial x_j}(1-\frac{1}{[1+(v_{\ell}^{\epsilon}-x_{\ell})^2]^{1/2}})|\}\cdot$$

$$\cdot\{1+2|\frac{\partial u_{\delta}^{k}}{\partial x_{\ell}}\frac{\partial v_j^{\epsilon}}{\partial x_j}+\frac{\partial^2 u_{\delta}^{k}}{\partial x_{\ell}\partial x_j}\frac{(v_{\ell}^{\epsilon}-x_{\ell})}{[1+(v_{\ell}^{\epsilon}-x_{\ell})^2]^{1/2}}+$$

$$+\frac{\partial u_{\delta}^{k}}{\partial x_j}[1-\frac{1}{(1+(v_j^{\epsilon}-x_j)^2)^{3/2}}]+$$

$$-\frac{\partial u_{\delta}^{\ell}}{\partial x_{\ell}}\frac{\partial v_{\ell}^{\epsilon}}{\partial x_j}[1-\frac{1}{(1+(v_{\ell}^{\epsilon}-x_{\ell})^2)^{3/2}}]|+|u_{\delta}^{k}|^{1/2}\}|u^{\epsilon,k}-u_{\delta}^{\epsilon,k}|F_{\delta}^{\epsilon,k}dx$$

$$\leq C(H_{\delta}^{k})^2E\int_{O}|v^{\epsilon}-x|^2dx+CE\int_{O}a_{ij}^{\epsilon}\frac{\partial v_{\ell}^{\epsilon}}{\partial x_j}\frac{\partial v_{\ell}^{\epsilon}}{\partial x_i}(\frac{\partial u_{\delta}^{k}}{\partial x_{\ell}})^2\cdot$$

$$\cdot(1-\frac{1}{[1+(v_{\ell}^{\epsilon}-x_{\ell})^2]^{1/2}})^2dx+CH_{\delta}^{\epsilon}(E\int_{O}|v^{\epsilon}-x|^2dx)^{1/2}$$

$$+C\{E\int_{O}a_{ij}^{\epsilon}\frac{\partial v_{\ell}^{\epsilon}}{\partial x_j}\frac{\partial v_{\ell}^{\epsilon}}{\partial x_i}(\frac{\partial u_{\delta}^{k}}{\partial x_{\ell}})^2(1-\frac{1}{[1+(v_{\ell}^{\epsilon}-x_{\ell})^2]^{1/2}})^2dx\}^{1/2}$$

and we use as in (4.18)

$$E \int_O a_{ij}^\epsilon \frac{\partial v_\ell^\epsilon}{\partial x_j} \frac{\partial v_\ell^\epsilon}{\partial x_i} \left(\frac{\partial u_\delta^k}{\partial x_\ell}\right)^2 \left(1 - \frac{1}{[1 + (v_\ell^\epsilon - x_\ell)^2]^{1/2}}\right)^2 dx$$

$$\leq E \int_O \left(a_{ij}^\epsilon \frac{\partial v_\ell^\epsilon}{\partial x_j} \frac{\partial v_\ell^\epsilon}{\partial x_i} - q_{\ell\ell}\right)\left(\frac{\partial u_\delta^k}{\partial x_\ell}\right)^2 \left(1 - \frac{1}{[1 + (v_\ell^\epsilon - x_\ell)^2]^{1/2}}\right)^2 dx$$

$$+ C \sum_\ell E \int_O \left(1 - \frac{1}{[1 + (v_\ell^\epsilon - x_\ell)^2]^{1/2}}\right)^2 dx.$$

4.4. Convergence proof. We deduce from (4.13) that

$$E \int_O a_{ij}^\epsilon \frac{\partial}{\partial x_j}(u^{\epsilon,k} - u_\delta^{\epsilon,k}) \frac{\partial}{\partial x_i}(u^{\epsilon,k} - u_\delta^{\epsilon,k}) \cdot$$

$$\cdot [1 + 2s(u^{\epsilon,k} - u_\delta^{\epsilon,k})^2 \, F_\delta^{\epsilon,k} dx +$$

$$+ (N + \beta)E \int_O (u^{\epsilon,k} - u_\delta^{\epsilon,k})^2 F_\delta^{\epsilon,k} dx$$

$$+ E < -\frac{\partial}{\partial x_i} q_{i\ell}, \frac{\partial u_\delta^k}{\partial x_\ell}(u^{\epsilon,k} - u_\delta^{\epsilon,k})F_\delta^{\epsilon,k} >$$

$$(4.19) \qquad - E \int_O q_{i\ell} \frac{\partial^2 u_\delta^k}{\partial x_\ell \partial x_i}(u^{\epsilon,k} - u_\delta^{\epsilon,k})F_\delta^{\epsilon,k} dx$$

$$+ NE \int_O u_\delta^k (u^{\epsilon,k} - u_\delta^{\epsilon,k})F_\delta^{\epsilon,k} dx \leq$$

$$\leq E \int_O \mathcal{H}(x, Du_\delta^k, u_\delta^k)(u^{\epsilon,k} - u_\delta^{\epsilon,k})F_\delta^{\epsilon,k} dx$$

$$+ NE \int_O (u^{\epsilon,k-1} - u_\delta^{\epsilon,k-1})(u^{\epsilon,k} - u_\delta^{\epsilon,k})F_\delta^{\epsilon,k} dx$$

$$+ NE \int_O u_\delta^{\epsilon,k-1}(u^{\epsilon,k} - u_\delta^{\epsilon,k})F_\delta^{\epsilon,k} dx + X_\delta^{\epsilon,k},$$

where

$$X_\delta^{\epsilon,k} \leq CE \int_O \left|D(u^{\epsilon,k} - u_\delta^{\epsilon,k})\right| \left\{ H_\delta^k |v^\epsilon - x| + \left[a_{ij}^\epsilon \frac{\partial v_\ell^\epsilon}{\partial x_j} \frac{\partial v_\ell^\epsilon}{\partial x_i}\left(\frac{\partial u_\delta^k}{\partial x_\ell}\right)^2 \cdot \right. \right.$$

$$\cdot \left(1 - \frac{1}{[1 + (v_\ell^\epsilon - x_\ell)^2]^{1/2}}\right)\bigg]^{1/2}\bigg\}[1 + 2s(u^{\epsilon,k} - u_\delta^{\epsilon,k})^2]F_\delta^{\epsilon,k}dx$$

$$+ CE\int_O |v^\epsilon - x||u^{\epsilon,k} - u_\delta^{\epsilon,k}|F_\delta^{\epsilon,k}dx$$

$$+ \bar{H}\frac{(\rho + \sigma)}{2}E\int_O |Du^{\epsilon,k} - Du_\delta^{\epsilon,k}|^2 F_\delta^{\epsilon,k}dx$$

$$+ \frac{\bar{H}}{2\rho}E\int_O |Du^{\epsilon,k} - Du_\delta^{\epsilon,k}|^2(u^{\epsilon,k} - u_\delta^{\epsilon,k})^2 F_\delta^{\epsilon,k}dx$$

$$+ \frac{\bar{H}}{\sigma}\left(1 + \bar{u} + 8C_1^2\bar{u}^2 + \frac{2C_1^2\bar{u}^2}{\alpha}\sum_\ell q_{\ell\ell}\right)E\int_O (u^{\epsilon,k} - u_\delta^{\epsilon,k})^2 F_\delta^{\epsilon,k}dx$$

$$+ C(H_\delta^k)^2 E\int_O |v^\epsilon - x|^2 dx + CH_\delta^k\left(E\int_O |v^\epsilon - x|^2 dx\right)^{1/2}$$

$$+ C\left|E\int_O (a_{ij}^\epsilon\frac{\partial v_\ell^\epsilon}{\partial x_j}\frac{\partial v_\ell^\epsilon}{\partial x_i} - q_{\ell\ell})(\frac{\partial u_\delta^k}{\partial x_\ell})^2\left(1 - \frac{1}{[1 + (v_\ell^\epsilon - x_\ell)^2]^{1/2}}\right)^2 dx\right|$$

$$+ C\left|E\int_O (a_{ij}^\epsilon\frac{\partial v_\ell^\epsilon}{\partial x_j}\frac{\partial v_\ell^\epsilon}{\partial x_i} - q_{\ell\ell})(\frac{\partial u_\delta^k}{\partial x_\ell})^2\left(1 - \frac{1}{[1 + (v_\ell^\epsilon - x_\ell)^2]^{1/2}}\right)^2 dx\right|^{\frac{1}{2}}$$

$$+ \left|E\int_O (H^\epsilon(x, \frac{\partial u_\delta^k}{\partial x_\ell}Dv_\ell^\epsilon, u_\delta^k) - \mathcal{H}(x, Du_\delta^k, u_\delta^k))(u^{\epsilon,k} - u_\delta^{\epsilon,k})F_\delta^{\epsilon,k}dx\right|.$$

We define now various constants. We choose

$$\rho = \sigma = \frac{\alpha}{2\bar{H}} \quad , \quad s = \frac{\bar{H}^2}{2\alpha}$$

$$N = \frac{2\bar{H}^2}{\alpha}\left(1 + \bar{u} + 8C_1^2\bar{u}^2 + 2C_1^2\frac{\bar{u}^2}{\alpha}\sum_\ell q_{\ell\ell}\right).$$

With these choices we deduce form (4.19) that

$$\frac{\alpha}{4}E\int_O |D(u^{\epsilon,k} - u_\delta^{\epsilon,k})|^2 F_\delta^{\epsilon,k}dx$$

(4.20)
$$+ \beta E\int_O (u^{\epsilon,k} - u_\delta^{\epsilon,k})^2 F_\delta^{\epsilon,k}dx + E < -\frac{\partial}{\partial x_i}(q_{ij}\frac{\partial u_\delta^k}{\partial x_j}) +$$

$$+ Nu_\delta^k - \mathcal{H}(x, Du_\delta^k, u_\delta^k) - Nu_\delta^{k-1}, (u^{\epsilon,k} - u_\delta^{\epsilon,k})F_\delta^{\epsilon,k} >$$

$$\leq NE\int_O (u^{\epsilon,k-1} - u_\delta^{\epsilon,k-1})(u^{\epsilon,k} - u_\delta^{\epsilon,k})F_\delta^{\epsilon,k}dx + Z_\delta^{\epsilon,k}$$

with the property

(4.21) $$Z_\delta^{\epsilon,k} \to 0 \text{ as } \epsilon \to 0, \ \forall \ k, \delta \text{ fixed.}$$

We have made use of the assumptions (1.17) and (3.8),(3.13) to get (4.21).

Now from the definition of u^k (see (4.1)), we have

$$|E < \mathcal{A}u_\delta^k + Nu_\delta^k - \mathcal{H}(x, Du_\delta^k, x_\delta^k) - Nu_\delta^{k-1}, (u^{\epsilon,k} - u_\delta^{\epsilon,k})F_\delta^{\epsilon,k} > |$$

$$= |E < \mathcal{A}(u_\delta^k - u^k) + N(u_\delta^k - u^k) - \mathcal{H}(x, Du_\delta^k, u_\delta^k) +$$

$$+ \mathcal{H}(x, Du^k, u^k) - N(u_\delta^{k-1} - u^{k-1}), (u^{\epsilon,k} - u_\delta^{\epsilon,k})F_\delta^{\epsilon,k} > |$$

$$\leq C(E \int_O |D(u_\delta^k - u^k)|^2 dx)^{1/2}(1 + (H_\delta^k)^2 E \int_O |v^\epsilon - x|^2 dx)^{1/2}$$

$$+ C(E \int_O (u_\delta^k - u^k)^2 dx)^{1/2} + CE \int_O |D(u_\delta^k - u^k)|^2 dx.$$

We then deduce from (4.20) that for k, δ fixed,

(4.22)
$$\limsup_{\epsilon \to 0}[E \int_O |Du^{\epsilon,k} - u_\delta^{\epsilon,k}|^2 dx + E \int_O (u^{\epsilon,k} - u_\delta^{\epsilon,k})^2 dx]$$

$$\leq C(E \int_O |D(u_\delta^k - u^k)|^2 dx)^{1/2} + CE \int_O |D(u_\delta^k - u^k)|^2 dx$$

$$+ C(E \int_O (u_\delta^k - u^k)^2 dx)^{1/2} + C \limsup_{\epsilon \to 0} E \int_O (u^{\epsilon,k-1} - u_\delta^{\epsilon,k-1})^2 dx.$$

From (4.22) we deduce easily that

(4.23)
$$\limsup_{\epsilon \to 0}[E \int_O |Du^{\epsilon,k} - \frac{\partial u^k}{\partial x_\ell} Dv_\ell^\epsilon|^2 dx + E \int_O (u^{\epsilon,k} - u^k)^2 dx]$$

$$\leq \limsup_{\epsilon \to 0} E \int_O (u^{\epsilon,k-1} - u_\delta^{\epsilon,k-1})^2 dx.$$

By induction in k we deduce

$$(4.24) \quad \limsup_{\epsilon \to 0} [E \int_O |Du^{\epsilon,k} - \frac{\partial u^k}{\partial x_\ell} Dv_\ell^\epsilon|^2 dx + E \int_O (u^{\epsilon,k} - u^k)^2 dx] = 0.$$

From this and the estimates (4.6),(4.2),(4.3) we deduce also

$$E \int_O (u^\epsilon - u)^2 dx \to 0 \quad \text{as} \quad \epsilon \to 0.$$

Redoing a calculation similar to that leading to (4.24), this time directly on (3.1) and (3.11), we can prove that

$$E \int_O |Du^\epsilon - \frac{\partial u}{\partial x_\ell} Dv_\ell^\epsilon|^2 dx \to 0.$$

The desired result has been proven.

References

[1] A.Bensoussan, G.Blankenship, *Controlled diffusions in a random medium*, to be published in Stochastics.

[2] A.Bensoussan, L.Boccardo, F.Murat, *Homogenization of Elliptic Equations with Principal Part not in Divergence Form...* Comm. Pure Appl. Math. **34** (1986), 769-805.

[3] A.Bensoussan, J.L.Lions, G.Papanicolau, *Asymptotic Analysis for Periodic Structures*, North Holland, Amsterdam, 1978.

[4] A.Bensoussan, L.Boccardo, F.Murat, *H convergence for quasi linear elliptic equations with quadratic growth*, to be published.

[5] L.Boccardo, F.Murat, J.P.Puel, *Résultats d'existence pour certains problèmes elliptiques quasilinéaires*, Ann. Scu. Norm. Sup. Pisa **11** (1984), 213-235.

[6] G.Dal Maso, L.Modica, *Non linear stochastic homogenization*, Ann. Mat. Pura e Appl. **144** (1986), 347-389.

[7] G.Dal Maso, L.Modica, *Non Linear stochastic homogenization and ergodic theory*, J. Reine und Ang. Math. **368** (1986), 28-49.

[8] E.De Giorgi, *Generalized limits in Calculus of Variations*, in F.Strocchi et al., Topics in Functional Analysis, 1980-81, Scuola Normale Superiore Pisa, 1981.

[9] G.Facchinetti, L.Russo, *Un caso unidimensionale di omogeneizzazione stocastica*, Boll. Un. Mat. It. **2-6** (1983), 159-170.

[10] V.V.Jurinskii, *Averaging an elliptic boudary value problem with random coefficients*, Siberian Math. J. **21** (1980), 470-482.

[11] C.M.Kozlov, *Averaging of Random Operators*, Math. USSR Sbornik **37** (1980), 167-180.

[12] L.Modica, *Omogeneizzazione con coefficienti casuali*, Atti Convegno su "Studio di Problemi-Limite dell'Analisi Funzionale", Bressanone, 1981, Pitagora, Bologna, 1982, 155-165.

[13] F.Murat, *Homogénéisation*, Cours Faculté Sciences Alger.

[14] F.Murat, L.Tartar, *Calcul des variations et homogénéisation*, Eyrolles, Paris 85, Collect. D.E.R. E.D.F., Lecture Notes of a Summer School on Homogeneization.

[15] G.C.Papanicolau, S.R.S. Varadhan, *Boundary Value problems with rapidly oscillating random coefficients*, Colloquia Mathematica Socrataties Jonos Bolyai **37** random Fields, Esytergom (Hungary), 1979.

[16] S.Spagnolo, *Sulla convergenza di soluzioni di equazioni paraboliche ed ellittiche*, Ann.Scu. Norm. Sup. Pisa **22** (1965), 571-597.

[17] L.Tartar, Cours Peccot, Collège de France.

[18] L.Tartar, *Remarks on homogenization*.

[19] L.Tartar, *Estimations finies de coefficients homogénéisés*, Research Notes in Math. **125**, Pitman 1985 (ed. by P.Kree, Colloque De Giorgi).

INRIA

Domaine de Voluceau - Rocquencourt

F-78153 LE CHESNAY CEDEX

L^∞ AND L^1 VARIATIONS ON
A THEME OF Γ-CONVERGENCE

LUCIO BOCCARDO

Dedicated to Ennio De Giorgi on his sixtieth birthday

1. Introduction. The theory of Γ-convergence is an important tool in Calculus of Variations, because the equicoercivity and the Γ-convergence of a sequence of functionals F_ε to F_o imply the weak convergence of minima $(u_\varepsilon \rightharpoonup u_o)$ and the convergence of $F_\varepsilon(u_\varepsilon)$ to $F_o(u_o)$. Unfortunately, in the general case, the Γ-convergence of F_ε to F_o do not imply the Γ-convergence of $(F_\varepsilon + G)$ to $(F_o + G)$. Thus, if we want study the convergence of the sequence u_ε, where u_ε is a minimum of F_ε over the convex set

$$K(\psi) = \{v \in W_o^{s,p}(\Omega) \ \ v \geq \psi \ \text{ a.e. in } \Omega\}$$

we must proof the $\Gamma-$convergence of

$$(F_\varepsilon + \delta(K(\psi))) \quad to \quad (F_o + \delta(K(\psi))).$$

The point of view we adopt is the minimization of functionals: thus no Euler equation will be really written.

Results of Γ-convergence for unilateral problems have already been obtained in several papers. Here we give a new simple proof, by means of techniques of [C-S] and [B-Ma] (section 3). In section 4 and 5 we give two convergence results of the minimum u_ε, when F_ε Γ-converges to F_o and G_ε converges to G_o in some sense, but $(F_\varepsilon + G_\varepsilon)(u_\varepsilon)$ do not converge to $(F_o + G_o)(u_o)$.

2. Let Ω be a bounded open set of $\mathbf{R}^N$. We shall consider in the sequel a sequence of Caratheodory functions $f_\varepsilon(x, s, \xi)$ (where $\varepsilon \geq 0, x \in \Omega, s \in \mathbf{R}, \xi \in \mathbf{R}^N$), convex with respect to ξ, which satisfy the growth conditions

$$(2.1) \qquad \alpha|\xi|^p \leq f_\varepsilon(x, s, \xi) \leq \beta(1 + |\xi|^p); \ \beta \geq \alpha > 0; \ p > 1.$$

For any $v \in W_o^{1,p}(\Omega)$, we define

$$(2.2) \qquad I_\varepsilon(v) = \int_\Omega f_\varepsilon(x, v, Dv)$$

We recall the following definition of Γ^--convergence.

Definition 2.1 *([D],[D-F],[B-Ma])- We shall say that I_ε Γ^-- converges to I_o whenever one has*

$$(2.3) \qquad v_\varepsilon \rightharpoonup v_o \quad \text{implies} \quad I_o(v_o) \leq \liminf I_\varepsilon(v_\varepsilon),$$

$$(2.4) \qquad \begin{array}{l} \text{for} \ \ v \in W_o^{1,p}(\Omega) \quad \text{there exists a sequence} \\ v_\varepsilon \quad \text{such that} \ \ v_\varepsilon \rightharpoonup v_o \quad \text{and} \ \ I_\varepsilon(v_\varepsilon) \to I_o(v_o) \,. \end{array}$$

The theory of Γ−convergence is an important tool in Calculus of Variations because the equicoercivity (like (2.1)) and the Γ−convergence of I_ε to I_o imply the weak convergence of the minima. Furthemore, in this context, we recall the following compactness result.

Theorem 2.2 *([D],[Sb],[B-D],[DM])-Consider a sequence I_ε. Then there exists a subsequence $I_{\varepsilon'}$, and a Caratheodory function $f_o(x, s, \xi)$, convex with respect to ξ, which satisfies the growth conditions (2.1), such that $I_{\varepsilon'}$ Γ^--converges to*

$$I_o(v) = \int_\Omega f_o(x, v, Dv)$$

Finally we recall the link between $\Gamma-$ convergence and $G-$convergence of differential operators ([Sp],[B-Ma]).

In the sequel, we shall consider also the problem of the minimization of I_ε over a convex set of the type

$$(2.5) \quad K(\psi) = \{v \in W_o^{1,p}(\Omega) : v \geq \psi \quad \text{a.e. in } \Omega , \psi \in W_0^{1,p}(\Omega)\}.$$

and we shall use the following definition.

Definition 2.3

$$\delta(K(\psi))[v] = \begin{cases} 0 & \text{if } v \in K(\psi) \\ +\infty & \text{if } v \notin K(\psi). \end{cases}$$

3. Unilateral Problems. Our objective is to prove the following

Theorem 3.1 *Under the assumptions* $(2.1),(2.2),(2.5),$

$$I_\varepsilon + \delta(K(\psi)) \xrightarrow{\Gamma^-} I_o + \delta(K(\psi)).$$

PROOF. The proof consists in 3 steps. *In the first step* we study the case

$$(3.1) \qquad\qquad \psi \in W_0^{1,s}(\Omega), \;\; s > p$$

and

$$u_o \geq \psi + t\varphi$$

where t is a positive real number and

$$\varphi \in \mathcal{D}(\Omega), \quad \varphi > 0 \;\; \text{in} \;\; \Omega, \; \|\varphi\| = 1.$$

By the Γ^--convergence of I_ε to I_o, there exists a sequence $\tilde{u}_\varepsilon$ such that

$$(3.3) \qquad\qquad \tilde{u}_\varepsilon \to u_o \;\; \text{in} \;\; W_o^{1,p}(\Omega) - \text{weak}$$

(3.4) $$I_\varepsilon(\tilde{u}_\varepsilon) \to I_o(u_o).$$

The functions u_ε defined by

$$u_\varepsilon = \max(\tilde{u}_\varepsilon, \psi)$$

belong to $K(\psi)$ and converge to u_o in $W_o^{1,p}(\Omega)$-weak.
    Note that

$$(\tilde{u}_\varepsilon - \psi) \to (u_o - \psi) \geq t\varphi \quad \text{in} \quad L^p(\Omega)$$

It is easy to see that

$$\text{meas } \{x \in \Omega : \tilde{u}_\varepsilon(x) \leq \psi(x)\} \to 0.$$

Define

$$\Omega_\varepsilon = \{x \in \Omega : \tilde{u}_\varepsilon(x) \leq \psi(x)\} = \{x \in \Omega : u_\varepsilon(x) = \psi(x)\}.$$

Therefore

$$I_\varepsilon(u_\varepsilon) = \int_{\Omega - \Omega_\varepsilon} f_\varepsilon(x, \tilde{u}_\varepsilon, D\tilde{u}_\varepsilon) + \int_{\Omega_\varepsilon} f_\varepsilon(x, \psi, D\psi) \leq$$

$$\leq \int_\Omega f_\varepsilon(x, \tilde{u}_\varepsilon, D\tilde{u}_\varepsilon) + \beta \int_{\Omega_\varepsilon} (1 + |\psi|^p + |D\psi|^p) \leq$$

$$\leq I_\varepsilon(\tilde{u}_\varepsilon) + c_1 [\text{meas } \Omega_\varepsilon]^{1-p/s}.$$

Then

$$I_o(u_o) \leq \liminf I_\varepsilon(u_\varepsilon) \leq \limsup I_\varepsilon(u_e) \leq \limsup I_\varepsilon(\tilde{u}_\varepsilon) +$$
$$\limsup c_1 \text{ meas } \{\Omega_\varepsilon\}^{1-p/s} = I_o(u_o),$$

that is

(3.5) $$I_\varepsilon(u_\varepsilon) \to I_o(u_o).$$

Remark that the proof of (2.3) is trivial.
    In the second step we consider the case

(3.6) $$u_o \geq \psi.$$

Note that the functions $v_{o,t}$ defined by

(3.7) $$v_{o,t} = u_o + t\psi$$

satisfy the inequality (3.2). Then, by the first step, there exist a sequence $v_{\varepsilon,t}$ such that

(3.8) $$v_{\varepsilon,t} \in K(\psi)$$

(3.9) $$v_{\varepsilon,t} \to v_{o,t} \quad \text{in} \quad W_o^{1,p}(\Omega) \quad \text{weak} \quad (\varepsilon \to 0)$$

(3.10) $$I_\varepsilon(v_{\varepsilon,t}) \to I_o(v_{o,t}) \qquad (\varepsilon \to 0).$$

Of course, we have also

(3.10) $$I_o(v_o, t) \to I_o(u_o) \qquad (t \to 0)$$

(3.11) $$v_{o,t} \to u_o \qquad (t \to 0).$$

Then, we can use a diagonal argument as in theorem 3.5 of [B-Ma] and we deduce again the existence of a sequence u_ε such that

(3.12) $$u_\varepsilon \to u_o$$

(3.13) $$I_\varepsilon(u_\varepsilon) \to I_o(u_o).$$

In the third step we assume only that the obstacle ψ belongs to $W_o^{1,p}(\Omega)$. By regularization, there exists a sequence of regular functions ψ_t such that $K(\psi_t)$ is not empty and

(3.14) $$\psi_t \in W_o^{1,s}(\Omega) , s > p, \ t > 0$$

(3.15) $$\psi_t \to \psi \quad \text{in} \quad W_o^{1,p}(\Omega).$$

Thus the function

$$v_{o,t} = u_o - \psi + \psi_t$$

belongs to $K(\psi_t)$. With the help of the above results, using the method of step 2, we deduce again (3.12) and (3.13) also in the general case.

Remark 3.2. We point out that, if ψ satisfies the hypothesis (3.1), then the sequence u_ε is bounded in $W_o^{1,r}(\Omega)$ for some $r > p$ ([B]).

4. A case of convergence of minima without Γ-convergence (L^∞). In this section we assume *also* the equi-uniformly convexity of I_ε. That is: there exists a continuous, strictly increasing real valued function such that

(4.1)
$$B : [0,\infty[\to [0,\infty[$$

(4.2)
$$B(0) = 0$$

(4.3)
$$B(\|v - w\|) + I_\varepsilon(\frac{v + w}{2}) \leq \frac{1}{2}I_\varepsilon(v) + \frac{1}{2}I_\varepsilon(w)$$
$$\text{for any } v, w \in W_o^{1,p}(\Omega)$$

(4.4)
$$I_\varepsilon(v) = \int_\Omega f_\varepsilon(x, Dv).$$

Note that if (4.3) holds for $\varepsilon > 0$ then it is easy to prove that (4.3) still holds for I_o.

Theorem 4.1. *Under the assumptions (4.1),(4.2),(4.3),(4.4), (4.5),(4.6),(4.7),*

(4.5)
$$I_\varepsilon \xrightarrow{\Gamma^-} I_o$$

(4.6)
$$\psi_\varepsilon \to \psi_o \quad in \quad W_o^{1,p}(\Omega) - weak$$

(4.7)
$$\psi_\varepsilon \to \psi_o \quad in \quad L^\infty(\Omega)$$

the sequence of the minima of the problem

$$\min\{I_\varepsilon(v), \quad v \in K(\psi_\varepsilon)\}$$

converges in $W_o^{1,p}$-*weak to* u_o, *where* u_o *minimizes* I_o *over* $K(\psi_o)$.

Remark 4.2. Note that also in the simple case $f_\varepsilon(x, Dv) = |Dv|^2$, for any ε, we don't have $I_\varepsilon(u_\varepsilon) \to I_o(u_o)$ (see [B-Mu]). This remark implies that $I_\varepsilon + \delta(K(\psi_\varepsilon))$ cannot Γ^--converge to $I_o + \delta(K(\psi_o))$.

Lemma 4.3. *Let* ψ_1, ψ_2 *be two obstacles such that* $\psi_1 - \psi_2 \in L^\infty(\Omega)$. *Consider a functional*

$$I(v) = \int_\Omega f(x, Dv)$$

which satisfies (4.1),(4.2),(4.3). *We define* u_j *the minimum of* I *over* $K(\psi_j)$. *Then*

(4.8) $$\|u_1 - u_2\|_{L^\infty} \le \|\psi_1 - \psi_2\|_{L^\infty} = K.$$

PROOF. By the definition of the minimum we have

$$\int_\Omega f(x, Du_1) \le \int_\Omega f(x, \frac{Du_1 - D(u_1 - u_2 - K)^+ + Du_1}{2})$$

which implies

(4.9)
$$\int_E f(x, Du_1) \le \int_E (x, D\frac{u_1 + u_2}{2})$$
$$\le \frac{1}{2}\int_E f(x, Du_1) + \frac{1}{2}\int_E f(x, Du_2),$$

where

$$E = \{x \in \Omega : (u_1 - u_2 - K) > 0\}.$$

We have also

$$\int_\Omega f(x, Du_2) \le \int_\Omega f(x, \frac{Du_2 + D(u_1 - u_2 - K)^+ + Du_2}{2}),$$

which implies

$$(4.10) \quad \int_E f(x, Du_2) \leq \int_E f(x, D\frac{u_1 + u_2}{2})$$
$$\leq \frac{1}{2}\int_E f(x, Du_1) + \frac{1}{2}\int_E f(x, Du_2).$$

The inequalities (4.9) and (4.10) yield

$$(4.11) \quad \int_E f(u, Du_1) + \int_E f(x, Du_2) = 2\int_E f(x, \frac{Du_1 + Du_2}{2}).$$

Now we use the inequality (4.3) with

$$v = \frac{1}{2}(u_1 - u_2 - K)^+ + \frac{1}{2}(u_1 + u_2)$$

and

$$w = \frac{1}{2}(u_1 + u_2) - \frac{1}{2}(u_1 - u_2 - K)^+$$

and we have

$$B(\|(u_1 - u_2 - K)^+\|) + \int_\Omega f(x, D\frac{u_1 + u_2}{2})$$
$$\leq \frac{1}{2}\int_\Omega f(x, \frac{1}{2}D(u_1 - u_2 - K)^+ + \frac{1}{2}D(u_1 + u_2))$$
$$+ \frac{1}{2}\int_\Omega f(x, \frac{1}{2}D(u_1 + u_2) - \frac{1}{2}D(u_1 - u_2 - K)^+) =$$
$$= \int_{\Omega - E} f(x, \frac{1}{2}D(u_1 + u_2)) + \frac{1}{2}\int_E f(x, Du_1) + \frac{1}{2}\int_E f(x, Du_2).$$

Now we use (4.11) and we deduce that

$$B(\|(u_1 - u_2 - K)^+\|) + I(\frac{u_1 + u_2}{2}) \leq I(\frac{u_1 + u_2}{2}),$$

which implies that

$$u_1 - u_2 - K \leq 0 \quad \text{a.e in} \quad \Omega.$$

In a like manner, it is possible to prove that

$$u_1 - u_2 \geq -K.$$

Proof of the Theorem 4.1. Define z_ε the minimum of I_ε on $K(\psi_o)$. By virtue of Theorem 3.1 z_ε converges to u_o in $W_0^{1,p}(\Omega)$-weak and Lemma 4.3 implies that

$$(u_\varepsilon - z_\varepsilon) \to 0 \quad \text{in} \quad L^\infty(\Omega).$$

Thus the sequence u_ε, which is bounded in $W_o^{1,p}(\Omega)$ because (4.6), is such that

$$\|u_\varepsilon - u_o\|_{L^p} \le \|u_\varepsilon - z_\varepsilon\|_{L^p} + \|z_\varepsilon - u_o\|_{L^p}.$$

This inequality together with Rellich's Theorem imply the convergence of u_ε to u_o in $W_o^{1,p}(\Omega)$-weak.

5. A case of convergence of minima without Γ-convergence (L^1). In this section we add

$$(5.1) \qquad B(\|v\|) = \int_\Omega |Dv|^p$$

to the assumptions of section 4 and we study an *other case of convergence of minima without Γ-convergence*. We assume that h_ε is a sequence such that

$$(5.2) \qquad h_\varepsilon \to h_o \quad \text{in} \quad W^{-1,p'}(\Omega)-\text{weak}$$

$$(5.3) \qquad h_\varepsilon \to h_o \quad \text{in} \quad L^1(\Omega)$$

and we consider u_ε, the minimum of the functional

$$(5.4) \qquad I_\varepsilon(v) - \int_\Omega fv, \quad \varepsilon \ge 0.$$

One has the following result of convergence of u_ε.

Theorem 5.1. *Under the assumptions (4.1),(4.2),(4.3),(4.4), (4.5),(5.1),(5.2),(5.3) the sequence u_ε converges in $W_o^{1,p}(\Omega)$-weak to u_o.*

We need the following Lemma.

Lemma 5.2. Let h_1, h_2 be in $W_o^{-1,p'}(\Omega)$ and let h_1, h_2 be in $L^1(\Omega)$. Consider a functional $I(v) - < h_i, v >$

(5.5)
$$I(v) = \int_\Omega f(x, Dv)$$

which satisifies (4.1),(4.2),(4.3),(5.1). We have the following estimate

(5.6)
$$\int_\Omega |DT_K(u_1 - u_2)|^p \le c_1 K \|h_1 - h_2\|_{L^1}.$$

PROOF. By definition of minimum we have

(5.7)
$$I(u_1) - < h_1, u_1 > \le I(u_1 - \frac{1}{2}T_K(u_1 - u_2)) -$$
$$< h_1, u_1 - \frac{1}{2}T_K(u_1 - u_2) >$$

and

(5.8)
$$I(u_2) - < h_2, u_2 > \le I(u_2 + \frac{1}{2}T_K(u_1 - u_2)) -$$
$$< h_2, u_2 + \frac{1}{2}T_K(u_1 - u_2) >$$

which imply

(5.10)
$$\frac{1}{2}I(u_1) + \frac{1}{2}I(u_2) \le \frac{1}{2}\int_{G_K} f(x, Du_1) +$$
$$+ \frac{1}{2}\int_{G_K} f(x, Du_2) + \int_{P_K} f(x, D\frac{u_1 + u_2}{2}) +$$
$$+ \frac{1}{4}\int_\Omega (h_1 - h_2)T_K(u_1 - u_2),$$

where

$$P_K = \{x \in \Omega : |u_1(x) - u_2(x)| \le K\}$$
$$G_K = \{x \in \Omega : |u_1(x) - u_2(x)| > K\}.$$

The uniform convexity yields

$$\gamma \int_\Omega |DT_K(u_1 - u_2)|^p + I(\frac{u_1 + u_2}{2}) \le$$

$$\le \frac{1}{2}I(\frac{u_1 + u_2}{2} + \frac{1}{2}T_K(u_1 - u_2))$$

(5.11)
$$+ \frac{1}{2}I(\frac{u_1 + u_2}{2} - \frac{1}{2}T_K(u_1 - u_2)) =$$

$$= \frac{1}{2}I(u_1) + \frac{1}{2}I(u_2) - \frac{1}{2}\int_{G_K} f(x, Du_1)$$

$$- \frac{1}{2}\int_{G_K} f(x, Du_2) + \int_{G_K} f(x, D\frac{u_1 + u_2}{2}).$$

The inequalities (5.10), (5.11) give the estimate (5.6).

Proof of Theorem 5.1. The sequence u_ε is bounded in $W_o^{1,p}(\Omega)$. Then for some subsequence (still denoted u_ε) and for some $u^* \in W_o^{1,p}(\Omega)$ we have

(5.12) $$u_\varepsilon \to u^* \quad \text{in} \quad W_o^{1,p}(\Omega) - \text{weak}.$$

Define ζ_ε the minimum of the functional $I_\varepsilon(v) - < h_o, v >$. By the definition of Γ^--convergence

(5.13) $$\zeta_\varepsilon \to u_o \quad \text{in} \quad W_o^{1,p}(\Omega) - \text{weak}.$$

Now from the previous Lemma follows that

(5.14) $$\|T_K(u_\varepsilon - \zeta_\varepsilon)\|_{W_o^{1,p}} \le c_2 K \|h_\varepsilon - h_o\|_{L^1}^{1/p},$$

which implies, by lower semicontinuity of the norm with respect to the weak convergence, that

$$\|T_K(u^* - u_o)\| = 0 \quad \text{for any } K > 0.$$

that is

(5.15) $$u^* = u_o$$

and all the sequence u_ε converges to u_o.

Remark 5.3. *In some sense the assumptions (5.2) and (5.3) can be seen as dual to the assumptions (4.6),(4.7).*

Acknowledgements. Parts of this work were done at the times the author was visiting the E.P.F. of Lausanne and was "Professeur associé " at the University of Besançon.

References

[A] H.Attouch, *Variational convergence for the functions and operators,* Appl. Math. Series, Pitman (1984).

[B.1] L.Boccardo, *An L^s estimate for the gradient of solutions of some nonlinear unilateral problems,* Ann. Mat. Pura Appl. **141** (1985) 277-287.

[B.2] L.Boccardo, *Homogénéisation pour une classe d'équations fortement non linéaires,* C.R.A.S. **306** (1988), 253-256.

[B-Ma] L.Boccardo, P.Marcellini, *Sulla convergenza delle soluzioni delle disequazioni variazionali,* Ann. Mat. Pura Appl. **110** (1976), 137-159.

[B-Mu] L.Boccardo, F.Murat, *Nouveaux resultats de convergence dans des problèmes unilateraux,* Res. Notes Math. **60** Pitman (1982), 64-85.

[B-D] G.Buttazzo, G.Dal Maso, *Γ-limits of integral functionals,* J.Anal. Math. **37** (1980), 145-185.

[C-S] L.Carbone, C.Sbordone, *Some properties of Γ-limits of integrals functionals,* Ann. Mat. Pura Appl. **122** (1979), 1-60.

[D-M] G.Dal Maso, L.Modica, *A general theory of variational functionals,* S.N.S. Pisa (1981), 149-221.

[D] E.De Giorgi, *Sulla convergenza di alcune successioni di integrali del tipo dell'area,* Rend. Mat. **8** (1975), 277-294.

[D-F] E.De Giorgi, T.Franzoni, *Su un tipo di convergenza variazionale,* Rend. Accad. Lincei **58** (1975), 842-850.

[Sb] C.Sbordone, *Su alcune applicazioni di un tipo di convergenza variazionale*, Ann. Sc. Norm. Sup. Pisa **2** (1975), 617-632.

[Sp] S.Spagnolo, *Sulla convergenza delle soluzioni di equazioni el-littiche e paraboliche*, Ann. Sc. Norm. Sup. Pisa **22** (1968), 575-597.

[T] L.Tartar, *Cours Peccot*, Collège France (1977).

Dipartimento di Matematica

Università di Roma I

Piazza A.Moro 2

I-00185 ROMA

ASYMPTOTICS FOR ELLIPTIC EQUATIONS INVOLVING CRITICAL GROWTH*

Haïm Brezis   Lambertus A. Peletier

Dedicated to Ennio De Giorgi on his sixtieth birthday

1. Introduction. Consider the problem

$$(1.1) \qquad (I) \qquad \begin{cases} -\Delta u - \lambda u = 3u^{5-\epsilon} & \text{in } \Omega \\ \quad u > 0 & \text{in } \Omega \\ \quad u = 0 & \text{on } \partial\Omega \end{cases}$$

where Ω is the unit ball in $\mathbf{R}^3$, $\lambda \geq 0$ and $\epsilon \geq 0$. It is well known that if $\epsilon > 0$, Problem (I) has a solution u_ϵ for any $\lambda < \lambda_1 = \pi^2$. On the other hand, if $\epsilon = 0$ Problem (I) has a solution if and only if $\pi^2/4 < \lambda < \pi^2$ (See [BN]).

In this paper we return to the question, first studied in [AP2], of the asymptotic behaviour of u_ϵ as $\epsilon \to 0$. There it was shown that if $\lambda = 0$, any solution u_ϵ of Problem (I) has the following limiting behaviour:

$$(1.2) \qquad \lim_{\epsilon \to 0} \epsilon\, u_\epsilon^2(0) = \frac{32}{\pi}$$

* This work was done whilst the first author held the Kloosterman Chair of the Mathematical Institute of the University of Leiden.

and at any $x \in \Omega \backslash \{0\}$:

$$(1.3) \qquad \lim_{\epsilon \to 0} \epsilon^{-1/2} u_\epsilon(x) = \frac{1}{4} \sqrt{\frac{\pi}{2}} \left(\frac{1}{|x|} - 1 \right).$$

We return to the study of u_ϵ for two reasons. First we shall give a different proof of (1.2) and (1.3), which is mainly based on PDE methods and – we hope – will enable us in due course to handle non-spherical domains. The second reason is that the method of [AP2], which readily extends to the case $0 < \lambda < \pi^2/4$, cannot be applied when $\lambda = \pi^2/4$. Here, a new phenomenon occurs, which was first discovered by Budd [Bu]. A formal computation, based on the method of matched asymptotic expansions, suggests that

$$(1.4) \qquad \lim_{\epsilon \to 0} \epsilon \, u_\epsilon^4(0) = 8\pi^2.$$

The method presented here does apply to the case $\lambda = \pi^2/4$ and provides a rigorous proof of (1.4).

As in [AP2], the method we use to prove (1.2) – (1.4) is based on estimating the different terms in the Pohozaev identity for Problem (I):

$$(1.5) \qquad \frac{3\epsilon}{6 - \epsilon} \int_\Omega u_\epsilon^{6-\epsilon} = J(u_\epsilon),$$

where

$$J(w) = \int_{\partial \Omega} (x, n) \left(\frac{\partial w}{\partial n} \right)^2 - 2\lambda \int_\Omega w^2.$$

Here n denotes the outward normal on $\partial \Omega$.

Writing $\mu = \left(u_\epsilon(0) \right)^{-2}$, we shall establish that

$$(1.6) \qquad \mu^{-1/2} u_\epsilon \to 4\pi G_\lambda \quad \text{as} \quad \epsilon \to 0,$$

where G_λ is the Green's function of $-\Delta - \lambda$, i.e. G_λ solves

$$(1.7) \qquad -\Delta G - \lambda G = \delta_0 \quad \text{in} \quad \Omega$$
$$(1.8) \qquad G = 0 \quad \text{on} \quad \partial\Omega$$

in which δ_0 is the Dirac mass centered at the origin. By elliptic regularity theory (1.6) implies that

$$(1.9) \qquad \frac{1}{\mu} J(u_\epsilon) \to J(4\pi G_\lambda).$$

On the other hand, we shall establish in Section 4 that

$$J(G_\lambda) = -g_\lambda(0)$$

so that

$$J(u_\epsilon) = -16\pi^2 g_\lambda(0)\mu + o(\mu) \quad \text{as} \quad \epsilon \to 0.$$

Here

(1.10)
$$g_\lambda(x) = G_\lambda(x) - \frac{1}{4\pi|x|},$$

i.e. g_λ is the regular part of the Green's function. Finally we check that

(1.11)
$$\int_\Omega u_\epsilon^{6-\epsilon} \to \frac{\pi^2}{4} \quad \text{as} \quad \epsilon \to 0$$

and so, putting (1.9) and (1.11) into (1.5) we obtain that

$$\frac{\epsilon}{\mu} \to -\frac{32}{\pi} \cdot 4\pi g_\lambda(0) \quad \text{as} \quad \epsilon \to 0$$

or, since

(1.12)
$$g_\lambda(x) = \frac{1}{4\pi|x|}\left\{\cos(\sqrt{\lambda}|x|) - \frac{\sin(\sqrt{\lambda}|x|)}{\tan\sqrt{\lambda}} - 1\right\},$$

this yields our first result:

Theorem 1. *Let u_ϵ be a solution of Problem (I).*
(a) If $\quad 0 \le \lambda \le \pi^2/4$, then

(1.13)
$$\lim_{\epsilon \to 0} \epsilon\, u_\epsilon^2(0) = \frac{32}{\pi}\frac{\sqrt{\lambda}}{\tan\sqrt{\lambda}}.$$

(b) If $\quad 0 \le \lambda < \pi^2/4$, then at any $x \ne 0$,

$$\lim_{\epsilon \to 0} \epsilon^{-1/2} u_\epsilon(x) = \left(\frac{\pi^3}{2} \cdot \frac{\tan\sqrt{\lambda}}{\sqrt{\lambda}}\right)^{1/2} G_\lambda(x),$$

where we define $\sqrt{\lambda}/\tan\sqrt{\lambda} = 1$ if $\lambda = 0$. If $\lambda = \pi^2/4$, the right hand side of (1.13) vanishes and all we can conclude is that

$$\epsilon\, u_\epsilon^2(0) = o(1) \quad \text{as} \quad \epsilon \to 0.$$

To obtain a precise estimate, we need a better global asymptotic approximation of u_ϵ than is given by (1.6). Hereby the following family of functions plays a central role

$$(1.14) \qquad U_\mu(x) = \left(\frac{\mu}{\mu^2 + |x|^2}\right)^{1/2};$$

it satisfies the equation

$$-\Delta u = 3u^5 \quad \text{in} \quad \mathbf{R}^3.$$

The function

$$\phi_\mu = U_\mu + 4\pi\mu^{1/2}g_\lambda$$

turns out to be the required approximation of u_ϵ. In Section 6 we shall establish that
(1.15)
$$J(u_\epsilon) = -16\pi^2 \, g_\lambda(0)\mu + 4\pi^2\mu^2 + O(\epsilon\mu|\log\mu| + \mu^3|\log\mu|) \quad \text{as} \quad \epsilon \to 0$$

Putting (1.15), together with (1.11), in (1.5) again and using the fact that $g_\lambda(0) = 0$ when $\lambda = \pi^2/4$ we obtain our second result.

Theorem 2. *Let u_ϵ be a solution of Problem (I) in which $\lambda = \pi^2/4$. Then*

(a) $$\lim_{\epsilon \to 0} \epsilon \, u_\epsilon^4(0) = 8\pi^2.$$

(b) $$\lim_{\epsilon \to 0} \epsilon^{-1/4}u_\epsilon(x) = (8\pi^2)^{-1/4}\frac{\cos(\frac{\pi}{2}|x|)}{|x|}, \quad x \neq 0.$$

An important ingredient in the proof of Theorem 2, which is of some interest in its own right, is a Pohozaev–type identity for the Green's function. This identity, which is valid for arbitrary bounded domains Ω in $\mathbf{R}^3$ is derived in Section 4. Some other integral identities involving the Green's function are also established in this section. The proof of Theorem 2 is subsequently given in Sections 5 and 6.

In Section 7, we consider a different, but related problem:

$$(II) \qquad \begin{cases} -\Delta u - \left(\dfrac{\pi^2}{4} + \epsilon\right)u = 3u^5 & \text{in } \Omega \\ u > 0 & \text{in } \Omega \\ u = 0 & \text{on } \partial\Omega, \end{cases}$$

where Ω is again the unit ball in $\mathbf{R}^3$. It is known [BN] that if $0 < \epsilon < 3\pi^2/4$, Problem (II) has a solution u_ϵ, and that, as with Problem (I), $u_\epsilon(0) \to \infty$ as $\epsilon \to 0$.

With the machinery built for dealing with Problem (I), we can now readily establish the asymptotic properties of u_ϵ both at the origin, and away from the origin, as $\epsilon \to 0$.

Theorem 3. *Let u_ϵ be a solution of Problem (II). Then*

(a) $$\lim_{\epsilon \to 0} \epsilon\, u_\epsilon^2(0) = \frac{\pi^3}{2}.$$

(b) $$\lim_{\epsilon \to 0} \epsilon^{-1/2} u_\epsilon(x) = 4\sqrt{2}\, G_{\pi^2/4}(x), \quad x \neq 0.$$

Many of the arguments and results in this paper continue to hold when the domain Ω is not a ball. They furnish insight in the behaviour of solutions u_ϵ of Problems (I) and (II) as $\epsilon \to 0$ when Ω is a general domain in $\mathbf{R}^3$ or $\mathbf{R}^N (N > 2)$ and enable us to formulate a number of conjectures about the behaviour. This is done in Section 8.

2. Preliminary bounds. Since uniqueness for Problem (I) is not known, we shall always take u_ϵ to be *any* solution of Problem (I). Because $u_\epsilon > 0$, it is known to be radially symmetric and decreasing [GNN].

As a first observation, note that if $0 \leq \lambda \leq \pi^2/4$, $u_\epsilon(0) \to \infty$ as $\epsilon \to 0$. For suppose to the contrary that there exists a sequence $\{\epsilon_n\}$, $\epsilon_n \to 0$ as $n \to \infty$ such that $u_{\epsilon_n}(0)$ remains bounded as $n \to \infty$. Then u_{ϵ_n} remains bounded in $L^\infty(\Omega)$ and, in view of the elliptic regularity theory applied to (1.1), u_{ϵ_n} remains bounded in $C^1(\bar{\Omega})$. So we can extract a subsequence, still denoted by $\{u_{\epsilon_n}\}$, which converges uniformly to a limit v, which satisfies

$$\begin{cases} -\Delta v - \lambda v = 3v^5, \ v \geq 0 & \text{in } \Omega \\ \qquad\qquad v = 0 & \text{on } \partial\Omega. \end{cases}$$

This implies that $v = 0$ because $0 \leq \lambda \leq \pi^2/4$ [BN].

On the other hand, we assert that

(2.1) $$\| u_\epsilon \|_\infty \geq \kappa > 0$$

for some constant $\kappa > 0$ independent of ϵ, which contradicts the conclusion drawn above. To prove (2.1), we multiply (1.1) by the principal eigenfunction ϕ_1 of $-\Delta$ (chosen positive) and integrate by parts. This leads to

$$(\lambda_1 - \lambda) \int_\Omega u_\epsilon \phi_1 = 3 \int_\Omega u_\epsilon^{5-\epsilon} \phi_1$$

$$\leq 3 \parallel u_\epsilon \parallel_\infty^{4-\epsilon} \int_\Omega u_\epsilon \phi_1,$$

where λ_1 is the principal eigenvalue of $-\Delta$, and thus

$$\parallel u_\epsilon \parallel_\infty^{4-\epsilon} \geq \frac{1}{3}(\lambda_1 - \lambda) \geq \frac{\pi^2}{4},$$

because $\lambda_1 = \pi^2$ and $\lambda \leq \pi^2/4$.

Whereas $u_\epsilon(0) \to \infty$ as $\epsilon \to 0$, $u_\epsilon(x) \to 0$ as $\epsilon \to 0$ at any point $x \neq 0$. This follows from the upper bound which we shall present next. Define the function

(2.2) $$W_\mu(x) = \left(\frac{\mu}{\mu^2 + \alpha|x|^2}\right)^{1/2},$$

where

(2.3) $$\alpha = \mu^{\epsilon/2} + \frac{\lambda}{3}\mu^2.$$

It satisfies the equation

(2.4) $$-\Delta u = \gamma^{-5} f(\gamma) u^5,$$

where $\gamma = \mu^{-1/2}$ and

(2.5) $$f(s) = \lambda s + 3s^{5-\epsilon}.$$

Lemma 2.1. *Let u_ϵ be any solution of Problem (I). Then*

$$u_\epsilon(x) \leq W_\mu(x) \qquad \text{for} \quad x \in \bar{\Omega},$$

where μ is given by

(2.6) $$\mu = \{u_\epsilon(0)\}^{-2}.$$

Remark. In view of the observations made at the beginning of this section, (2.6) implies that $\mu \to 0$ as $\epsilon \to 0$. Thus, we shall always think of μ as a small quantity.

Remark. It follows from Lemma 2.1 that if $x \neq 0$,

$$u_\epsilon(x) = O(\mu^{1/2-\epsilon/2}) \quad \text{as} \quad \mu \to 0.$$

For the proof of Lemma 2.1 we refer to [AP1, Lemma 1(iii)].

The following elliptic estimates will be needed. Consider the problem

$$(P) \qquad \begin{cases} -\Delta u - \lambda u = f & \text{in } \Omega \\ \qquad\qquad u = b & \text{on } \partial\Omega, \end{cases}$$

where Ω is a bounded domain in $\mathbf{R}^3$, with smooth boundary $\partial\Omega$ and $\lambda < \lambda_1$, λ_1 being the principal eigenvalue of $-\Delta$ in Ω.

Lemma 2.2. *Let u be a solution of Problem (P) in which $\lambda < \lambda_1$. Then*

$$\| u \|_{W^{1,q}(\Omega)} + \| \nabla u \|_{C^{0,\alpha}(\partial\Omega)} \leq$$
$$\leq C(\| f \|_{L^1(\Omega)} + \| f \|_{L^\infty(\omega)} + \| b \|_{C^{2,\alpha}(\partial\Omega)}),$$

for any $q < 3/2$, any $\alpha \in (0,1)$ and any neighbourhood ω of the boundary $\partial\Omega$ in Ω.

Remark. As a consequence of Lemma 2.2. we have

(i) $\quad \| u \|_{L^2(\Omega)} + \| \nabla u \|_{L^2(\partial\Omega)} \leq$

(2.7) $\qquad\qquad \leq C(\| f \|_{L^1(\Omega)} + \| f \|_{L^\infty(\omega)} + \| b \|_{C^{2,\alpha}(\partial\Omega)}).$

(ii) If $\{f_n\}$ is a bounded sequence in $L^1(\Omega)$ and in $L^\infty(\omega)$, and $\{b_n\}$ is a bounded sequence in $C^{2,\alpha}(\partial\Omega)$, then the corresponding sequence of solutions $\{u_n\}$ has a compact closure in $L^2(\Omega)$, whilst the sequence $\{\nabla u_n\}$, restricted to $\partial\Omega$, has a compact closure in $L^2(\partial\Omega)$.

Proof of Lemma 2.2. First, in view of the classical Schauder estimates, we may always assume that $b = 0$. Next, we claim that for all $q < 3/2$,

$$\| u \|_{W^{1,q}(\Omega)} \leq C \| f \|_{L^1(\Omega)}.$$

This follows easily, by duality, from the fact that if v satisfies

$$\begin{cases} -\Delta v - \lambda v = f_0 + \sum_{i=1}^{3} \dfrac{\partial}{\partial x_i} f_i & \text{in } \Omega \\ \quad\quad v = 0 & \text{on } \partial\Omega, \end{cases}$$

where $f_0, f_1, f_2 \in L^p(\Omega)$, then

$$\| v \|_{L^\infty(\Omega)} \le C_p \sum_{i=0}^{3} \| f_i \|_{L^p(\Omega)}$$

for any $p > 3$ (see e.g. [GT], Theorem 8.15); in other words $(-\Delta - \lambda)^{-1}$ maps $W^{-1,p}(\Omega)$ into $L^\infty(\Omega)$ and, by duality, it also maps $L^1(\Omega)$ into $W_0^{1,p'}(\Omega)$ with $1/p + 1/p' = 1$.

Next, we claim that

$$\| \nabla u \|_{C^{0,\alpha}(\partial\Omega)} \le C(\| f \|_{L^1(\Omega)} + \| f \|_{L^\infty(\omega)}).$$

Indeed let χ denote the characteristic function of ω and write $f = f_1 + f_2$ with $f_1 = f\chi$ and $f_2 = f(1 - \chi)$. For $i = 1, 2$, let u_i be the solutions of the problems

$$\begin{aligned} -\Delta u_i - \lambda u_i &= f_i & \text{in} \quad \Omega \\ u_i &= 0 & \text{on} \quad \partial\Omega, \end{aligned}$$

so that $u = u_1 + u_2$.

By the L^p regularity theory (see e.g. [GT], Chapter 9) we have

$$\| u_1 \|_{W^{2,q}(\Omega)} \le C \| f_1 \|_{L^\infty(\Omega)} = C \| f \|_{L^\infty(\omega)}$$

for any $q < \infty$, and consequently

$$\| u_1 \|_{C^{1,\alpha}(\bar{\Omega})} \le C \| f \|_{L^\infty(\omega)}$$

for any $\alpha < 1$. On the other hand we have, as above,

$$\| u_2 \|_{W^{1,q}(\Omega)} \le C \| f_1 \|_{L^1(\Omega)} \le C \| f \|_{L^1(\Omega)}$$

and thus, by the Sobolev imbedding,

$$\| u_2 \|_{L^2(\Omega)} \le C \| f \|_{L^1(\Omega)} .$$

Finally, we note that u_2 satisfies

$$\begin{cases} -\Delta u_2 - \lambda u_2 = 0 & \text{in } \omega \\ \qquad\qquad u_2 = 0 & \text{on } \partial\Omega. \end{cases}$$

It follows from standard elliptic regularity that

$$\| u_2 \|_{C^{1,\alpha}(\overline{\omega'})} \leq C \| u_2 \|_{L^2(\omega)}$$

for any neighbourhood ω' of $\partial\Omega$, strictly smaller than ω. In particular we have

$$\| \nabla u_2 \|_{C^{0,\alpha}(\partial\Omega)} \leq C \| f \|_{L^1(\Omega)} .$$

3. A first estimate. We shall proceed in two steps. First we shall show that when $0 \leq \lambda \leq \pi^2/4$,

$$(3.1) \qquad\qquad \epsilon \leq C\mu^{1-\epsilon/2},$$

where C is some positive constant, and then we shall establish the desired limits.

To prove (3.1), note that (1.5) implies that

$$(3.2) \qquad\qquad \frac{3\epsilon}{6-\epsilon} \int_\Omega u_\epsilon^{6-\epsilon} \leq \int_{\partial\Omega} (x,n)\left(\frac{\partial u_\epsilon}{\partial n}\right)^2.$$

In the next two lemmas, we shall estimate the two integrals in (3.2).

Lemma 3.1. *Let u_ϵ be a solution of Problem (I) in which $\lambda < \pi^2$. Then*

$$\int_{\partial\Omega} (x,n)\left(\frac{\partial u_\epsilon}{\partial n}\right)^2 \leq C\mu^{1-\epsilon/2},$$

where C is some positive constant.

PROOF. According to Lemma 2.2 it is sufficient to estimate the right hand side of (1.1) in $L^1(\Omega)$ and $L^\infty(\omega)$, where ω is some neighbourhood of $\partial\Omega$ in Ω.
By Lemma 2.1

$$\int_\Omega u_\epsilon^{5-\epsilon} \leq \int_\Omega W_\mu^{5-\epsilon}$$

$$= 4\pi\mu^{(5-\epsilon)/2} \int_0^1 \frac{r^2 dr}{(\mu^2 + ar^2)^{(5-\epsilon)/2}}.$$

If we now set $r = \mu\alpha^{-1/2}s$, and observe that $\alpha \geq \mu^{\epsilon/2}$, we obtain

(3.3)
$$\int_\Omega u_\epsilon^{5-\epsilon} \leq 4\pi\mu^{1/2-\epsilon/4} \int_0^\infty \frac{s^2 ds}{(1+s^2)^{(5-\epsilon)/2}}$$
$$\leq C\mu^{1/2-\epsilon/4}$$

for some positive constant C, because the integral is convergent. On the other hand, let ω be a neighbourhood of $\partial\Omega$ in Ω, which does not contain $x = 0$. Then, by Lemma 2.1, $u_\epsilon \leq C\mu^{1/2-\epsilon/4}$ in ω and hence

(3.4)
$$u_\epsilon^{5-\epsilon} \leq C\mu^{(5-\epsilon)(1/2-\epsilon/4)} \quad \text{in} \quad \omega.$$

Putting (3.3) and (3.4) into (2.7), the desired estimate follows.

Lemma 3.2. *We have*

$$\liminf_{\epsilon\to 0} \int_\Omega u_\epsilon^{6-\epsilon} \geq \frac{\pi^2}{4}.$$

PROOF. We first prove that

(3.5)
$$\liminf_{\epsilon\to 0} \int_\Omega u_\epsilon^{6-\epsilon} > 0.$$

Suppose, to the contrary, that there exists a subsequence $\{\epsilon_n\}, \epsilon_n \to 0$ as $n \to \infty$, but still denoted by ϵ, such that

(3.6)
$$\lim_{\epsilon\to 0} \int_\Omega u_\epsilon^{6-\epsilon} = 0.$$

Multiplying equation (1.1) by u_ϵ we deduce from (3.6), since $\lambda < \pi^2$, that

$$u_\epsilon \to 0 \quad \text{as} \quad \epsilon \to 0 \quad \text{in} \quad H^1(\Omega).$$

This implies by the Sobolev injection that

(3.7)
$$u_\epsilon \to 0 \quad \text{as} \quad \epsilon \to 0 \quad \text{in} \quad L^6(\Omega).$$

Next, we multiply (1.1) by u_ϵ^5 and integrate. This yields

$$\frac{5}{9} \int_\Omega |\nabla u_\epsilon^3|^2 = \lambda \int_\Omega u_\epsilon^6 + 3 \int_\Omega u_\epsilon^{10-\epsilon},$$

and hence, using Sobolev's injection again, in view of (3.7),

$$(3.8) \qquad \alpha \parallel u_\epsilon \parallel_{18}^6 \le \int_\Omega u_\epsilon^{10-\epsilon} + o(1) \quad \text{as} \quad \epsilon \to 0,$$

where α is some positive number. However, using Hölder's inequality, we find that

$$(3.9) \qquad \int_\Omega u_\epsilon^{10-\epsilon} \le \ \parallel u_\epsilon \parallel_{18}^6 \parallel u_\epsilon \parallel_{6-3\epsilon/2}^{4-\epsilon} = o(1) \parallel u_\epsilon \parallel_{18}^6 .$$

Combining (3.8) and (3.9) we find that

$$(3.10) \qquad \parallel u_\epsilon \parallel_{18} \to 0 \quad \text{as} \quad \epsilon \to 0.$$

Going back to (1.1) and using standard elliptic regularity theory, we see that (3.10) implies that

$$\parallel u_\epsilon \parallel_{H^2(\Omega)} \to 0 \quad \text{as} \quad \epsilon \to 0$$

and thus, since $N = 3$,

$$\parallel u_\epsilon \parallel_\infty \to 0 \quad \text{as} \quad \epsilon \to 0.$$

This contradicts the observation made at the beginning of Section 2, that $u_\epsilon(0) \to \infty$ as $\epsilon \to 0$. Thus (3.5) is established .

To complete the proof of Lemma 3.2, we multiply (1.1) by u_ϵ, use Sobolev's inequality and the observation that

$$\int_\Omega u_\epsilon^2 \le \int_\Omega W_\mu^2 \le 4\pi\mu^{1-\epsilon/2}.$$

This yields

$$S \parallel u_\epsilon \parallel_6^2 \le 3 \int_\Omega u_\epsilon^{6-\epsilon} + o(1) \quad \text{as} \quad \epsilon \to 0,$$

where S is the best Sobolev constant. On the other hand, by Hölder's inequality,

$$\parallel u_\epsilon \parallel_{6-\epsilon} \le \parallel u_\epsilon \parallel_6 (1 + o(1)) \quad \text{as} \quad \epsilon \to 0.$$

So

$$(3.11) \quad \left(S + o(1)\right) \parallel u_\epsilon \parallel_{6-\epsilon}^2 \leq 3 \parallel u_\epsilon \parallel_{6-\epsilon}^{6-\epsilon} + o(1) \quad \text{as} \quad \epsilon \to 0.$$

Set

$$L = \liminf_{\epsilon \to 0} \int_\Omega u_\epsilon^{6-\epsilon}.$$

We have shown that $L > 0$. If $L = \infty$, the lemma is proved, so suppose that $L < \infty$. Then, if we let $\epsilon \to 0$ through a sequence chosen so that $\parallel u_\epsilon \parallel_{6-\epsilon}^{6-\epsilon} \to L$, we deduce from (3.11) that

$$SL^{1/3} \leq 3L$$

or

$$L \geq (\frac{1}{3}S)^{3/2} = \frac{\pi^2}{4},$$

(see for instance [T]). This finishes the proof.

If we use the Lemmas 3.1 and 3.2 in (3.2), we obtain the following first estimate.

Lemma 3.3. *Suppose $0 \leq \lambda \leq \pi^2/4$. Then there exists a constant $C > 0$ such that*

$$\epsilon \leq C\mu^{1-\epsilon/2}$$

for ϵ sufficiently small.

Corollary 3.4. *Suppose $0 \leq \lambda \leq \pi^2/4$. Then*

$$|\mu^\epsilon - 1| = O(\mu|\log \mu|) \qquad \text{as} \quad \epsilon \to 0.$$

We now return to the Pohozaev identity

$$(3.12) \qquad \frac{3\epsilon}{6-\epsilon} \int_\Omega u_\epsilon^{6-\epsilon} = \int_{\partial\Omega} (x, n)\left(\frac{\partial u_\epsilon}{\partial n}\right)^2 - 2\lambda \int_\Omega u_\epsilon^2$$

for a precise estimate. With Lemma 3.2 it is easy to establish the limit of the first integral.

Lemma 3.5. *We have*

$$\lim_{\epsilon \to 0} \int_\Omega u_\epsilon^{6-\epsilon} = \frac{\pi^2}{4}.$$

PROOF. By Lemma 2.1

$$\int_\Omega u_\epsilon^{6-\epsilon} \le \int_\Omega W_\mu^{6-\epsilon}$$

$$\le 4\pi \mu^{-\epsilon/4} \int_0^\infty \frac{s^2 ds}{(1+s^2)^{3-(\epsilon/2)}}$$

$$\to \frac{\pi^2}{4} \qquad \text{as} \quad \epsilon \to 0$$

by Corollary 3.4. Remembering Lemma 3.2, this completes the proof.

To estimate the right hand side of (3.12), we investigate the behaviour of u_ϵ as $\epsilon \to 0$.

Lemma 3.6. *We have*

$$\mu^{1/2} u_\epsilon(\mu x) \to U_1(x) = \frac{1}{(1+|x|^2)^{1/2}} \qquad \text{as} \quad \epsilon \to 0$$

uniformly on $\mathbf{R}^3$, *where* u_ϵ *is extended by* 0 *outside* $B_{1/\mu}$.

PROOF. We use a scaling argument, and define the family of functions

$$(3.13) \qquad v_\epsilon(x) = \mu^{1/2} u_\epsilon(\mu x).$$

They satisfy

$$(3.14) \qquad -\Delta v_\epsilon - \lambda \mu^2 v_\epsilon = 3\mu^{\epsilon/2} v_\epsilon^{5-\epsilon} \quad \text{on} \quad B_{1/\mu}$$
$$v_\epsilon(0) = 1,$$

and, in addition, by Lemma 2.1,

$$v_\epsilon(x) \le \mu^{1/2} W_\mu(\mu x) \le \frac{1}{(1+\mu^{\epsilon/2}|x|^2)^{1/2}} \qquad \text{on} \quad B_{1/\mu}.$$

Thus, in view of Corollary 3.4, the family $\{v_\epsilon\}$ is bounded. The elliptic regularity theory implies that $\{v_\epsilon\}$ is equicontinuous on every compact subset of $B_{1/\mu}$. Hence, by Arzéla–Ascoli there exists a sequence, also denoted by v_ϵ, which converges to some function V uniformly on compact sets, but since $v_\epsilon(x) \to 0$ as $x \to \infty$, uniformly with respect to ϵ, also on $\mathbf{R}^3$. Taking the limit in (3.13), using Corollary 3.4 again, we conclude that V satisfies

$$-\Delta V = 3V^5 \quad \text{in} \quad \mathbf{R}^3$$
$$V(0) = 1.$$

In addition, V depends only on $|x|$. Thus $V = U_1$. Since V is determined uniquely it follows that the entire family v_ϵ converges to U_1 as $\epsilon \to 0$.

We use this limit to estimate the right hand side of (1.1).

Lemma 3.7. *We have*

$$\lim_{\epsilon \to 0} \mu^{-1/2} \int_\Omega u_\epsilon^{5-\epsilon} = \frac{4\pi}{3}.$$

PROOF. Defining v_ϵ as in (3.13), we can write the integral as

$$\mu^{-1/2} \int_{B_1} u_\epsilon^{5-\epsilon}(x)dx = \mu^{-3+\epsilon/2} \int_{B_1} v_\epsilon^{5-\epsilon}\left(\frac{x}{\mu}\right)dx$$
$$= \mu^{\epsilon/2} \int_{B_{1/\mu}} v_\epsilon^{5-\epsilon}(y)dy.$$
$$\to \int_{\mathbf{R}^3} U_1^5(y)dy \qquad \text{as} \quad \epsilon \to 0$$
$$= \frac{4\pi}{3}.$$

(The last integral can easily be computed by recalling that $-\Delta U_1 = 3U_1^5$ on $\mathbf{R}^3$).

Remark. Because at any $x \neq 0$,

$$\mu^{-1/2}u_\epsilon^{5-\epsilon}(x) \leq \frac{C}{|x|^5}\mu^2 \to 0 \qquad \text{as} \quad \epsilon \to 0,$$

Lemma 3.7 implies that

(3.15) $$\mu^{-1/2} u_\epsilon^{5-\epsilon} \to \frac{4\pi}{3} \delta_0 \qquad \text{as} \quad \epsilon \to 0,$$

where δ_0 is the Dirac mass centered at the origin.

Now, define the function

$$w_\epsilon = \mu^{-1/2} u_\epsilon.$$

It satisfies

$$\begin{cases} -\Delta w_\epsilon - \lambda w_\epsilon = 3\mu^{-1/2} u_\epsilon^{5-\epsilon} & \text{in } \Omega \\ \qquad\qquad\quad w_\epsilon = 0 & \text{on } \partial\Omega. \end{cases}$$

By Lemma 2.2 and the subsequent Remark, it follows from (3.15) that

(3.16) $$w_\epsilon \to 4\pi G_\lambda \quad \text{as} \quad \epsilon \to 0 \quad \text{in} \quad L^2(\Omega),$$

and also uniformly away from the origin. Here G_λ is the Green's function defined by (1.7) and (1.8). In addition, restricted to $\partial\Omega$ we have

$$\nabla w_\epsilon \to 4\pi \nabla G_\lambda \qquad \text{as} \quad \epsilon \to 0 \quad \text{in} \quad L^2(\partial\Omega).$$

Thus, writing the right hand side of (3.12) again as $J(u_\epsilon)$, we conclude that

$$\frac{1}{\mu} J(u_\epsilon) \to 16\pi^2 J(G_\lambda) \quad \text{as} \quad \epsilon \to 0.$$

We now use a result, which will be proved in the next section:

Lemma 3.8. *Let G_λ be the Green's function defined by (1.7) and (1.8). Then for any domain $\Omega \subset \mathbf{R}^3$ with smooth boundary $\partial\Omega$,*

$$\int_{\partial\Omega} (x, n) \left(\frac{\partial G_\lambda}{\partial n} \right)^2 - 2\lambda \int_\Omega G_\lambda^2 = -g_\lambda(0),$$

where

$$g_\lambda(x) = G_\lambda(x) - \frac{1}{4\pi|x|}.$$

Thus, we find that the right hand side of (3.12) behaves asymptotically as

$$(3.17) \qquad J(u_\epsilon) = 16\pi^2 g_\lambda(0)\mu[1 + o(1)] \quad \text{as} \quad \epsilon \to 0.$$

Remembering the behaviour of the integral on the left hand side of (3.12), given in Lemma 3.5, we arrive at the final result

$$(3.18) \qquad \lim_{\epsilon \to 0} \frac{\epsilon}{\mu} = -\frac{32}{\pi} \cdot 4\pi g_\lambda(0).$$

This completes part (a) of the proof of the first part of Theorem 1. The proof of part (b) follows immediately from (3.16) and (3.18) and the observation that

$$-4\pi g_\lambda(0) = \frac{\sqrt{\lambda}}{\tan \sqrt{\lambda}} > 0 \qquad \text{for} \quad \lambda < \pi^2/4.$$

4. A Pohozaev-type identity for the Green's function.

Throughout this section, Ω is an *arbitrary* bounded domain in $\mathbf{R}^N$ ($N > 2$) with smooth boundary $\partial\Omega$, and λ_1 the first eigenvalue of $-\Delta$ in Ω with Dirichlet boundary conditions.

Let $G_\lambda(x,y)$ be the Green's function of $-\Delta - \lambda$ in Ω, i.e. for any $y \in \Omega$, $G_\lambda(\cdot,y)$ satisfies

$$(4.1) \qquad -\Delta G_\lambda - \lambda G_\lambda = \delta_y \qquad \text{in} \quad \Omega$$

$$(4.2) \qquad \qquad G_\lambda = 0 \qquad \text{on} \quad \partial\Omega,$$

where δ_y denotes the Dirac mass concentrated at $x = y$, and let $g_\lambda(x,y)$ denote the regular part of G_λ,

$$(4.3) \qquad g_\lambda(x,y) = G_\lambda(x,y) - \frac{1}{(N-2)\sigma_N|x-y|^{N-2}},$$

where σ_N denotes the area of the unit sphere in $\mathbf{R}^N$.

Remark. When $\lambda \neq 0$, then if $N = 3$, $g_\lambda(\cdot,y) \in C(\Omega)$, but if $N \geq 4$, $g_\lambda(x,y)$ has no finite limit as $x \to y$. Of course, if $\lambda = 0$, $g_0(\cdot,y)$ is a smooth function for any $N > 2$.

Conforming with the notation of the previous sections, we shall write $G_\lambda(x) = G_\lambda(x, 0)$ and $g_\lambda(x) = g_\lambda(x, 0)$ if $0 \in \Omega$.

Theorem 4.1. *Suppose $N = 3$, and let $G_\lambda(\cdot, y)$ be defined by (4.1), (4.2) and $g_\lambda(\cdot, y)$ by (4.3). Then*

$$\int_{\partial\Omega} (x - y, n) \left(\frac{\partial G_\lambda}{\partial n}(x, y) \right)^2 - 2\lambda \int_\Omega G_\lambda^2(x, y) = -g_\lambda(y, y).$$

In the proof of Theorem 4.1 we shall use the following Lemma about the linear problem

$$(4.4) \qquad\qquad -\Delta u - \lambda u = f(x) \qquad \text{in} \quad \Omega$$
$$(4.5) \qquad\qquad\qquad u = 0 \qquad \text{on} \quad \partial\Omega.$$

Lemma 4.2. *Let u be a solution of (4.4), (4.5) in which $\lambda < \lambda_1$ and $f \in L^\infty(\Omega)$. Then*

$$(4.6) \quad \int_{\partial\Omega} (x, n) \left(\frac{\partial u}{\partial n} \right)^2 - 2\lambda \int_\Omega u^2 = - \int_\Omega f\{(N - 2)u + 2(x, \nabla u)\}.$$

PROOF. Observe that since $f \in L^\infty$, $u \in W^{2,p}$ for any $p > 1$, whence the integrals in (4.6) are all well defined.
We first multiply (4.4) by u and integrate over Ω. This yields, in view of the boundary condition,

$$(4.7) \qquad\qquad \int_\Omega |\nabla u|^2 - \lambda \int_\Omega u^2 = \int_\Omega fu.$$

Next, we multiply (4.4) by $(x, \nabla u)$ and integrate over Ω to obtain

$$- \int_{\partial\Omega} (x, n) \left(\frac{\partial u}{\partial n} \right)^2 + \int_\Omega (\nabla u, \nabla(x, \nabla u)) - \lambda \int_\Omega u(x, \nabla u)$$
$$= \int_\Omega f(x, \nabla u),$$

which we can write as

$$- \int_{\partial\Omega} (x, n) \left(\frac{\partial u}{\partial n} \right)^2 + \int_\Omega |\nabla u|^2 + \frac{1}{2} \int_\Omega (x, \nabla(|\nabla u|^2)) +$$
$$- \frac{1}{2}\lambda \int_\Omega (x, \nabla u^2) = \int_\Omega f(x, \nabla u)$$

or

$$(4.8) \quad -\frac{1}{2}\int_{\partial\Omega}(x,n)\Big(\frac{\partial u}{\partial n}\Big)^2 + \Big(1 - \frac{N}{2}\Big)\int_{\Omega}|\nabla u|^2 + \\ + \lambda\frac{N}{2}\int_{\Omega}u^2 = \int_{\Omega}f(x,\nabla u).$$

If we now use (4.7) to eliminate the second integral on the left hand side in (4.8) we end up with the required identity.

Proof of Theorem 4.1. Without loss of generality we set $y = 0$. Define the family of functions

$$\delta_\rho = \frac{1}{|B_\rho|}\chi_{B_\rho}, \quad \rho > 0$$

with $|B_\rho| = \frac{4}{3}\pi\rho^3$. Plainly the functions δ_ρ converge weakly to δ_0 as $\rho \to 0$.
Let v_ρ be the solution of the equation

$$-\Delta v = \delta_\rho \quad \text{in} \quad \mathbf{R}^3$$

such that $v_\rho(x) \to 0$ as $|x| \to \infty$. It is readily verified that

$$v_p(x) = \begin{cases} -\dfrac{r^2}{8\pi\rho^3} + \dfrac{3}{8\pi\rho} & \text{if } 0 < r < \rho \\ \dfrac{1}{4\pi r} & \text{if } \rho \le r < \infty . \end{cases}$$

Let u_ρ be the solution of the problem

$$\begin{cases} -\Delta u - \lambda u = \delta_\rho & \text{in } \Omega \\ u = 0 & \text{on } \partial\Omega \end{cases}$$

so that by Lemma 4.2,

$$(4.9) \quad \int_{\partial\Omega}(x,n)\Big(\frac{\partial u_\rho}{\partial n}\Big)^2 - 2\lambda\int_{\Omega}u_\rho^2 = -\int_{\Omega}\delta_\rho\{u_\rho + 2(x,\nabla u_\rho)\}.$$

Now let $\rho \to 0$. Then by Lemma 2.2 and the subsequent Remark, $u_\rho \to G_\lambda$ and the *left hand side* of (4.9) converges to

$$(4.10) \quad \int_{\partial\Omega}(x,n)\Big(\frac{\partial G_\lambda}{\partial n}\Big)^2 - 2\lambda\int_{\Omega}G_\lambda^2.$$

To evaluate the limit of the *right hand side* we write it as

$$(4.11) \quad -\int_\Omega \delta_\rho\{(u_\rho - v_\rho) + 2(x, \nabla(u_\rho - v_\rho))\} - \int_\Omega \delta_\rho\{v_\rho + 2(x, \nabla v_\rho)\}.$$

It follows by direct computation that

$$(4.12) \qquad \int_\Omega \delta_\rho\{v_\rho + 2(x, \nabla v_\rho)\} = 0.$$

On the other hand,

$$\begin{cases} -\Delta(u_\rho - v_\rho) - \lambda(u_\rho - v_\rho) = \lambda v_\rho & \text{in } \Omega \\ u_\rho - v_\rho = -v_\rho & \text{on } \partial\Omega \end{cases}$$

and hence, since $v_\rho \to 1/4\pi r$ as $\rho \to 0$ in $L^2(\Omega)$, $u_\rho - v_\rho \to g_\lambda$ as $\rho \to 0$ in $H^2(\Omega)$, and hence uniformly in $\bar\Omega$, where g_λ is the solution of the problem

$$-\Delta g - \lambda g = \frac{\lambda}{4\pi r} \quad \text{in} \quad \Omega$$

$$g = -\frac{1}{4\pi r} \quad \text{on} \quad \partial\Omega,$$

i.e. g_λ is the regular part of the Green's function. Therefore

$$(4.13) \qquad \int_\Omega \delta_\rho(u_\rho - v_\rho) \to g_\lambda(0) \quad \text{as} \quad \rho \to 0.$$

Finally,

$$(4.14) \qquad \left| \int_\Omega \delta_\rho(x, \nabla(u_\rho - v_\rho)) \right| \leq \frac{3}{4\pi\rho^2} \int_{B_\rho} |\nabla(u_\rho - v_\rho)|.$$

But $\nabla(u_\rho - v_\rho)$ is bounded in $H^1(\Omega)$, and hence in $L^6(\Omega)$. So, by Hölder's inequality,

$$\int_{B_\rho} |\nabla(u_\rho - v_\rho)| \leq C\|\nabla(u_\rho - v_\rho)\|_6 \rho^{5/2}$$

for some positive constant C, which implies by (4.14) that

$$(4.15) \qquad \left| \int_\Omega \delta_\rho(x, \nabla(u_\rho - v_\rho)) \right| = O(\rho^{1/2}) \qquad \text{as} \quad \rho \to 0.$$

So, (4.12), (4.13) and (4.15) together imply that the right hand side of (4.9), given by (4.11), tends to

(4.16) $-g_\lambda(0)$

as $\rho \to 0$. Thus we conclude that letting $\rho \to 0$ in (4.9) yields the desired identity.

In the next two theorems we establish Pohozaev–type identities for the Green's function G_λ, when $\lambda = 0$, in arbitrary domains in $\mathbf{R}^N (N > 2)$. The second of these will be used in Section 8. For convenience we write $G(x,y) = G_0(x,y)$ and $g(x,y) = g_0(x,y)$.

Theorem 4.3.   *We have, for every $y \in \Omega$,*

$$\int_{\partial\Omega} (x - y, n)\left(\frac{\partial G}{\partial n}(x,y)\right)^2 dx = -(N - 2)g(y,y),$$

where $n = n(x)$ denotes the outward normal to $\partial\Omega$ at x.

PROOF.   As in the proof of Theorem 4.1 we may assume that $y = 0$ and consider the family of functions

$$\delta_\rho = \frac{1}{|B_\rho|}\chi_{B_\rho}$$

with $|B_\rho| = \sigma_N \rho^N/N$. Let v_ρ be the solution of the equation

$$-\Delta v = \delta_\rho \qquad \text{in} \quad \mathbf{R}^N$$

such that $v_\rho(x) \to 0$ as $|x| \to \infty$. It is readily verified that

$$v_\rho(x) = \begin{cases} -\dfrac{r^2}{2\sigma_N\rho^N} + \dfrac{N}{2\sigma_N(N-2)\rho^{N-2}} & \text{if } 0 < r < \rho \\[2mm] \dfrac{1}{(N-2)\sigma_N r^{N-2}} & \text{if } \rho \le r < \infty. \end{cases}$$

Let u_ρ be the solution of the problem

$$\begin{cases} -\Delta u = \delta_\rho & \text{in } \Omega \\ u = 0 & \text{on } \partial\Omega, \end{cases}$$

so that, by Lemma 4.2

$$\int_{\partial\Omega} (x,n)\left(\frac{\partial u_\rho}{\partial n}\right)^2 = -\int_\Omega \delta_\rho\{(N-2)u_\rho + 2(x,\nabla u_\rho)\}.$$

Now let $\rho \to 0$. Clearly $u_\rho \to G$ in $C^k(\omega)$ for any integer k and any neighbourhood ω of $\partial\Omega$ which does not contain 0, and thus the left hand side converges to

$$\int_{\partial\Omega} (x,n)\left(\frac{\partial G}{\partial n}\right)^2.$$

To evaluate the right hand side we write it as

$$-\int_\Omega \delta_\rho\{(N-2)(u_\rho - v_\rho) + 2(x,\nabla(u_\rho - v_\rho))\}$$
$$-\int_\Omega \delta_\rho\{(N-2)v_\rho + 2(x,\nabla v_\rho)\}.$$

It follows by direct computation that

$$\int_\Omega \delta_\rho\{(N-2)v_\rho + 2(x,\nabla v_\rho)\} = 0.$$

On the other hand,

$$-\Delta(u_\rho - v_\rho) = 0 \quad \text{in } \Omega$$
$$u_\rho - v_\rho = -v_\rho \quad \text{on } \partial\Omega$$

and hence, since $v_\rho(x) \to 1/\{(N-2)\sigma_N r^{N-2}\}$ as $\rho \to 0$, $u_\rho - v_\rho \to g$ as $\rho \to 0$ in $C^k(\overline{\Omega})$ for any integer k. Therefore

$$\int_\Omega \delta_\rho(u_\rho - v_\rho) \to g(0,0)$$

and

$$\int_\Omega \delta_\rho(x,\nabla(u_\rho - v_\rho)) \to 0$$

as $\rho \to 0$.

Theorem 4.4. *We have for every* $y \in \Omega$,

$$\int_{\partial\Omega} \left(\frac{\partial G}{\partial n}(x, y)\right)^2 n(x)dx = -\nabla\phi(y),$$

where $\phi(y) = g(y, y)$ *and* $n(x)$ *denotes the outward normal to* $\partial\Omega$ *at* x.

PROOF.   Without loss of generality we may assume that $y = 0$. We use the same notation as in the proof of Theorem 4.1. Multiplying the equation

$$-\Delta u_\rho = \delta_\rho \quad \text{in} \quad \Omega$$

through by $\partial u_\rho / \partial x_i$ and integrating over Ω we obtain

(4.17) $$-\frac{1}{2}\int_{\partial\Omega}\left(\frac{\partial u_\rho}{\partial n}\right)^2 (n, e_i) = \int_\Omega \delta_\rho \frac{\partial u_\rho}{\partial x_i} dx,$$

where e_i denotes the unit vector along the x_i-axis. But

$$\int_\Omega \delta_\rho \frac{\partial u_\rho}{\partial x_i} dx = \int_\Omega \delta_\rho \frac{\partial}{\partial x_i}(u_\rho - v_\rho)dx + \int_\Omega \delta_\rho \frac{\partial v_\rho}{\partial x_i} dx$$

and, since the last integral is zero by symmetry, we have

(4.18) $$\int_\Omega \delta_\rho \frac{\partial u_\rho}{\partial x_i} dx = \int_\Omega \delta_\rho \frac{\partial}{\partial x_i}(u_\rho - v_\rho)dx.$$

On the other hand $(u_\rho - v_\rho)(x) \to g(x, 0)$ in $C^k(\overline{\Omega})$ for any integer k, as $\rho \to 0$ and thus from (4.18) we deduce that

$$\lim_{\rho\to 0}\int_\Omega \delta_\rho \frac{\partial u_\rho}{\partial x_i} = \frac{\partial}{\partial x_i}g(x, 0)\Big|_{x=0}.$$

Passing to the limit in (4.17) we are led to

$$-\frac{1}{2}\int_\Omega \left(\frac{\partial G}{\partial n}(x, 0)\right)^2 (n, e_i) = \frac{\partial}{\partial x_i}g(x, 0).$$

Finally, we differentiate the relation $\phi(x) = g(x, x)$ and we find

$$\frac{\partial\phi}{\partial x_i}(0) = \frac{\partial g}{\partial x_i}(0, 0) + \frac{\partial g}{\partial y_i}(0, 0) = 2\frac{\partial g}{\partial x_i}(0, 0)$$

by the symmetry of the function $g(x, y)$.

5. A good approximation for u_ϵ. To study the properties of u_ϵ as $\epsilon \to 0$ when $\lambda = \pi^2/4$, we need a better estimate for $J(u_\epsilon)$ than is given in (3.17). Thus it is necessary to establish with greater precision the behaviour of $u_\epsilon(x)$ as $\epsilon \to 0$. Writing as before

$$(5.1) \qquad \mu = (u_\epsilon(0))^{-2},$$

We shall prove in this section that in all of Ω, u_ϵ can be very well approximated – in some adequate norms – by the function

$$(5.2) \qquad \phi_\mu = U_\mu + 4\pi\sqrt{\mu}g_\lambda,$$

where g_λ is the regular part of the Green's function defined in (1.10) and $0 \le \lambda \le \pi^2/4$. Consider the "error term" η defined by

$$(5.3) \qquad \eta = u_\epsilon - \phi_\mu.$$

Then an easy calculation shows that

$$(5.4) \qquad -\Delta\eta - \lambda\eta = f \quad \text{in} \quad \Omega$$
$$(5.5) \qquad \eta = b \quad \text{on} \quad \partial\Omega,$$

where

$$(5.6) \qquad f = \lambda\left(U_\mu - \frac{\sqrt{\mu}}{|x|}\right) + 3\left(u_\epsilon^{5-\epsilon} - U_\mu^5\right),$$
$$(5.7) \qquad b = -(U_\mu + 4\pi\sqrt{\mu}g_\lambda).$$

Recall, moreover, from Section 3 that

$$(5.8) \qquad \epsilon = O(\mu) \quad \text{as} \quad \mu \to 0.$$

Our main estimate for η will be the following.

Proposition 5.1. *Let η be defined by (5.3). Then*

$$\|\eta\|_{L^2(\Omega)} + \|\nabla\eta\|_{L^2(\partial\Omega)} \le C(\mu^{5/2}|\log\mu| + \epsilon\mu^{1/2}|\log\mu|),$$

where $\nabla\eta$ denotes the full gradient of η, and not just the tangential gradient.

The proof of Proposition 5.1 is based on Lemma 2.2 which requires estimates for f in $L^1(\Omega)$ and $L^\infty(\omega)$, where ω is a suitable neighbourhood of $\partial\Omega$ and for b in $C^{2,\alpha}(\partial\Omega)$. We shall now derive these bounds in succession.

Lemma 5.2. *Let $\alpha \in (0,1)$. Then*

$$\|b\|_{C^{2,\alpha}(\partial\Omega)} = O(\mu^{5/2}) \quad as \quad \mu \to 0.$$

PROOF. On $\partial\Omega$, $G_\lambda = 0$, and so

$$b(x) = -\sqrt{\mu}\Big(\frac{1}{\sqrt{\mu^2 + |x|^2}} - \frac{1}{|x|}\Big).$$

Since $|x|$ is bounded away from zero on $\partial\Omega$, it readily follows that

$$\|b\|_{C^k(\partial\Omega)} = O(\mu^{5/2}) \quad as \quad \mu \to 0$$

for any $k \geq 0$.

Lemma 5.3. *Let ω be a neighbourhood of $\partial\Omega$ in Ω, which does not contain the origin. Then*

$$\|f\|_{L^\infty(\omega)} = O(\mu^{5/2}) \quad as \quad \mu \to 0.$$

PROOF. As in the proof of Lemma 5.2, one proves that the first term in (5.6) is $O(\mu^{5/2})$ as $\mu \to 0$. As to the second term, plainly

$$\|U_\mu\|_{L^\infty(\omega)} \leq C\sqrt{\mu},$$

where C is some generic constant, and, since by Lemma 2.1 $u_\epsilon \leq W_\mu$, we have

$$\|u_\epsilon\|_{L^\infty(\omega)} \leq |W_\mu|_{L^\infty(\omega)} \leq C\sqrt{\mu}$$

Thus

$$\|U_\mu^5\|_{L^\infty(\omega)} \leq C\mu^{5/2}$$

and

$$\|u_\epsilon^5\|_{L^\infty(\omega)} \leq C\mu^{5/2}\mu^{-\epsilon/2} \leq C\mu^{5/2}$$

in view of (5.8).

To estimate f in $L^1(\Omega)$, we split it into three pieces:

$$f = f_1 + f_2 + f_3,$$

where

(5.9) $$f_1 = \lambda\left(U_\mu - \frac{\sqrt{\mu}}{|x|}\right),$$

(5.10) $$f_2 = 3(u_\epsilon^{5-\epsilon} - W_\mu^{5-\epsilon}),$$

(5.11) $$f_3 = 3(W_\mu^{5-\epsilon} - U_\mu^5).$$

Lemma 5.4. *Let f_1 be given by (5.9). Then*

$$\|f_1\|_{L^1(\Omega)} = O(\mu^{5/2}|\log\mu|) \quad as \quad \mu \to 0.$$

PROOF. By the expression for U_μ, we have

$$\int_\Omega |f_1| = 4\pi\lambda\sqrt{\mu} \int_0^1 \left|\frac{1}{r} - \frac{1}{\sqrt{\mu^2 + r^2}}\right| r^2 \, dr,$$

or, if we set $r = \mu s$,

$$\int_\Omega |f_1| = 4\pi\lambda\mu^{5/2} \int_0^{1/\mu} \left|\frac{1}{s} - \frac{1}{\sqrt{1+s^2}}\right| s^2 \, ds$$

$$\leq C\mu^{5/2} \int_0^{1/\mu} \frac{s \, ds}{1 + s^2}$$

$$< C\mu^{5/2}|\log\mu| \quad \text{as} \quad \mu \to 0.$$

The splitting of the second term in the function f into f_2 and f_3 was inspired by the fact that $u_\epsilon \leq W_\mu$ in Ω. Thus $f_2 \leq 0$ and to estimate it we merely need a lower bound for u_ϵ. This bound is given in the following Lemma. It is very similar to a bound given in [AP3], but for completeness, we give the proof in the Appendix.

Lemma 5.5. *Let u_ϵ be any solution of (1.1) in which $0 \leq \lambda < \pi^2$. Then, for $x \in \overline{\Omega}$,*

$$u_\epsilon(x) \geq W_\mu(x)\left(1 - \frac{\lambda}{2}|x|^2\right) - \frac{1}{\sqrt{\mu}}[\{\sqrt{\mu}W_\mu(x)\}^{-\epsilon} - 1].$$

Lemma 5.6. *Let f_2 be given by (5.10). Then for μ small*

$$\|f_2\|_{L^1(\Omega)} \leq C(\epsilon\mu^{1/2}|\log\mu| + \mu^{5/2}|\log\mu|),$$

where C is some positive constant.

PROOF. By Lemmas 2.1 and 5.5

$$
\begin{aligned}
|f_2| &\leq CW_\mu^{4-\epsilon}(W_\mu - u_\epsilon) \\
(5.12) \qquad &\leq C\mu^{-1/2}W_\mu^{4-\epsilon}[\{\sqrt{\mu}W_\mu\}^{-\epsilon} - 1] + C|x|^2W_\mu^{5-\epsilon},
\end{aligned}
$$

where, like throughout this proof, C is some positive generic constant.
Observe that

$$\{\sqrt{\mu}W_\mu(x)\}^{-\epsilon} = \left(\frac{\mu^2 + \alpha|x|^2}{\mu^2}\right)^{\epsilon/2},$$

where $\alpha = \mu^{\epsilon/2} + \frac{1}{3}\lambda\mu^2$. Hence, because $|x| \leq 1$,

$$
\begin{aligned}
\{\sqrt{\mu}W_\mu(x)\}^{-\epsilon} - 1 &\leq (2\mu^{-2})^{\epsilon/2} - 1 \\
(5.13) \qquad &= C\epsilon|\log\mu|
\end{aligned}
$$

in view of (5.8). Also, note that since $\alpha \geq \mu^{\epsilon/2}$ and $\mu^{\epsilon/2} < 1$,

$$(5.14) \qquad W_\mu(x) \leq \left(\frac{\mu}{\mu^2 + \mu^{\epsilon/2}|x|^2}\right)^{1/2} < \mu^{-\epsilon/4}U_\mu(x).$$

Thus, (5.13) and (5.14), as well as (5.8) enable us to simplify (5.12) to the following bound:

$$
\begin{aligned}
(5.15) \qquad |f_2| &\leq C\epsilon\mu^{-1/2}|\log\mu|U_\mu^{4-\epsilon} + C|x|^2U_\mu^{5-\epsilon} \\
&\leq C\epsilon\mu^{-1/2}|\log\mu|U_\mu^4 + C|x|^2U_\mu^5
\end{aligned}
$$

since $U_\mu^{-\epsilon} \leq C$. However

$$\int_\Omega U_\mu^4 = 4\pi\mu^2 \int_0^1 \frac{r^2 dr}{(\mu^2 + r^2)^2}$$
$$= 4\pi\mu \int_0^{1/\mu} \frac{s^2 ds}{(1+s^2)^2},$$

where we have set $r = \mu s$. Because the last integral is convergent if we let μ tend to zero, we may conclude that

(5.16) $$\int_\Omega U_\mu^4 \leq C\mu.$$

Similarly, we obtain

$$\int_\Omega U_\mu^5 |x|^2 = 4\pi\mu^{5/2} \int_0^{1/\mu} \frac{s^4 ds}{(1+s^2)^{5/2}}$$
(5.17) $$\leq C\mu^{5/2} |\log \mu|.$$

Thus, integrating (5.15) over Ω, and using (5.16) and (5.17), we end up with the desired bound

$$\int_\Omega |f_2| \leq C(\epsilon\mu^{1/2} |\log \mu| + \mu^{5/2} |\log \mu|).$$

Lemma 5.7. *Let f_3 be given by (5.11). Then for μ small*

$$\|f_3\|_{L^1(\Omega)} \leq C(\epsilon\mu^{1/2}|\log \mu| + \mu^{5/2}),$$

where C is some positive constant.

PROOF. We split f_3 as follows:

$$f_3 = 3(W_\mu^{5-\epsilon} - U_\mu^{5-\epsilon}) + 3(U_\mu^{5-\epsilon} - U_\mu^5)$$
$$= f_{31} + f_{32}$$

in an obvious notation. By the Mean Value Theorem

$$|f_{31}| \leq C\mu^{(5-\epsilon)/2} \frac{|\mu^{\epsilon/2} - 1| + \lambda\mu^2}{(\mu^2 + \mu^{\epsilon/2}|x|^2)^{(7-\epsilon)/2}} \cdot |x|^2,$$

and hence, because $\epsilon = O(\mu)$ by (5.8),

$$|f_{31}| \leq C\mu^{5/2}(\epsilon|\log\mu| + \mu^2)\frac{|x|^2}{(\mu^2 + |x|^2)^{7/2}}.$$

This yields upon integration over Ω,

$$(5.18) \qquad \int_\Omega |f_{31}| \leq C(\epsilon\mu^{1/2}|\log\mu| + \mu^{5/2}).$$

As to the second term,

$$\begin{aligned}
|f_{32}| &= 3U_\mu^5|U_\mu^{-\epsilon} - 1| \\
&= 3U_\mu^5\left|\left(\frac{\mu^2 + |x|^2}{\mu}\right)^{\epsilon/2} - 1\right| \\
&\leq C\epsilon|\log\mu|U_\mu^5,
\end{aligned}$$

and so

$$(5.19) \qquad \int_\Omega |f_{32}| \leq C\epsilon\mu^{1/2}|\log\mu|.$$

Together (5.18) and (5.19) yield the required estimate .

From Lemmas 5.4, 5.6 and 5.7 we finally conclude that for μ small

$$\int_\Omega |f| \leq C(\epsilon\mu^{1/2}|\log\mu| + \mu^{5/2}|\log\mu|),$$

C being some positive constant. This estimate, together with earlier estimates for f near $\partial\Omega$ and b on $\partial\Omega$ given in Lemmas 5.3 and 5.2 suffice to prove Proposition 5.1 with the help of Lemma 2.2.

The following two bounds for ϕ_μ will prove useful later.

Lemma 5.8. *Let ϕ_μ be given by (5.2). Then*

$$(a) \qquad \int_\Omega \phi_\mu^2 \;\leq\; C\mu,$$

$$(b) \qquad \int_{\partial\Omega} |\nabla\phi_\mu|^2 \;\leq\; C\mu,$$

where C is some positive constant.

PROOF. Both estimates follow immediately from the definition of U_μ and the fact that λ does not depend on μ.

6. The main results. We now return again to Pohozaev's identity for Problem (I):

$$(6.1) \qquad \frac{3\epsilon}{6-\epsilon} \int_\Omega u_\epsilon^{6-\epsilon} = \int_{\partial\Omega} (x,n)\left(\frac{\partial u_\epsilon}{\partial n}\right)^2 - 2\lambda \int_\Omega u_\epsilon^2,$$

and use the approximation

$$(6.2) \qquad u_\epsilon = \phi_\mu + \eta,$$

where

$$(6.3) \qquad \phi_\mu = U_\mu + 4\pi\sqrt{\mu} g_\lambda,$$

which was discussed in the previous section to obtain an expansion for the right hand side $J(u_\epsilon)$ of (6.1).

Theorem 6.1. *We have the expansion, as $\epsilon \to 0$*

$$J(u_\epsilon) = -16\phi^2 g_\lambda(0)\mu + 4\pi^2\lambda\mu^2 + O(\epsilon\mu|\log\mu| + \mu^3|\log\mu|).$$

The proof of Theorem 6.1 is given in two lemmas.

Lemma 6.2. *Let ϕ_μ be defined by (6.3). Then*

$$J(u_\epsilon) = J(\phi_\mu) + R_\epsilon,$$

where

$$|R_\epsilon| \le C(\epsilon\mu|\log\mu| + \mu^3|\log\mu|),$$

in which C is some positive constant.

PROOF. By (6.2),

$$R_\epsilon = J(\phi_\mu + \eta) - J(\phi_\mu)$$
$$= \int_{\partial\Omega} (x,n)\left\{2\frac{\partial\phi_\mu}{\partial n}\cdot\frac{\partial\eta}{\partial\eta} + \left(\frac{\partial\eta}{\partial\eta}\right)^2\right\} - 2\lambda\int_\Omega (2\phi_\mu\eta + \eta^2)$$

and hence, by Cauchy– Schwarz's inequality and Lemma 5.8

$$|R_\epsilon| \le C\{\mu^{1/2}(\|\nabla\eta\|_{L^2(\partial\Omega)} + \|\eta\|_{L^2(\Omega)})$$
$$+ \|\nabla\eta\|^2_{L^2(\partial\Omega)} + \|\eta\|^2_{L^2(\Omega)}\}.$$

If we now use Proposition 5.1 to estimate η, we arrive at the required bound.

The function ϕ_μ is explicitly given in terms of the parameter μ, and closely related to the Green's function G_λ. Indeed

$$\phi_\mu(x) = 4\pi\sqrt{\mu}\Big(\frac{1}{4\pi\sqrt{\mu^2 + |x|^2}} + g_\lambda(x)\Big)$$
$$\simeq 4\pi\sqrt{\mu}\Big(\frac{1}{4\pi|x|} + g_\lambda(x)\Big)$$
$$= 4\pi\sqrt{\mu}G_\lambda(x).$$

In the next Lemma we shall exploit this relationship, and our knowledge of $J(G_\lambda)$ from Theorem 4.1, to determine the first two terms of the asymptotic expansion of $J(\phi_\mu)$ as $\mu \to 0$.

Lemma 6.3.   *Let ϕ_μ be given by (6.3). Then, as $\mu \to 0$,*

$$J(\phi_\mu) = -16\pi^2 g_\lambda(0)\mu + 4\lambda\pi^2\mu^2 + O(\mu^3|\log\mu|).$$

PROOF.   Recall that

(6.4) $$\phi_\mu = 4\pi\sqrt{\mu}G_\lambda + \Big(U_\mu - \frac{\sqrt{\mu}}{|x|}\Big)$$

Hence,

$$\frac{\partial\phi_\mu}{\partial n} = 4\pi\sqrt{\mu}\frac{\partial G_\lambda}{\partial n} + \sqrt{\mu}\frac{\partial}{\partial n}\Big(\frac{1}{\sqrt{\mu^2 + |x|^2}} - \frac{1}{|x|}\Big),$$

and therefore, since $\Omega = B_1$,

$$\frac{\partial\phi_\mu}{\partial n}\Big|_{\partial\Omega} = 4\pi\sqrt{\mu}\frac{\partial G_\lambda}{\partial n}\Big|_{\partial\Omega} + O(\mu^{5/2}) \quad \text{as} \quad \mu \to 0.$$

Thus
(6.5)
$$\int_{\partial\Omega}(x,n)\Big(\frac{\partial\phi_\mu}{\partial n}\Big)^2 = 16\pi^2\mu\int_{\partial\Omega}(x,n)\Big(\frac{\partial G_\lambda}{\partial n}\Big)^2 + O(\mu^3) \quad \text{as} \quad \mu \to 0.$$

Next, by (6.3) we can write the second integral in $J(\phi_\mu)$ as

$$\int_\Omega \phi_\mu^2 = \int_\Omega \frac{\mu}{\mu^2 + |x|^2} + 8\pi\mu\int_\Omega \frac{g_\lambda}{\sqrt{\mu^2 + |x|^2}} + 16\pi^2\mu\int_\Omega g_\lambda^2$$

$$= 16\pi^2\mu\int_\Omega G_\lambda^2 + X_1 + X_2,$$

where
$$X_1 = 8\pi\mu\int_\Omega\Big(\frac{1}{\sqrt{\mu^2 + |x|^2}} - \frac{1}{|x|}\Big)g_\lambda$$

and
$$X_2 = \mu\int_\Omega\Big(\frac{1}{\mu^2 + |x|^2} - \frac{1}{|x|^2}\Big).$$

Plainly, because $g_\lambda \in L^\infty(\Omega)$,

$$|X_1| \le C\mu\int_0^1\Big|\frac{1}{\sqrt{\mu^2 + r^2}} - \frac{1}{r}\Big|r^2dr$$

$$\le C\mu^3|\log\mu|.$$

On the other hand

$$X_2 = -4\pi\mu^3\int_0^1 \frac{dr}{\mu^2 + r^2} = -2\pi^2\mu^2 + O(\mu^3).$$

Thus

$$(6.6) \quad \int_\Omega \phi_\mu^2 = 16\pi^2\mu\int_\Omega G_\lambda^2 - 2\pi^2\mu^2 + O(\mu^3|\log\mu|) \quad \text{as} \quad \mu \to 0.$$

Putting (6.5) and (6.6) together, we obtain

$$J(\phi_\mu) = 16\pi^2\mu J(G_\lambda) + 4\pi^2\lambda\mu^2 + O(\mu^3|\log\mu|) \quad \text{as} \quad \mu \to 0.$$

Since $J(G_\lambda) = -g_\lambda(0)$ according to Theorem 4.1, the proof is complete.

Plainly, Lemma's 6.2 and 6.3 yield Theorem 6.1.

We are now ready to prove Theorem 2. We have $\lambda = \pi^2/4$, and so $g_\lambda(0) = 0$. This implies, according to Theorem 6.1, that $J(u_\epsilon) = O(\mu^2)$ as $\epsilon \to 0$. Therefore, we divide (6.1) by μ^2 and let ϵ tend to zero. Remembering the limit of the integral on the left hand side of (6.1), given in Lemma 3.5, we obtain

$$(6.7) \qquad \frac{\pi^2}{8} \lim_{\epsilon \to 0} \frac{\epsilon}{\mu^2} = \pi^4,$$

or, since $\mu = (u_\epsilon(0))^{-2}$,

$$\lim_{\epsilon \to 0} \epsilon u_\epsilon^4(0) = 8\pi^2$$

which is the content of Part (a).
As to Part (b), we recall from (3.16) that

$$\mu^{-1/2} u_\epsilon(x) \to 4\pi G_{\pi^2/4}(x) \quad \text{as} \quad \epsilon \to 0$$

in $L^2(\Omega)$, and pointwise for $x \neq 0$. If we now use (6.7) to eliminate μ, we find the limit

$$\epsilon^{-1/4} u_\epsilon(x) \to 2^{5/4} \sqrt{\pi} G_{\pi^2/4}(x) \quad \text{as} \quad \epsilon \to 0$$

in $L^2(\Omega)$, which is the content of Part (b).

7. A related problem. Consider the problem

$$(II) \qquad \begin{cases} -\Delta u - \left(\dfrac{\pi^2}{4} + \epsilon\right) u = 3u^5 & \text{in } \Omega \\ \qquad\quad u > 0 & \text{in } \Omega \\ \qquad\quad u = 0 & \text{on } \partial\Omega, \end{cases}$$

where Ω is the unit ball in $\mathbf{R}^3$ and ϵ a small positive number. As in Problem (I), Problem (II) has a solution u_ϵ if $\epsilon > 0$ (and small enough), but it has no solution if $\epsilon = 0$ [BN]. It is again our objective to study the behaviour of u_ϵ as $\epsilon \to 0$, and specifically, to prove Theorem 3.

It is readily shown, as in Section 2, that $u_\epsilon(0) \to \infty$ as $\epsilon \to 0$, and from [AP1] one has the upper bound

$$(7.1) \qquad u_\epsilon(x) \le W_\mu(x) = \left(\frac{\mu}{\mu^2 + \alpha|x|^2}\right)^{1/2} \quad \text{in } \overline{\Omega},$$

where

$$(7.2) \qquad \mu = (u_\epsilon(0))^{-2}$$

and

$$(7.3) \qquad \alpha = 1 + \frac{1}{3}\left(\frac{\pi^2}{4} + \epsilon\right)\mu^2.$$

One also has from [AP3] the lower bound

$$(7.4) \qquad u_\epsilon(x) \ge W_\mu(x)\left\{1 - \frac{1}{2}\left(\frac{\pi^2}{4} + \epsilon\right)|x|^2\right\}.$$

The two bounds (7.1) and (7.4) relate $u_\epsilon(x)$ via the parameter μ to its central value $u_\epsilon(0)$. To relate μ to ϵ we use Pohozaev's identity again. For Problem (II) it becomes

$$(7.5) \qquad J(u_\epsilon) = \int_{\partial\Omega}(x,n)\left(\frac{\partial u_\epsilon}{\partial n}\right)^2 - 2\left(\frac{\pi^2}{4} + \epsilon\right)\int_\Omega u_\epsilon^2 = 0.$$

We now estimate the two terms in $J(u_\epsilon)$ by finding a good approximation for u_ϵ. As before we use for this purpose the function

$$\phi = U_\mu + 4\pi\sqrt{\mu}g_{(\pi^2/4)+\epsilon},$$

where U_μ has been defined by (1.14) and g_λ in (1.10). To see that ϕ is indeed a good approximation, we note that the remainder term

$$\eta = u_\epsilon - \phi,$$

is a solution of the problem

$$(7.6) \qquad -\Delta\eta - \left(\frac{\pi^2}{4} + \epsilon\right)\eta = f$$

$$(7.7) \qquad \eta = b,$$

where

$$f = \left(\frac{\pi^2}{4} + \epsilon\right)\left(U_\mu - \frac{\sqrt{\mu}}{|x|}\right) + 3\left(u_\epsilon^5 - U_\mu^5\right)$$
$$b = -(U_\mu + 4\pi\sqrt{\mu}g_{(\pi^2/4)+\epsilon}).$$

and hence that it can be estimated by

$$(7.8) \qquad \|\eta\|_{L^2(\Omega)} + \|\nabla\eta\|_{L^2(\partial\Omega)} \le C\mu^{5/2}|\log\mu|.$$

The proof of (7.8) is very similar to that of Proposition 5.1 (actually it is a little simpler) and we therefore omit it.

It follows from (7.8) that $J(u_\epsilon)$ can be expressed as

$$(7.9) \qquad\qquad J(u_\epsilon) = J(\phi) + R_\epsilon,$$

where

$$(7.10) \qquad\qquad R_\epsilon = O(\mu^3|\log\mu|) \quad \text{as} \quad \epsilon \to 0.$$

For the proof of (7.9) and (7.10) we refer to Lemma 6.2.

If we now use Lemma 6.3 to evaluate $J(\phi)$, setting $\lambda = (\pi^2/4) + \epsilon$, we conclude from (7.9) that

$$-16\pi^2 g_{(\pi^2/4)+\epsilon}(0)\mu + 4\pi^2\left(\frac{\pi^2}{4} + \epsilon\right)\mu^2 = O(\mu^3|\log\mu|) \quad \text{as} \quad \epsilon \to 0$$

or, if we divide by $-16\pi^2\mu$,

$$(7.11) \quad g_{(\pi^2/4)+\epsilon}(0) = \frac{1}{4}\left(\frac{\pi^2}{4} + \epsilon\right)\mu + O(\mu^2|\log\mu|) \quad \text{as} \quad \epsilon \to 0.$$

However, $g_\lambda(0)$ is given by

$$g_\lambda(0) = -\frac{1}{4\pi}\sqrt{\lambda}\cotan\sqrt{\lambda},$$

and so, if $\lambda = (\pi^2/4) + \epsilon$,

$$(7.12) \qquad g_{(\pi^2/4)+\epsilon}(0) = \frac{\epsilon}{8\pi} + O(\epsilon^2) \quad \text{as} \quad \epsilon \to 0.$$

Inserting (7.12) into (7.11) we conclude first that $\mu = O(\epsilon)$ as $\epsilon \to 0$ and subsequently that

$$(7.13) \qquad \frac{\epsilon}{\mu} = \frac{\pi^3}{2} + O(\epsilon|\log\epsilon|) \quad \text{as} \quad \epsilon \to 0.$$

As to the limiting behaviour of u_ϵ as $\epsilon \to 0$, we recall that by (7.8),

$$\mu^{-1/2}u_\epsilon \to 4\pi G_{\pi^2/4} \quad \text{as} \quad \epsilon \to 0$$

in $L^2(\Omega)$, and pointwise away from the origin. Hence, by (7.13)

$$\epsilon^{-1/2}u_\epsilon(x) \to 4\sqrt{2/\pi}G_{\pi^2/4} \quad \text{as} \quad \epsilon \to 0.$$

This completes the proof of Theorem 3.

8. Conjectures for general domains. We shall now formulate various conjectures for general domains in $\mathbf{R}^N$. They are motivated partly by the results in the previous sections and partly by some recent results of [R1,2]. We shall also present evidence in support of these conjectures.

CONJECTURE 1.   Let $\Omega \subset \mathbf{R}^N, N \geq 3$, be a bounded domain with smooth boundary. Let u_ϵ be a solution of

$$(8.1) \qquad -\Delta u_\epsilon = N(N-2)u_\epsilon^{p-\epsilon} \quad \text{in} \quad \Omega$$
$$(8.2) \qquad u_\epsilon > 0 \quad \text{in} \quad \Omega$$
$$(8.3) \qquad u_\epsilon = 0 \quad \text{on} \quad \partial\Omega,$$

where $p = (N+2)/(N-2)$.
We denote again by $G(x,y) = G_0(x,y)$ the Green's function of $-\Delta$ and by $g(x,y) = g_0(x,y)$ its regular part i.e.

$$g(x,y) = G(x,y) - \frac{1}{(N-2)\sigma_N|x-y|^{N-2}},$$

where σ_N is the area of the unit sphere in $\mathbf{R}^N$:

$$\sigma_N = \frac{2\pi^{N/2}}{\Gamma(N/2)}.$$

Recall that $g(x, y)$ is smooth on $\Omega \times \Omega$.

Conjecture 1. *Assume u_ϵ is a minimizing sequence for the Sobolev inequality, i.e.*

$$(8.4) \qquad \frac{\int_\Omega |\nabla u_\epsilon|^2}{\|u_\epsilon\|_{p+1}^2} = S_N + o(1),$$

where S_N is the best Sobolev constant in $\mathbf{R}^N$:

$$S_N = \pi N(N-2)\left\{\frac{\Gamma(N/2)}{\Gamma(N)}\right\}^{2/N}.$$

Then

$$\lim_{\epsilon \to 0} \epsilon \|u_\epsilon\|_{L^\infty}^2 = 2\sigma_N^2 \left[\frac{N(N-2)}{S_N}\right]^{N/2} |g|,$$

where g is a critical value of the function $\phi(x) = g(x, x)$ i.e. $g = \phi(x_0)$ for some point $x_0 \in \Omega$ such that $\nabla \phi(x_0) = 0$.

Evidence. Recall that Pohozaev's identity says that if u is a solution of the problem

$$\begin{cases} -\Delta u = f(u) & \text{in } \Omega, \\ u = 0 & \text{on } \partial\Omega \end{cases}$$

then,

$$(8.5) \quad \left(1 - \frac{N}{2}\right)\int_\Omega uf(u) + N\int_\Omega F(u) = \frac{1}{2}\int_{\partial\Omega}(x - y, n)\left(\frac{\partial u}{\partial n}\right)^2,$$

where $F(u) = \int_0^u f(t)dt$ and y is any point in $\mathbf{R}^N$. Applying this identity to $(8.1) - (8.3)$ we obtain

$$\frac{N(N-2)^3}{2N - \epsilon(N-2)}\epsilon\int_\Omega u_\epsilon^{p+1-\epsilon} = \int_{\partial\Omega}(x - y, n)\left(\frac{\partial u_\epsilon}{\partial n}\right)^2,$$

or

$$(8.6) \qquad \frac{1}{2}(N-2)^3\epsilon\|u_\epsilon\|_{p+1}^{p+1} \simeq \int_{\partial\Omega}(x - y, n)\left(\frac{\partial u_\epsilon}{\partial n}\right)^2.$$

If we multiply (8.1) by u_ϵ, integrate over Ω and use (8.4) we obtain

$$\|u_\epsilon\|_{p+1}^{p-1} \simeq \frac{S_N}{N(N-2)},$$

and therefore

$$\|u_\epsilon\|_{p+1}^{p+1} \simeq \left[\frac{S_N}{N(N-2)}\right]^{N/2}.$$

In view of (8.6) we thus find that

$$(8.7) \qquad \int_{\partial\Omega} (x-y,n)\left(\frac{\partial u_\epsilon}{\partial n}\right)^2 \simeq \frac{1}{2}(N-2)^3 \left[\frac{S_N}{N(N-2)}\right]^{N/2} \epsilon.$$

On the other hand we know that u_ϵ "concentrates" around some point x_0 (see [S], [L]) and near x_0 we have

$$(8.8) \qquad u_\epsilon(x) \simeq \frac{\mu^{(N-2)/2}}{(\mu^2 + |x-x_0|^2)^{(N-2)/2}}$$

for some appropriate $\mu = \mu_\epsilon$ which tends to 0 as $\epsilon \to 0$. Then, we have

$$(8.9) \qquad \|u_\epsilon\|_{L^\infty(\Omega)} \simeq \mu^{-(N-2)/2}.$$

From (8.8) we deduce that, near $x = x_0$,

$$N(N-2)u_\epsilon^{p-\epsilon} \simeq N(N-2)\frac{\mu^{(N+2)/2}}{(\mu^2 + |x-x_0|^2)^{(N+2)/2}}$$

$$= \mu^{(N-2)/2}\frac{N(N-2)\mu^2}{(\mu^2 + |x-x_0|^2)^{(N+2)/2}}$$

$$(8.10) \qquad \simeq \mu^{(N-2)/2}K_N\delta_{x_0},$$

where

$$K_N = N(N-2)\int_{\mathbf{R}^N} \frac{dx}{(1+|x|^2)^{(N+2)/2}} = N(N-2)\int_{\mathbf{R}^N} U^p(x)dx$$

and

$$U(x) = \frac{1}{(1+|x|^2)^{(N-2)/2}}.$$

Since U satisfies the equation $-\Delta U = N(N-2)U^p$ we see that

$$N(N-2)\int_{\mathbf{R}^N} U^p dx = \int_{\mathbf{R}^N}(-\Delta U) = -\lim_{r\to\infty} r^{N-1}\sigma_N U'(r)$$
$$= (N-2)\sigma_N,$$

and therefore

$$(8.11) \qquad\qquad K_N = (N-2)\sigma_N.$$

Going back to (8.1) – (8.3) we conclude that, globally on Ω,

$$(8.12) \qquad\qquad u_\epsilon(x) \simeq \mu^{(N-2)/2} K_N G(x,x_0)$$

and therefore
$$(8.13)$$
$$\int_{\partial\Omega}(x-y,n)\left(\frac{\partial u_\epsilon}{\partial n}\right)^2 dx \simeq \int_{\partial\Omega}(x-y_0,n)\mu^{N-2}K_N^2\left(\frac{\partial G}{\partial n}(x,x_0)\right)^2 dx.$$

Recall that (see Theorem 4.3)

$$\int_{\partial\Omega}(x-x_0,n)\left(\frac{\partial G}{\partial n}(x,x_0)\right)^2 dx = -(N-2)g(x_0,x_0).$$

Hence, if we put $y = x_0$ in (8.13) we obtain using (8.11)

$$(8.14) \quad \int_{\partial\Omega}(x-x_0,n)\left(\frac{\partial u_\epsilon}{\partial n}\right)^2 dx = -\mu^{N-2}\sigma_N^2(N-2)^3 g(x_0,x_0).$$

Putting (8.7) and (8.14) together we are led to

$$\frac{1}{2}(N-2)^3\left[\frac{S_N}{N(N-2)}\right]^{N/2}\epsilon \simeq -\mu^{(N-2)}\sigma_N^2(N-2)^3 g(x_0,x_0).$$

Consequently, using (8.9), we conjecture that

$$\|u_\epsilon\|^2_{L^\infty(\Omega)} \simeq \mu^{-(N-2)} \simeq \frac{2}{\epsilon}\sigma_N^2\left[\frac{N(N-2)}{S_N}\right]|g(x_0,x_0)|$$

since $g(x,y) < 0$ (by the maximum principle).

Finally, we claim that the point of concentration x_0 is a critical point of the function $\phi(x) = g(x, x)$. First, note that in Pohozaev's identity (8.5) the point y is arbitrary and thus we have

$$(8.15) \qquad \int_{\partial\Omega} \left(\frac{\partial u_\epsilon}{\partial n}\right)^2 n dx = 0.$$

From (8.12) we deduce that

$$(8.16) \qquad \int_{\partial\Omega} \left(\frac{\partial G}{\partial n}(x, x_0)\right)^2 n(x) dx = 0.$$

To complete the argument it now suffices to apply Theorem 4.4.

Remark. Conjecture 1 is consistent with the results obtained in [AP2] when Ω is a ball.

CONJECTURE 2. Let $\Omega \subset \mathbf{R}^N, N \geq 4$, be a bounded domain with smooth boundary. Let u_ϵ be a solution of

$$\begin{cases} -\Delta u_\epsilon = N(N-2)u_\epsilon^p + \epsilon u_\epsilon & \text{in } \Omega, \\ u_\epsilon > 0 & \text{in } \Omega \\ u_\epsilon = 0 & \text{on } \partial\Omega, \end{cases}$$

where $p = (N+2)/(N-2)$.

Conjecture 2. *Assume $\{u_\epsilon\}$ is a minimizing sequence for the Sobolev inequality. Then*

$$\lim_{\epsilon\to 0} \epsilon\|u_\epsilon\|_{L^\infty(\Omega)}^{2(N-4)/N-2)} = \frac{(N-2)^3\sigma_N}{2a_N}|g| \qquad \text{if } N > 4$$

$$\lim_{\epsilon\to 0} \epsilon\log\|u_\epsilon\|_{L^\infty(\Omega)} = 4\sigma_4|g| \qquad \text{if } N = 4,$$

where g is a critical value of the function $\phi(x) = g(x, x)$ and

$$a_N = \int_0^\infty \frac{r^{N-1}dr}{(1+r^2)^{N-2}} = 2\frac{N-1}{N-4}\frac{\{\Gamma(N/2)\}^2}{\Gamma(N)}.$$

Evidence. From Pohozaev's identity we obtain

$$(8.17) \qquad \epsilon\int_\Omega u_\epsilon^2 = \frac{1}{2}\int_{\partial\Omega}(x-y, n)\left(\frac{\partial u_\epsilon}{\partial n}\right)^2$$

for any point $y \in \mathbf{R}^N$. As above, we have (8.8) at the point of concentration x_0 and thus

$$(8.18) \qquad \int_\Omega u_\epsilon^2 \simeq \mu^2 \int_{\mathbf{R}^N} \frac{dx}{(1+|x|^2)^{N-2}} = \mu^2 \sigma_N a_N \quad \text{if} \quad N > 4$$

and

$$(8.19) \qquad \int_\Omega u_\epsilon^2 \simeq \mu^2 |\log \mu| \sigma_4 \quad \text{if} \quad N = 4.$$

On the other hand, we have as in Conjecture 1,

$$\int_{\partial\Omega} (x - x_0, n)\left(\frac{\partial u_\epsilon}{\partial n}\right)^2 \simeq \int_{\partial\Omega} (x - x_0, n)\mu^{N-2} K_N^2 \left(\frac{\partial G}{\partial n}(x, x_0)\right)^2 dx$$
$$(8.20) \qquad\qquad = -\mu^{N-2}(N-2)^3 \sigma_N^2 g(x_0, x_0).$$

Putting together (8.17) with $y = x_0$, (8.18), (8.19) and (8.20) we are led to Conjecture 2. The argument for showing that x_0 as a critical point of ϕ is the same as in Conjecture 1. Part of this programme has been made rigorous by O. Rey [R2].

Remark.   Conjecture 2 is consistent with the results for the ball in [AP3].

CONJECTURE 3.   Let $\Omega \subset \mathbf{R}^3$ be a bounded domain with smooth boundary. Let u_ϵ be the solution of

$$\begin{cases} -\Delta u_\epsilon = 3u_\epsilon^{5-\epsilon} + \lambda u_\epsilon & \text{in } \Omega, \\ u_\epsilon > 0 & \text{in } \Omega, \\ u_\epsilon = 0 & \text{on } \partial\Omega. \end{cases}$$

As before, we denote by $G_\lambda(x, y)$ the Green's function of $-\Delta - \lambda$ on Ω and by $g_\lambda(x, y)$ its regular part. It is not difficult to see that $\phi_\lambda(x) = g_\lambda(x, x)$ is smooth on Ω

Conjecture 3.   *(i) Assume $\phi_\lambda(x) \leq 0$ on Ω and $\{u_\epsilon\}$ is a minimizing sequence for the Sobolev inequality. Then*

$$\lim_{\epsilon \to 0} \epsilon \|u_\epsilon\|_{L^\infty(\Omega)}^2 = 128|g_\lambda|,$$

where g_λ is a critical value of the function ϕ_λ, i.e $g_\lambda = \phi_\lambda(x_0)$ for some point $x_0 \in \Omega$ such that $\nabla\phi_\lambda(x_0) = 0$.
(ii) If $\phi_\lambda(x_0) = 0$, then

$$\lim_{\epsilon \to 0} \epsilon \|u_\epsilon\|_{L^\infty(\Omega)}^4 = 32\lambda.$$

Remark.   This would be consistent with Theorems 1 and 2.

Appendix.

Let u_ϵ be a positive radial solution of

$$(A.1) \qquad -\Delta u - \lambda u = 3u^{5-\epsilon} \quad \text{in} \quad \Omega$$
$$(A.2) \qquad u(0) = \mu^{-1/2}.$$

Then, according to [AP1], we have the *upper bound*

$$(A.3) \qquad u_\epsilon(x) \leq W_\mu(x),$$

where

$$(A.4) \qquad W_\mu(x) = \left(\frac{\mu}{\mu^2 + \alpha|x|^2} \right)^{1/2}$$

and

$$(A.5) \qquad \alpha = \mu^{\epsilon/2} + \frac{1}{3}\lambda\mu^2.$$

In addition we have the *lower bound*:

Lemma 5.5. *We have*

$$(A.6) \quad u_\epsilon(x) \geq W_\mu(x)\left(1 - \frac{\lambda}{2}|x|^2\right) - \mu^{-1/2}[\{\mu^{1/2}W_\mu(x)\}^{-\epsilon} - 1].$$

PROOF. We follow [AP3], exploiting the radial symmetry of u_ϵ. Setting

$$t = \frac{1}{|x|} \quad \text{and} \quad y(t) = u_\epsilon(x),$$

we transform (A.1), (A.2) to

$$(A.7) \qquad y'' + t^{-4}f(y) = 0$$

$$(A.8) \qquad y(t) \to \gamma \quad \text{as} \quad t \to \infty,$$

where f is given by

$$f(s) = \lambda s + 3s^{5-\epsilon}$$

and $\gamma = \mu^{-1/2}$. Rephrasing (A.3) we have

$(A.9)$ $\qquad\qquad y(t) \leq z(t) \qquad$ for $\quad 1 \leq t < \infty$,

where $z(t) = W_\mu(x)$, and is given explicitly by

$$z(t) = \gamma t\left(t^2 + \frac{1}{3}\frac{f(\gamma)}{\gamma}\right)^{-1/2}.$$

Note that z is a solution of the problem

$(A.10)$ $\qquad\qquad z'' + t^{-4}\gamma^{-5}f(\gamma)z^5 = 0$

$(A.11)$ $\qquad\qquad z(t) \to \gamma \quad$ as $\quad t \to \infty$.

We now integrate the differential equation (A.7) for y twice. That yields, if we use (A.8),

$$y(t) = \gamma - \int_t^\infty (s-t)s^{-4}f(y(s))ds$$

$$\geq \gamma - \int_t^\infty (s-t)s^{-4}f(z(s))ds$$

$(A.12)$ $\qquad\qquad = \gamma - \lambda I_1 - 3I_2,$

where

$$I_1 = \int_t^\infty (s-t)s^{-4}z(s)ds$$

$$I_2 = \int_t^\infty (s-t)s^{-4}z^{5-\epsilon}(s)ds.$$

By the concavity and positivity of z, we have

$$\frac{z(s)}{s} < \frac{z(t)}{t} \quad \text{if} \quad s > t,$$

and hence we can estimate I_1 by

$(A.13)$ $\qquad\qquad I_1 < \dfrac{z(t)}{t}\displaystyle\int_t^\infty (s-t)s^{-3}ds = \dfrac{z(t)}{2t^2}.$

To estimate I_2, we note that since z is increasing,

$$z^{-\epsilon}(s) < z^{-\epsilon}(t) \quad \text{if} \quad s > t$$

and so

$$(A.14) \qquad I_2 < z^{-\epsilon}(t) \int_t^\infty (s-t)s^{-4}z^5(s)ds.$$

If we integrate the differential equation (A.10) for z twice and use (A.11), we obtain

$$z(t) = \gamma - \gamma^{-5}f(\gamma) \int_t^\infty (s-t)s^{-4}z^5(s)ds.$$

Using this relation to eliminate the integral from (A.14) we arrive at the estimate

$$I_2 < z^{-\epsilon}(t)\frac{\gamma - z(t)}{\gamma^{-5}f(\gamma)}.$$

This yields, remembering the definition of f,

$$(A.15) \qquad I_2 < \frac{1}{3}\{\gamma - z(t)\}\left(\frac{\gamma}{z(t)}\right)^\epsilon.$$

Putting the estimates (A.13) and (A.15) for I_1 and I_2 into (A.12) finally yields the lower bound

$$y(t) > z(t)\left(1 - \frac{\lambda}{2t^2}\right) - \{\gamma - z(t)\}\left[\left(\frac{\gamma}{z(t)}\right)^\epsilon - 1\right]$$
$$< z(t)\left(1 - \frac{\lambda}{2t^2}\right) - \gamma\left[\left(\frac{\gamma}{2(t)}\right)^\epsilon - 1\right]$$

because $\gamma > z(t)$. This reads, in terms of the original variables,

$$u_\epsilon(x) \geq W_\mu(x)\left(1 - \frac{\lambda}{2}|x|^2\right) - \mu^{-1/2}[\{\mu^{1/2}W_\mu(x)\}^{-\epsilon} - 1],$$

i.e. the lower bound we set out to prove.

References

[AP1] F.V.Atkinson, L.A.Peletier, *Emder–Fowler equations involving critical exponents*, Nonlinear Anal. TMA **10** (1986), 755-776.

[AP2] F.V.Atkinson, L.A.Peletier, *Elliptic equations with nearly critical growth*, J. Diff. Equ. **70** (1987), 349-365.

[AP3] F.V.Atkinson, L.A.Peletier, *Large solutions of elliptic equations involving critical exponents*, to appear in Asymptotic Analysis **1** (1988).

[BN] H.Brezis, L.Nirenberg, *Positive solutions of nonlinear elliptic equations involving critical Sobolev exponents*, Comm. Pure Appl. Math. XXXVI (1983), 437-477.

[Bu] C.Budd, *Semilinear elliptic equations with near critical growth rates*, Proc. Roy. Soc. Edinburgh **107A** (1987), 249-270.

[GNN] B.Gidas, W.-M.Ni, L.Nirenberg, *Symmetry and related properties via the maximum principle*, Comm. Math. Phys. **68** (1979), 209-243.

[GT] D.Gilbarg, N.Trudinger, *Elliptic partial differential equations of second order*, Grundl. math. wiss. #224, Springer Verlag, 1977.

[L] P.L.Lions, *The concentration–compactness principle in the calculus of variations, the limit case*, Rev. Mat. Iberoamericano **1** (1985), 45-121 and 145-201.

[R1] O.Rey, *Le rôle de la fonction de Green dans une équation elliptique non linéaire avec l'exposant critique de Sobolev*, C.R. Acad. Sci. Paris **305** (1987), 591-594.

[R2] O.Rey, *The role of the Green's function in a nonlinear elliptic equation involving the critical Sobolev exponent*, to appear.

[S] M.Struwe, *A global compactness result for elliptic boundary value problems involving limiting nonlinearities*, Math. Z. **187** (1984), 511-517.

[T] G.Talenti, *Best constants in Sobolev inequality*, Annali di Mat. **110** (1976), 353-372.

Département de Mathématiques
Université Paris VI
4 Place Jussieu
F-75230 PARIS, cedex 05

Mathematical Institute
University of Leiden
The Nederlands

ASYMPTOTIC BEHAVIOUR FOR DIRICHLET PROBLEMS IN DOMAINS BOUNDED BY THIN LAYERS

Giuseppe Buttazzo   Gianni Dal Maso   Umberto Mosco

Dedicated to Ennio De Giorgi on his sixtieth birthday

1. Introduction. Let $\Omega \subset \mathbf{R}^n$ be a bounded Lipschitz domain surrounded along its boundary by a layer Σ_ϵ of maximum thickness ϵ.

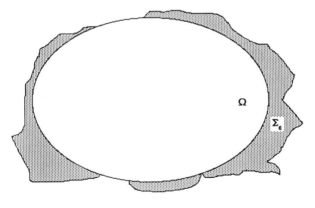

Figure 0

Let us consider the Laplace operator $-\Delta$ in Ω and the operator $-\epsilon\Delta$ in the layer Σ_ϵ. Given a function $g \in L^2(\mathbf{R}^n)$, let u_ϵ be the solution of the equations

$$-\Delta u_\epsilon = g \quad \text{in} \quad \Omega \quad , \quad -\epsilon\Delta u_\epsilon = g \quad \text{in} \quad \Sigma_\epsilon \, ,$$

satisfying the Dirichlet condition $u_\epsilon = 0$ on the boundary of $\Omega_\epsilon = \Omega \cup \Sigma_\epsilon$ and the natural transmission conditions on $\Omega_\epsilon \cap \partial\Omega$ (see figure 0).

Without any further assumption on the geometry and the regularity of Σ_ϵ, is it possible to describe and characterize the asymptotic behaviour of u_ϵ as $\epsilon \to 0$?

Asymptotic properties of this kind have been investigated by several authors, with different degrees of generality and under various regularity assumptions on the data (see for instance [1], [5], [9], [10]).

On the other hand, similar questions have been also raised in connection with boundary value problems in highly perturbed domains, and an extensive literature is by now available on this subject.

For Dirichlet problems, a general approach to the theory has been recently developed in [4], [8], [13], [15], [16]. The basic feature of this approach is that, without demanding any property on the varying domains, it still allows to characterize the asymptotic behaviour of the solutions, by relying on suitable variational compactness and density properties of a class of Borel measures: Namely, the class $\mathcal{M}_o$ of all nonnegative Borel measures on $\mathbf{R}^n$ not charging polar sets, but possibly $+\infty$ on large subsets of $\mathbf{R}^n$.

In the present paper we answer the question raised at the beginning in full generality, by reframing it in terms of the relaxed Dirichlet problems introduced in [15]. By defining the measure

$$\mu^\epsilon(B) = \begin{cases} +\infty & \text{if } B \cap \partial\Omega^\epsilon \text{ has positive capacity,} \\ 0 & \text{otherwise,} \end{cases}$$

our initial problem can be given a variational formulation as follows:

$$\begin{cases} u_\epsilon \in H^1(\mathbf{R}^n) \cap L^2(\mathbf{R}^n, \mu^\epsilon) \\ \int_\Omega Du_\epsilon Dv dx + \epsilon \int_{\Sigma_\epsilon} Du_\epsilon Dv \, dx + \int_{\bar\Omega_\epsilon} u_\epsilon v d\mu^\epsilon = \int_{\Omega_\epsilon} gv dx \\ \text{for every } v \in H^1(\mathbf{R}^n) \cap L^2(\mathbf{R}^n, \mu^\epsilon). \end{cases}$$

Here, $H^1(\mathbf{R}^n)$ denotes the usual Sobolev space, $L^2(\mathbf{R}^n, \mu^\epsilon)$ the space of all Borel functions which are square integrable with respect to μ^ϵ, and the integral term in $d\mu^\epsilon$ is unambiguously determined because functions of $H^1(\mathbf{R}^n)$ can be defined up to sets of capacity zero.

Therefore, the Dirichlet condition $u_\epsilon = 0$ on $\partial\Omega^\epsilon$ is prescribed in the capacity sense, and u_ϵ can be taken to be equal to zero on $\mathbf{R}^n - \bar{\Omega}^\epsilon$.

The first basic result of this paper is that there exist a subsequence $(\Omega^{\epsilon'})$ and a measure $\mu \in \mathcal{M}_o$ supported by $\partial\Omega$, both independent of g, such that the corresponding solutions $u_{\epsilon'}$ converge strongly in $L^2(\mathbf{R}^n)$ to a function u, which belongs to $H^1(\Omega) \cap L^2(\bar{\Omega}, \mu)$, is a solution of the equation

$$-\Delta u = g \quad \text{in} \quad \Omega,$$

and satisfies a boundary condition of Robin-type formally written as

$$(1.1) \qquad \frac{\partial u}{\partial \nu} + \mu u = 0 \quad \text{on} \quad \partial\Omega,$$

where ν denotes the outer unit normal to Ω. The rigorous variational meaning of (1.1) is that of the natural boundary condition associated with the minimum problem:

$$\min\left\{ \int_\Omega |Du|^2 dx + \int_{\partial\Omega} u^2 d\mu - 2\int_\Omega gu \, dx \right\}$$

of which the limit function u is the (unique) solution.

The second basic result is the characterization of the limit measure μ in terms of suitable asymptotic capacities associated with Σ_ϵ. We recall that the evaluation of μ by using capacitary techniques was already obtained in [9] in the particular case of periodic oscillating boundaries.

The first result is based on compactness propeties of variational type for the class $\mathcal{M}_o$, which are described in detail in Section 4, and on the corresponding convergence properties of the solutions, described in Section 5. The second result is based on the additional properties established in Sections 3, 6, and 7, which provide a general procedure for the identification of an arbitrary asymptotic measure $\mu \in \mathcal{M}_o$.

2. Notation and preliminary results. In this section we fix the notation and collect all results from previous papers we shall use in the sequel. Let n be an integer with $n \geq 2$.

1. For every bounded open set $U \subset \mathbf{R}^n$ and for every compact set $K \subset U$ the *capacity* of K with respect to U is defined by

$$\text{cap}(K,U) = \inf\left\{\int_U |Dv|^2 dx : v \in C_o^\infty(U), v \geq 1 \quad \text{on} \quad K\right\};$$

the definition is extended to open sets $V \subset U$ by

$$\text{cap}(V,U) = \sup\left\{\text{cap}(K,U) : K \subset V, K \quad \text{compact}\right\},$$

and to Borel sets $B \subset U$ by

$$\text{cap}(B,U) = \inf\left\{\text{cap}(V,U) : V \supset B, V \quad \text{open}\right\}.$$

We say that a Borel set $B \subset \mathbf{R}^n$ has *capacity zero* if $\text{cap}(B \cap U, U) = 0$ for every bounded open set $U \subset \mathbf{R}^n$. If a property $P(x)$ holds for all $x \in B$ except for a set $B_o \subset B$ with capacity zero, then we say that $P(x)$ holds *quasi everywhere* on B (q.e. on B). We say that a Borel set $A \subset \mathbf{R}^n$ is *quasi open* if for every bounded open set $U \subset \mathbf{R}^n$ and for every $\epsilon > 0$ there exists an open set $V \subset U$ such that $\text{cap}((A \cap U)\Delta V, U) < \epsilon$, where Δ denotes the symmetric difference between sets.

We say that a function $f : B \to \overline{\mathbf{R}}$ is *quasi continuous* on B if for every bounded open set $U \subset \mathbf{R}^n$ and for every $\epsilon > 0$ there exists an open set $V \subset U$, with $\text{cap}(V, U) < \epsilon$, such that the restriction of f to $(B \cap U) - V$ is continuous.

It is well known that a bounded set $B \subset \mathbf{R}^n$ has capacity zero (resp. B is quasi open or f is quasi continuous on B) if and only if the above conditions are satisfied for one (hence for all) bounded open set $U \subset \mathbf{R}^n$ with $B \subset U$.

2. For every open set $U \subset \mathbf{R}^n$ we denote by $H^1(U)$ the usual Sobolev space of all functions in $L^2(U)$ with first order distribution derivatives in $L^2(U)$, and by $H_o^1(U)$ the closure of $C_o^\infty(U)$ in $H^1(U)$.

For every $a, b \in \mathbf{R}$ we set $a \wedge b = \min(a, b), a \vee b = \max(a, b), a^+ = a \vee 0, a^- = (-a) \vee 0$. It is well known that $u \wedge v$ and $u \vee v$

belong to $H^1(U)$ (resp. $H^1_o(U)$) whenever u, v belong to $H^1(U)$ (resp. $H^1_o(U)$).

Let U be an open subset of $\mathbf{R}^n$, let S be a subset of $\bar{U}$, and let $u \in H^1(U)$. We say that $u = 0$ on S *in the sense of* $H^1(U)$ if there exist a sequence (u_h) converging to u in $H^1(U)$ and a sequence (V_h) of open neighbourhoods of S in $\mathbf{R}^n$ such that $u_h = 0$ a.e. on $U \cap V_h$ for every $h \in \mathbf{N}$. If $u, v \in H^1(U)$, we say that $u = v$ on S *in the sense of* $H^1(U)$ if $u - v = 0$ on S in the sense of $H^1(U)$ (compare with [24], Definition 1.2, and [27], Definition 1.1). It follows immediately from the definition that the set of all functions $u \in H^1(U)$ such that $u = 0$ on S in the sense of $H^1(U)$ is a closed linear subspace of $H^1(U)$, and that $H^1_o(U)$ is the set of all functions $u \in H^1(U)$ such that $u = 0$ on ∂U in the sense of $H^1(U)$.

Let U be an open subset of $\mathbf{R}^n$ and let $\partial_* U$ be the Lipschitz part of the boundary of U, defined as the set of all points $x \in \partial U$ for which there exists an open neighbourhood V of x and a Lipschitz map ϕ from V onto an open neighbourhood W of 0, with a Lipschitz inverse, such that $\phi(x) = 0$ and $\phi(V \cap U) = \{y \in W : y_n < 0\}$, where y_n denotes the n-th coordinate of y.

For every $x \in \mathbf{R}^n$ and for every $r > 0$ we set

$$B_r(x) = \{y \in \mathbf{R}^n : |x - y| < r\}$$

and for every Borel set $B \subset \mathbf{R}^n$ we denote by $|B|$ its Lebesgue measure. It is well known that for every $u \in H^1(\Omega)$ the limit

$$(2.1) \qquad \tilde{u}(x) = \lim_{r \to 0} \frac{1}{|U \cap B_r(x)|} \int_{U \cap B_r(x)} u(y) \, dy$$

exists and is finite q.e. on $U \cup \partial_* U$. Moreover, the function $\tilde{u}$ defined q.e. by (2.1) is quasi-continuous on $U \cup \partial_* U$. Finally, if (u_h) converges to u strongly in $H^1(U)$, then there exists a subsequence (u_{h_k}) of (u_h) such that $(\tilde{u}_{h_k})$ converges to $\tilde{u}$ q.e. on $U \cup \partial_* U$.

A proof of these facts can be found in [22] for the interior points, and the study of the boundary points can be reduced easily to this case by means of the following extension theorem (see [2], Theorem 4.32).

Proposition 2.1. *Let U, V, W be bounded open subsets of $\mathbf{R}^n$ such that $V \subset\subset W$ and $W \cap \partial U = W \cap \partial_* U$. Then for every*

$u \in H^1(W \cap U)$ *there exists* $v \in H^1(V)$ *such that* $u = v$ *q.e. on* $V \cap U$.

The following proposition illustrates the relationships between the property "$u = 0$ on S in the sense of $H^1(U)$" and the property "$\tilde{u} = 0$ q.e. on S" (see [20] and [23]).

Proposition 2.2 *Let* U *be an open subset of* $\mathbf{R}^n$, *let* S *be a closed subset of* U, *and let* $u \in H^1(U)$. *Then* $u = 0$ *on* S *in the sense of* $H^1(U)$ *if and only if* $\tilde{u} = 0$ *q.e. on* S.

More generally, if S is a closed subset of $U \cup \partial_* U$, then $u = 0$ on S in the sense of $H^1(U)$ if and only if $\tilde{u} = 0$ q.e. on S. Indeed we can reduce easily the problem to the case of Proposition 2.2 by means of Proposition 2.1.

Let U be a bounded open subset of $\mathbf{R}^n$. The definition of $\tilde{u}$ allows to express the capacity of an arbitrary Borel set $B \subset U$ as the solution of a minimum problem. Indeed we have (see [22], Section 10)

$$\mathrm{cap}(B, U) = \inf\{ \int_U |Dv|^2 dx : v \in H^1_o(U), \tilde{v} \geq 1 \quad \text{q.e. on} \quad B\}.$$

3. By a *Borel measure* we mean a non-negative countably additive set function defined on the Borel σ-algebra of $\mathbf{R}^n$ and with values in $[0, +\infty]$. By a *Radon measure* we mean a Borel measure which is finite on every compact set. If μ is a Borel measure and $f : \mathbf{R}^n \to [0, +\infty]$ is a Borel function, we denote by $f\mu$ the Borel measure defined by

$$(f\mu)(B) = \int_B f d\mu$$

for every Borel set $B \subset \mathbf{R}^n$.

Following [15] we denote by $\mathcal{M}_o$ the class of all Borel measures μ such that $\mu(B) = 0$ for every Borel set $B \subset \mathbf{R}^n$ with capacity zero. Note that the measures of the class $\mathcal{M}_o$ are not required to be regular of σ-finite, as the following examples show (see [15]).
  i) the n-dimensional Lebesgue measure and the $(n-1)$-dimensional Hausdorff measure belong to $\mathcal{M}_o$;

ii) if $\mu \in \mathcal{M}_o$ and $f : \mathbf{R}^n \to [0, +\infty]$ is a Borel function, then the Borel measure $f\mu$ belongs to $\mathcal{M}_o$;

iii) for every Borel set $A \subset \mathbf{R}^n$, the Borel measure ∞_A defined below belongs to $\mathcal{M}_o$:

$$(2.2) \qquad \infty_A(B) = \begin{cases} 0 & \text{if } B \cap A \text{ has capacity zero,} \\ +\infty & \text{otherwise.} \end{cases}$$

Following [13], Section 3, we denote by $\mathcal{M}_o^*$ the class of all measures $\mu \in \mathcal{M}_o$ such that

$$\mu(B) = \inf\{\mu(A) : A \quad \text{quasi open}, \quad B \subset A\}$$

for every Borel set $B \subset \mathbf{R}^n$. For every $\mu \in \mathcal{M}_o$ there exists a unique measure $\mu^* \in \mathcal{M}_o^*$ such that

$$(2.3) \qquad \int_U \tilde{u}^2 d\mu^* = \int_U \tilde{u}^2 d\mu$$

for every open set $U \subset \mathbf{R}^n$ and every $u \in H^1(U)$. The measure μ^* is defined by

$$\mu^*(B) = \inf\{\mu(A) : A \quad \text{quasi open}, \quad B \subset A\}$$

for every Borel set $B \subset \mathbf{R}^n$. In Section 3 we shall associated to every $\mu \in \mathcal{M}_o$ a suitable set function (called μ-capacity); those associated with measures $\mu^* \in \mathcal{M}_o^*$ enjoy additional regularity properties (see Proposition 3.3).

4. In the following we shall use the notion of Γ-convergence. We recall here only the definition and the main properties of Γ-limits (see [18]). For further information we refer to [3], [6], [7], [17], and to the references quoted there.

Let X be a metric space, let $(F_\epsilon)_{\epsilon > 0}$ be a family of functions from X into $\overline{\mathbf{R}}$, and let F be a function from X into $\overline{\mathbf{R}}$. We say that (F_ϵ) Γ-converges to F in X as $\epsilon \to 0$ if

(a) for every $u \in X$ and for every family (u_ϵ) converging to u in X (as $\epsilon \to 0$) we have

$$F(u) \leq \liminf_{\epsilon \to 0} F_\epsilon(u_\epsilon)$$

(b) for every $u \in X$ there exists a family (u_ϵ) converging to u in X (as $\epsilon \to 0$) such that

$$F(u) = \limsup_{\epsilon \to 0} F_\epsilon(u_\epsilon).$$

Similarly, we define Γ-convergence for sequences F_{ϵ_h} as $\epsilon_h \to 0$. The following compactness theorem holds (see [19], Proposition 3.1).

Proposition 2.3. *Assume X is a separable metric space. For every family $(F_\epsilon)_{\epsilon > 0}$ there exists a subfamily $(F_{\epsilon'})$ which Γ-converges in X to a lower semicontinuous functional F as $\epsilon' \to 0$.*

The following property will be used in the sequel (see [14], Proposition 1.15(c)).

Proposition 2.4. *Assume X is a separable metric space. The family (F_ϵ) Γ-converges to F in X as $\epsilon \to 0$ if and only if every Γ-convergent subfamily $(F_{\epsilon'})$ Γ-converges to F in X as $\epsilon' \to 0$.*

3. Some properties of the variational μ-capacities. In this section we associate with every measure $\mu \in \mathcal{M}_o$ a family of increasing set functions, called μ-capacities, which are obtained as solutions of some variational problems associated with μ. We show that all the relevant information concerning the measure μ is contained in its μ-capacity. In particular, μ can be completely reconstructed from the knowledge of its μ-capacity on a suitable family of subsets of $\mathbf{R}^n$.

Throughout the paper we denote by Ω a fixed *bounded open subset* of $\mathbf{R}^n$ with a *Lipschitz boundary* and by L a fixed *elliptic operator* of the form

$$(3.1) \qquad Lu = -\sum_{i,j=1}^{n} D_i(a_{ij}(x)D_j u)$$

where $a_{ij} = a_{ji} \in L^\infty(\mathbf{R}^n)$ and, for suitable constants $0 < \Lambda_1 \leq \Lambda_2$,

$$(3.2) \qquad \Lambda_1 |\xi|^2 \leq \sum_{i,j=1}^{n} a_{ij}(x)\xi_i\xi_j \leq \Lambda_2 |\xi|^2$$

for almost every $x \in \mathbf{R}^n$ and for every $\xi \in \mathbf{R}^n$. For every $\epsilon \geq 0$ we consider the operator

$$(3.3) \qquad L^\epsilon u = - \sum_{i,j=1}^n D_i(a_{ij}^\epsilon(x)D_j u),$$

where

$$(3.4) \qquad a_{ij}^\epsilon(x) = \begin{cases} a_{ij}(x) & \text{if } x \in \Omega, \\ \epsilon a_{ij}(x) & \text{if } x \in \mathbf{R}^n - \Omega. \end{cases}$$

We denote by $a(x,\xi)$ and $a^\epsilon(x,\xi)$ the quadratic forms associated to the matrices (a_{ij}) and (a_{ij}^ϵ), i.e.

$$(3.5) \qquad a(x,\xi) = \sum_{i,j=1}^n a_{ij}(x)\xi_j\xi_i$$

$$(3.6) \qquad a^\epsilon(x,\xi) = \sum_{i,j=1}^n a_{ij}^\epsilon(x)\xi_j\xi_i$$

for every $x \in \mathbf{R}^n$ and $\xi \in \mathbf{R}^n$.

Let $\mu \in \mathcal{M}_o$ and let $\epsilon \geq 0$. For every bounded open set $U \subset \mathbf{R}^n$ and for every Borel set $B \subset\subset U$ (i.e. $\bar{B}$ compact and $\bar{B} \subset U$) we define the μ-capacity of B in U relative to the operator L^ϵ by (see [16])

$$c^\epsilon(\mu,B,U) = \inf\left\{ \int_U a^\epsilon(x,Du)dx + \int_B \tilde{u}^2 d\mu : u - 1 \in H_o^1(U) \right\}.$$

It is easy to prove that $c^\epsilon(\mu,B,U) \leq \mu(B)$ and $c^\epsilon(\mu,B,U) \leq k\mathrm{cap}(B,U)$, where $k = \max\{1,\Lambda_2\}$. Moreover the set function $c^\epsilon(\mu,\cdot,U)$ is increasing, continuous along increasing sequences, and finitely subadditive (hence countably subadditive). If L is the Laplace operator $-\Delta$ and μ is the measure ∞_A defined in (2.2), then

$$c^1(\mu,B,U) = \mathrm{cap}(B \cap A, U).$$

If $\epsilon > 0$, then the infimum in the definition of $c^\epsilon(\mu,B,U)$ is achieved by the lower semicontinuity and the coerciveness of the functional to be minimized. In the case $\epsilon = 0$ we have the following result, provided that the support of μ (denoted by spt μ) is contained in $\bar{\Omega}$.

Proposition 3.1. *Suppose that $\mu \in \mathcal{M}_o$ and spt $\mu \subset \bar{\Omega}$. Then for every bounded open set $U \subset \mathbf{R}^n$ and for every Borel set $B \subset\subset U$ we have*

$$c^o(\mu, B, U) = \min \left\{ \int_{U \cap \Omega} a(x, Du)dx + \int_{B \cap \bar{\Omega}} \tilde{u}^2 d\mu : u \in K \right\}$$

where K is the set of all $u \in H^1(U \cap \Omega)$ such that $u = 1$ on $\partial(U \cap \Omega) - U$ in the sense of $H^1(U \cap \Omega)$ (see fig. 1).

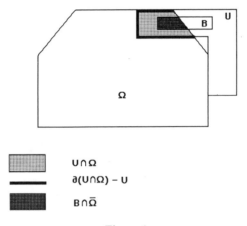

U∩Ω
∂(U∩Ω) – U
B∩Ω̄

Figure 1

The previous proposition is an immediate consequence of the following lemma.

Lemma 3.2. *Let U and V be two bounded open subsets of $\mathbf{R}^n$ with $V \subset\subset U$ and let $u \in H^1(U \cap \Omega)$ with $u = 0$ on $\partial(U \cap \Omega) - U$ in the sense of $H^1(U \cap \Omega)$. Then there exists a sequence (u_h) in $H_o^1(U)$ such that $u_h|_{U \cap \Omega}$ converges to u in $H^1(U \cap \Omega)$ and $u_h = u$ on $V \cap \Omega$ for every $h \in N$.*

PROOF. By hypothesis there exist a sequence (v_h) converging to u in $H^1(U \cap \Omega)$ and a sequence (V_h) of open neighbourhoods of

$(U \cap \Omega) - U$ in $\mathbf{R}^n$ such that $v_h = 0$ a.e. on $U \cap \Omega \cap V_h$ for every $h \in \mathbf{N}$. We extend v_h to Ω by setting $v_h = 0$ on $\Omega - U$. In this way $v_h \in H^1(\Omega)$, and, since Ω has a Lipschitz boundary, there exists $w_h \in H^1(\mathbf{R}^n)$ such that $w_h = v_h$ on Ω. Let ϕ_h be a sequence in $C^1_o(U)$ with $\phi_h = 1$ on $(\overline{U \cap \Omega}) - V_h$ and let $\phi \in C^1_o(U)$ with $\phi = 1$ on V. Set

$$u_h = \phi u + (1 - \phi)\phi_h w_h.$$

Then $u_h \in H^1_o(U), u_h = u$ on $V \cap \Omega$, and $(u_h|_{U\cap\Omega})$ converges to u in $H^1(U \cap \Omega)$.

The following proposition shows that the μ^*-capacity and the μ-capacity agree on every open set.

Proposition 3.3. *Let* $\mu \in \mathcal{M}_o$, *let* $\epsilon \geq 0$, *and let* U *be a bounded open subset of* $\mathbf{R}^n$. *If* $\epsilon = 0$ *assume that* $\mathrm{spt}\mu \subset \bar{\Omega}$. *Then*

$$c^\epsilon(\mu^*, V, U) = c^\epsilon(\mu, V, U)$$

for every open set $V \subset\subset U$ *and*

(3.7) $\qquad c^\epsilon(\mu^*, B, U) = \inf\{c^\epsilon(\mu, V, U) : V \quad \text{open}, \quad B \subset V\}$

for every Borel set $B \subset\subset U$.

PROOF. A proof in the case $\epsilon > 0$ can be found in [13], Sections 2 and 3, and it can be adapted to the case $\epsilon = 0$ by using Proposition 3.1.

The following theorem allows us to obtain μ from $c^\epsilon(\mu, B, U)$ even if μ is not σ-finite.

Theorem 3.4. *Let* $\mu \in \mathcal{M}_o$, *let* $\epsilon \geq 0$, *let* U *be a bounded open subset of* $\mathbf{R}^n$, *and let* B *be a Borel set with* $B \subset\subset U$.
(a) If $\epsilon > 0$, *then*

(3.8) $$\mu(B) = \sup \sum_{i \in I} c^\epsilon(\mu, B_i, U)$$

    where the supremum is taken over all finite Borel partitions
    $(B_i)_{i \in I}$ *of* B.

(b) *If* $\epsilon = 0$ *and* $\mathrm{spt}\,\mu \subset \bar{\Omega}$, *then for every* $x \in \bar{\Omega}$ *there exists an open*
    neighbourhood $W(x)$ *of* x *such that (3.8) holds whenever*
    $B \subset\subset U \subset W(x)$. *In the case* $x \in \Omega$ *we can take* $W(x) = \Omega$.

    A proof of the previous theorem is given in [13], Theorem 4.3, in the case $\epsilon > 0$ or $U \subset \Omega$. This proof can be easily adapted to the case $\epsilon = 0$ and $x \in \partial\Omega$ by replacing the derivation theorem for μ-capacities with its boundary version (see [8], Theorem 4.1).

    In the rest of this section we shall consider the special case $\mathrm{spt}\,\mu \subset \partial\Omega$ and we introduce a different notion of μ-capacity, denoted by $b_\delta^o(\mu, B, U)$, which will be used in Section 5 to characterize the limit of a sequence of Dirichlet problems in domains surrounded by thin layers.

    Let $(\Omega_\delta)_{\delta > 0}$ be a family of open subsets of Ω such that

$$(3.9) \qquad\qquad \Omega_{\delta_1} \subset\subset \Omega_{\delta_2} \quad \text{for} \quad \delta_1 > \delta_2,$$

and

$$(3.10) \qquad\qquad \Omega = \bigcup_{\delta > 0} \Omega_\delta.$$

    For every bounded open set $U \subset \mathbf{R}^n$ we put $U_\delta^o = U \cap (\Omega - \bar{\Omega}_\delta)$. Given $\mu \in \mathcal{M}_o$, for every Borel set $B \subset U$ we define the boundary μ-capacity of B in U, relative to the operator L, by
(3.11)

$$b_\delta^o(\mu, B, U) = \min\left\{ \int_{U_\delta^o} a(x, Du)dx + \int_{B \cap \partial\Omega} \tilde{u}^2 d\mu : u \in K_\delta^o \right\}$$

where K_δ^o is the set of all functions $u \in H^1(U_\delta^o)$ such that $u = 1$ on $U \cap \partial\Omega_\delta$ in the sense of $H^1(U_\delta^o)$ (see fig.2). The minimum in (3.11) is clearly achieved by the lower semicontinuity and the coerciveness of the functional. It is easy to see that the set function $b_\delta^o(\mu, \cdot, U)$ is increasing, continuous along increasing sequences, and finitely subadditive (hence countably subadditive). Moreover
(3.12)   if B_1, B_2 are Borel subsets of U, then

$$b_\delta^o(\mu, B_1 \cup B_2, U) + b_\delta^o(\mu, B_1 \cap B_2, U) \leq b_\delta^o(\mu, B_1, U) + b_\delta^o(\mu, B_2, U);$$

(3.13)   if B is a Borel subset of U and V is a bounded open subset
of $\mathbf{R}^n$ with $B \cap \partial\Omega \subset V$ and $V \cap U \cap \partial\Omega_\delta = \emptyset$, then

$$b_\delta^o(\mu, B, U) \leq \Lambda_2 \mathrm{cap}(B \cap \partial\Omega, V);$$

(3.14)   if B is a Borel subset of U, then

$$b_\delta^o(\mu, B, U) = b_\delta^o(\mu, B \cap \partial\Omega, U) \leq \mu(B \cap \partial\Omega).$$

These properties can be proved by adapting the arguments of
Theorem 2.9 of [13]. Finally, we remark that the function $\delta \to$
$b_\delta^o(\mu, B, U)$ is decreasing.

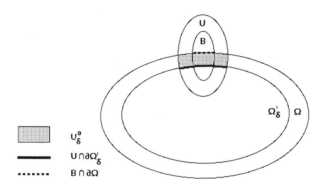

Figure 2

Lemma 3.5. *Let* $\mu \in \mathcal{M}_o$ *and let* U *be a bounded open subset
of* $\mathbf{R}^n$. *The following properties hold:*
(a) if B_1, B_2 *are Borel subsets of* U *and* $\mathrm{dist}(B_1, B_2) = \sigma > 0$, *then
for every* $\eta \in]0, 1[$

$$b_\delta^o(\mu, B_1, U) + b_\delta^o(\mu, B_2, U) \leq \frac{1}{1-\eta} b_\delta^o(\mu, B_1 \cup B_2, U) + \frac{4\Lambda_2|U_\delta^o|}{\eta\sigma^2};$$

(b) if B *is a Borel subset of* U *and* $\mathrm{dist}(B, \mathbf{R}^n - U) = \sigma > 0$, *then
for every* $\eta \in]0, 1[$

$$c^0(\mu, B \cap \partial\Omega, U) \leq (1-\eta)^{-1} b_\delta^o(\mu, B, U) + \Lambda_2 \eta^{-1} \sigma^{-2} |U_\delta^o|;$$

(c) if B is a Borel subset of U, then

$$b_\delta^o(\mu^*, B, U) = \inf\{b_\delta^o(\mu^*, V, U) : V \quad \text{open}, \quad B \subset V\}.$$

PROOF. Property (c) can be obtained by adapting the proof of Theorem 3.5 of [13].

Let us prove (a). Let B_1 and B_2 be two Borel subsets of U with $\text{dist}(B_1, B_2) = \sigma > 0$. There exists a function $u \in H^1(U_\delta^o)$ such that

$$b_\delta^o(\mu, B_1 \cup B_2, U) = \int_{U_\delta^o} a(x, Du)dx + \int_{(B_1 \cup B_2) \cap \partial\Omega} \tilde{u}^2 d\mu$$

and $u = 1$ on $U \cap \partial\Omega_\delta$ in the sense of $H^1(U_\delta^o)$. By a truncation argument we can prove that $0 \le u \le 1$ a.e. on U_δ^o. Since $\text{dist}(B_1, B_2) = \sigma > 0$, for every $\epsilon > 0$ there exist two functions $\phi_1, \phi_2 \in C_o^1(\mathbf{R}^n)$ such that $\phi_1 \phi_2 = 0$ on $\mathbf{R}^n$, $\phi_i = 1$ in a neighbourhood of B_i, $0 \le \phi_i \le 1$ on $\mathbf{R}^n$, and $|D\phi_i| \le (2 + \epsilon)\sigma^{-1}$ on $\mathbf{R}^n$ for $i = 1, 2$. Let $V_i = \{x \in \mathbf{R}^n : \phi_i(x) > 0\}$ and $v_i = 1 + \phi_i(u - 1)$. Then $v_i = 1$ on $U \cap \partial\Omega_\delta$ in the sense of $H^1(U_\delta^o)$, hence

$$b_\delta^o(\mu, B_i, U) \le \int_{U_\delta^o} a(x, Dv_i)dx + \int_{B_i \cap \partial\Omega} \tilde{v}_i^2 d\mu =$$

$$= \int_{V_i \cap U_\delta^o} a(x, Dv_i)dx + \int_{B_i \cap \partial\Omega} \tilde{u}^2 d\mu.$$

For every $\eta \in \ ]0, 1[$ we have by convexity

$$a(x, Dv_i) \le (1 - \eta)^{-1}\phi_i^2 a(x, Du) + \eta^{-1}(u - 1)^2 a(x, D\phi_i)$$
$$\le (1 - \eta)^{-1} a(x, Du) + \eta^{-1}\Lambda_2(2 + \epsilon)^2\sigma^{-2},$$

hence, using the fact that $V_1 \cap V_2 = \emptyset$, we obtain

$$b_\delta^o(\mu, B_1, U) + b_\delta^o(\mu, B_2, U) \le (1 - \eta)^{-1}\int_{U_\delta^o} a(x, Du)dx$$

$$+ \int_{(B_1 \cup B_2) \cap \partial\Omega} \tilde{u}^2 d\mu + (2 + \epsilon)^2\Lambda_2\eta^{-1}\sigma^{-2}|U_\delta^o| =$$
$$= (1 - \eta)^{-1}b_\delta^o(\mu, B_1 \cup B_2, U) + (2 + \epsilon)^2\Lambda_2\eta^{-1}\sigma^{-2}|U_\delta^o|.$$

The conclusion follows now by taking the limit as $\epsilon \to 0$.

Let us prove (b). Let B a Borel of U with $\mathrm{dist}(B, \mathbf{R}^n - U) = \sigma > 0$. There exists a function $u \in H^1(U^o_\delta)$ such that

$$b^o_\delta(\mu, B, U) = \int_{U^o_\delta} a(x, Du)dx + \int_{B \cap \partial\Omega} \tilde{u}^2 d\mu$$

and $u = 1$ on $U \cap \partial\Omega_\delta$ in the sense of $H^1(U^o_\delta)$. Therefore we can extend u to a function of $H^1(U \cap \Omega)$ still denoted by u, by setting $u = 1$ on $U \cap \bar{\Omega}_\delta$. By a truncation argument we can prove that $0 \le u \le 1$ a.e. on $U \cap \Omega$.

Since $\mathrm{dist}(B, \mathbf{R}^n - U) = \sigma > 0$, for every $\eta > 0$ there exists a function $\phi \in C^1_o(U)$ such that $\phi = 1$ in a neighbourhood of B, $|D\phi| \le (1 + \eta)\sigma^{-1}$ on U, and $0 \le \phi \le 1$ on U. Let $v = 1 + \phi(u - 1)$. Then $v = 1$ on $\partial(U \cap \Omega) - U$ in the sense of $H^1(U \cap \Omega)$, hence

$$c^o(\mu, B, U) \le \int_{U \cap \Omega} a(x, Dv)dx + \int_{B \cap \partial\Omega} \tilde{v}^2 d\mu$$
$$= \int_{U^o_\delta} a(x, Dv)dx + \int_{B \cap \partial\Omega} \tilde{u}^2 d\mu.$$

By arguing as in the proof of (a) we obtain

$$c^o(\mu, B, U) \le (1 - \eta)^{-1}b^o_\delta(\mu, B, U) + (1 + \epsilon)^{-2}\sigma^{-2}|U^o_\delta|.$$

The conclusion follows now by taking the limit as $\epsilon \to 0$.

Theorem 3.6. *If $\mu \in \mathcal{M}_o$, then*

$$\mu(B \cap \partial\Omega) = \sup\{b^o_\delta(\mu, B, U) : \delta > 0\}$$

for every bounded Borel set $B \subset \mathbf{R}^n$ and for every bounded open set $U \subset \mathbf{R}^n$ with $B \subset U$.

PROOF. Let $\mu \in \mathcal{M}_o$. For every bounded open set $U \subset \mathbf{R}^n$ and for every Borel set $B \subset U$ we define

(3.15) $\qquad b^o(\mu, B, U) = \sup_{\delta > 0} b^o_\delta(\mu, B, U) = \lim_{\delta \to 0} b^o_\delta(\mu, B, U),$

so we have to prove that

(3.16) $b^o(\mu, B, U) = \mu(B \cap \partial\Omega)$

for every Borel set $B \subset U$.

The properties of the functions b^o_δ yield that the set function $b^o(\mu, \cdot, U)$ is increasing and countably subadditive. Moreover, if B_1 and B_2 are Borel subsets of U with $\text{dist}(B_1, B_2) > 0$, then Lemma 3.5(a) implies

$$b^o(\mu, B_1, U) + b^o(\mu, B_2, U) \leq b^o(\mu, B_1 \cup B_2, U)$$

(recall that $|U^o_\delta| \to 0$ as $\delta \to 0$), therefore, $b^o(\mu, \cdot, U)$ is a Borel measure on U by the Carathéodory criterion (see [21], 2.3.2(9)). By (3.14) we have

(3.17) $b^o(\mu, B, U) = b^o(\mu, B \cap \partial\Omega, U) \leq \mu(B \cap \partial\Omega)$

for every Borel set $B \subset U$. Since $B \to b^o(\mu, B, U)$ and $B \to \mu(B \cap \partial\Omega)$ are Borel measures supported by $\partial\Omega$, it is enough to prove (3.16) locally, i.e. in a neighbourhood of each point $x \in U \cap \partial\Omega$. Let us fix $x \in U \cap \partial\Omega$, let $W(x)$ be the open neighbourhood of x given by Theorem 3.4(b), and let $V(x) = U \cap W(x)$. We shall prove (3.16) for every Borel set $B \subset\subset V(x)$. By Lemma 3.5(b) for such sets we have

$$c^o(\mu, B \cap \partial\Omega, V(x)) \leq b^o(\mu, B, V(x)) \leq b^o(\mu, B, U).$$

Therefore Theorem 3.4(b) implies that for every Borel set $B \subset\subset V(x)$

(3.18)
$$\mu(B \cap \partial\Omega) = \sup \sum_{i \in I} c^o(\mu, B_i, V(x))$$
$$\leq \sup \sum_{i \in I} b^o(\mu, B_i, U) = b^o(\mu, B \cap \partial\Omega, U)$$

where the supremum is taken over all finite Borel partitions $(B_i)_{i \in I}$ of $B \cap \partial\Omega$. The conclusion follows now from (3.17) and (3.18).

4. A compactness result. For every $\epsilon > 0$ let $F^\epsilon : L^2(\mathbf{R}^n) \to [0, +\infty]$ be the functional defined by

$$(4.1) \qquad F^\epsilon(u) = \begin{cases} \int_{\mathbf{R}^n} a^\epsilon(x, Du)dx & \text{if } u \in H^1(\mathbf{R}^n), \\ +\infty & \text{elsewhere on } L^2(\mathbf{R}^n), \end{cases}$$

and let $F^o : L^2(\mathbf{R}^n) \to [0, +\infty]$ be the functional defined by

$$(4.2) \qquad F^o(u) = \begin{cases} \int_\Omega a^o(x, Du)dx & \text{if } u|_\Omega \in H^1(\Omega), \\ +\infty & \text{elsewhere on } L^2(\mathbf{R}^n), \end{cases}$$

where $a^\epsilon(x, \xi)$ and $a^o(x, \xi)$ are the quadratic forms defined in (3.5) and (3.6).

Let $(\mu^\epsilon)_{\epsilon > 0}$ be a family of measures of the class $\mathcal{M}_o$ and let

$$d_\epsilon = \sup\{\text{dist}(x, \Omega) : x \in \text{spt}\mu^\epsilon\}$$

where $\text{spt}\mu^\epsilon$ denotes the support of μ^ϵ. We assume that

$$(4.3) \qquad \lim_{\epsilon \to 0} d_\epsilon = 0.$$

With each measure μ^ϵ we associate the following functional defined on $L^2(\mathbf{R}^n)$:

$$(4.4) \qquad M^\epsilon(u) = \begin{cases} \int_{\mathbf{R}^n} \tilde{u}^2 d\mu^\epsilon & \text{if } u \in H^1(\mathbf{R}^n), \\ +\infty & \text{elsewhere on } L^2(\mathbf{R}^n). \end{cases}$$

Given a measure $\mu \in \mathcal{M}_o$ supported by $\bar{\Omega}$, we consider the functional

$$(4.5) \qquad M^o(u) = \begin{cases} \int_{\bar{\Omega}} \tilde{u}^2 d\mu & \text{if } u \in H^1(\mathbf{R}^n), \\ +\infty & \text{elsewhere on } L^2(\mathbf{R}^n). \end{cases}$$

The main result of this section is the following compactness theorem.

Theorem 4.1. *There exist a subsequence* $(F^{\epsilon'} + M^{\epsilon'})$ *and* $\mu \in \mathcal{M}_o^*$ *supported by* $\bar{\Omega}$, *such that the family* $(F^{\epsilon'} + M^{\epsilon'})$ Γ*-converges to* $F^o + M^o$ *in* $L^2(\mathbf{R}^n)$ *as* $\epsilon' \to 0$.

PROOF. We prove first that (F^ϵ) Γ-converges to F^o in $L^2(\mathbf{R}^n)$ as $\epsilon \to 0$. Condition (a) in the definition of Γ-convergence follows

from the inequality $F^o \leq F^\epsilon$ and from the lower semicontinuity of
F^o. It is enough to prove condition (b) when $u|_\Omega \in H^1(\Omega)$. Since
Ω has a Lipschitz boundary, there exists a family (u_ϵ) in $H^1(\mathbf{R}^n)$
converging to u in $L^2(\mathbf{R}^n)$ such that $u_\epsilon = u$ on Ω and

$$\lim_{\epsilon \to 0} \epsilon \int_{\mathbf{R}^n} |Du_\epsilon|^2 dx = 0.$$

Since

$$F^\epsilon(u_\epsilon) \leq F^o(u) + \epsilon \Lambda_2 \int_{\mathbf{R}^n - \Omega} |Du_\epsilon|^2 dx,$$

condition (b) follows.

By Theorem 2.3 there exists a subsequence $(F^{\epsilon'} + M^{\epsilon'})$ Γ-con-
verging in $L^2(\mathbf{R}^n)$ as $\epsilon' \to 0$ to a lower semicontinuous functional
$G : L^2(\mathbf{R}^n) \to [0, +\infty]$. Since (F^ϵ) Γ-converges to F^o in $L^2(\mathbf{R}^n)$ as
$\epsilon \to 0$, we have $G \geq F^o$. We define for every $u \in L^2(\mathbf{R}^n)$

$$(4.6) \qquad M(u) = \begin{cases} G(u) - F^o(u) & \text{if } u|_\Omega \in H^1(\Omega), \\ +\infty & \text{elsewhere on } L^2(\mathbf{R}^n). \end{cases}$$

Let us prove that

$$(4.7) \qquad\qquad\qquad M(u) = M(v)$$

for every $u, v \in L^2(\mathbf{R}^n)$ with $u = v$ a.e. on Ω. Fix u, v as required;
we may assume $u|_\Omega = v|_\Omega \in H^1(\Omega)$. Let (u_ϵ) be a family in $H^1(\mathbf{R}^n)$
converging to u in $L^2(\mathbf{R}^n)$ as $\epsilon \to 0$ such that

$$(4.8) \qquad F^o(u) + M(u) = \lim_{\epsilon \to 0}[F^\epsilon(u_\epsilon) + M^\epsilon(u_\epsilon)]$$

and let (v_ϵ) be a family in $H^1(\mathbf{R}^n)$ converging to v in $L^2(\mathbf{R}^n)$ as
$\epsilon \to 0$ such that

$$(4.9) \qquad\qquad \lim_{\epsilon \to 0} \epsilon \int_{\mathbf{R}^n} |Dv_\epsilon|^2 dx = 0.$$

Let $w_\epsilon = \phi_\epsilon u_\epsilon + (1 - \phi_\epsilon) v_\epsilon$, where (ϕ_ϵ) is a sequence in $C_o^1(\mathbf{R}^n)$
with $0 \leq \phi_\epsilon \leq 1$ on $\mathbf{R}^n$, $\phi_\epsilon = 1$ on $\bar{\Omega} \cap \text{spt}\mu_\epsilon$, $\epsilon|D\phi_\epsilon|^2 \leq 2\epsilon^{1/2}$, and
$\phi_\epsilon(x) = 0$ whenever $\text{dist}(x, \Omega) > d_\epsilon + \epsilon^{1/4}$. Then w_ϵ converges to

v in $L^2(\mathbf{R}^n)$, hence condition (a) of the definition of Γ-convergence implies

$$(4.10) \qquad F^o(v) + M(v) \le \liminf_{\epsilon \to 0}[F^\epsilon(w_\epsilon) + M^\epsilon(w_\epsilon)].$$

Since $\tilde{w}^\epsilon = \tilde{u}^\epsilon$ q.e. on $\operatorname{spt}\mu^\epsilon$, we have

$$(4.11) \qquad M^\epsilon(w^\epsilon) = M^\epsilon(u^\epsilon).$$

Moreover, by convexity, for every $\eta > 0$ we have

$$a^\epsilon(x, Dw_\epsilon) \le \frac{\phi_\epsilon}{1-\eta} a^\epsilon(x, Du_\epsilon) +$$

$$+ \frac{1-\phi_\epsilon}{1-\eta} a^\epsilon(x, Dv_\epsilon) + \frac{(u_\epsilon - v_\epsilon)^2}{\eta} a^\epsilon(x, D\phi_\epsilon),$$

hence

$$(4.12) \qquad \begin{aligned} F^\epsilon(w_\epsilon) &\le \frac{1}{1-\eta} F^\epsilon(u_\epsilon) + \frac{\Lambda_2}{1-\eta}\epsilon \int_{\mathbf{R}^n} |Dv_\epsilon|^2 dx \\ &+ \Lambda_2 \frac{\epsilon^{1/2}}{\eta} \int_{\mathbf{R}^n - \Omega} (u_\epsilon - v_\epsilon)^2 dx. \end{aligned}$$

From (4.8), (4.9), (4.10), (4.11), (4.12) it follows that

$$F^o(v) + M(v) \le \frac{1}{1-\eta}[F^o(u) + M(u)].$$

Since $F^o(v) = F^o(u)$, as $\eta \to 0$ we obtain $M(v) \le M(u)$. By changing the role of u and v we obtain (4.7).

To conclude the proof of the theorem we have to show that there exists $\mu \in \mathcal{M}_o$ with $\operatorname{spt}\mu \subset \bar{\Omega}$, such that $M = M^o$ on $L^2(\mathbf{R}^n)$, where M^o is the functional defined in (4.5). To this aim it is enough to prove that there exists $\mu \in \mathcal{M}_o$ such that

$$(4.13) \qquad M(u) = \int_{\mathbf{R}^n} \tilde{u}^2 d\mu$$

for every $u \in H^1(\mathbf{R}^n)$. In fact, by (2.3) it is not restrictive to assume that $\mu \in \mathcal{M}_o^*$. Moreover, from (4.7) and (4.13) it follows that

$$M(u) = \int_{\mathbf{R}^n} \tilde{v}^2 d\mu$$

for every $u \in L^2(\mathbf{R}^n)$ with $u|_\Omega \in H^1(\Omega)$ and for every $v \in H^1(\mathbf{R}^n)$ which extends $u|_\Omega$. Therefore $\mathrm{spt}\mu \subset \bar\Omega$ and

$$M(u) = \int_{\bar\Omega} \tilde{u}^2 d\mu.$$

The proof of (4.13) will be obtained in the following seven lemmas.

Lemma 4.2. *Let* $u, v \in H^1(\mathbf{R}^n)$ *with* $0 \leq u \leq v$ *a.e. on* $\mathbf{R}^n$. *Then* $M(u) \leq M(v)$.

PROOF. By the definition of Γ-convergence there exist two families (u_ϵ) and (v_ϵ) converging in $L^2(\mathbf{R}^n)$ as $\epsilon \to 0$ to u and v respectively, such that

$$F^o(u) = \lim_{\epsilon \to 0} F^\epsilon(u_\epsilon)$$

and

$$F^o(v) + M(v) = \lim_{\epsilon \to 0}[F^\epsilon(v_\epsilon) + M^\epsilon(v_\epsilon)].$$

Since $u \geq 0$ and $v \geq 0$, we may assume $u_\epsilon \geq 0$ and $v_\epsilon \geq 0$. Then the relations

$$F^\epsilon(u_\epsilon \wedge v_\epsilon) + F^\epsilon(u_\epsilon \vee v_\epsilon) = F^\epsilon(u_\epsilon) + F^\epsilon(v_\epsilon)$$

$$M^\epsilon(u_\epsilon \wedge v_\epsilon) \leq M^\epsilon(v_\epsilon)$$

imply

$$F^o(u) + M(u) + F^o(v) \leq \liminf_{\epsilon \to 0}[F^\epsilon(u_\epsilon \wedge v_\epsilon) + M^\epsilon(u_\epsilon \wedge v_\epsilon)] +$$
$$+ \liminf_{\epsilon \to 0} F^\epsilon(u_\epsilon \vee v_\epsilon) \leq \liminf_{\epsilon \to 0}[F^\epsilon(u_\epsilon) + F^\epsilon(v_\epsilon) + M^\epsilon(v_\epsilon)] =$$
$$= F^o(u) + F^o(v) + M(v),$$

which concludes the proof of the lemma.

Let H_M be the class of all quasi-continuous Borel functions $f : \mathbf{R}^n \to \mathbf{R}$ such that there exists $u \in H^1(\mathbf{R}^n)$ with $\tilde{u} = f$ q.e. on $\mathbf{R}^n$ and $M(u) < +\infty$. Since the function $u \in H^1(\mathbf{R}^n)$ associated to $f \in$

H_M is unique, M can be defined on H_M by setting $M(f) = M(u)$. For every $f, g \in H_M$ we define

$$B(f,g) = \frac{1}{2}[M(f+g) - M(f) - M(g)].$$

Lemma 4.3. *The class H_M is a Riesz space, B is a symmetric bilinear form on H_M, and $M(f) = B(f, f)$ for every $f \in H_M$.*

PROOF. From Proposition V of [26] it follows that for every $u, v \in H^1(\mathbf{R}^n)$ and $t \in \mathbf{R}$ the functional M satisfies the conditions

$$(4.14) \qquad M(u) \geq 0, \qquad M(0) = 0, \qquad M(tu) = t^2 M(u);$$

$$(4.15) \qquad M(u+v) + M(u-v) = 2[M(u) + M(v)].$$

Hence H_M is a vector space. To prove that H_M is a Riesz space, it is enough to show that $M(u^+) < +\infty$ for every $u \in H^1(\mathbf{R}^n)$ with $M(u) < +\infty$. For, let $u \in H^1(\mathbf{R}^n)$ with $M(u) < +\infty$, and let $u_\epsilon \to u$ in $L^2(\mathbf{R}^n)$ be such that

$$F^o(u) + M(u) = \lim_{\epsilon \to 0}[F^\epsilon(u_\epsilon) + M^\epsilon(u_\epsilon)].$$

Since (u_ϵ^+) converges to u^+ in $L^2(\mathbf{R}^n)$ as $\epsilon \to 0$, we have

$$F^o(u^+) + M(u^+) \leq \liminf_{\epsilon \to 0}[F^\epsilon(u_\epsilon^+) + M^\epsilon(u_\epsilon^+)]$$
$$\leq \lim_{\epsilon \to 0}[F^\epsilon(u_\epsilon) + M^\epsilon(u_\epsilon)] = F^o(u) + M(u) < +\infty,$$

hence $M(u^+) < +\infty$.

Finally, the fact that B is a symmetric bilinear form on H_M follows from (4.14) and (4.15) by the standard algebraic manipulations which show that any norm which satisfies the parallelogram identity comes from a scalar product (see for instance [28], Chapter 1, Section 5, Theorem 1).

Lemma 4.4. *For every $f, g \in H_M$ with $f \geq 0$ and $g \geq 0$ we have $B(f, g) \geq 0$.*

PROOF. Since B is a bilinear form, it is

$$B(f,g) = \lim_{t \to 0} \frac{M(f + tg) - M(f)}{2t},$$

so that the assertion follows immediately from Lemma 4.2.

Lemma 4.5. *Let* $u \in H^1(\mathbf{R}^n)$ *and let* (u_h) *be an increasing sequence of nonnegative functions of* $H^1(\mathbf{R}^n)$ *such that* $(\tilde{u}_h)$ *converges to* $\tilde{u}$ *q.e. on* $\mathbf{R}^n$. *Then* $M(u) = \lim_{h \to \infty} M(u_h)$.

PROOF. The inequality $M(u) \geq \limsup_{h \to \infty} M(u_h)$ follows from Lemma 4.2. By Lemma 1.6 of [12] there exists an increasing sequence (v_h) converging to u strongly in $H^1(\mathbf{R}^n)$ such that $0 \leq v_h \leq u_h$ a.e. on $\mathbf{R}^n$. Hence

$$M(u) \leq \liminf_{h \to \infty} M(v_h) \leq \liminf_{h \to \infty} M(u_h)$$

by Lemma 4.2 and by the lower semicontinuity of M in $H^1(\mathbf{R}^n)$ (recall that the functional G in (4.6) is lower semicontinuous on $L^2(\mathbf{R}^n)$).

Lemma 4.6. *Let* $f, g \in H_M$ *with* $f \geq 0$, *and let* (g_h) *be an increasing sequence in* H_M *such that* $g = \sup\{g_h : h \in \mathbf{N}\}$. *Then*

$$B(f,g) = \lim_{h \to \infty} B(f, g_h).$$

PROOF. It is not restrictive to assume $g \geq 0$ and $g_h \geq 0$. By Schwarz inequality we have

$$|B(f, g_h - g)|^2 \leq M(f)M(g_h - g),$$

so that the lemma is proved if we show that $\lim_{h \to \infty} M(g_h - g) = 0$. By (4.15) we have

$$M(g_h - g) = 2M(g) + 2M(g_h) - M(g_h + g),$$

so that, using Lemma 4.5

$$\lim_{h \to \infty} M(g_h - g) = 4M(g) - M(2g) = 0.$$

Lemma 4.7. *Let* $f, g \in H_M$ *with* $|f| \wedge |g| = 0$. *Then* $B(f, g) = 0$.

PROOF. It is not restrictive to assume $f \geq 0$ and $g \geq 0$. In this case, by Lemma 4.4 we have $B(f, g) \geq 0$. To prove the opposite inequality, by the definition of B, we have to prove that

$$M(u + v) \leq M(u) + M(v)$$

for every $u, v \in H^1(\mathbf{R}^n)$ with $u \geq 0, v \geq 0$ and $u \wedge v = 0$ a.e. on $\mathbf{R}^n$. Let us fix u and v as required. By the definition of Γ-convergence, there exist two families (u_ϵ) and (v_ϵ) of non-negative functions converging in $L^2(\mathbf{R}^n)$ as $\epsilon \to 0$ to u and v respectively, such that

$$F^o(u) + M(u) = \lim_{\epsilon \to 0} [F^\epsilon(u_\epsilon) + M^\epsilon(u_\epsilon)]$$

$$F^o(v) + M(v) = \lim_{\epsilon \to 0} [F^\epsilon(v_\epsilon) + M^\epsilon(v_\epsilon)].$$

Since $(u_\epsilon \vee v_\epsilon)$ converges to $u \vee v = u + v$ in $L^2(\mathbf{R}^n)$ as $\epsilon \to 0$, we have

$$F^o(u + v) + M(u + v) \leq \liminf_{\epsilon \to 0} [F^\epsilon(u_\epsilon \vee v_\epsilon) + M^\epsilon(u_\epsilon \vee v_\epsilon)] \leq$$

$$\leq \liminf_{\epsilon \to 0} [F^\epsilon(u_\epsilon) + M^\epsilon(u_\epsilon) + F^\epsilon(v_\epsilon) + M^\epsilon(v_\epsilon)] =$$

$$= F^o(u) + M(u) + F^o(v) + M(v).$$

Since $|u| \wedge |v| = 0$, we have $F^o(u+v) = F^o(u) + F^o(v)$, and the proof is achieved.

Lemma 4.8. *There exists a measure* $\mu \in \mathcal{M}_o$ *such that*

(4.16)
$$M(u) = \int_{\mathbf{R}^n} \tilde{u}^2 d\mu$$

for every $u \in H^1(\mathbf{R}^n)$.

PROOF. By Lemmas 4.3, 4.4, 4.6, 4.7, H_M is a Riesz space which satisfies the Stone condition (8.2), and the bilinear form B is symmetric, positive, local, and continuous on monotone sequences (see the definitions in the Appendix). It is easy to see that the monotone class $\hat{H}_M$ generated by H_M still satisfies condition (8.2). Therefore by Proposition 8.6 and Theorem 8.7 of the Appendix, the extension $\hat{B}$ of B to $\hat{H}_M \times \hat{H}_M$ can be represented in the form

$$(4.17) \qquad \hat{B}(f,g) = \int_{\mathbf{R}^n} fg \, d\mu$$

for every $f, g \in \hat{H}_M$, where μ is the measure on the δ − ring $\mathcal{A} = \{A \subset \mathbf{R}^n : 1_A \in \hat{H}_M\}$ defined by

$$\mu(A) = \hat{B}(1_A, 1_A).$$

Note that every Borel set A of capacity zero belongs to $\mathcal{A}$ and $\mu(A) = 0$. Let $\mathcal{A}_\sigma$ be the σ-ring generated by $\mathcal{A}$. Then μ can be extended to a measure defined on the Borel σ-field of $\mathbf{R}^n$, still denoted by μ, such that $\mu(A) = +\infty$ whenever $A \notin \mathcal{A}_\sigma$. From (4.17) it follows

$$(4.18) \qquad M(f) = \int_{\mathbf{R}^n} f^2 d\mu$$

for every $f \in H_M$. Our goal is to extend (4.18) to all functions $u \in H^1(\mathbf{R}^n)$. Fix $u \in H^1(\mathbf{R}^n)$. If $M(u) < +\infty$, there exists $f \in H_M$ such that $\tilde{u} = f$ q.e. on $\mathbf{R}^n$, hence (4.16) follows from (4.18). Suppose now

$$(4.19) \qquad M(u) = +\infty.$$

To prove (4.16) we argue by contradiction. Assume that $\int_{\mathbf{R}^n} \tilde{u}^2 d\mu < +\infty$; then, for every $\epsilon > 0$ it is

$$\mu(\{|\tilde{u}| > \epsilon\}) \le \epsilon^{-2} \int_{\mathbf{R}^n} \tilde{u}^2 d\mu < +\infty,$$

so that the set $A = \{|\tilde{u}| > \epsilon\}$ belongs to $\mathcal{A}_\sigma$. Therefore, by a monotone class argument, there exists an increasing sequence (g_h) in H_M such that $\sup_h g_h = +\infty$ on A. Define

$$f_h = g_h \wedge (|\tilde{u}| - \epsilon)^+;$$

then $f_h \in H_M$ and $\sup_h f_h = (|\tilde{u}| - \epsilon)^+$, so that by Lemma 4.5

$$M((|u| - \epsilon)^+) = \sup_h M(f_h) = \sup_h \int_{\mathbf{R}^n} f_h^2 d\mu = \int_{\mathbf{R}^n} |(|\tilde{u}| - \epsilon)^+|^2 d\mu.$$

By Lemma 4.5, passing to the limit as $\epsilon \to 0$ we get

$$M(|u|) = \int_{\mathbf{R}^n} \tilde{u}^2 d\mu < +\infty.$$

By Lemma 4.2 and by (4.15) we obtain

$$M(u) \le 2[M(u^+) + M(u^-)] \le 4M(|u|) < +\infty$$

which contradicts (4.19).

5. Convergence of resolvents.

In this section we introduce the resolvent operator R_λ^ϵ associated with the measures μ^ϵ, and we study their convergence as $\epsilon \to 0$.

Let (μ^ϵ) be a family of measures of the class $\mathcal{M}_o$ satisfying (4.3), let μ be a measure of the class $\mathcal{M}_o$ supported by $\bar{\Omega}$, and let $F^\epsilon, F^o, M^\epsilon, M^o$ be the functionals on $L^2(\mathbf{R}^n)$ defined by (4.1), (4.2), (4.4), (4.5). In this section we shall prove that $(F^\epsilon + M^\epsilon)$ Γ-converges to $F^o + M^o$ in $L^2(\mathbf{R}^n)$ as $\epsilon \to 0$ if and only if for a given $\lambda > 0$ and for every $g \in L^2(\mathbf{R}^n)$ the family (u_ϵ) of the solutions of the problems formally written as

$$(5.1) \qquad \begin{cases} L^\epsilon u_\epsilon + \mu^\epsilon u_\epsilon + \lambda u_\epsilon = g & \text{in } \mathbf{R}^n \\ u_\epsilon \in H^1(\mathbf{R}^n) \end{cases}$$

converges in $L^2(\mathbf{R}^n)$ to the function u_o defined by

$$(5.2) \qquad u_o = \begin{cases} w_o & \text{in } \Omega \\ \frac{1}{\lambda} g & \text{in } \mathbf{R}^n - \Omega, \end{cases}$$

where $w_o \in H^1(\Omega)$ is the solution of the problem formally written as

$$(5.3) \qquad \begin{cases} L w_o + \mu w_o + \lambda w_o = g & \text{in } \Omega \\ \frac{\partial w_o}{\partial \nu_a} + \mu w_o = 0 & \text{on } \partial\Omega, \end{cases}$$

ν_a being the conormal vector to $\partial\Omega$ relative to the operator L defined in (3.1).

According to Definition 2.2 of [16] we say that u_ϵ is a variational solution of (5.1) if $u_\epsilon \in H^1(\mathbf{R}^n)$, $\tilde{u}_\epsilon \in L^2(\mathbf{R}^n, \mu^\epsilon)$, and

(5.4)
$$\int_{\mathbf{R}^n} [\sum_{i,j=1}^n a_{ij}^\epsilon D_j u_\epsilon D_i v] dx +$$
$$+ \int_{\mathbf{R}^n} \tilde{u}_\epsilon \tilde{v} d\mu^\epsilon + \lambda \int_{\mathbf{R}^n} u_\epsilon v dx = \int_{\mathbf{R}^n} gv dx$$

for every $v \in H^1(\mathbf{R}^n)$ with $\tilde{v} \in L^2(\mathbf{R}^n, \mu^\epsilon)$. It is easy to see that for every $g \in L^2(\mathbf{R}^n)$ the solution u_ϵ of (5.4) exists and is unique, and coincides with the solution of the minimum problem

(5.5)
$$\min \{ \int_{\mathbf{R}^n} [\sum_{i,j=1}^n a_{ij}^\epsilon D_j u D_i u] dx + \int_{\mathbf{R}^n} \tilde{u}^2 d\mu^\epsilon +$$
$$+ \lambda \int_{\mathbf{R}^n} u^2 dx - 2 \int_{\mathbf{R}^n} gu dx : u \in L^2(\mathbf{R}^n) \}.$$

We define the resolvent operator $R_\lambda^\epsilon : L^2(\mathbf{R}^n) \to L^2(\mathbf{R}^n)$ by setting $R_\lambda^\epsilon(g) = u_\epsilon$.

We say that w_o is a variational solution of (5.3) if $w_o \in H^1(\Omega)$, $\tilde{w}_o \in L^2(\bar{\Omega}, \mu)$ and

(5.6)
$$\int_\Omega [\sum_{i,j=1}^n a_{ij} D_j w_o D_i v] dx +$$
$$+ \int_{\bar{\Omega}} \tilde{w}_o \tilde{v} d\mu + \lambda \int_\Omega uv dx = \int_\Omega gv dx$$

for every $v \in H^1(\Omega)$ with $\tilde{v} \in L^2(\bar{\Omega}, \mu)$. Again, for every $g \in L^2(\mathbf{R}^n)$ such a solution exists and is unique, and it is obtained as the restriction of the (unique) minimum point of the problem (5.5) with $\epsilon = 0$. Let us note that this minimizer is given by (5.2). For $\epsilon = 0$ the resolvent operator $R_\lambda^o : L^2(\mathbf{R}^n) \to L^2(\mathbf{R}^n)$ is defined by $R_\lambda^0(g) = u_o$.

According to (5.4) and (5.6), the solutions of (5.1) and (5.3) are not taken in the usual distributional sense, because of the constraints $\tilde{v} \in L^2(\mathbf{R}^n, \mu^\epsilon)$ and $\tilde{v} \in L^2(\bar{\Omega}, \mu)$ on the test functions (see [15], Section 3).

For every $\epsilon \geq 0$ and $\lambda > 0$ we denote by $m_\lambda^\epsilon(g)$ the minimum value of the problem (5.5), and by $Y_\lambda^\epsilon(g)$ the Moreau-Yosida approximation of $F^\epsilon + M^\epsilon$ given by

(5.7)
$$Y_\lambda^\epsilon(g) = \min \left\{ F^\epsilon(u) + M^\epsilon(u) + \lambda \int_{\mathbf{R}^n} |u - g|^2 dx : u \in L^2(\mathbf{R}^n) \right\}.$$

We are now in a position to state the following theorem.

Theorem 5.1. *For every $\lambda > 0$ the following conditions are equivalent:*
(a) $F^\epsilon + M^\epsilon$ *Γ-converges to $F^o + M^o$ in $L^2(\mathbf{R}^n)$ as $\epsilon \to 0$;*
(b) $R_\lambda^\epsilon(g)$ *converges to $R_\lambda^o(g)$ strongly in $L^2(\mathbf{R}^n)$ (as $\epsilon \to 0$) for every $g \in L^2(\mathbf{R}^n)$;*
(c) $R_\lambda^\epsilon(g)$ *converges to $R_\lambda^o(g)$ weakly in $L^2(\mathbf{R}^n)$ (as $\epsilon \to 0$) for every $g \in L^2(\mathbf{R}^n)$;*
(d) $m_\lambda^\epsilon(g)$ *converges to $m_\lambda^o(g)$ (as $\epsilon \to 0$) for every $g \in L^2(\mathbf{R}^n)$;*
(e) $Y_\lambda^\epsilon(g)$ *converges to $Y_\lambda^o(g)$ (as $\epsilon \to 0$) for every $g \in L^2(\mathbf{R}^n)$.*

To prove the theorem we need the following lemma.

Lemma 5.2. *Assume that $F^\epsilon + M^\epsilon$ Γ-converges to $F^o + M^o$ in $L^2(\mathbf{R}^n)$ as $\epsilon \to 0$. Then*

(5.8)
$$F^o(u) + M^o(u) \leq \liminf_{\epsilon \to 0} [F^\epsilon(u_\epsilon) + M^\epsilon(u_\epsilon)]$$

for every $u \in L^2(\mathbf{R}^n)$ and for every family (u_ϵ) converging to u weakly in $L^2(\mathbf{R}^n)$.

PROOF. We may assume that the right-hand side of (5.8) is finite, that the lower limit is a limit, and that $u_\epsilon \in H^1(\mathbf{R}^n)$ for every $\epsilon > 0$. Then $(u_\epsilon|_\Omega)$ is bounded in $H^1(\Omega)$, hence $u|_\Omega \in H^1(\Omega)$ and $u_\epsilon|_\Omega$ converges to $u|_\Omega$ weakly in $H^1(\Omega)$. Since Ω has a Lipschitz boundary, there exists $w \in H^1(\mathbf{R}^n)$ such that $w = |u|$ on Ω and $w \geq 0$ on $\mathbf{R}^n$. For every $\epsilon > 0$ we define

$$\psi_\epsilon(x) = \left[1 - \frac{\text{dist}(x, \Omega)}{\sqrt{\epsilon}} \right]^+$$

and $w_\epsilon = (w+1)\psi_\epsilon$. Then $w_\epsilon \in H^1(\mathbf{R}^n), w_\epsilon|_{\mathbf{R}^n-\Omega}$ converges to 0 in $L^2(\mathbf{R}^n - \Omega)$, and

$$(5.9) \qquad \lim_{\epsilon \to 0} \epsilon \int_{\mathbf{R}^n-\Omega} |Dw_\epsilon|^2 dx = 0.$$

Let us define

$$v = \begin{cases} u & \text{on } \Omega, \\ 0 & \text{on } \mathbf{R}^n - \Omega, \end{cases}$$

and

$$v_\epsilon = \begin{cases} -w_\epsilon & \text{where } u_\epsilon < -w_\epsilon, \\ u_\epsilon & \text{where } |u_\epsilon| \le w_\epsilon, \\ w_\epsilon & \text{where } u_\epsilon > w_\epsilon. \end{cases}$$

Then $v_\epsilon \in H^1(\mathbf{R}^n)$, $|v_\epsilon - v| \le |u_\epsilon - u|$ a.e. on Ω, and $|v_\epsilon - v| \le 2w_\epsilon$ a.e. on $\mathbf{R}^n - \Omega$, hence v_ϵ converges to v in $L^2(\mathbf{R}^n)$ as $\epsilon \to 0$. Therefore condition (a) in the definition of Γ-convergence (Section 2) implies

$$(5.10) \quad F^o(u) + M^o(u) = F^o(v) + M^o(v) \le \liminf_{\epsilon \to 0}[F^\epsilon(v_\epsilon) + M^\epsilon(v_\epsilon)].$$

For every $\epsilon > 0$ we have $\tilde{v}_\epsilon^2 \le \tilde{u}_\epsilon^2$ q.e. on $\mathbf{R}^n$, hence

$$(5.11) \qquad\qquad M^\epsilon(v_\epsilon) \le M^\epsilon(u_\epsilon).$$

Moreover, setting $B_\epsilon = \{x \in \Omega : |u_\epsilon(x)| > w(x) + 1\}$, we have

$$F^\epsilon(v_\epsilon) \le \int_\Omega a(x, Du_\epsilon)dx + \Lambda_2 \int_{B_\epsilon} |Dw|^2 dx+$$

$$(5.12) \quad + \epsilon \int_{\mathbf{R}^n-\Omega} a(x, Du_\epsilon)dx + \epsilon\Lambda_2 \int_{\mathbf{R}^n-\Omega} |Dw_\epsilon|^2 dx =$$

$$= F^\epsilon(u_\epsilon) + \Lambda_2 \int_{B_\epsilon} |Dw|^2 dx + \epsilon\Lambda_2 \int_{\mathbf{R}^n-\Omega} |Dw_\epsilon|^2 dx.$$

Since $|u_\epsilon|$ converges to w strongly in $L^2(\Omega)$, the Lebesgue measure of B_ϵ tends to 0 as $\epsilon \to 0$, therefore (5.8) follows easily from (5.9), (5.10), (5.11), (5.12).

Proof of Theorem 5.1. Lemma 5.2 implies that if $(F^\epsilon + M^\epsilon)$ Γ-converges to $F^o + M^o$ in $L^2(\mathbf{R}^n)$ as $\epsilon \to 0$, then $(F^\epsilon + M^\epsilon)$ converges to $F^0 + M^0$ in $L^2(\mathbf{R}^n)$ according to Definition 1.4 of [25]. Therefore the equivalence of (a), (b), and (e) can be obtained as a consequence of the abstract result proved in [3], Theorem 3.26. A more direct proof of this fact and of the implications (b) $\Rightarrow$ (c) $\Rightarrow$ (d) $\Rightarrow$ (e) can be obtained by adapting the proof of Proposition 2.9 of [4], by replacing the weak convergence in $H_o^1(D)$ with the weak convergence in $L^2(\mathbf{R}^n)$.

We now study the resolvent operators $\hat{R}_\lambda^\epsilon$ associated with Dirichlet problems in a family of bounded open domains Ω^ϵ shrinking to Ω. Let (μ^ϵ) be a family in $\mathcal{M}_o$ as stated at the beginning of this section, and let (Ω^ϵ) be a family of bounded open domains of $\mathbf{R}^n$ such that

$$(5.13) \qquad \begin{cases} \Omega \subset \Omega^\epsilon \\ \sup\{\text{dist}(x,\Omega) : x \in \Omega^\epsilon\} \to 0 \quad \text{as } \epsilon \to 0 \end{cases}$$

and

$$(5.14) \qquad \mu^\epsilon \geq \infty_{\partial\Omega^\epsilon},$$

where $\infty_{\partial\Omega^\epsilon}$ is the measure defined in (2.2).

For a given $g \in L^2(\mathbf{R}^n)$ and $\epsilon > 0$ we consider the Dirichlet problem formally written as

$$(5.15) \qquad \begin{cases} L^\epsilon w_\epsilon + \mu^\epsilon w_\epsilon + \lambda w_\epsilon = g & \text{in } \Omega^\epsilon, \\ w_\epsilon = 0 & \text{on } \partial\Omega^\epsilon, \end{cases}$$

where $\lambda \geq 0$. The variational solution w_ϵ is defined similarly to the previous case (5.1) by replacing $\mathbf{R}^n$ with Ω^ϵ and the space $H^1(\mathbf{R}^n)$ with $H_o^1(\Omega^\epsilon)$. Such a solution exists and is unique, and it is obtained by problem (5.5) by replacing again $\mathbf{R}^n$ with Ω^ϵ and $H^1(\mathbf{R}^n)$ with $H_o^1(\Omega^\epsilon)$. As before, we define the resolvent operators $R_\lambda^\epsilon : L^2(\mathbf{R}^n) \to L^2(\mathbf{R}^n)$ by setting

$$\hat{R}_\lambda^\epsilon(g) = \begin{cases} w_\epsilon & \text{on } \Omega^\epsilon, \\ 0 & \text{on } \mathbf{R}^n - \Omega^\epsilon. \end{cases}$$

Let us now consider again problem (5.3) ($\epsilon = 0$) with $\lambda > 0$. Let w_o be the variational solution of (5.3); we define the resolvent operator $\hat{R}_\lambda^o : L^2(\mathbf{R}^n) \to L^2(\mathbf{R}^n)$ by setting

$$(5.16) \qquad \hat{R}_\lambda^o(g) = u_o \quad \text{where } u_o = \begin{cases} w_o & \text{on } \Omega, \\ 0 & \text{on } \mathbf{R}^n - \Omega. \end{cases}$$

Note that w_o is the solution of the minimum problem

$$(5.17) \quad \hat{m}_\lambda^o(g) = \min\{\int_\Omega a(x, Dw)dx + \int_{\bar{\Omega}} \tilde{w}^2 d\mu +$$
$$+ \lambda \int_\Omega w^2 dx - 2 \int_\Omega gw dx : w \in H_0^1(\Omega)\}.$$

Note that problem (5.3) has a solution also in the case $\epsilon = \lambda = 0$, provided the minimum in (5.17) is attained. This happens, for instance, when there exists a constant $c > 0$ such that

$$(5.18) \quad \Omega^\epsilon \subset \{x \in \mathbf{R}^n : \text{dist}(x, \Omega) < c\epsilon\}$$

and $(F^\epsilon + M^\epsilon)$ Γ-converges to $F^o + M^o$ in $L^2(\mathbf{R}^n)$ as $\epsilon \to 0$. In fact the following result holds, where H^{n-1} denotes the $(n-1)$-dimensional Hausdorff measure on $\mathbf{R}^n$.

Proposition 5.3. *Suppose that $(F^\epsilon + M^\epsilon)$ Γ-converges to $F^o + M^o$ in $L^2(\mathbf{R}^n)$ as $\epsilon \to 0$. If (5.13), (5.14), and (5.18) hold, then for every $u \in H^1(\Omega)$*

$$(5.19) \quad \int_{\bar{\Omega}} \tilde{u}^2 d\mu \geq k \int_{\partial\Omega} \tilde{u}^2 dH^{n-1}$$

where $k > 0$ is a constant depending only on $c, \lambda, n,$ and Ω.

To prove the proposition we need the following Poincaré inequalities.

Lemma 5.4. *Assume (5.13) and (5.18). Then there exists a constant $k > 0$ depending only on $c, n,$ and Ω, such that*

$$\int_{\partial\Omega} \tilde{u}^2 dH^{n-1} \leq k\epsilon \int_{\Omega^\epsilon - \Omega} |Du|^2 dx;$$

$$\int_{\Omega^\epsilon - \Omega} u^2 dx \leq k\epsilon^2 \int_{\Omega^\epsilon - \Omega} |Du|^2 dx;$$

$$\int_\Omega u^2 dx \leq k\left\{\int_\Omega |Du|^2 dx + \epsilon \int_{\Omega^\epsilon - \Omega} |Du|^2 dx\right\}$$

for every $\epsilon > 0$ and for every $u \in H_o^1(\Omega^\epsilon)$.

PROOF. This result is proved in [1], Theorem III.3 under the additional hypothesis that Ω has a $C^{1,1}$ boundary. The extension to the Lipschitz continuous case can be obtained by a smooth approximation of the domain Ω or by a Lipschitz change of variables which maps, locally, Ω into a half space.

Proof of Proposition 5.3. Let $u \in L^2(\mathbf{R}^n)$ with $M^o(u) < +\infty$ and let (u_ϵ) be a family in $H^1(\mathbf{R}^n)$ converging to u in $L^2(\mathbf{R}^n)$ such that

$$(5.20) \qquad F^o(u) + M^o(u) = \lim_{\epsilon \to 0}[F^\epsilon(u_\epsilon) + M^\epsilon(u_\epsilon)].$$

Since $M^o(u) < +\infty$, we may assume that $M^\epsilon(u_\epsilon) < +\infty$ for every $\epsilon > 0$, hence $\tilde{u}_\epsilon = 0$ q.e. on $\partial\Omega^\epsilon$, which implies $u_\epsilon|_{\Omega^\epsilon} \in H_o^1(\Omega^\epsilon)$ by Proposition 2.2. By lower semicontinuity we have

$$F^o(u) = \int_\Omega a(x, Du)dx \le \liminf_{\epsilon \to 0} \int_\Omega a(x, Du_\epsilon)dx,$$

so Lemma 5.4 and (5.20) imply

$$\int_{\bar\Omega} \tilde{u}^2 d\mu \ge \liminf_{\epsilon \to 0} \epsilon \int_{\Omega^\epsilon - \Omega} a(x, Du_\epsilon)dx \ge$$

$$\ge \liminf_{\epsilon \to 0} \epsilon\Lambda_1 \int_{\Omega^\epsilon - \Omega} |Du_\epsilon|^2 dx \ge \frac{\Lambda_1}{k} \int_{\partial\Omega} \tilde{u}^2 dH^{n-1},$$

which proves (5.19).

Theorem 5.5. *Assume (5.13) and (5.14). For every $\lambda > 0$ the following conditions are equivalent:*
(a) $F^\epsilon + M^\epsilon$ Γ-converges to $F^o + M^o$ in $L^2(\mathbf{R}^n)$ as $\epsilon \to 0$;
(b) $\hat{R}_\lambda^\epsilon(g)$ converges to $\hat{R}_\lambda^o(g)$ strongly in $L^2(\mathbf{R}^n)$ (as $\epsilon \to 0$) for every $g \in L^2(\mathbf{R}^n)$;
(c) $\hat{R}_\lambda^\epsilon(g)$ converges to $\hat{R}_\lambda^o(g)$ weakly in $L^2(\mathbf{R}^n)$ (as $\epsilon \to 0$) for every $g \in L^2(\mathbf{R}^n)$;
(d) $\hat{m}_\lambda^\epsilon(g)$ converges to $\hat{m}_\lambda^o(g)$ (as $\epsilon \to 0$) for every $g \in L^2(\mathbf{R}^n)$.
The same result holds also for $\lambda = 0$ if (5.18) is satisfied.

PROOF. It is enough to adapt the proof of Proposition 2.9 of [4], by replacing the functionals G_n and G defined before (2.34) by

$$\hat{G}^\epsilon(v) = \begin{cases} F^\epsilon(v) + M^\epsilon(v) + \lambda \int_{\Omega^\epsilon} v^2 dx & \text{if } v = 0 \text{ a.e. on } \mathbf{R}^n - \Omega^\epsilon, \\ +\infty & \text{otherwise in } L^2(\mathbf{R}^n), \end{cases}$$

$$\hat{G}(v) = \begin{cases} F^o(v) + M^o(v) + \lambda \int_\Omega v^2 dx & \text{if } v = 0 \text{ a.e. on } \mathbf{R}^n - \Omega, \\ +\infty & \text{otherwise in } L^2(\mathbf{R}^n), \end{cases}$$

and by using the weak convergence in $L^2(\mathbf{R}^n)$ instead of the weak convergence in $H_o^1(D)$. In the case $\lambda = 0$, the compactness of the family $(\hat{R}_\lambda^\epsilon(g))_{\epsilon>0}$ in the strong topology of $L^2(\mathbf{R}^n)$ can be obtained by using the Poincaré inequalities of Lemma 5.4 as in the proof of Theorem III.3 of [1].

6. Thin layers around the boundary. Let (Ω^ϵ) be a family of bounded open subsets of $\mathbf{R}^n$ such that

$$(6.1) \qquad\qquad \Omega \subset \Omega^\epsilon$$

for every $\epsilon > 0$, and let $d_\epsilon = \sup\{\text{dist}(x, \Omega) : x \in \Omega^\epsilon\}$. We assume that

$$(6.2) \qquad\qquad \lim_{\epsilon \to 0} d_\epsilon = 0.$$

In this section we shall determine some conditions on Ω^ϵ which imply the convergence, as $\epsilon \to 0$, of the solutions w_ϵ of the Dirichlet problem

$$(6.3) \qquad \begin{cases} L^\epsilon w_\epsilon + \lambda w_\epsilon = g & \text{in } \Omega^\epsilon, \\ w_\epsilon \in H_o^1(\Omega^\epsilon), \end{cases}$$

to the solution w_o of the problem formally written as

$$(6.4) \qquad \begin{cases} Lw_o + \lambda w_o = g & \text{in } \Omega, \\ \frac{\partial w_o}{\partial \nu_a} + \mu w_o = 0 & \text{on } \partial\Omega, \end{cases}$$

where $\lambda > 0$, $g \in L^2(\mathbf{R}^n)$, and μ is a measure of the class $\mathcal{M}_o^*$ supported by $\partial\Omega$.

To study this problem on a fixed function space, for every $\epsilon \geq 0$ we consider the functions

$$u_\epsilon = \begin{cases} w_\epsilon & \text{on } \Omega^\epsilon \\ 0 & \text{on } \mathbf{R}^n - \Omega^\epsilon, \end{cases}$$

where w_ϵ is the weak solution of (6.3) for $\epsilon > 0$, $\Omega^o = \Omega$, and w_o is the weak solution of (6.4) according to (5.6), i.e. $w_o \in H^1(\Omega), \tilde{w}_o \in L^2(\partial\Omega, \mu)$, and

$$\int_\Omega [\sum_{i,j=1}^n a_{ij} D_j w_o D_i v]\, dx +$$

$$+ \int_{\partial\Omega} \tilde{w}_o \tilde{v} d\mu + \lambda \int_\Omega w_o v dx = \int_\Omega g v dx$$

for every $v \in H^1(\Omega)$ with $\tilde{v} \in L^2(\partial\Omega, \mu)$.

In order to use the results of Section 5, we introduce the measures $\mu^\epsilon = \infty_{\partial\Omega^\epsilon}$, i.e.

$$(6.5) \qquad \mu^\epsilon(B) = \begin{cases} 0 & \text{if } B \cap \partial\Omega^\epsilon \text{ has capacity zero,} \\ +\infty & \text{otherwise,} \end{cases}$$

and the corresponding functionals M^ϵ defined by (4.4). Moreover, given a measure μ of the class $\mathcal{M}_o$ supported by $\partial\Omega$, we denote by M^o the corresponding functional defined in (4.5). By Theorem 5.5 the family (u_ϵ) converges to u_o in $L^2(\mathbf{R}^n)$ as $\epsilon \to 0$ for every right-hand side $g \in L^2(\mathbf{R}^n)$ if and only if $F^\epsilon + M^\epsilon$ Γ-converges to $F^o + M^o$ in $L^2(\mathbf{R}^n)$ as $\epsilon \to 0$. Moreover, this result can be extended to the case $\lambda = 0$ if (5.18) is satisfied.

Therefore, this section is devoted to find necessary and sufficient conditions on (Ω^ϵ) in order that $F^\epsilon + M^\epsilon$ Γ-converges to $F^o + M^o$. More precisely, we prove a theorem which allows us to calculate the measure μ in terms of the limit, as $\epsilon \to 0$, of the capacities of some subsets of $\partial\Omega^\epsilon$ with respect to the operators L^ϵ.

To determine the Γ-limit of the family $(F^\epsilon + M^\epsilon)$ we fix a family $(\Omega_\delta)_{\delta>0}$ of open subsets of Ω which satisfies conditions (3.9) and (3.10). Let $\epsilon > 0$ and $\delta > 0$. For every bounded open set $W \subset \mathbf{R}^n$ we put $W_\delta^\epsilon = W \cap (\Omega^\epsilon - \bar{\Omega}_\delta)$ and we define

$$(6.6) \qquad b_\delta^\epsilon(W) = \min \{ \int_{W_\delta^\epsilon} a^\epsilon(x, Du) dx : u \in K_\delta^\epsilon \},$$

where K_δ^ϵ is the set of all functions $u \in H^1(W_\delta^\epsilon)$ such that $u = 0$ on $W \cap \partial \Omega^\epsilon$ and $u = 1$ on $W \cap \partial \Omega_\delta$ in the sense of $H^1(W_\delta^\epsilon)$ (see fig. 3). The minimum in (6.6) is clearly achieved by the lower semicontinuity and the coerciveness of the functional (recall that $\epsilon > 0$).

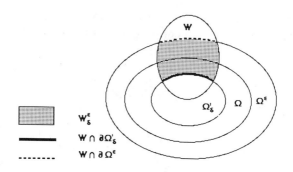

Figure 3

For every pair V, U of bounded open sets with $V \subset U$ we define

$$(6.7) \qquad\qquad b_\delta^\epsilon(V, U) = b_\delta^\epsilon(W)$$

with $W = V \cup (U \cap \Omega)$ (see fig. 4), and

$$(6.8) \qquad\qquad \beta_\delta'(V, U) = \liminf_{\epsilon \to 0} b_\delta^\epsilon(V, U),$$

$$(6.9) \qquad\qquad \beta_\delta''(V, U) = \limsup_{\epsilon \to 0} b_\delta^\epsilon(V, U).$$

We will now define the set functions $\hat{\beta}'(B, U)$ and $\hat{\beta}''(B, U)$ for arbitrary Borel subsets $B \subset U$. We first consider the inner regularizations $\hat{\beta}_\delta'(\cdot, U)$ and $\hat{\beta}_\delta''(\cdot, U)$ of $\beta_\delta'(\cdot, U)$ and $\beta_\delta''(\cdot, U)$, defined for every open set $V \subset U$ by

$$(6.10) \qquad \hat{\beta}_\delta'(V, U) = \sup\{\beta_\delta'(V', U) : V' \text{ open}, \ V' \subset\subset V\},$$

$$(6.11) \qquad \hat{\beta}_\delta''(V, U) = \sup\{\beta_\delta''(V', U) : V' \text{ open}, \ V' \subset\subset V\}.$$

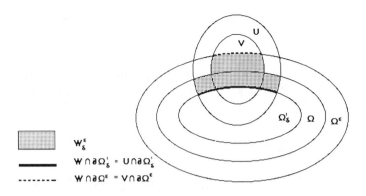

Figure 4

The definition of $\hat{\beta}'_\delta$ and $\hat{\beta}''_\delta$ is extended to arbitrary Borel sets $B \subset U$ by

$$(6.12) \qquad \hat{\beta}'_\delta(B,U) = \inf\{\hat{\beta}'_\delta(V,U) : V \quad \text{open}, \quad B \subset V\}$$

$$(6.13) \qquad \hat{\beta}''_\delta(B,U) = \inf\{\hat{\beta}''_\delta(V,U) : V \quad \text{open}, \quad B \subset V\}.$$

Finally we define

$$(6.14) \qquad \hat{\beta}'(B,U) = \sup\{\hat{\beta}'_\delta(B,U) : \delta > 0\}$$

$$(6.15) \qquad \hat{\beta}''(B,U) = \sup\{\hat{\beta}''_\delta(B,U) : \delta > 0\}$$

for every Borel set $B \subset U$. It is possible to prove that the functions $b^\epsilon_\delta, \beta'_\delta, \beta''_\delta, \hat{\beta}'_\delta, \hat{\beta}''_\delta$ are decreasing in δ. We shall prove the following theorem.

Theorem 6.1. *Assume (6.1) and (6.2). Then $F^\epsilon + M^\epsilon$ Γ-converges to $F^o + M^o$ in $L^2(\mathbf{R}^n)$ as $\epsilon \to 0$ if and only if $\hat{\beta}' = \hat{\beta}''$. In this case the measure μ associated with M^o is supported by $\partial\Omega$, and*

$$(6.16) \qquad \mu^*(B) = \hat{\beta}'(B,U) = \hat{\beta}''(B,U)$$

for every bounded open set $U \subset \mathbf{R}^n$ *and for every Borel set* $B \subset U$.

To prove the theorem we shall show that $F^\epsilon + M^\epsilon$ Γ-converges to $F^o + M^o$ in $L^2(\mathbf{R}^n)$ as $\epsilon \to 0$ if and only if for every $\delta > 0$ and for every open set $V \subset U$ we have

$$\hat{\beta}'_\delta(V,U) = \hat{\beta}''_\delta(V,U) = b^o_\delta(\mu, V, U),$$

where $b^o_\delta(\mu, V, U)$ is the boundary μ-capacity defined in (3.11).

Lemma 6.2. *If* $F^\epsilon + M^\epsilon$ Γ-*converges to* $F^o + M^o$ *in* $L^2(\mathbf{R}^n)$ *as* $\epsilon \to 0$, *then* $\mathrm{spt}\mu \subset \partial\Omega$.

PROOF. By Theorem 4.1 we have $\mathrm{spt}\mu \subset \bar{\Omega}$; moreover, for every $u \in H^1_o(\Omega)$

$$\int_{\mathbf{R}^n} \tilde{u}^2 d\mu \leq \liminf_{\epsilon \to 0}[F^\epsilon(u) + M^\epsilon(u)] - F^o(u),$$

and, since $M^\epsilon(u) = 0$ for all $\epsilon > 0$, we get

$$\int_{\mathbf{R}^n} \tilde{u}^2 d\mu = 0,$$

for every $u \in H^1_o(\Omega)$, so that $\mu(\Omega) = 0$.

Lemma 6.3. *Suppose that* $F^\epsilon + M^\epsilon$ Γ-*converges to* $F^o + M^o$ *in* $L^2(\mathbf{R}^n)$ *as* $\epsilon \to 0$. *Let* U *and* V *be two bounded open subsets of* $\mathbf{R}^n$ *with* $V \subset U$. *Then*

(6.17)
$$\int_{U \cap \Omega} a(x, Du)dx + \int_{V \cap \bar{\Omega}} \tilde{u}^2 d\mu \leq$$
$$\leq \liminf_{\epsilon \to 0} \left\{ \int_U a^\epsilon(x, Du_\epsilon)dx + \int_V \tilde{u}^2_\epsilon d\mu^\epsilon \right\}$$

for every family (u_ϵ) *in* $H^1(U)$ *which converges weakly in* $L^2(U)$ *to a function* u *such that* $u|_{U \cap \Omega} \in H^1(U \cap \Omega)$.

PROOF. Wihout loss of generality we may assume that the right-hand side of (6.17) is finite and that the lower limit is a limit, so there exists $c \in \mathbf{R}$ such that

$$(6.18) \qquad \int_{U \cap \Omega} |Du_\epsilon|^2 dx + \epsilon \int_{U-\Omega} |Du_\epsilon|^2 dx \le c$$

for every $\epsilon > 0$. This implies that $u_\epsilon|_{U \cap \Omega}$ converges to $u|_{U \cap \Omega}$ weakly in $H^1(U \cap \Omega)$, and $\sqrt{\epsilon}u_\epsilon|_{U-\bar{\Omega}}$ converges to 0 weakly in $H^1(U - \bar{\Omega})$ as $\epsilon \to 0$.

Let K be a compact subset of V, and let ϕ be a function of $C_o^1(V)$ with $0 \le \phi \le 1$ on V and $\phi = 1$ on a neighbourhood of K. We define $v = \phi u$ and $v_\epsilon = \phi u_\epsilon$. Then v_ϵ converges to v weakly in $L^2(\mathbf{R}^n)$, thus by Lemma 5.2

$$(6.19) \qquad F^o(v) + M^o(v) \le \liminf_{\epsilon \to 0}[F^\epsilon(v_\epsilon) + M^\epsilon(v_\epsilon)].$$

Let us denote by $\hat{a}(x, \xi, \eta)$ the bilinear form associated with $a(x, \xi)$, defined by

$$\hat{a}(x, \xi, \eta) = \sum_{i,j=1}^n a_{ij}(x)\xi_j\eta_i$$

for every $x, \xi, \eta \in \mathbf{R}^n$. Then

$$F^o(v) + M^o(v) =$$
$$(6.20) \qquad = \int_{V \cap \Omega} a(x, D\phi)u^2 dx + 2\int_{V \cap \Omega} \hat{a}(x, D\phi, Du)\phi u dx +$$
$$+ \int_{V \cap \Omega} a(x, Du)\phi^2 dx + \int_{V \cap \bar{\Omega}} \phi^2 \tilde{u}^2 d\mu$$

and

$$F^\epsilon(v_\epsilon) + M^\epsilon(v_\epsilon) = \int_{V \cap \Omega} a(x, D\phi)u_\epsilon^2 dx +$$
$$+ 2\int_{V \cap \Omega} \hat{a}(x, D\phi, Du_\epsilon)\phi u_\epsilon dx + \int_{V \cap \Omega} a(x, Du_\epsilon)\phi^2 dx$$
$$(6.21)$$
$$+ \int_{V \cap \bar{\Omega}} \phi^2 \tilde{u}_\epsilon^2 d\mu^\epsilon + \epsilon \int_{V-\bar{\Omega}} a(x, Du_\epsilon)\phi^2 dx$$
$$+ 2\epsilon \int_{V-\bar{\Omega}} \hat{a}(x, D\phi, Du_\epsilon)\phi u_\epsilon dx + \epsilon \int_{V-\bar{\Omega}} a(x, D\phi)u_\epsilon^2 dx.$$

Since $u_\epsilon|_{V\cap\Omega}$ converges to $u|_{V\cap\Omega}$ weakly in $H^1(V\cap\Omega)$ as $\epsilon \to 0$, the first two terms in the right-hand side of (6.21) converge to the corresponding terms in (6.20) as $\epsilon \to 0$. Since $\sqrt{\epsilon}u_\epsilon|_{V-\bar{\Omega}}$ converges to 0 weakly in $H^1(V-\bar{\Omega})$ as $\epsilon \to 0$, the last two terms in the right-hand side of (6.21) converge to 0 as $\epsilon \to 0$. Therefore (6.19), (6.20), and (6.21) imply that

(6.22)
$$\int_{U\cap\Omega} a(x,Du)\phi^2 dx + \int_{K\cap\bar{\Omega}} \tilde{u}^2 d\mu \le$$
$$\le \liminf_{\epsilon\to 0} \left\{ \int_U a^\epsilon(x,Du_\epsilon)\phi^2 dx + \int_V \tilde{u}_\epsilon^2 d\mu^\epsilon \right\}.$$

By lower semicontinuity we also have

(6.23) $$\int_{U\cap\Omega} a(x,Du)(1-\phi^2)dx \le$$
$$\le \liminf_{\epsilon\to 0} \int_{U\cap\Omega} a(x,Du_\epsilon)(1-\phi^2)dx$$
$$\le \liminf_{\epsilon\to 0} \int_U a^\epsilon(x,Du_\epsilon)(1-\phi^2)dx.$$

By adding (6.22) and (6.23), and by taking the limit as $K \uparrow V$ we obtain (6.17).

Lemma 6.4. *Suppose that $F^\epsilon + M^\epsilon$ Γ-converges to $F^o + M^o$ in $L^2(\mathbf{R}^n)$ as $\epsilon \to 0$. Let $\delta > 0$ and let U be a bounded open subset of $\mathbf{R}^n$. Then*
$$b^o_\delta(\mu,V,U) \le \beta'_\delta(V,U)$$
for every open set $V \subset U$.

PROOF. Let V be an open subset of U, and let $W = V\cup(U\cap\Omega)$. Possibly passing to a subsequence we may assume that

(6.24) $$\lim_{\epsilon\to 0} b^\epsilon_\delta(V,U) \le \beta'_\delta(V,U) < +\infty.$$

For every $\epsilon > 0$ there exists $w_\epsilon \in H^1(W^\epsilon_\delta)$ such that

(6.25) $$b^\epsilon_\delta(V,U) = \int_{W^\epsilon_\delta} a^\epsilon(x,Dw_\epsilon)dx,$$

$w_\epsilon = 0$ on $W \cap \partial\Omega^\epsilon = V \cap \partial\Omega^\epsilon$, and $w_\epsilon = 1$ on $W \cap \partial\Omega_\delta = U \cap \partial\Omega_\delta$ in the sense of $H^1(W_\delta^\epsilon)$. By a truncation argument we can prove that $0 \le w_\epsilon \le 1$ a.e. on W_δ^ϵ. Let

$$W_\delta = W - \bar{\Omega}_\delta = (W - \Omega^\epsilon) \cup W_\delta^\epsilon.$$

Since $w_\epsilon = 0$ on $W \cap \partial\Omega^\epsilon$ in the sense of $H^1(W_\delta^\epsilon)$, we can extend w_ϵ to a function of $H^1(W_\delta)$, still denoted by w_ϵ, by setting $w_\epsilon = 0$ on $W - \Omega^\epsilon$. It is easy to see that $\tilde{w}_\epsilon = 0$ q.e. on $W \cap \partial\Omega^\epsilon = W_\delta \cap \partial\Omega^\epsilon$. Since (w_ϵ) is bounded in $L^\infty(W_\delta)$, by extracting a subsequence we may assume that (w_ϵ) converges weakly in $L^2(W_\delta)$ to a function $w \in L^2(W_\delta)$. Let

$$U_\delta^o = U \cap (\Omega - \bar{\Omega}_\delta) = W_\delta^\epsilon \cap (\Omega - \bar{\Omega}_\delta),$$

let $u_\epsilon = w_\epsilon|_{U_\delta^o}$, and let $u = w|_{U_\delta^o}$. By (6.24) and (6.25) the family (u_ϵ) is bounded in $H^1(U_\delta^o)$, hence $u \in H^1(U_\delta^o)$ and (u_ϵ) converges to u weakly in $H^1(U_\delta^o)$. Since $u_\epsilon = 1$ on $U \cap \partial\Omega_\delta$ in the sense of $H^1(U_\delta^o)$, we have also $u = 1$ on $U \cap \partial\Omega_\delta$ in the sense of $H^1(U_\delta^o)$. Since $\tilde{w}_\epsilon = 0$ q.e. on $W_\delta \cap \partial\Omega^\epsilon$, by (6.24), (6.25), and by Lemma 6.3 we have

$$b_\delta^o(\mu, V, U) \le \int_{U_\delta^o} a(x, Du)dx + \int_{V \cap \partial\Omega} \tilde{u}^2 d\mu$$

$$\le \int_{W_\delta \cap \Omega} a(x, Dw)dx + \int_{W_\delta \cap \bar{\Omega}} \tilde{w}^2 d\mu$$

$$\le \liminf_{\epsilon \to 0} \{ \int_{W_\delta} a^\epsilon(x, Dw_\epsilon)dx + \int_{W_\delta} \tilde{w}_\epsilon^2 d\mu^\epsilon \} =$$

$$= \liminf_{\epsilon \to 0} \int_{W_\delta^\epsilon} a^\epsilon(x, Dw_\epsilon)dx = \beta_\delta'(V, U),$$

which concludes the proof of the lemma.

Lemma 6.5. *Suppose that $F^\epsilon + M^\epsilon$ Γ-converges to $F^o + M^o$ in $L^2(\mathbf{R}^n)$ as $\epsilon \to 0$. Let U, V be two bounded open subsets of $\mathbf{R}^n$ with $V \subset U$ and let K be a compact subset of V. Then for every $u \in H^1(U)$ there exists a family (u_ϵ) in $H^1(U)$ converging to u in $L^2(U)$ such that*

$$\int_{U \cap \Omega} a(x, Du)dx + \int_{V \cap \bar{\Omega}} \tilde{u}^2 d\mu \ge$$

$$\ge \limsup_{\epsilon \to 0} \{ \int_U a^\epsilon(x, Du_\epsilon)dx + \int_K \tilde{u}_\epsilon^2 d\mu^\epsilon \}$$

and $u_\epsilon - u \in H^1_o(U)$ *for every $\epsilon > 0$.*

PROOF. It is enough to adapt the proof of [13], Lemma 5.6.

Lemma 6.6. *Under the hypotheses of Lemma 6.4 we have*

$$b^o_\delta(\mu, V, U) \geq \beta''_\delta(V', U)$$

for every pair V, V' of open sets with $V' \subset\subset V \subset U$.

PROOF. Let V, V' be two open sets with $V' \subset\subset V \subset U$. By (3.11) there exists $u \in H^1(U^o_\delta)$ such that

(6.26) $$b^o_\delta(\mu, V, U) = \int_{U^o_\delta} a(x, Du)dx + \int_{V\cap\partial\Omega} \tilde{u}^2 d\mu$$

and $u = 1$ on $U \cap \partial\Omega_\delta$ in the sense of $H^1(U^o_\delta)$. Let V_1 and V_2 be two open sets with $V' \subset\subset V_1 \subset\subset V_2 \subset\subset V$, let $W_1 = V_1 - \bar{\Omega}_{\delta/2}$, and let $W_2 = (V_2 - \bar{\Omega}_\delta) \cup U^o_\delta = (V_2 - \Omega) \cup U^o_\delta$. Since Ω has a Lipschitz boundary, by Proposition 2.1 there exists $w \in H^1(W_2)$ such that $w = u$ on U^o_δ. Since $W_1 \subset\subset W_2$, by Lemma 6.5 there exists a family (w_ϵ) in $H^1(W_2)$ converging to w in $L^2(W_2)$ such that $w_\epsilon - w \in H^1_o(W_2)$ for every $\epsilon \to 0$ and

(6.27)
$$\limsup_{\epsilon\to 0}\{ \int_{W_2} a^\epsilon(x, Dw_\epsilon)dx + \int_{W_1} \tilde{w}^2_\epsilon d\mu^\epsilon \} \leq$$
$$\leq \int_{W_2\cap\Omega} a(x, Dw)dx + \int_{W_2\cap\bar{\Omega}} \tilde{w}^2 d\mu.$$

By Lemma 6.2 we have

(6.28)
$$\int_{W_2\cap\Omega} a(x, Dw)dx + \int_{W_2\cap\bar{\Omega}} \tilde{w}^2 d\mu =$$
$$= \int_{U^o_\delta} a(x, Dw)dx + \int_{V_2\cap\partial\Omega} \tilde{w}^2 d\mu \leq$$
$$\leq \int_{U^o_\delta} a(x, Du)dx + \int_{V\cap\partial\Omega} \tilde{u}^2 d\mu =$$
$$= b^o_\delta(\mu, V, U) < +\infty,$$

thus (6.27) implies that

$$\int_{W_1} \tilde{w}_\epsilon^2 d\mu^\epsilon < +\infty$$

for ϵ sufficiently small. By (6.5) we have $\tilde{w}_\epsilon = 0$ q.e. on $W_1 \cap \partial\Omega^\epsilon = V_1 \cap \partial\Omega_\epsilon$, hence $w_\epsilon = 0$ on $\bar{V}' \cap \partial\Omega^\epsilon$ in the sense of $H^1(W_1)$ by Proposition 2.2. Let $W' = V' \cup (U \cap \Omega)$, let $(W')_\delta^\epsilon = W' \cap (\Omega^\epsilon - \bar{\Omega}_\delta)$, and let $u_\epsilon = w_\epsilon|_{(W')_\delta^\epsilon}$. Then $u_\epsilon = 0$ on $W' \cap \partial\Omega^\epsilon = V' \cap \partial\Omega^\epsilon$ and $u_\epsilon = u = 1$ on $W' \cap \partial\Omega_\delta = U \cap \partial\Omega_\delta$ in the sense of $H^1((W')_\delta^\epsilon)$, therefore

$$(6.29) \qquad \begin{aligned} b_\delta^\epsilon(V',U) &\leq \int_{(W')_\delta^\epsilon} a^\epsilon(x, Du_\epsilon)dx \\ &\leq \int_{W_2} a^\epsilon(x, Dw_\epsilon)dx + \int_{W_1} \tilde{w}_\epsilon^2 d\mu^\epsilon \end{aligned}$$

for ϵ sufficiently small. The conclusion follows now from (6.27), (6.28), (6.29).

Proof of Theorem 6.1. By Proposition 2.4 and by the compactness Theorem 4.1 we may suppose that $F^\epsilon + M^\epsilon$ Γ-converges in $L^2(\mathbf{R}^n)$ as $\epsilon \to 0$, and we have to identify the Γ-limit by proving (6.16). So we assume that the Γ-limit has the form $F^o + M^o$, where M^o is given by (4.5) and μ is a measure of the class $\mathcal{M}_o^*$ supported by $\bar{\Omega}$. By Lemma 6.2 we have spt$\mu \subset \partial\Omega$.

It remains to prove (6.16). Let U be a bounded open subset of $\mathbf{R}^n$ and let $\delta > 0$. By using Lemma 6.4 and the fact that $b_\delta^o(\mu, \cdot, U)$ is continuous along increasing sequences we have

$$b_\delta^o(\mu, V, U) \leq \hat{\beta}_\delta'(V, U)$$

for every open set $V \subset U$. By Lemma 6.6 we have also

$$\hat{\beta}_\delta''(V, U) \leq b_\delta^o(\mu, V, U),$$

hence

$$\hat{\beta}_\delta'(V, U) = \hat{\beta}_\delta''(V, U) = b_\delta^o(\mu, V, U)$$

for every open set $V \subset U$. By Lemma 3.5 (c) we have

$$\hat{\beta}_\delta'(B, U) = \hat{\beta}_\delta''(B, U) = b_\delta^o(\mu, B, U)$$

for every Borel set $B \subset U$. Therefore (6.16) follows from Theorem 3.6 and from the fact that $\text{spt}\mu \subset \partial\Omega$.

7. The uniform case. In this section we suppose, in addition to (6.1) and (6.2), that there exists $c > 0$ such that

$$(7.1) \qquad \Omega^\epsilon \supset \{x \in \mathbf{R}^n : \text{dist}(x, \Omega) < c\epsilon\}.$$

In this case the limit measure μ can be computed by a simpler procedure.

Let $\epsilon > 0$. For every bounded open set $V \subset \mathbf{R}^n$ we put $V_o^\epsilon = V \cap (\Omega^\epsilon - \bar{\Omega})$ and we define

$$(7.2) \qquad \begin{aligned} b_o^\epsilon(V) &= \min\{\int_{V_o^\epsilon} a^\epsilon(x, Du)dx : u \in K_o^\epsilon\} \\ &= \epsilon \min\{\int_{V_o^\epsilon} a(x, Du)dx : u \in K_o^\epsilon\}, \end{aligned}$$

where K_o^ϵ is the set of all functions $u \in H^1(V_o^\epsilon)$ such that $u = 0$ on $V \cap \partial\Omega^\epsilon$ and $u = 1$ on $V \cap \partial\Omega$ in the sense of $H^1(V_o^\epsilon)$ (see fig. 5). The minimum in (7.2) is clearly achieved by the lower semicontinuity and the coerciveness of the functional.

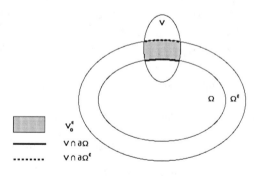

Figure 5

For every bounded open set $V \subset \mathbf{R}^n$ we define

$$(7.3) \qquad \beta_o'(V) = \liminf_{\epsilon \to 0} b_o^\epsilon(V),$$

$$(7.4) \qquad \beta_o''(V) = \limsup_{\epsilon \to 0} b_o^\epsilon(V).$$

We now define the set functions $\hat\beta_o'(B)$ and $\hat\beta_o''(B)$ for arbitrary Borel sets B setting

$$(7.5) \qquad \hat\beta_o'(V) = \sup\{\beta_o'(V') : V' \quad \text{open}, \quad V' \subset\subset V\}$$

$$(7.6) \qquad \hat\beta_o''(V) = \sup\{\beta_o''(V') : V' \quad \text{open}, \quad V' \subset\subset V\},$$

and, for every bounded Borel set $B \subset \mathbf{R}^n$,

$$(7.7) \qquad \hat\beta_o'(B) = \inf\{\hat\beta_o'(V) : B \quad \text{bounded open set}, \quad B \subset\subset V\}$$

$$(7.8) \qquad \hat\beta_o''(B) = \inf\{\hat\beta_o''(V) : V \quad \text{bounded open set}, \quad B \subset\subset V\}.$$

Theorem 7.1. *Assume (6.1), (6.2), and (7.1). Then $F^\epsilon + M^\epsilon$ Γ-converges to $F^o + M^o$ in $L^2(\mathbf{R}^n)$ as $\epsilon \to 0$ if and only if $\hat\beta_o' = \hat\beta_o''$. In this case the measure μ associated with M^o is supported by $\partial\Omega$ and*

$$(7.9) \qquad \mu(B) = \hat\beta_o'(B) = \hat\beta_o''(B)$$

for every bounded Borel set $B \subset \mathbf{R}^n$.

To prove the theorem we need the following lemmas.

Lemma 7.2. *Assume (6.1), (6.2), (7.1). Then the set function $\hat\beta_o''$ is subadditive.*

PROOF. It is enough to prove that

$$(7.10) \qquad \beta_o''(U' \cup V) \le \beta_o''(U) = \beta_o''(V)$$

for every triple U, U', V of bounded open sets with $U' \subset\subset U$. Let us fix U, U', V as required and let $\phi \in C_o^1(U)$ with $0 \le \phi \le 1$ on

U and $\phi = 1$ on U'. For every $\epsilon > 0$ let $U_o^\epsilon = U \cap (\Omega^\epsilon - \bar{\Omega})$ and $V_o^\epsilon = V \cap (\Omega^\epsilon - \bar{\Omega})$. By (7.2) there exist $u_\epsilon \in H^1(U_o^\epsilon)$ and $v_\epsilon \in H^1(V_o^\epsilon)$ such that

$$b_o^\epsilon(U) = \epsilon \int_{U_o^\epsilon} a(x, Du_\epsilon)dx,$$

$$b_o^\epsilon(V) = \epsilon \int_{V_o^\epsilon} a(x, Dv_\epsilon)dx,$$

$u_\epsilon = 0$ on $U \cap \partial\Omega^\epsilon$, $u_\epsilon = 1$ on $U \cap \partial\Omega$ in the sense of $H^1(U_o^\epsilon)$, $v_\epsilon = 0$ on $V \cap \partial\Omega^\epsilon$, and $v_\epsilon = 1$ on $V \cap \partial\Omega$ in the sense of $H^1(V_o^\epsilon)$. By a truncation argument we can prove that $0 \leq u_\epsilon \leq 1$ on U_o^ϵ and $0 \leq v_\epsilon \leq 1$ on V_o^ϵ. Let $W = U' \cup V$, let $W_o^\epsilon = W \cap (\Omega^\epsilon - \bar{\Omega})$, and let $w_\epsilon = \phi u_\epsilon + (1-\phi)v_\epsilon$. Then $w_\epsilon \in H^1(W_o^\epsilon)$, $w_\epsilon = 0$ on $W \cap \partial\Omega^\epsilon$, $w_\epsilon = 1$ on $W \cap \partial\Omega$, hence

$$b_o^\epsilon(W) \leq \epsilon \int_{W_o^\epsilon} a(x, Dw_\epsilon)dx.$$

By convexity, for every $\eta \in]0,1[$ we have

$$a(x, Dw_\epsilon) \leq \frac{\phi}{1-\eta}a(x, Du_\epsilon) + \frac{1-\phi}{1-\eta}a(x, Dv_\epsilon) + \frac{1}{\eta}(u_\epsilon - v_\epsilon)^2 a(x, D\phi),$$

hence

$$b_o^\epsilon(W) \leq \frac{\epsilon}{1-\eta}\int_{U_o^\epsilon} a(x, Du_\epsilon)dx + \frac{\epsilon}{1-\eta}\int_{V_o^\epsilon} a(x, Dv_\epsilon)dx$$

$$+ \frac{\epsilon}{\eta}\Lambda_2 \int_{W_o^\epsilon} |D\phi|^2 dx = \frac{1}{1-\eta}b_o^\epsilon(U) + \frac{1}{1-\eta}b_o^\epsilon(V)$$

$$+ \frac{\epsilon}{\eta}\Lambda_2 \int_{W_o^\epsilon} |D\phi|^2 dx.$$

By letting $\epsilon \to 0$ first, and then $\eta \to 0$, we obtain (7.10).

Lemma 7.3. *Assume (6.1), (6.2), (7.1). Then there exists a constant k, depending on Ω, Λ_2 and c, such that*

$$\hat{\beta}_o''(B) \leq k$$

for every bounded Borel set $B \subset \mathbf{R}^n$.

PROOF. Since $\hat{\beta}''_o$ is an increasing set function, it is enough to prove that

(7.11) $$\hat{\beta}''_o(U) < +\infty$$

for a given bounded open set $U \subset \mathbf{R}^n$ with $\bar{\Omega} \subset U$. For every $\epsilon > 0$ let $u_\epsilon : \mathbf{R}^n \to \mathbf{R}$ be defined by

$$u_\epsilon(x) = \left[1 - \frac{1}{c\epsilon}\mathrm{dist}(x,\Omega)\right]^+.$$

By (7.1) we have $u_\epsilon = 0$ on $U \cap \partial\Omega^\epsilon$ and $u_\epsilon = 1$ on $\partial\Omega$ in the sense of $H^1(U_o^\epsilon)$. By (7.2) we have

$$b_o^\epsilon(U) \leq \int_{U_o^\epsilon} a^\epsilon(x, Du_\epsilon)dx \leq$$

$$\leq \epsilon \int_{\Omega^\epsilon - \Omega} a(x, Du_\epsilon)dx \leq \epsilon\Lambda_2 \left(\frac{1}{c\epsilon}\right)^2 |\Omega^\epsilon - \Omega|.$$

Since Ω has a Lipschitz boundary, there exists a constant c_1 such that $|\Omega^\epsilon - \Omega| \leq c_1\epsilon$, hence there exists a constant c_2 such that

$$b_o^\epsilon(U) \leq c_2$$

for every $\epsilon > 0$, which implies (7.11).

Proof of Theorem 7.1. As in Theorem 6.1 we may assume that $F^\epsilon + M^\epsilon$ Γ-converges to $F^o + M^o$ in $L^2(\mathbf{R}^n)$ as $\epsilon \to 0$, where M^o is given by (4.5) and μ is a measure of the class $\mathcal{M}_o^*$ supported by $\partial\Omega$, and we have only to identify M^o and μ by proving (7.9).

By Theorem 6.1 we have

(7.12) $$\mu^*(B) = \hat{\beta}'(B, U) = \hat{\beta}''(B, U)$$

for every bounded open set $U \subset \mathbf{R}^n$ and for every Borel set $B \subset U$. By (7.12), to prove (7.9) it is enough to show that $\mu = \mu^*$ and

(7.13) $$\hat{\beta}'(B, U) \leq \hat{\beta}'_o(B)$$

(7.14) $$\hat{\beta}''_o(B) \leq \hat{\beta}''(B, U)$$

for every bounded open set $U \subset \mathbf{R}^n$ and for every Borel set $B \subset U$.

It is easy to see that

$$b_\delta^\epsilon(V, U) \le b_o^\epsilon(V)$$

for every $\epsilon > 0$, $\delta > 0$ and for every pair U, V of bounded open sets with $V \subset U \subset \mathbf{R}^n$. By the definition of $\hat{\beta}'$ and $\hat{\beta}_o'$ this inequality implies (7.13). By Lemma 7.3 and by (7.12) and (7.13) we have

$$\mu(B) \le \mu^*(B) \le k < +\infty$$

for every Borel set $B \subset \mathbf{R}^n$. This implies that μ is bounded and therefore $\mu^* = \mu$.

It remains to prove (7.14). Since $\hat{\beta}''$ is subadditive (see Lemma 7.2) and $\hat{\beta}''(\cdot, U)$ is a Borel measure on U supported by $U \cap \partial\Omega$ (see (7.12)), it is enough to prove (7.14) locally, i.e. when U is a small neighbourhood of a point $x_o \in \partial\Omega$. Since Ω has a Lipschitz boundary, by a Lipschitz change of variables we can reduce the problem to the case $x_o = 0$, $U = Q = \{x \in \mathbf{R}^n : |x_i| < 1/2 \text{ for } i = 1, ..., n\}$, and $U \cap \Omega = \{x \in Q : x_n < 0\}$. By (7.1) there exists a constant $k > 0$, depending on c and on the Lipschitz constant of the change of coordinates, such that

$$(7.15) \qquad Q \cap \Omega^\epsilon \supset Q^\epsilon,$$

where $Q^\epsilon = \{x \in Q : x_n < k\epsilon\}$. By (3.9) and (3.10) there exists a decreasing function $\eta :]0, +\infty[\to]0, +\infty[$ converging to 0 as $\delta \to 0$, such that

$$Q \cap \Omega_\delta \supset Q_\delta$$

where $Q_\delta = \{x \in Q : x_n < -\eta(\delta)\}$ (see fig. 6 and fig. 7).

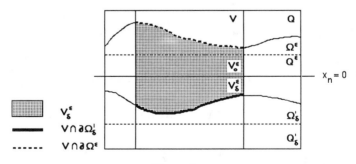

Figure 6

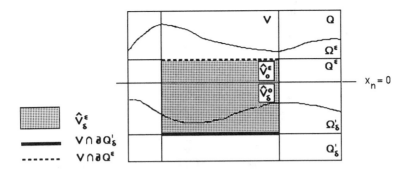

Figure 7

Since $\hat{\beta}''(\cdot, Q)$ is a bounded Borel measure on Q by (7.12), to prove (7.14) it is enough to show that

$$\beta_o''(V') \leq \hat{\beta}''(V, Q)$$

for every pair V, V' of open sets with $V' \subset\subset V \subset Q$, and since $\beta_\delta''(V, V) \leq \beta_\delta''(V, Q)$ for every $\delta > 0$, it is enough to prove that

$$(7.16) \qquad \hat{\beta}_o''(V') \leq \sup\{\beta_\delta''(V', V') : \delta > 0\}$$

for every open set $V' \subset Q$ which can be expressed as a finite union of rectangles of the form $]a_1, b_1[\times...\times]a_n, b_n[$ with a_n and b_n different from 0. Let V' be such a set, let

$$A = \{\hat{x} \in \mathbf{R}^{n-1} : (\hat{x}, 0) \in V'\},$$

and let V be the cylinder defined by

$$V = \{x \in Q : \hat{x} \in A\} = A\times]-1/2, 1/2[$$

where we denote by $\hat{x}$ the projection of x on $\mathbf{R}^{n-1}$ defined by $\hat{x} = (x_1, ..., x_{n-1})$, so that $x = (\hat{x}, x_n)$.

For every $\epsilon > 0$ and every $\delta > 0$ we set $V_\delta^\epsilon = V' \cap (\Omega^\epsilon - \bar{\Omega}_\delta)$ and $\hat{V}_\delta^\epsilon = V' \cap (Q^\epsilon - Q_\delta)$. By the special form of the set V' (in particular, by the hypothesis that a_n and b_n are different from 0) we have $V' \cap (\bar{\Omega}^\epsilon - Q_\delta) = V \cap (\bar{\Omega}^\epsilon - Q_\delta)$ for every ϵ and δ sufficiently small. Therefore we have

$$V_\delta^\epsilon = V \cap (\Omega^\epsilon - \Omega_\delta) \qquad \hat{V}_\delta^\epsilon = V \cap (Q^\epsilon - Q_\delta) = A\times]-\eta(\delta), k\epsilon[$$
$$b_\delta^\epsilon(V') = b_\delta^\epsilon(V) \qquad b_o^\epsilon(V') = b_o^\epsilon(V)$$

for every ϵ and δ sufficiently small, so (7.16) is equivalent to

$$(7.17) \qquad \beta_o''(V) \le \sup\{\beta_\delta''(V,V) : \delta > 0\}.$$

For ϵ and δ sufficiently small we denote by v_δ^ϵ the unique function in $H^1(V_\delta^\epsilon)$ such that

$$(7.18) \qquad b_\delta^\epsilon(V) = \int_{V_\delta^\epsilon} a^\epsilon(x, Dv_\delta^\epsilon)dx,$$

with $v_\delta^\epsilon = 0$ on $V \cap \partial\Omega^\epsilon$, and $v_\delta^\epsilon = 1$ on $V \cap \partial\Omega_\delta$ in the sense of $H^1(V_\delta^\epsilon)$ (see fig. 6). To estimate $b_\delta^\epsilon(V)$ from below we consider the solution w_δ^ϵ of the minimum problem

$$\min\left\{\int_{\hat{V}_\delta^\epsilon} a^\epsilon(x, Du)dx : u \in \hat{K}_\delta^\epsilon\right\},$$

where $\hat{K}_\delta^\epsilon$ is the set of all functions $u \in H^1(\hat{V}_\delta^\epsilon)$ such that $u = 0$ on $V \cap \partial Q^\epsilon$ and $u = 1$ on $V \cap \partial Q_\delta$ in the sense of $H^1(\hat{V}_\delta^\epsilon)$. By a truncation argument we can prove that $0 \le v_\delta^\epsilon \le 1$ a.e. on V_δ^ϵ and $0 \le w_\delta^\epsilon \le 1$ a.e. on $\hat{V}_\delta^\epsilon$, therefore, by the comparison principles for elliptic equations we have

$$(7.19) \qquad w_\delta^\epsilon \le v_\delta^\epsilon \quad \text{a.e. on} \quad \hat{V}_\delta^\epsilon - \bar{\Omega}_\delta = V_\delta^\epsilon \cap Q^\epsilon.$$

Let $\hat{V}_\delta^o = V \cap (\Omega - \bar{Q}_\delta) = A\times] - \eta(\delta), 0[$ and let k the constant occurring in the definition of Q^ϵ in (7.15). Let us prove that

$$(7.20) \qquad \int_{\hat{V}_\delta^o} |Dw_\delta^\epsilon|^2 dx \le \frac{\Lambda_2}{k\Lambda_1}.$$

Let u^ϵ be the function defined by

$$u^\epsilon(x) = 1 \wedge \{1 - \frac{x_n}{k\epsilon}\}^+.$$

Then $u^\epsilon \in H^1(\hat{V}_\delta^\epsilon)$, $u^\epsilon = 0$ on $V \cap \partial Q^\epsilon$, and $u^\epsilon = 1$ on $V \cap \partial Q_\delta$ in the sense of $H^1(\hat{V}_\delta^\epsilon)$, hence the minimality of w_δ^ϵ implies

$$\Lambda_1 \int_{\hat{V}_\delta^o} |Dw_\delta^\epsilon|^2 dx \le \int_{\hat{V}_\delta^o} a(x, Dw_\delta^\epsilon)dx$$

$$\le \int_{\hat{V}_\delta^\epsilon} a^\epsilon(x, Dw_\delta^\epsilon)dx \le \int_{\hat{V}_\delta^\epsilon} a^\epsilon(x, Du^\epsilon)dx$$

$$\le \epsilon\Lambda_2 \int_{\hat{V}_\delta^\epsilon} |Du^\epsilon|^2 dx \le \frac{\Lambda_2}{k},$$

which proves (7.20).

Let $V_o^\epsilon = V \cap (\Omega^\epsilon - \bar{\Omega})$ and $\hat{V}_o^\epsilon = V \cap (Q^\epsilon - \bar{\Omega}) = A \times]0, k\epsilon[$. For every $t \in]0, 1[$ we consider the function $z_\delta^\epsilon \in H_o^1(\hat{V}^\epsilon)$ defined by

$$(7.21) \qquad z_\delta^\epsilon(\hat{x}, x_n) = \frac{1 - w_\delta^\epsilon(\hat{x}, \sigma x_n)}{1 - t} \qquad (\text{where} \quad \sigma = -\frac{\eta(\delta)}{k\epsilon}).$$

Then $z_\delta^\epsilon = 0$ on $V \cap \partial Q^\epsilon$ in the sense of $H^1(\hat{V}_o^\epsilon)$, so we can extend z_δ^ϵ to a function of $H^1(V_o^\epsilon)$ by setting $z_\delta^\epsilon = 0$ on $V_o^\epsilon - \hat{V}_o^\epsilon$. Let u_δ^ϵ be the function on $H^1(V_o^\epsilon)$ defined by

$$u_o^\epsilon = \left[\left(\frac{1}{t} v_\delta^\epsilon \right) \vee z_\delta^\epsilon \right] \wedge 1.$$

Then $u_o^\epsilon = 0$ on $V \cap \partial\Omega^\epsilon$ in the sense of $H^1(V_o^\epsilon)$. Moreover, (7.19) and (7.21) imply that $u_\delta^\epsilon = 1$ on $V \cap \partial\Omega = A \times \{0\}$ in the sense of $H^1(V_o^\epsilon)$. Therefore by (7.2) and (7.18) we have

$$(7.22) \qquad \begin{aligned} b_o^\epsilon(V) &\leq \epsilon \int_{V_o^\epsilon} a(x, Du_\delta^\epsilon)dx \leq t^{-2} \int_{V_\delta^\epsilon} a^\epsilon(x, Dv_\delta^\epsilon)dx \\ &+ \epsilon \int_{\hat{V}_o^\epsilon} a(x, Dz_\delta^\epsilon)dx \leq t^{-2} b_\delta^\epsilon(V) + \epsilon\Lambda_2 \int_{\hat{V}_o^\epsilon} |Dz_\delta^\epsilon|^2 dx. \end{aligned}$$

Let us estimate the last integral. By the definition (7.21) of z_δ^ϵ and by (7.20) we get

$$\int_{\hat{V}_\delta^\epsilon} |Dz_\delta^\epsilon|^2 dx = (1-t)^{-2} \int_A \{ \sum_{i=1}^{n-1} \int_0^{k\epsilon} |D_i w_\delta^\epsilon(\hat{x}, \sigma x_n)|^2 dx_n +$$

$$+ \int_0^{k\epsilon} |D_n w_\delta^\epsilon(\hat{x}, \sigma x_n)|^2 \sigma^2 dx_n \} d\hat{x} =$$

$$= (1-t)^{-2} \int_A \{ \frac{1}{|\sigma|} \sum_{i=1}^{n-1} \int_{-\eta(\delta)}^0 |D_i w_\delta^\epsilon(\hat{x}, x_n)|^2 dx_n +$$

$$+ |\sigma| \int_{-\eta(\delta)}^0 |D_n w_\delta^\epsilon(\hat{x}, x_n)|^2 dx_n \} d\hat{x} \leq$$

$$\leq (n-1)(1-t)^{-2} \frac{k\epsilon}{\eta(\delta)} \int_{\hat{V}_\delta^o} |Dw_\delta^\epsilon|^2 dx +$$

$$+ (1-t)^{-2} \frac{\eta(\delta)}{k\epsilon} \int_{\hat{V}_\delta^\epsilon} |Dw_\delta^\epsilon|^2 dx \leq (n-1)(1-t)^2 \frac{\Lambda_2}{\Lambda_1} \frac{\epsilon}{\eta(\delta)} +$$

$$+ (1-t)^{-2} \frac{\Lambda_2}{\Lambda_1} \frac{\eta(\delta)}{\epsilon} k^{-2},$$

thus (7.22) implies

$$b_o^\epsilon(V) \le t^{-2}b_\delta^\epsilon(V) + (n-1)(1-t)^{-2}\frac{\Lambda_2^2}{\Lambda_1}\frac{\epsilon^2}{\eta(\delta)} + (1-t)^{-2}\frac{\Lambda_2^2}{\Lambda_1}\eta(\delta)k^{-2}.$$

By taking the limit first as $\epsilon \to 0$, then as $\delta \to 0$, and finally as $t \to 1$ we obtain (7.17), which concludes the proof of the theorem.

The measure μ given by Theorem 7.1 has been explicitely computed in some cases of particular interest. Take $a(x,\xi) = |\xi|^2$, $\Omega^\epsilon = \bar{\Omega} \cup \Sigma^\epsilon$, and denote by $\nu(\sigma)$ the outer normal versor at the point $\sigma \in \partial\Omega$. The case

$$\Sigma^\epsilon = \{\sigma + t\nu(\sigma) : \sigma \in \partial\Omega, 0 < t < \epsilon d(\sigma)\}$$

has been considered in [1] when d is a continuous striclty positive function on $\partial\Omega$. Then we have

$$\mu = \frac{1}{d(\sigma)} \cdot H^{n-1}|_{\partial\Omega}$$

where H^{n-1} denotes the $(n-1)$-dimensional Hausdorff measure. The case (take for simplicity the dimension n equal to 2)

$$\Sigma^\epsilon = \left\{\sigma + t\nu(\sigma) : \sigma \in \partial\Omega, 0 < t < \epsilon d\left(\frac{s(\sigma)}{s_\epsilon}\right)\right\}$$

has been considered in [9], where $s(\sigma)$ represents the curvilinear abscissa on $\partial\Omega$, d is a periodic function, and $s_\epsilon \to 0$ with $\frac{\epsilon}{s_\epsilon} \to k$. In this case

$$\mu = c_k \cdot H^{n-1}|_{\partial\Omega},$$

where the constant c_k is given by (set for brevity $L = |\partial\Omega|$)

$$c_k = \min\left\{\frac{1}{L}\int_0^L ds \int_0^{d(s)} \left(k^2\left|\frac{\partial v}{\partial s}\right|^2 + \left|\frac{\partial v}{\partial t}\right|^2\right)dt : \right.$$
$$\left. v(s,0) = 1, \quad v(s,d(s)) = 0, \quad v(0,t) = v(L,t)\right\}.$$

8. Appendix. In all this section, X denotes a fixed set. If $u : X \to \bar{R}$ is a function, we denote by u^- and u^+ the negative and positive parts of u respectively:

$$u^- = (-u) \vee 0 \qquad u^+ = u \vee 0.$$

If F is a class of functions from X into $\bar{\mathbf{R}}$ we set

$$F^- = \{u^- : u \in F\} \qquad F^+ = \{u^+ : u \in F\}.$$

We recall now some standard definitions in measure theory. A *Riesz space* on X is a vector space R of functions from X into $\mathbf{R}$ such that

$$u, v \in R \Rightarrow u \wedge v \in R \quad \text{and} \quad u \vee v \in R.$$

A *monotone class* on X is a family S of functions from X into $\mathbf{R}$ such that

i) if (u_h) is an increasing sequence in S having a majorant in S, then $u = \sup_h u_h$ belongs to S;

ii) if (u_h) is a decreasing sequence in S having a minorant in S, then $u = \inf_h u_h$ belongs to S.

Remark that if R is a Riesz space, then the monotone class generated by R (i.e. the smallest monotone class containing R) is still a Riesz space, which will be denoted by $\hat{R}$.

If R is a Riesz space, a linear form $L : R \to \mathbf{R}$ is said to be *positive* if $L(u) \geq 0$ whenever $u \in R^+$, and *continuous on monotone sequences* if $L(u_h)$ tends to zero whenever $u_h \downarrow 0$ in R.

Analogously, if R_1, R_2 are Riesz spaces, a bilinear form $B : R_1 \times R_2 \to \mathbf{R}$ is said to be positive if $B(u, v) \geq 0$ whenever $u \in R_1^+$ and $v \in R_2^+$, and *continuous on monotone sequences* if for every $u \in R_1$ and $v \in R_2$ the linear forms $B(u, \cdot)$ and $B(\cdot, v)$ are continuous on monotone sequences. It is easy to see that, if B is positive and continuous on monotone sequences, then $B(u_h, v_h)$ tends to zero whenever $u_h \downarrow 0$ in R_1 and $v_h \downarrow 0$ in R_2.

For linear forms on Riesz spaces the following Daniell's extension theorem is classical (see for instance [11], Chapter 3).

Theorem 8.1. *Let R be a Riesz space and let $\hat{R}$ be the monotone class generated by R. Let $L_o : R \to \mathbf{R}$ be a positive linear form continuous on monotone sequences. Then, there exists a unique positive linear form $L : \hat{R} \to R$ continuous on monotone sequences, such that $L = L_o$ on R.*

An analogous result for bilinear forms holds.

Theorem 8.2. *Let R_1 and R_2 be two Riesz spaces and let $\hat{R}_1$ and $\hat{R}_2$ be the monotone classes generated by R_1 and R_2 respectively. Let $B_o : R_1 \times R_2 \to \mathbf{R}$ be a positive bilinear form continuous on monotone sequences. Then, there exists a unique bilinear form $B : \hat{R}_1 \times \hat{R}_2 \to \mathbf{R}$ positive and continuous on monotone sequences, such that $B = B_o$ on $R_1 \times R_2$.*

PROOF. For every $v \in R_2^+$ the linear form $B_o(\cdot, v)$ on R_1 is positive and continuous on monotone sequences. By Theorem 8.1 there exists a unique linear form $B_1(\cdot, v)$ on $\hat{R}_1$ which is positive, continuous on monotone sequences, and extends $B_o(\cdot, v)$. For every $v \in R_2$ we set

$$B_1(\cdot, v) = B_1(\cdot, v^+) - B_1(\cdot, v^-).$$

In this way, we have defined a positive bilinear form $B_1 : \hat{R}_1 \times R_2 \to \mathbf{R}$. Since each element of $\hat{R}_1$ is between two elements of R_1, it is not difficult to prove that B_1 is continuous on monotone sequences.

For every $u \in \hat{R}_1^+$ the linear form $B_1(u, \cdot)$ on R_2 is positive and continuous on monotone sequences. Again by Theorem 8.1 there exists a unique linear form $B(u, \cdot)$ on $\hat{R}_2$ which is positive, continuous on monotone sequences, and extends $B_1(u, \cdot)$. For every $u \in \hat{R}_1$ we set

$$B(u, \cdot) = B(u^+, \cdot) - B(u^-, \cdot).$$

In this way, we have defined a positive bilinear form on $\hat{R}_1 \times \hat{R}_2$. As before, we can prove that B is continuous on monotone sequences and that B is the unique extension of B_o.

Definition 8.3. *Let R be a Riesz spaces; we say a bilinear form $B : R \times R \to \mathbf{R}$ is local if $B(u, v) = 0$ whenever $|u| \wedge |v| = 0$.*

Remark 8.4. It is easy to see that B is local if and only if $B(u, v) = 0$ whenever $u \wedge v = 0$.

Proposition 8.5. *Assume the bilinear form $B : R \times R \to \mathbf{R}$ is symmetric. Then B is local if and only if*

$$(8.1) \qquad B(u \vee v, u \wedge v) = B(u, v) \quad \text{for every} \quad u, v \in R.$$

PROOF. If (8.1) holds, Remark 8.4 implies that B is local. On the contrary, assume that B is local and let $u, v \in R$; since $|u - u \wedge v| \wedge |v - u \wedge v| = 0$ we have

$$B(u - u \wedge v, v - u \wedge v) = 0,$$

so that

$$
\begin{aligned}
B(u \vee v, u \wedge v) &= B(u + v - u \wedge v, u \wedge v) \\
&= B(u \wedge v + (u - u \wedge v) + (v - u \wedge v), u \wedge v) \\
&= B(u \wedge v, u \wedge v) + B(u - u \wedge v, u \wedge v) \\
&\quad + B(v - u \wedge v, u \wedge v) + B(u - u \wedge v, v - u \wedge v) \\
&= B(u, u \wedge v) + B(u, v - u \wedge v) = B(u, v).
\end{aligned}
$$

Proposition 8.6. *Let R be a Riesz space, let $\hat{R}$ be the monotone class generated by R, let $B_o : R \times R \to \mathbf{R}$ be a positive bilinear form continuous on monotone sequences, and let $B : \hat{R} \times \hat{R} \to \mathbf{R}$ be the extension of B_o given by Theorem 8.2. Assume B_o is symmetric and local; then B is symmetric and local.*

PROOF. It follows easily from Proposition 8.5 by a standard monotone classes argument.

Theorem 8.7. *Let S be a monotone Riesz space such that*

(8.2) $\qquad\qquad u \wedge 1 \in S \quad \text{whenever} \quad u \in S,$

and let $B : S \times S \to \mathbf{R}$ be a symmetric positive bilinear form which is local and continuous on monotone sequences. Denote by $\mathcal{E}$ the family

$$\mathcal{E} = \{E \subset X : 1_E \in S\}$$

and set for every $E \in \mathcal{E}$

$$\mu(E) = B(1_E, 1_E).$$

Then, $\mathcal{E}$ is a δ-ring, μ is a measure on $\mathcal{E}$, S is a subset of $L^2(X, \mathcal{E}, \mu)$, and

$$B(u, v) = \int uv \, d\mu \quad \text{for every} \quad u, v \in S.$$

PROOF. We divide the proof into several steps.

Step 1. $u \in S$, $t > 0 \Rightarrow \{u > t\} \in \mathcal{E}$.
In fact, by (8.2), $u \wedge t \in S$ whenever $u \in S$ and $t > 0$, so that

$$(u - t)^+ = u - u \wedge t \in S.$$

Then, $1_{\{u > t\}} = \sup \{h(u - t)^+ \wedge 1 : h \in \mathbf{N}\}$ belongs to S, being majorized by the function $\frac{2}{t}(u - \frac{t}{2})^+ \in S$.

Step 2. $E \in \mathcal{E}$, $u \in S \Rightarrow u \cdot 1_E \in S$.
It is enough to consider only the case $u \in S^+$. In this case it is

$$u \cdot 1_E = \sup \{u \wedge (h \cdot 1_E) : h \in \mathbf{N}\}.$$

Step 3. $E, F \in \mathcal{E}$ with $E \subset F \Rightarrow \mu(E) = B(1_E, 1_F)$.
In fact, by using the locality of B, it is $B(1_E, 1_{F-E}) = 0$, so that

$$\mu(E) = B(1_E, 1_E) = B(1_E, 1_E) + B(1_E, 1_{F-E}) = B(1_E, 1_F).$$

Step 4. μ is a measure on the δ-ring $\mathcal{E}$.
$\mathcal{E}$ is a δ-ring because S is a monotone Riesz space. The finite additivity of μ follows easily from Step 3. To prove that μ is a measure on $\mathcal{E}$, it is enough to observe that

$$\mu(E) = B(1_E, 1_E) = \sup_h B(1_{E_h}, 1_E) = \sup_h B(1_{E_h}, 1_{E_h}) = \sup_h \mu(E_h)$$

for every sequence $E_h \uparrow E$.

Step 5. $B(u, 1_E) = \int_E u \, d\mu$ whenever $u \in S$ and $E \in \mathcal{E}$.

Let $E \in \mathcal{E}$; it is enough to consider only the case $u \in S^+$. By Steps 2, 3, and 4, and by the locality of B, the equality

$$(8.3) \qquad\qquad B(u, 1_E) = \int_E u \, d\mu$$

holds for every $\mathcal{E}$-simple function $u \in S^+$. Since every $u \in S^+$ is the limit of an increasing sequence of $\mathcal{E}$-simple functions in S^+, equality (8.3) holds for every $u \in S^+$.

Step 6. $B(u, v) = \int uv \, d\mu$ for every $u, v \in S$.
It is enough to consider only the case $u, v \in S^+$. Let $u \in S^+$; by Step 5 the equality

$$(8.4) \qquad\qquad B(u, v) = \int uv \, d\mu$$

holds for every $\mathcal{E}$-simple function $v \in S^+$. The same argument of Step 5 yields that (8.4) holds for every $v \in S^+$.

Acknowledgements. The authors have been partially supported by National Research Projects of Ministero della Pubblica Istruzione. The third author wishes to thank Institut für Angewandte Mathematik and SFB-256 of Bonn University for hospitality and support.

References

[1] E.Acerbi, G.Buttazzo, *Reinforcement problems in the calculus of variations*, Ann. Inst. H.Poincaré Anal. Non Linéaire, 4 (1986), 273-284.

[2] R.A.Adams, *Sobolev Spaces*, Academic Press, New York, 1975.

[3] H.Attouch, *Variational Convergence for Functions and Operators*, Pitman, London, 1984.

[4]   J.R.Baxter, G.Dal Maso, U.Mosco, *Stopping times and Γ-convergence*, Trans. Amer. Math. Soc. **303** (1987), 1-38

[5]   H.Brezis, L.Caffarelli, A.Friedman, *Reinforcement problems for elliptic equations and variational inequalities*, Ann. Mat. Pura Appl. **123** (1980), 219-246.

[6]   G.Buttazzo, *Su una definizione generale dei Γ-limiti*, Boll. Un. Mat. Ital. **14-B** (1977), 722-744.

[7]   G.Buttazzo, G.Dal Maso, *Γ-limits of integral functionals*, J.Analyse Math. **37** (1980), 145-185.

[8]   G.Buttazzo, G.Dal Maso, U.Mosco, *A derivation theorem for capacities with respect to a Radon measure*, J. Funct. Anal. **71** (1987), 263-278.

[9]   G.Buttazzo, R.V.Kohn, *Reinforcement by a thin layer with oscillating thickness*, Appl. Math. Optim. **16** (1987), 247-261.

[10] L.Caffarelli, A.Friedman, *Reinforcement problems in elastoplasticity*, Rocky Mountain J. Math **10** (1980), 155-184.

[11] C.Costantinescu, K.Weber, A.Sontag, *Integration Theory, Vol. 1: Measure and Integral*, Wiley, New York 1985.

[12] G.Dal Maso, *On the integral representation of certain local functionals*, Ricerche Mat. **32** (1983), 85-131.

[13] G.Dal Maso, *Γ-convergence and μ-capacities*, Ann Sc. Norm. Sup. Pisa Cl. Sci. (4) **14** (1987), 423-464

[14] G.Dal Maso, L.Modica, *Nonlinear stochastic homogenization*, Ann. Mat. Pura Appl. **144** (1986), 347-389.

[15] G.Dal Maso, U.Mosco, *Wiener's criterion and Γ-convergence*, Appl. Math. Optim. **15** (1987), 15-63.

[16] G. Dal Maso, U. Mosco, *Wiener criteria and energy decay for relaxed Dirichlet problems*, Arch. Rational Mech. Anal. **95** (1986), 345-387.

[17] E.De Giorgi, *G-operators and Γ-convergence*. Proceedings of the "International Congress of Mathematicians", Warsaw 1983, 1175-1191, North Holland, Amsterdam, 1984.

[18] E.De Giorgi, T.Franzoni, *Su un tipo di convergenza variazionale*, Atti Accad. Naz. Lincei Rend. Cl. Sci. Fis. Mat. Natur. **58** (1975), 842-850.

[19] E.De Giorgi, T.Franzoni, *Su un tipo di convergenza variazionale*, Rend. Sem. Mat. Brescia **3** (1979), 63-101.

[20] J.Deny, *Les potentiels d'énergie finie*, Acta Math. **82** (1950), 107-183.

[21] H.Federer, *Geometric Measure Theory*, Springer-Verlag, Berlin, 1969.

[22] H.Federer, W.Ziemer, *The Lebesgue set of a function whose distribution derivatives are p-th power summable*, Indiana Univ. Math. J. **22** (1972), 139-158.

[23] L.I.Hedberg, *Nonlinear potentials and approximations in the mean by analytic functions*, Math. Z. **129** (1972), 299-319.

[24] W.Littman, G.Stampacchia, H.F.Weinberger, *Regular points for elliptic equations with discontinuous coefficients*, Ann. Sc. Norm. Sup. Pisa Cl. Sci. **17** (1963), 41-77.

[25] U.Mosco, *Convergence of convex sets and solutions of variational inequalities*, Adv. in Math. **3** (1969), 510-585.

[26] C.Sbordone, *Su alcune applicazioni di un tipo di convergenza variazionale*, Ann. Sc. Norm. Sup. Pisa Cl. Sci. **2** (1975), 617-638.

[27] G.Stampacchia, *Le problème de Dirichlet pour les équations elliptiques du second ordre à coefficients discontinus*, Ann. Inst. Fourier (Grenoble) **15** (1965), 189-258.

[28] K.Yosida, *Functional Analysis*, Springer-Verlag, Berlin, 1980.

Dipartimento di Matematica
Università di Ferrara
Via Machiavelli 35
I-44100 FERRARA

SISSA
Strada Costiera 11
I-34014 TRIESTE

Dipartimento di Matematica
Università "La Sapienza"
I-00185 ROMA

FUNDAMENTAL INTERIOR ESTIMATES FOR A CLASS OF SECOND ORDER ELLIPTIC OPERATORS

SERGIO CAMPANATO

Dedicated to Ennio De Giorgi on his sixtieth birthday

1. Notations and introduction. Let Ω be an open set in $\mathbf{R}^n$, $n \geq 2$, let $x = (x_1, \ldots, x_n)$ denote a point of Ω and $N > 1$ be an integer, $p = (p^1, \ldots, p^n)$, with $p^i \in \mathbf{R}^N$, denotes a generic vector of $\mathbf{R}^{nN}$. For a vector $u : \Omega \to \mathbf{R}^N$ we set $Du = (D_1 u, \ldots, D_n u)$. If $a(x) : \Omega \to \mathbf{R}^{nN}$ is a vector we set

$$\operatorname{div} a(x) = \sum_{i=1}^{n} D_i a^i(x).$$

Let q be a real number ≥ 2 and if $\xi \in \mathbf{R}^k, k \geq 1$, we set

(1.1)
$$V(\xi) = (1 + \|\xi\|^2)^{1/2}$$
$$W(\xi) = V^{(q-2)/2}(\xi)\xi.$$

Let

$$a(x, u, p) : \Omega \times \mathbf{R}^N \times \mathbf{R}^{nN} \to \mathbf{R}^{nN}$$

be a vector, strictly monotone in p, with non-linearity q [1]. In particular,

(1.2)
$$a(x, u, 0) = 0$$
$$\|a(x, u, p)\| \leq M V^\beta(u) V^{q-2-\beta}(p)\|p\|$$

where
$$M > 0 \quad \text{and} \quad 0 \leq \beta \leq q - 2.$$

Particulary important are the limit cases $\beta = 0$ and $\beta = q - 2$ in which cases we have

(1.3)
$$\|a(x, u, p)\| \leq M V^{q-2}(p)\|p\|$$

(1.4)
$$\|a(x, u, p)\| \leq M V^{q-2}(u)\|p\|.$$

The vector $a(x, u, p)$, in the case (1.3), is totally non-linear and in the case (1.4) the vector $a(x, u, p)$ is quasi-linear. In this case, if $p \to a(x, u, p)$ is of class C^1, we can write

$$a(x, u, p) = V^{q-2}(u)A(x, u)p$$

where $A(x, u)$ is an $nN \times nN$ matrix, which is bounded and elliptic

(1.5)
$$\|A(x, u)\| \leq M$$
$$(A(x, u)\xi|\xi) \geq \nu\|\xi\|^2$$

$\forall \xi \in \mathbf{R}^{nN}$ and $M \geq \nu > 0$.

It is known that the study of $\mathcal{L}^{q,\lambda}$ -regularity, and in particular, the Hölder regularity, of the solutions u of the differential system

(1.6)
$$-\text{div } a(x, u, Du) = b(x, u, Du) \quad \text{in } \Omega$$

makes it necessary to preliminarily obtain the so called "fundamental estimates" for the solutions of the basic system associated to the operator (1.6).

In the case of (1.3), the basic system associated is of the following type

(1.7)
$$\text{div } a(Du) = 0 \quad \text{in } \Omega$$

where $a(p)$ is a vector in $\mathbf{R}^{nN}$ which satisfies the above mentioned hypothesis for the vector a. In particular,

$$a(0) = 0 \quad \text{and} \quad \|a(p)\| \leq M V^{q-2}(p)\|p\|.$$

The solutions of the system (1.7) are vectors $u \in H^{1,q}(\Omega)$ such that

$$(1.8) \qquad \int_\Omega (a(Du)|D\phi)dx = 0, \ \forall \phi \in H_0^{1,q}(\Omega)$$

In the case (1.4), the basic system associated is of the type

$$(1.9) \qquad \begin{aligned} &\text{div } V^{q-2}(u)ADu = 0 \text{ in } \Omega \\ &W(u) \in H^{1,2}(\Omega) \end{aligned}$$

where A is a constant $nN \times nN$ elliptic matrix.

A solutionof the system (1.9) is a vector u such that

$$(1.10) \qquad \begin{aligned} &W(u) \in H^{1,2}(\Omega) \\ &\int_\Omega V^{q-2}(u)(ADu|D\phi)dx = 0, \quad \forall \phi \in {}^*H_0^{1,2}(\Omega) \end{aligned}$$

where we have set [1]

$$(1.11) \qquad \begin{aligned} {}^*H_0^{1,2}(\Omega) = \{\phi \in \mathbf{H}^{1,2}(\Omega) : V^{(q-2)/2}(u)D\phi \in L^2(\Omega), \\ V^{(q-2)/2}(u)\phi \in L^2(\Omega)\} \quad [1]. \end{aligned}$$

The fundamental estimates are concerned either with the vector Du or with the vector u.

For a system of the type (1.7) these estimates are well known in the literature (see for example, [1]) and are of the following kind:

---

[1] See also [3]. One can define in a similar way the space ${}^*H^{1,2}(\Omega)$ and it is evident that we have

$$W(u) \in H^{1,2}(\Omega) \Rightarrow u \in {}^*H^{1,2}(\Omega)$$

For each ball $B(\sigma) \subset\subset \Omega$ and $\forall t \in (0,1)$

$$(1.12) \qquad \int_{B(t\sigma)} \|W(Du)\|^2 dx \leq ct^\lambda \int_{B(\sigma)} \|W(Du)\|^2 dx$$

where $\lambda = \min\{2 + \epsilon, n\}$.

$$(1.13)$$

$$\int_{B(t\sigma)} \|u\|^q dx \leq ct^{\lambda_0} \left\{ \int_{B(\sigma)} \|u\|^q dx + \sigma^q \int_{B(\sigma)} \|W(Du)\|^2 dx \right\}$$

where $\lambda_0 = \min\{q + \lambda, n\}$.

The fundamental estimates are not yet known in the case of systems of the type (1.9) although one can guess what kind of estimates they would have to be.

In this paper we shall obtain the fundamental estimates for the vector u in the particular case of systems (1.9) wherein $A = I_{nN}$.

2. A preliminary result. The exponents λ and λ_o that occur in the fundamental estimates (1.12) and (1.13), depend on the structure of the basic system and are $\leq n$. For example, if the basic system (1.7) is linear

$$a(Du) = ADu$$

where A is a constant $nN \times nN$ elliptic matrix, it is known for a long time that $\lambda = \lambda_o = n$; but, in general, it follows from the definition given in (1.13) that $\lambda_o = n$ only when n is sufficiently small

$$n \leq q + \lambda \qquad (2).$$

We do not know even now whether the same phenomenon holds good also for the fundamental estimates related to the basic system (1.9).

There has been considerable interest in determining when does one have $\lambda_0 = n$ in the fundamental estimate for u. In fact, it is related to the possibility of proving a maximum principle by the method of Cannarsa.

---

[2] This is certainly the case if $n=2,3,4$.

We shall now present a procedure which might be useful in certain circumstances.

We recall the following propositions:

Lemma 2.1. *if $y(\sigma)$, of class C^1, is a solution of the inequation*

$$(2.1) \qquad \sigma y'(\sigma) \geq \alpha y(\sigma), \qquad 0 < \sigma \leq \sigma_o$$

then $\forall \sigma \in (0, \sigma_0]$ and $\forall t \in (0,1)$ we have

$$(2.2) \qquad y(t\sigma) \leq t^\alpha y(\sigma).$$

Infact, setting

$$Y(\sigma) = \sigma^{-\alpha} y(\sigma)$$

the inequation (2.1) becomes

$$Y'(\sigma) \geq 0.$$

Let $G(x)$ be a non negative function on Ω of class $H^{1,1}_{\text{loc}}(\Omega)$. Suppose that, for every ball $B(\sigma) \subset\subset \Omega$ and for any function $\theta(x) \in C^\infty(\overline{B(\sigma)})$, non negative in $B(\sigma)$ and zero on $\partial B(\sigma)$, we have

$$(2.3) \qquad \int_{B(\sigma)} (DG(x)|D\theta(x))dx \leq 0$$

then one can prove the following lemma.

Lemma 2.2. *For each ball $B(\sigma) = B(x^o, \sigma) \subset\subset \Omega$ and $\forall t \in (0,1)$ we have the estimate*

$$(2.4) \qquad \int_{B(t\sigma)} G(x)dx \leq t^n \int_{B(\sigma)} G(x)dx$$

Infact, choosing

$$\theta(x) = \sigma^2 - \|x - x^o\|^2$$

and denoting by $\nu(x)$ the external normal vector at the point $x \in \partial B(\sigma)$ we obtain from (2.3)

$$\int_{\partial B(\sigma)} G(x)\frac{d\theta(x)}{d\nu(x)}dx \leq \int_{B(\sigma)} G(x)\Delta\theta(x)dx$$

and hence

$$(2.5) \qquad \sigma \frac{d}{d\sigma} \int_{B(\sigma)} G(x)dx \geq n \int_{B(\sigma)} G(x)dx.$$

From this, in view of Lemma 2.1, (2.4) follows.

As a consequence of the estimate (2.4) we obtain the fact that $G \in L^\infty_{\text{loc}}(\Omega)$ and we have, $\forall B(\sigma) \subset\subset \Omega$,

$$(2.6) \qquad \sup_{B(\sigma/2)} G \leq c(n) \fint_{B(\sigma)} G dx.$$

It follows from this the following fact of a general nature

Lemma 2.3. *If the function $G(x) > 0$ in Ω satisfies the estimate (2.4), then $\forall q > 1$, for any ball $B(\sigma) \subset\subset \Omega$ and $\forall t \in (0,1)$ we have*

$$(2.7) \qquad \int_{B(t\sigma)} G^q dx \leq c(n,q)t^n \int_{B(\sigma)} G^q dx \qquad [3].$$

In fact, if $t \geq 1/2$ then (2.7) is trivially true. If, instead $0 < t < 1/2$, then recalling that $G \in L^\infty(B(\sigma))$ and that (2.5) holds we get

$$\int_{B(t\sigma)} G^q dx \leq \left[\sup_{B(\sigma/2)} G \right]^{q-1} \int_{B(t\sigma)} G dx$$

$$\leq c(n)^{q-1} \left[\int_{B(\sigma)} G dx \right]^{q-1} \cdot t^n \int_{B(\sigma)} G dx$$

$$\leq c(n)^{q-1} t^n \int_{B(\sigma)} G^q dx.$$

3. Harmonic vectors. We observe that if G is of class $C^2(\Omega)$ then the condition (2.3) is equivalent to assuming that $\Delta G \geq 0$ in Ω.

---

[3] In other words, if the estimate (2.4) holds for a certain exponent q_0 then it holds also for $q \geq q_0$.

Let $u \in H^{1,2}(\Omega)$ be a harmonic vector $\Omega \to \mathbf{R}^N$

$$\int_\Omega (Du|D\phi)dx = 0, \ \forall \phi \in H_0^{1,2}(\Omega).$$

Then, as is well known, $u \in C^\infty(\Omega)$. Let us fix a $q \geq 2$ and set $G(x) = \|u(x)\|^q$.

In view of the fact that $\Delta u = 0$ in Ω we trivially have $\Delta G \geq 0$ in Ω and hence we have, by the Lemma 2.2, the estimate

$$(3.1) \qquad \int_{B(t\sigma)} \|u\|^q dx \leq t^n \int_{B(\sigma)} \|u\|^q dx$$

$\forall B(\sigma) \subset\subset \Omega$ and $\forall t \in (0,1)$.

This estimate obviously holds also for all the derivatives $D^\alpha u$.

The above result is well known in literature when $q = 2$.

4. A basic quasi-linear system.

Consider the basic quasi-linear system

$$(4.1) \qquad \operatorname{div} V^{q-2}(u)Du = 0 \ \text{ in } \ \Omega.$$

Let $u : \Omega \to \mathbf{R}^N$ be a solution of the system (4.1) in the sense described in §1:

$$W(u) \in H^{1,2}(\Omega)$$

$$(4.2) \qquad \int_\Omega V^{q-2}(u)(Du|D\phi)dx = 0, \ \forall \phi \in {}^*H_0^{1,2}(\Omega).$$

We shall prove the following fundamental estimate for the vector u.

Theorem 4.1 *If u is a solution of the system (4.1) then, $\forall B(\sigma) \subset\subset \Omega$ and $\forall t \in (0,1)$ we have*

$$(4.3) \qquad \int_{B(t\sigma)} \|W(u)\|^2 dx \leq t^n \int_{B(\sigma)} \|W(u)\|^2 dx.$$

PROOF. Since $G(x) = \|W(u)\|^2 \in H^{1,1}(\Omega)$ it is enough, by Lemma 2.2, to show that $G(x)$ satisfies the condition (2.3).

Let us fix $B(\sigma) \subset\subset \Omega$ and the function $\theta(x) \in C^\infty(\overline{B(\sigma)})$ such that $\theta \geq 0$ on $B(\sigma)$ and $\theta = 0$ on $\partial B(\sigma)$. We observe that, since $W(u) \in H^{1,2}(\Omega)$, it follows that

$$(4.4) \qquad \theta u \in {}^*H_0^{1,2}(\Omega) \text{ and also } \frac{\theta u}{V^2(u)} \in {}^*H_0^{1,2}(\Omega) .$$

On the other hand

$$D_i\|W(u)\|^2 = qV^{q-2}(u|D_iu) - (q-2)V^{q-4}(u)(u|D_iu), \quad i = 1,\ldots,n$$

from which, in view of the hypothesis (4.2) and of (4.4), we easily see that

$$(4.5) \qquad \begin{aligned} \int_{B(\sigma)} &(D\|W(u)\|^2|D\theta)dx = \\ &= -q\int_{B(\sigma)} \theta V^{q-2}(u)\|Du\|^2dx \\ &+ (q-2)\int_{B(\sigma)} \theta V^{q-2}(u)(Du|D(\frac{u}{V^2}))dx \leq \\ &\leq \int_{B(\sigma)} \theta V^{q-2}(u)\|Du\|^2(\frac{q-2}{V^2(u)} - q)dx \leq 0 \,{}^{(4)}. \end{aligned}$$

The estimate (4.3) is thus proved.

A consequence of the estimate (4.3) is the fact that

$$W(u) \in L^\infty_{\text{loc}}(\Omega) .$$

Since we have, from the system (4.1),

$$\text{div } V^{q-2}(u)Du_k = 0, \qquad k = 1,\ldots,N \text{ in } \Omega$$

each function u_k is, in the ball $B(\sigma) \subset\subset \Omega$, a solution of a linear second order elliptic equation with coefficients belonging to $L^\infty(B(\sigma))$; we have, by a well known result of De Giorgi [2], the following

---

[4] We note that $(q-2)/V^2(u) - q < 0$.

Lemma 4.1. *If u ia a solution of the system (4.1) then u is Hölder continuous of a certain exponent* $\alpha \in (0,1)$.

It now follows that the system (4.1) can be considered as a linear system with Hölder continuous coefficients. Hence Du is also Hölder continuous in Ω.

From the system (4.1) and the linear theory it follows that the second derivatives of u exist locally and if, $s = 1, \ldots, n$ then we have

$$(4.6) \quad \operatorname{div} V^{q-2}(u)DD_s u = -(q-2)\Sigma_i D_i[V^{q-4}(u)(u|D_s u)D_i u]$$

The vectors

$$F^i = V^{q-4}(u)(u|D_s u)D_i u$$

occurring on the right hand side, are Hölder continuous in Ω. Hence, by the linear theory applied to the system (4.6), it follows that the vectors $D_s Du$, $s = 1, \ldots, n$ are also Hölder continuous in Ω. One can thus prove, by an iterative procedure, that any solution u the system (4.1) belongs to $C^\infty(\Omega)$.

References

[1] S.Campanato, *Qualche risultato recente per sistemi differenziali in ipotesi di monotonia,* Boll. U.M.I. **2-A** (1988), 27-57.

[2] E.De Giorgi, *Sulla differenzibilità e l'analiticità delle estremali degli integrali multipli regolari,* Mem. Acc. Sci. Torino (1957), 25-43.

[3] E.Giusti, *Regolarità parziale di sistemi ellittici quasi-lineari di ordine arbitrario,* Annali Sc. Norm. Sup. Pisa **23** (1969), 115-141.

Dipartimento di Matematica

Università di Pisa

Via Buonarroti 2

I-56127 PISA

Γ-CONVERGENCE OF INTEGRAL FUNCTIONALS DEFINED ON VECTOR-VALUED FUNCTIONS

Luciano Carbone   Riccardo De Arcangelis

Dedicated to Ennio De Giorgi on his sixtieth birthday

0. Introduction. Let (U, τ) be a topological space satisfying the first countability axiom and let $F_h, h = 1, 2, ...,$ be real functionals defined on U. Let us assume that there exists a functional F_∞ defined on U and verifying for every $u \in U$ the following properties:

i) for every $v_h \xrightarrow{\tau} u$ it results $F_\infty(u) \leq \liminf_{h \to \infty} F_h(v_h)$,

ii) there exists $u_h \xrightarrow{\tau} u$ such that $F_\infty(u) \geq \limsup_{h \to \infty} F_h(u_h)$,

then we will say that F_∞ is the $\Gamma^-(\tau)$–limit of the sequence (F_h) and we will write:

$$(0.1) \qquad F_\infty(u) = \Gamma^-(\tau) \lim_{\substack{h \to \infty \\ v \to u}} F_h(v).$$

This notion of convergence has been introduced by E.De Giorgi in [DG1], [DG F].

It is particulary useful in Calculus of Variations because of the following property: if we assume that there exist $u_h, h = 1, 2, ...,$

such that $F_h(u_h) = \min_{v \in U} F_h(v)$ and $u_h \to u_\infty$ then it results $F_\infty(u_\infty) = \min_{v \in U} F_\infty(v)$.

Besides, as a particular case if $F_h = F_1$ for every h, the functional F_∞ given by (0.1) is the lower semicontinuous envelope in the topology τ of F_1, i.e. the greatest lower semicontinuous functional less than or equal to F_1 (usually denoted by $sc^-(\tau)F_1$).

Therefore the Γ^--convergence of functionals of the type

$$(0.2) \qquad F_h(\Omega; u) = \int_\Omega f_h(x; Du),$$

where Ω is a bounded open set of R^n, $u : R^n \to R^m$, $m \geq 1$ and Du denotes the $n \times m$ matrix having as columns the gradients of the components of u, has been partucularly studied.

In this case U is generally the space of the functions with components in $W_{loc}^{1,p}(R^n)$ and τ is the weak topology of $(W_{loc}^{1,p}(R^n))^m$ or the strong one of $(L_{loc}^p(R^n))^m$.

In this study the problems of the existence of the functional F_∞ and of its characterization as a variational functional like those in (0.2) are of pre-eminent interest.

If $m = 1$ these problems have been studied by several authors (see e.g. [DG1], [DG S], [C S], [B DM 1]) and the theory looks exaustive enaugh. In this study the notion of convexity appears naturally. Anyway there exist already some remarkable differences between the coercive and non coercive cases.

If $m > 1$ the case of a single functional and in which τ is the weak topology of $W^{1,p}(\Omega; R^m)$ has been studied in several papers (see [A F], [A B F], [D], [M], [Me], [B]).

It has been proved that the functional in (0.1) is still of the type $\int_\Omega \bar{f}(x; Du)$, $\bar{f}(x; .)$ being the greatest quasiconvex function less than or equal to $f(x; .)$.

In the case of topologies of the type $L^p(\Omega; R^m)$ the above problems have been studied in [F1] for coercive functionals.

The general case of non coercive functionals is still open.

Contrary to what happens in the coercive case the functionals in (0.1) relative to the topologies $L^1(\Omega; R^m)$ and $L^\infty(\Omega; R^m)$ may be different and the lower semicontinuity in $L^\infty(\Omega; R^m)$ of a functional is linked to some geometric properties of the integrand (see [A B F], [A B]).

In this paper we want to deal with the previous problem in the case of topologies $L^p(\Omega; R^m)$, $1 \leq p \leq +\infty$, using intermediate hypotheses between coerciveness and convexity.

More precisely consider a sequence of Caratheodory functions

$$(0.4) \qquad f_h : (x; z) \in R^n \times R^{nm} \mapsto f_h(x; z) \in [0, +\infty[$$

and let $l \in N$ such that $1 \leq l \leq m$.

For every $z \in R^{nm}$ let us set $z = (\hat{z}, \tilde{z})$ with $\hat{z} \in R^{nl}$, $\tilde{z} \in R^{n(m-1)}$.

We will suppose, for example, that the partial functions $f_h(x; \hat{z}, .)$ are convex, that the $f_h(x; ., \tilde{z})$ coercive, i.e. verifying $|\hat{z}|^p \leq f_h(x; \hat{z}, \tilde{z})$, and that $f_h(x; z) \leq \Lambda(1 + |z|^p)$.

We will prove that there exist a nonnegative function f_∞ defined on $R^n \times R^{nm}$ and a subsequence (f_{h_k}) of (f_h) for which the Γ^--limit in the $L^p(\Omega; R^m)$ topology of the sequence of functionals $(\int_\Omega f_{h_k}(x; Du))$ exists and is equal to $\int_\Omega f_\infty(x; Du)$ for every bounded open set Ω of R^n and every u with locally Lipschitz components on R^n (Theorem 2.4).

The above convexity hypotheses can be related to the notion of rank-one convexity (see [D]).

For $l = m$ we reobtain some of the results of [F1].

Our results are used in order to prove some relaxation properties, i.e. representation formulas for the lower semicontinuous envelope of a functional.

Lastly some examples are discussed.

1. Definitions and preliminaries. We recall the following fundamental results about Γ-convergence theory proved in [DG F].

Theorem 1.1. *Let (F_h) be a sequence of functionals defined on U. Then:*

1) *if there exists the functional $\Gamma^-(\tau) \lim_{\substack{h \to \infty \\ v \to u}} F_h(v)$ on U it is τ-lower semicontinuous on U;*

2) *there exists a subsequence (F_{h_k}) of (F_h) such that there exists the*

$$\Gamma^-(\tau) \lim_{\substack{h \to \infty \\ v \to u}} F_{h_k}(v) \quad \text{for every } u \in U$$

Let $n, m, l \in N$ with $1 \leq l \leq m$. For every $n \times m$ matrix z we denote with $\hat{z}$ the $n \times l$ matrix formed by the first l columns of z and with $\tilde{z}$ the one formed by the last $m - l$ ones. We will write $z = (\hat{z}, \tilde{z})$. In this paper we will consider a sequence of Carathéodory functions

$$(1.1) \qquad f_h : (x; z) \in R^n \times R^{nm} \mapsto f_h(x; z) \in [0, +\infty[$$

verifying for a fixed l

$$(1.2) \qquad f_h(x; \hat{z}, .) \quad convex \ for \ a.a. \ x \ in \ R^n, \ \hat{z} \ in \ R^{nl}$$

$$(1.3) \qquad |\hat{z}|^p \leq f_h(x; \hat{z}, \tilde{z}) \quad for \ a.a \ x \ in \ \mathbf{R}^n, \ z \ in \ R^{nm}, \ p \geq 1.$$

Definition 1.2. *Let p and q be extended real numbers with $1 \leq p < +\infty$, $1 \leq q \leq +\infty$.*

We say that a sequence of functions (f_h) as in (1.1) satisfies a growth condition of order (p, q) if:

$$(1.4) \qquad f_h(x; \hat{z}, \tilde{z}) \leq b_h(x) + \Lambda(|\hat{z}|^p + |\tilde{z}|^q) \qquad if \ q < +\infty$$

$$(1.5) \qquad f_h(x; \hat{z}, \tilde{z}) \leq b_h(x) + \Phi(\tilde{z})(1 + |\hat{z}|^p) \qquad if \ q = +\infty$$

for a.a. x in R^n, z in R^{nm},
where $\Lambda \geq 0$, Φ is a finite function on $R^{n(m-l)}$ and b_h are functions in $L^1_{loc}(R^n)$ such that there exists b in $L^1_{loc}(R^n)$ verifying:

$$(1.6) \qquad \int_A b_h \, dx \ \rightarrow \ \int_A b \, dx$$

for every bounded open set A of R^n.

From (1.2) and (1.5) it is not restrictive to assume that Φ is a convex and finite function, hence continuous on $R^{n(m-l)}$.
We further observe that (1.4) implies (1.5).

For every measurable set E we denote with $|E|$ the Lebesgue measure of E.

$Lip_{loc}(R^n; R^m)$ will be the set of the functions with values in R^m and with locally Lipschitz components. If no ambiguity occurs we will set $Lip_{loc} = Lip_{loc}(R^n; R^m)$.

If $u \in Lip_{loc}$ we will denote with $D\hat{u}$ the $n \times l$ matrix formed by the first l columns of the $n \times m$ matrix Du and with $D\tilde{u}$ the one formed by the last $m - l$.

As usual we will set $Du = (D\hat{u}, D\tilde{u}) = (Du^1, ..., Du^m)$.

With an abuse of notation we will write

$$\hat{u} = (u^1, ..., u^l), \quad \tilde{u} = (u^{l+1}, ..., u^m), \quad u = (\hat{u}, \tilde{u}).$$

For semplicity we will denote with the symbols $L^p(\Omega) \times L^q(\Omega)$ the topology $L^p(\Omega; R^l) \times L^q(\Omega; R^{m-l})$, $1 \leq p, q \leq +\infty$. Finally we will denote with $L_o^p(\Omega; R^k)$, $1 \leq p \leq +\infty$, the topology induced by the extended metric

$$d(u, v) = \begin{cases} \|u - v\|_{L^p(\Omega; R^m)} & \text{if spt } (u - v) \subset\subset \Omega \\ +\infty & \text{otherwise.} \end{cases}$$

Lastly we state a direct consequence of a theorem of G. Buttazzo and G. Dal Maso (see [B DM 2]).

Theorem 1.3. *Let F be a functional defined for every bounded open set of R^n and every u in Lip_{loc} satisfying the following hypotheses:*

1) *F is local, i.e. for every bounded open set A, u and v in Lip_{loc} such that $u = v$ a.e. in A it results $F(A; u) = F(A; v)$;*

2) *for every $u \in Lip_{loc}$ $F(.; u)$ is the restriction to the set of all bounded open sets of R^n of a Borel measure;*

3) *for every bounded open set A, u in Lip_{loc},*

$$0 \leq F(A; u) \leq \int_\Omega b(x) + \int_\Omega \Phi(D\tilde{u}) \ (1 + |D\hat{u}|^p)$$

with b in $L_{loc}^1(R^n)$, Φ as in (1.5);

4) *for every bounded open set A, u in Lip_{loc}, $c \in R^m$ it results $F(A; u + c) = F(A; u)$;*

5) *for every bounded open set A the functional $F(A;.)$ is sequentially lower semicontinuous in the topology weak* $W^{1,\infty}(\Omega, R^m)$.*

Then there exists a measurable function $f : R^n \times R^{nm} \to [0, +\infty[$ such that:

(α) *for every bounded open set A, u in Lip_{loc}*

$$F(A; u) = \int_A f(x; Du);$$

(β) $f(x;.)$ *is quasiconvex for a.a. x in R^n;*

(γ) f *satisfies (1.5) written with b instead of b_h.*

2. The integral representation theorem. Let f_h be a sequence of functions as in (1.1), let us define the functionals

$$F_h(\Omega; u) = \int_\Omega f_h(x; Du)$$

for every $u \in Lip_{loc}$ and Ω bounded open set of R^n.

The following result holds.

Lemma 2.1. *Let (F_h) be a sequence of functions satisfying (1.1) and a growth condition of order (p,q).*

Let $r, s \in [1, +\infty]$ and assume that there exists the limit

$$F_o^{(r,s)}(\Omega; u) = \Gamma^-\left(L_o^r(\Omega) \times L_o^s(\Omega)\right) \lim_{\substack{h \to +\infty \\ v \to u}} F_h(\Omega; v)$$

for every Ω bounded open set of R^n u in Lip_{loc}.

Then

$$F_o^{(r,s)}(\Omega; u) \le \liminf_k F_o^{(r,s)}(\Omega_k; u)$$

for every Ω bounded open set of R^n, u in Lip_{loc} and every sequence of open sets such that $|\partial\Omega_k| = 0$, $\Omega_k \subseteq \Omega_{k+1} \subseteq \Omega$ and $|\Omega - \Omega_k| \to 0$.

PROOF. Assume that $q = +\infty$ the case $q < +\infty$ can be treated analagously.

Fix $k \in N$ and let $(u_h) \subseteq Lip_{loc}$ such that

$$u_h \to u \text{ in } L_o^r(\Omega_k) \times L_o^s(\Omega_k)$$

$$F_o^{(r,s)}(\Omega_k; u) \geq \limsup_h F_h(\Omega_k; u_h).$$

Let us define the functions

$$v_h(x) = \begin{cases} u_h(x) & \text{if } x \in \Omega_k \\ u(x) & \text{if } x \in \Omega - \Omega_k \end{cases}.$$

Then
$$v_h \to u \text{ in } L_o^r(\Omega) \times L_o^s(\Omega) \text{ and}$$

$$F_o^{(r,s)}(\Omega; u) - F_o^{(r,s)}(\Omega_k; u) \leq$$
$$\liminf_h F_h(\Omega; v_h) - \limsup_h F_h(\Omega_k; u_h) \leq$$
$$\int_{\Omega - \Omega_k} b \, dx + \int_{\Omega - \Omega_k} \Phi(D\tilde{u})(1 + |D\hat{u}|^p).$$

The thesis now follows taking the limit as $k \to +\infty$ in the above inequality. ∎

We now prove the following fundamental result.

Theorem 2.2. *Let Ω be a bounded open set of R^n and let (f_h) be a sequence of functions satisfying (1.1), (1.2), (1.3) and a growth condition of order (p,q).*
Let $r \in [1, +\infty]$, $s \in [q, +\infty]$ and assume that in $u \in Lip_{loc}$ there exists

(2.1) $$F^{(r,s)}(\Omega; u) = \Gamma^-(L^r(\Omega) \times L^s(\Omega)) \lim_{\substack{h \to \infty \\ v \to v}} F_h(\Omega; v).$$

Then in u there exists also

(2.2) $$F_o^{(r,s)}(\Omega; u) = \Gamma^-(L_o^r(\Omega) \times L_o^s(\Omega)) \lim_{\substack{h \to \infty \\ v \to u}} F_h(\Omega; v)$$

and it results

(2.3) $$F_o^{(r,s)}(\Omega; u) = F^{(r,s)}(\Omega; u).$$

PROOF. We divide the proof into some steps: at first we will consider the case in which $q = +\infty$ and Ω has Lipschitz boundary, then the general case.

(I) CASE $q = +\infty$, Ω WITH LIPSCHITZ BOUNDARY.

(I.A) CONSTRUCTION OF SUITABLE APPROXIMATING SEQUENCES.

Let $(F_{h'})$ be an arbitrary subsequence of (F_h). By Theorem 1.1 there exists a subsequence $(F_{h''})$ of $(F_{h'})$ such that there exists

$$G_o(\Omega; u) = \Gamma^-(L_o^r(\Omega) \times L_o^s(\Omega)) \lim_{\substack{h'' \to \infty \\ v \to u}} F_{h''}(\Omega; v).$$

Since we are going to prove that $G_o(\Omega; u) = F^{(r,s)}(\Omega; u)$ and since the subsequence $(F_{h'})$ is arbitrary it will not be restrictive to assume the existence of the limit in (2.2).

We have obviously:

(2.4) $$F^{(r,s)}(\Omega; u) \leq F_o^{(r,s)}(\Omega; u).$$

Let $(u_h) \subseteq Lip_{loc}$ be such that:

(2.5) $u_h \to u$ in $L^r(\Omega) \times L^s(\Omega)$ (recall that $s = +\infty$)

(2.6) $+\infty > F^{(r,s)}(\Omega; u) \geq \limsup_h F_h(\Omega; u_h).$

Since Ω has Lipschitz boundary by (2.5), (1.3), (2.6) and Rellich's theorem it follows that if $\bar{r} = \max\{r, p\}$

(2.7) $$u_h \to u \text{ in } L^{\bar{r}}(\Omega) \times L^s(\Omega).$$

Let $B_o \subset\subset \Omega$ be an open set such that $|\partial B_o| = 0$, let $\delta = \text{dist}(B_o, \partial\Omega)$ and let $\nu \in N$.

For every $j \in \{1, ..., \nu\}$ let us define the sets:

$$B_j = \{x \in \Omega : \text{dist}(x, B_o) < \frac{j\delta}{\nu}\}$$

and let ϕ_j be functions satisfying

(2.8) $\quad \begin{cases} \phi_j \in C_o^1(B_j), \ 0 \le \phi_j \le 1, \ \phi_j(x) = 1 \ \text{if} \ x \in B_{j-1} \\ \|D\phi_j\|_{L^\infty(R^n)} \le \frac{\nu+1}{\delta}. \end{cases}$

Define further, for every $j = 1, ..., \nu - 1$, the functions ψ_j as $\psi_j = \phi_{j+1}$ and let, for every $t \in]0, 1[$, $j = 1, ..., \nu-1$, γ_j^t be functions in $C^1(R^n)$ verifying

(2.9) $\quad \begin{cases} 1 \le \gamma_j^t \le \frac{1}{1-t} \ , \ \gamma_j^t(x) = 1 \ \text{if} \ x \in B_j \ , \ \gamma_j^t(x) = \frac{1}{1-t} \ \text{if} \ x \notin B_{j+1} \\ \|D\gamma_j^t\|_{L^\infty(R^n)} \le \frac{t}{1-t} \frac{\nu+1}{\delta}. \end{cases}$

For every $j \in \{1, ..., \nu - 1\}$, $t \in]0, 1[$ let us define the functions $v_{h,j}^t = (\hat{v}_{h,j}^t, \tilde{v}_{h,j}^t)$ as

$$\hat{v}_{h,j}^t = \psi_j \hat{u}_h + (1 - \psi_j)\hat{u}$$

$$\tilde{v}_{h,j}^t = (1 - t)\gamma_j^t(\phi_j \tilde{u}_h + (1 - \phi_j)\tilde{u}).$$

By (2.7) it soon follows that:

$$v_{h,j}^t \to (\hat{u}, \ (1 - t)\gamma_j^t \tilde{u}) \ \text{in} \ L_o^{\bar{r}}(\Omega) \times L_o^s(\Omega) \ \text{as} \ h \to +\infty$$

for every $j \in \{1, ..., \nu - 1\}$, $t \in]0, 1[$. Hence for every $j \in \{1, ..., \nu - 1\}$ we deduce by (2.6):

$$F_o^{(r,s)}(\Omega; \hat{u}, (1 - t)\gamma_j^t \tilde{u}) \le \liminf_h F_h(\Omega; v_{h,j}^t) \le$$

$$\le \limsup_h F_h(\Omega; u_h) + \limsup_h \{F_h(\Omega_j, v_{h,j}^t) - F_h(\Omega; u_h)\} \le$$

$$\le F^{(r,s)}(\Omega; u) + \limsup_h \{F_h(\Omega, v_{h,j}^t) - F_h(\Omega; u_h)\}.$$

(I.B) ESTIMATES.

We now estimate the right hand side of (2.10). We have:

$$(2.11) \qquad F_h(\Omega; v_{h,j}^t) - F_h(\Omega; u_h) =$$

$$= \int_{B_{j-1}} \{f_h(x; D\hat{u}_h, (1-t)D\tilde{u}_h) - f_h(x; D\hat{u}_h, D\tilde{u}_h)\} +$$

$$+ \int_{B_j - B_{j-1}} \{f_h(x; D\hat{u}_h, (1-t)(\phi_j D\tilde{u}_h + (1-\phi_j)D\tilde{u} + D\phi_j(\tilde{u}_h - \tilde{u})) +$$

$$- f_h(x; D\hat{u}_h, D\tilde{u}_h)\} +$$

$$+ \int_{B_{j+1} - B_j} \{f_h(x; D\hat{v}_{h,j}^t, (1-t)D(\gamma_j^t \tilde{u})) - f_h(x; D\hat{u}_h, D\tilde{u}_h)\} +$$

$$+ \int_{\Omega - B_{j+1}} \{f_h(x; D\hat{u}, D\tilde{u}) - f_h(x; D\hat{u}_h, D\tilde{u}_h)\} = \alpha_h + \beta_h + \gamma_h + \delta_h.$$

Let us estimate separately each of the terms in the right hand side of (2.11).

By (1.2) and (1.5) we deduce that:

$$(2.12) \qquad \begin{aligned} \alpha_h &\leq t \int_{B_{j-1}} f_h(x; D\hat{u}_h, \tilde{0}) \leq \\ &\leq t \int_{B_{j-1}} b_h \, dx + t \int_{B_{j-1}} \Phi(\tilde{0})(1 + |D\hat{u}_h|^p). \end{aligned}$$

By (1.2),(1.5),(2.8) and (2.9) we have:

$$(2.13) \quad \beta_h \leq t \int_{B_j - B_{j-1}} f_h(x; D\hat{u}_h, \frac{1-t}{t} D\phi_j(\tilde{u}_h - \tilde{u})) +$$

$$+ (1-t) \int_{B_j - B_{j-1}} f_h(x; D\hat{u}_h, \phi_j D\tilde{u}_h + (1-\phi_j)D\tilde{u}) +$$

$$- \int_{B_j - B_{j-1}} f_h(x; D\hat{u}_h, D\tilde{u}_h) \leq$$

$$\leq t \int_{B_j - B_{j-1}} b_h \, dx + t \int_{B_j - B_{j-1}} \Phi(\frac{1-t}{t} D\phi_j(\tilde{u}_h - \tilde{u}))(1 + |D\hat{u}_h|^p) +$$

$$+ \int_{B_j - B_{j-1}} \phi_j f_h(x; D\hat{u}_h, D\tilde{u}_h) + \int_{B_j - B_{j-1}} (1 - \phi_j) f_h(x; D\hat{u}_h, D\tilde{u}) +$$

$$- \int_{B_j - B_{j-1}} f_h(x; D\hat{u}_h, D\tilde{u}_h)$$

$$\leq t \int_{B_j - B_{j-1}} b_h \, dx + t \int_{B_j - B_{j-1}} \Phi(\frac{1-t}{t} D\phi_j(\tilde{u}_h - \tilde{u}))(1 + |D\hat{u}_h|^p) +$$

$$+ \int_{B_j - B_{j-1}} b_h \, dx + \int_{B_j - B_{j-1}} \Phi(D\tilde{u})(1 + |D\hat{u}_h|^p).$$

In order to estimate γ_h let us observe that by (1.2) and (1.5) it follows:

(2.14)
$$\gamma_h \leq t \int_{B_{j+1} - B_j} f_h(x; D\hat{v}_{h,j}^t, \tilde{0}) +$$

$$+ \int_{B_{j+1} - B_j} f_h(x; D\hat{v}_{h,j}^t, D(\gamma_j^t \tilde{u})) \leq$$

$$\leq t \int_{B_j - B_{j-1}} b_h + t \int_{B_{j+1} - B_j} \Phi(\tilde{0})(1 + |D\hat{v}_{h,j}^t|^p) +$$

$$+ \int_{B_{j+1} - B_j} b_h \, dx + \int_{B_{j+1} - B_j} \Phi(D(\gamma_j^t \tilde{u}))(1 + |D\hat{v}_{h,j}^t|^p).$$

Finally by (1.5) it follows:

(2.15)
$$\delta_h \leq \int_{\Omega - B_{j+1}} b_h \, dx + \int_{\Omega - B_{j+1}} \Phi(D\tilde{u})(1 + |D\hat{u}|^p).$$

(I.c) PASSAGE TO THE LIMIT.

We now fix $t \in]0,1[$, $\nu \in N$ and $\epsilon > 0$; then there exists $\bar{h} = \bar{h}(\epsilon, \nu, t)$ such that, by virtue of (2.8), (2.10) and (2.12)÷(2.15), it follows for every $h \geq \bar{h}$:

(2.16)

$$F_0^{(r,s)}(\Omega; \hat{u}, (1-t)\gamma_j^t \tilde{u}) \leq F^{(r,s)}(\Omega; \hat{u}, \tilde{u}) +$$

$$+ t \int_\Omega b_h \, dx + \int_{\Omega - B_o} b_h \, dx + t \int_\Omega \Phi(\tilde{0})(1 + |D\hat{u}_h|^p)$$

$$+ t \int_{B_j - B_{j-1}} \Phi(\frac{1-t}{t} D\phi_j(\tilde{u}_h - \tilde{u}))(1 + |D\hat{u}_h|^p)$$

$$+ \int_{B_{j+1} - B_j} \Phi(\tilde{0})(1 + 3^{p-1}(|D\hat{u}_h|^p + |D\hat{u}|^p + \left|\frac{\nu+1}{\delta}\right|^p |\hat{u}_h - \hat{u}|^p))$$

$$+ \int_{B_{j+1} - B_j} \Phi(D(\gamma_j^t \tilde{u}))(1 + 3^{p-1}(|D\hat{u}_h|^p + |D\hat{u}|^p + \left|\frac{\nu+1}{\delta}\right|^p |\hat{u}_h - \hat{u}|^p))$$

$$+ \int_{B_j - B_{j-1}} \Phi(D\tilde{u})(1 + |D\hat{u}_h|^p)$$

$$+ \int_{\Omega - B_o} \Phi(D\tilde{u})(1 + |D\hat{u}|^p) + \epsilon$$

for every $j \in \{1, ..., \nu - 1\}$.

Let us define now, for every $\nu \in N$, the functions

$$\Theta^\nu(x) = \begin{cases} D\phi_j(x) & \text{if } x \in B_j - B_{j-1} \quad j = 1, ..., \nu - 1 \\ 0 & \text{otherwise} \end{cases}$$

and, for every $t \in]0, 1[, \nu \in N$

$$M_\Phi(t, \nu) = \max_{1 \leq j \leq \nu - 1} \sup_{\bar{\Omega}} \Phi(D(\gamma_j^t \tilde{u})).$$

Obviously for every $\nu \in N$ it results:

(2.17)

$$\limsup_{t \to 0} M_\Phi(t, \nu) \leq M_\Phi = \sup\{\Phi(\tilde{z}) : |\tilde{z}| \leq \|\tilde{u}\|_{W^{1,\infty}(\Omega, R^{m-1})}\}.$$

With these notations in mind we deduce by taking the averages in both memebers of (2.16) that if $h \geq \bar{h}$:

(2.18)
$$\frac{1}{\nu-1}\sum_{j=1}^{\nu-1} F_o^{(r,s)}(\Omega;\hat{u},(1-t)\gamma_j^t\tilde{u}) \le F^{(r,s)}(\Omega;\hat{u},\tilde{u})+$$

$$+t\int_\Omega b_h\ dx + \int_{\Omega-B_o} b_h\ dx + t\int_\Omega \Phi(\tilde{0})(1+|D\hat{u}_h|^p)$$

$$+\frac{t}{\nu-1}\int_{\Omega-B_o}\Phi(\frac{1-t}{t}\Theta^\nu(\tilde{u}_h-\tilde{u}))(1+|D\hat{u}_h|^p)$$

$$+\frac{t}{\nu-1}\int_{\Omega-B_o}\Phi(\tilde{0})(1+3^{p-1}(|D\hat{u}_h|^p+|D\hat{u}|^p+\left|\frac{\nu+1}{\delta}\right|^p|\hat{u}_h-\hat{u}|^p))$$

$$+\frac{1}{\nu-1}M_\Phi(t,\nu)\int_{\Omega-B_o}(1+3^{p-1}(|D\hat{u}_h|^p+|D\hat{u}|^p+|\frac{\nu+1}{\delta}|^p|\hat{u}_h-\hat{u}|^p))$$

$$+\frac{1}{\nu-1}\int_{\Omega-B_o}\Phi(D\tilde{u})(1+|D\hat{u}_h|^p)$$

$$+\int_{\Omega-B_o}\Phi(D\tilde{u})(1+|D\hat{u}|^p)+\epsilon.$$

If we observe now that by (1.3) and (2.6) it follows:

$$\limsup_h \|D\hat{u}_h\|^p_{L^p(\Omega;R^{nl})} \le F^{(r,s)}(\Omega;u)$$

by virtue of (1.5) and (2.7) we can pass to the limit as $h \to +\infty$ in (2.18).
We get:

(2.19)
$$\frac{1}{\nu-1}\sum_{j=1}^{\nu-1} F_o^{(r,s)}(\Omega;\hat{u},(1-t)\gamma_j^t\tilde{u}) \le F^{(r,s)}(\Omega;\hat{u},\tilde{u})+$$

$$+t\int_\Omega b\ dx + \int_{\Omega-B_o} b\ dx + t\Phi(\tilde{0})(|\Omega-B_o|+F^{(r,s)}(\Omega;u))$$

$$+\frac{t}{\nu-1}\Phi(\tilde{0})(|\Omega-B_o|+F^{(r,s)}(\Omega,u))$$

$$+\frac{t}{\nu-1}\Phi(\tilde{0})(|\Omega\setminus B_o|3^{p-1}(F^{(r,s)}(\Omega;u)+\int_{\Omega-B_o}|D\hat{u}|^p))$$

$$+ \frac{1}{\nu - 1} \|\Phi(D\tilde{u})\|_{L^{\infty}(\Omega)}(|\Omega - B_o| + F^{(r,s)}(\Omega; u))$$

$$+ \int_{\Omega - B_o} \Phi(D\tilde{u})(1 + |D\hat{u}|^p) + \frac{1}{\nu - 1} M_{\Phi}(t, \nu)(|\Omega - B_o|$$

$$+ 3^{p-1}(F^{(r,s)}(\Omega; u) + \int_{\Omega - B_o} |D\hat{u}|^p)) + \epsilon.$$

At this point if we observe that for every $j \in \{1, ..., \nu - 1\}$

$$(1 - t)\gamma_j^t \tilde{u} \to \tilde{u} \quad \text{in } L_o^s(\Omega; R^{m-l}) \text{ as } t \to 0$$

and that $F_o^{(r,s)}(\Omega; \hat{u}, .)$ is lower semicontinuous in the topology $L_o^s(\Omega; R^{m-l})$ (see Theorem 1.1) we can pass to the limit as $t \to 0$ in (2.19); by (2.17) we get:

(2.20)

$$F_o^{(r,s)}(\Omega; \hat{u}, \tilde{u}) \le F^{(r,s)}(\Omega; \hat{u}, \tilde{u}) + \int_{\Omega - B_o} b \, dx +$$

$$+ \frac{1}{\nu - 1} \|\Phi(D\tilde{u})\|_{L^{\infty}(\Omega)}(|\Omega - B_o| + F^{(r,s)}(\Omega; u))$$

$$+ \frac{1}{\nu - 1} M_{\Phi}(|\Omega - B_o| + 3^{p-1}(F^{(r,s)}(\Omega; u) + \int_{\Omega - B_o} |D\hat{u}|^p))$$

$$+ \int_{\Omega - B_o} \Phi(D\tilde{u})(1 + |D\hat{u}|^p) + \epsilon.$$

If $\nu \to +\infty$ by (2.20) it follows:

(2.21)
$$F_o^{(r,s)}(\Omega; \hat{u}, \tilde{u}) \le F^{(r,s)}(\Omega; \hat{u}, \tilde{u}) + \int_{\Omega - B_o} b \, dx +$$

$$+ \int_{\Omega - B_o} \Phi(D\tilde{u})(1 + |D\hat{u}|^p) + \epsilon.$$

Finally if $|\Omega - B_o| \to 0$ and $\epsilon \to 0$ by (2.21) we get:

(2.22) $$F_o^{(r,s)}(\Omega; \hat{u}, \tilde{u}) \le F^{(r,s)}(\Omega; \hat{u}, \tilde{u}),$$

hence, by (2.22) and (2.24), the thesis follows.

(II) CASE WITH NO ASSUMPTIONS ON q AND Ω.

(II.A) CASE $q = +\infty$, NO ASSUMPTIONS ON Ω.

If Ω has not Lipschitz boundary let (Ω_k) be a sequence of open sets with Lipschitz boundary such that:

$$(2.23) \qquad \Omega_k \subseteq \Omega_{k+1} \subseteq \Omega$$

$$(2.24) \qquad |\Omega - \Omega_k| \to 0 \quad \text{as } k \to +\infty$$

Let $(F_{h'})$ be an arbitrary subsequence of (F_h).

By 2) of Theorem 1.1, by virtue of a diagonal process, there exists a subsequence $(F_{h''})$ of $(F_{h'})$ such that there exist the limits:

$$F_o^{(r,s)}(\Omega_k; u) = \Gamma^-(L_o^r(\Omega_k) \times L_o^s(\Omega_k)) \lim_{\substack{h'' \to \infty \\ v \to u}} F_{h''}(\Omega_k; v)$$

$$F^{(r,s)}(\Omega; u) = \Gamma^-(L^r(\Omega_k) \times L^s(\Omega_k)) \lim_{\substack{h'' \to \infty \\ v \to u}} F_{h''}(\Omega_k; v)$$

for every $k \in N$.

Moreover recall that:

$$(2.25) \qquad F^{(r,s)}(\Omega_k; u) \leq F^{(r,s)}(\Omega; u) \quad \text{for every } k \in N.$$

By (2.22), Lemma 2.1 and (2.25) it results:

$$(2.26) \quad F_o^{(r,s)}(\Omega; u) \leq \liminf_k F_o^{(r,s)}(\Omega_k; u) \leq \limsup_k F^{(r,s)}(\Omega; u) \leq$$

$$\leq F^{(r,s)}(\Omega; u).$$

Hence (2.3) follows by (2.26) recalling that obviously one has

$$F^{(r,s)}(\Omega; u) \leq F_o^{(r,s)}(\Omega; u).$$

(II.B) CASE $q < +\infty$, NO ASSUMPTIONS ON Ω.

In this case the proof works exactly as in the case $q = +\infty$ till (2.13) because the integral

$$(2.27) \qquad \int_{B_j - B_{j-1}} \Phi(\frac{1-t}{t} D\phi_j(\tilde{u}_h - \tilde{u}))(1 + |D\hat{u}_h|^p)$$

does not allow, in general, the passage to the limit as $h \to +\infty$ as in (2.19) if $\tilde{u}_h \to \tilde{u}$ in $L^s((\Omega; R^{m-l})$ with $q \le s < +\infty$.

To do this we replace (2.27) and the derived estimates with

$$\int_{B_J - B_{J-1}} \Lambda(|\frac{1-t}{t} D\phi_j(\tilde{u}_h - \tilde{u})|^q + |D\hat{u}_h|^p),$$

and this is the reason for which we require, if $q < +\infty$, to be satisfied a growth condition as in (1.4) instead of (1.5).

With these changes in mind the proof proceeds as in the previous case. ∎

It is well known that the growth conditions and Theorem 2.2 imply, by the use of standard techniques (see [DG1], [C S]), the existence of the limit in (2.1) for every bounded open set of R^n and u in Lip_{loc}.

Moreover it is soon proved that for every u in Lip_{loc} and for every couple of disjoint open sets A and B it results:

$$F_o^{(r,s)}(A \cup B; u) \le F_o^{(r,s)}(A; u) + F_o^{(r,s)}(B; u).$$

On the other side it is well known that

$$F^{(r,s)}(A \cup B; u) \ge F^{(r,s)}(A; u) + F^{(r,s)}(B; u).$$

At this point, by the use of standard techniques (see [DM]), it can be proved the following result.

Proposition 2.3. *Let (f_h) be a sequence of functions satisfying (1.1), (1.2), (1.3) and a growth condition of order (p,q).*

Then for every $r \in [1, +\infty]$, $s \in [q, +\infty]$ there exists a subsequence (f_{h_k}) such that there exists the functional

$$F^{(r,s)}(\Omega; u) = \Gamma^-(L^r(\Omega) \times L^s(\Omega)) \lim_{\substack{k \to \infty \\ v \to u}} F_{h_k}(\Omega, v)$$

for every bounded open set Ω of R^n, u in Lip_{loc}.

Moreover, for every u in Lip_{loc}, the functional $F^{(r,s)}(.;u)$ is the restriction to the set of all bounded open sets of R^n of an absolutely continuous Borel measure.

With these results in mind, by using Theorem 1.3, we are able to prove the integral representation theorem.

Theorem 2.4. *Let (f_h) be a sequence of functions satisfying (1.1), (1.2), (1.3) and a growth condition of order (p, q).*

Then for every $r \in [1, +\infty]$, $s \in [q, +\infty]$ there exist a subsequence (f_{h_k}) and a Carathéodory function

$$f_\infty^{(r,s)} : R^n \times R^{nm} \to [0, +\infty[$$

such that:

a) for a.a. x in R^n $f_\infty^{(r,s)}(x;.)$ is quasiconvex;

b) (2.28) $\begin{cases} |\hat{z}|^p \leq f_\infty^{(r,s)}(x;z) \leq b(x) + \Phi(\tilde{z})(1 + |\hat{z}|^p) & \text{if } q = +\infty \\ |\hat{z}|^p \leq f_\infty^{(r,s)}(x;z) \leq b(x) + \Lambda(|\tilde{z}|^q + |\hat{z}|^p) & \text{if } q < +\infty \end{cases}$

c) $\int_\Omega f_\infty^{(r,s)}(x; Du) = \Gamma^-(L^r(\Omega) \times L^s(\Omega)) \lim_{\substack{k \to \infty \\ v \to u}} \int_\Omega f_{h_k}(x; Dv) =$

(2.29) $= \Gamma^-(L_o^r(\Omega) \times L_o^s(\Omega)) \lim_{\substack{k \to \infty \\ v \to u}} \int_\Omega f_{h_k}(x; Dv)$

for every bounded open set Ω of R^n, u in Lip_{loc}.

PROOF. The hypotheses of Theorem 1.3 are satisfied. In fact (1) and (3) are obvious, (2) follows from Propostion 2.3, (4) follows from the fact that such property holds for the functionals F_h and (5) follows from the $L^r(\Omega) \times L^s(\Omega)$ lower semicontinuity of the functional in the left hand side of the first equality in (2.29) (see 1 of Theorem 1.1).

Therefore (α), (β) and (γ) of Theorem 1.3, together with Theorem 2.2, imply (a) and (c).

(b) follows from (1.4), (1.5) and the $L^r(\Omega; R^m)$ lower semicontinuity of the functional $\int_\Omega |D\hat{u}|^p$. ∎

We now apply our results in order to study the relaxation properties of a functional.

Given a Carathéodory function

$$f : (x; z) \in R^n \times R^{nm} \mapsto f(x; z) \in [0, +\infty[$$

we say that f satisfies a growth condition of order (p, q) if (1.4) or (1.5) are satisfied by f instead of f_h, in this case we will assume that $b_h = b$ for every h.

We will also assume the analogous of (1.2) and (1.3):

(2.30) $\qquad f(x; \hat{z}, .)$ convex for a.a. x in R^n, $\hat{z} \in R^{nm}$;

(2.31) $\qquad |\hat{z}|^p \leq f(x; \hat{z}, \hat{z})$ for a.a x in R^n, $z \in R^{nm}$, $p \geq 1$.

By Theorem 2.4 it follows that:

Corollary 2.5. *Let f be a function satisfying (2.29), (2.30), (2.31) and a growth condition of order (p, q).*

Then for every $r \in [1, +\infty]$, $s \in [q, +\infty]$ there exists a Carathéodory function

$$f_\infty^{(r,s)} : R^n \times R^{nm} \to [0, +\infty[$$

such that:
(a) for a.a. x in R^n $f_\infty^{(r,s)}(x; .)$ is quasiconvex;

(b) $\qquad \begin{cases} |\hat{z}|^p \leq f_\infty^{(r,s)}(x; z) \leq b(x) + \Phi(\tilde{z})(1 + |\hat{z}|^p) & \text{if } q = +\infty \\ |\hat{z}|^p \leq f_\infty^{(r,s)}(x; z) \leq b(x) + \Lambda(|\tilde{z}|^q + |\hat{z}|^p) & \text{if } q < +\infty \end{cases}$

(c) $\qquad \displaystyle\int_\Omega f_\infty^{(r,s)}(x; Du) = sc^-(L^r(\Omega) \times L^s(\Omega)) \int_\Omega f(x; Du) =$

$$= sc^-(L_o^r(\Omega) \times L_o^r(\Omega)) \int_\Omega f(x, ; Du)$$

for every bounded open set Ω of R^n, u in Lip_{loc}.

3. Some examples. Let us open this section with a well known property of the jacobian determinants that we prove for sake of completeness.

Lemma 3.1. *Let Ω be a bounded open set of R^n and let u and v be functions in $H_{loc}^{1,n}(R^n, R^n)$ such that $u - v$ is in $H_o^{1,n}(\Omega; R^n)$. Then*

$$\int_\Omega det\, Du = \int_\Omega det\, Dv.$$

PROOF. Let us first prove that if $\phi \in C^1(R^n; R^n)$ and if its i-th component ϕ^i is in $C_o^1(\Omega)$ then $\int_\Omega det\, D\phi = 0$.

To this aim let us recall that for every ϕ in $C^1(R^n; R^n)$ and for every $i \in \{1, ..., n\}$ it results:

(3.1)
$$\det D\phi = \frac{\partial(\phi^1, ..., \phi^n)}{\partial(x_1, ..., x_m)} =$$
$$= \sum_{j=1}^n (-1)^{i+j} \frac{\partial}{\partial x_j}\left(\phi^i \frac{\partial(\phi^1, ..., \phi^{i-1}, \phi^{i+1}, ..., \phi^n)}{\partial(x_1, ..., x_{j-1}, x_{j+1}, ..., x_n)}\right).$$

Then, recalling that $\phi^i = 0$ on $\partial\Omega$, by (3.1) and the divergence theorem it follows:

(3.2)
$$\int_\Omega det D\phi = \int_{\partial\Omega} \sum_{j=1}^n (-1)^{i+j} \phi^i \frac{\partial(\phi^1, ..., \phi^{i-1}, \phi^{i+1}, ..., \phi^n)}{\partial(x_1, ..., x_{j-1}, x_{j+1}, ..., x_n)} \nu_j d\sigma = 0.$$

Now if $\phi \in H_{loc}^{1,n}(R^n; R^n)$ with $\phi^i \in H_o^{1,n}(\Omega)$ let $(\phi_k) \subseteq C^1(R^n; R^n)$ be such that $\phi_k \to \phi$ in $H^{1,n}(\Omega; R^n)$ and $\phi_k^i \in C_o^1(\Omega)$.

By (3.2) it follows:

(3.3)
$$\int_\Omega det D\phi = \lim_k \int_\Omega det D\phi_k = 0.$$

Finally let u and v be in $H_{loc}^{1,n}(R^n; R^n)$ such that $u - v$ is in $H_o^{1,n}(\Omega; R^n)$.

By an interated use of (3.3) we deduce:

$$\int_\Omega \det[D(u^1 - v^1),\ Du^2, ..., Du^n] = 0,$$

$$\int_\Omega \det[Dv^1,\ D(u^2 - v^2), ..., Du^n] = 0$$

(3.4)

$$\cdots \ \cdots \ \cdots$$

$$\int_\Omega \det[Dv^1,\ Dv^2, ..., D(u^n - v^n)] = 0.$$

At this point the thesis follows directly by (3.4). ∎

We now discuss some examples.

Example 3.2. Let $n = m = 2$, $l = 1$ and let us consider the function:

$$f(z^1, z^2) = |\det[z^1, z^2]| + |z^1|, \quad z \in R^4$$

This function satisfies a growth condition of order $(1, +\infty)$, (2.30) and (2.31) with $p = 1$.

Consider the functional

$$F(u) = \int_\Omega f(Du^1, Du^2).$$

By the use of the functions considered in [A B F] Example IV.1 it can be proved that F is not $L^1(\Omega; R^2)$ lower semicontinuous in $u(x, y) = (x, y)$.

Nevertheless F is $L_o^1(\Omega; R^2)$ lower semicontinuous in $u(x, y) = (x, y)$.

To prove this let $(u_h) \subseteq Lip_{loc}$ such that $u_h \to u$ in $L_o^1(\Omega; R^2)$.

Thanks to the convexity of the function $t \to |t|$ we have:

(3.5) $|\det Du_h| \geq |\det Du| + (\det Du_h - \det Du)\mathrm{sg}(\det Du),$

hence, recalling that $\det Du = 1$ and integrating both members of (3.5) we deduce:

(3.6) $\liminf\limits_h \int_\Omega f(Du_h) \geq \liminf\limits_h \int_\Omega (\det Du_h - \det Du) + \int_\Omega f(Du).$

Hence, recalling that $u_h = u$ on $\partial\Omega$, by Lemma 3.1 we deduce our result.

Observe that the inequality

$$sc^-(L^1(\Omega; R^2))F(u) > F(u)$$

is linked to the fact that the thesis of Theorem 2.2 is not true in general if $q = +\infty$ and $s < q$.   ∎

Example 3.3. Let $n = m = 2$, $l = 1$. Let us consider the function

$$g(z^1, z^2) = |\det[z^1, z^2] - 1|^2.$$

Observe that the function g satisfies a growth condition of order $(2, +\infty)$ but is not coercive, hence the thesis of Theorem 2.2 can be no more true.

In fact the functional

$$G(u) = \int_\Omega g(Du)$$

is not $L^\infty(\Omega; R^2)$ lower semicontinuous (see [A B F] Example IV.2), nevertheless it is $L_o^\infty(\Omega; R^2)$ lower semicontinuous.

To prove this let $(u_h) \subseteq Lip_{loc}$ be such that $u_h \rightharpoonup u$ in $L_o^\infty(\Omega; R^2)$, then by Lemma 3.1 we deduce

$$\liminf_h \int_\Omega |\det Du_h - 1|^2 \geq \liminf_h \int_\Omega |\det Du_h|^2 + |\Omega| +$$

$$- 2 \limsup_h \int_\Omega \det Du_h =$$

$$= \liminf_h \int_\Omega |\det Du_h|^2 + |\Omega| - 2 \int_\Omega \det Du \geq$$

$$\geq \int_\Omega |\det Du|^2 + |\Omega| - 2 \int_\Omega \det Du = \int_\Omega |\det Du - 1|^2.$$

We further observe that in the study of the L_o^∞ lower semi-continuity of integral functionals $\int_\Omega f(\det Du)$, with f convex, the hypothesis

$$f(0) = \min_{t \in R} f(t)$$

is no longer necessary (see [A B], [A B F], [F 2]).   ■

References

[A B]   E.Acerbi, G.Buttazzo, *Semicontinuous envelopes of polyconvex integrals*, Proceed. Royal Soc. of Edinburgh **96 A** (1984), 51-54.

[A B F] E.Acerbi, G.Buttazzo, N.Fusco, *Semicontinuity and relaxation for integrals depending on vector-valued functions*, J. Math. Pures et Appl. **62** (1983), 371-387.

[A F]   E.Acerbi, N.Fusco, *Semicontinuity problems in the Calculus of Variations*, Arch. Rational Mech. Anal. **86** (1984), 125-145.

[A]   H.Attouch, *Convergence problems for functions and operators*, Pitman Appl. Math. Ser., Boston (1984).

[B]   J.M.Ball, *Convexity conditions and existence theorems in nonlinear elasticity*, Arch. Rational Mech. Anal. **63** (1977), 337-403.

[B M]   J.M.Ball, F.Murat, $W^{1,p}$*-quasiconvexity and variational problems for multiple integrals*, J. Funct. Anal. **58** (1984), 225-253.

[B DM 1] G.Buttazzo, G.Dal Maso, *Γ-limits of integral functionals*, J. Analyse Math. **37** (1980), 145-185.

[B DM 2] G.Buttazzo, G.Dal Maso, *Integral representation and relaxation of local functionals*, Nonlinear Anal. **9** (1985), 515-532.

[C S]   L.Carbone, C.Sbordone, *Some properties of Γ-limits of integral functionals*, Ann. Mat. Pura Appl. **122** (1979), 1-60.

[D]   B.Dacorogna, *Weak continuity and weak lower semicontinuity of nonlinear functionals*, Springer-Varlag, Berlin (1982).

[DM]   G.Dal Maso, *Alcuni teoremi sui Γ-limiti di misure*, Boll. Un. Mat. Ital. **15-B** (1979), 182-192.

[DM M]   G.Dal Maso, L.Modica, *A general theory of variational functionals*, "Topics in Functional Analysis 1980/81", Quaderno della Scuola Normale Superiore, Pisa (1982), 149-221.

[DG 1]   E.De Giorgi, *Sulla convergenza di alcune successioni di integrali del tipo dell'area*, Rend. Mat. **8** (1975), 277-294.

[DG 2] E.De Giorgi, *Convergence problems for functionals and operators,* Proceed. Int. Meeting on "Recent Methods in Nonlinear Analysis" Rome, May 8-12 1978, Pitagora ed., Bologna (1979), 131-188.

[DG F] E.De Giorgi, T.Franzoni, *Su un tipo di convergenza variazionale,* Rend. Sem. Mat. Brescia **3** (1979), 63-101.

[DG S] E.De Giorgi, S.Spagnolo, *Sulla convergenza degli integrali dell'energia per operatori ellittici del secondo ordine,* Boll. Un. Mat. Ital. **8** (1973), 391-411.

[F 1] N.Fusco, *On the convergence of integral functionals depending on vector-valued functions,* Ricerche Mat. **32** (1983), 321-339.

[F 2] N.Fusco, *Remarks on the relaxation of integrals of the Calculus of Variations,* in "Systems of Nonlinear Partial Differential Equations", ed. by J.M.Ball, D.Reidel Publishing Company, Dordrecht (1983), 401-408.

[Ma] P.Marcellini, *Approximation of quasiconvex functions and lower semicontinuity of multiple integrals,* Manuscripta Math. 51 (1985), 1-28.

[M S] P.Marcellini, C.Sbordone, *Semicontinuity problems in the Calculus of Variations,* Nonlin. Anal. (1980), 241-257.

[Me] N.G.Meyers *Quasi-convexity and lower-semicontinuity of multiple variational integrals of any order,* Trans. Am. Soc. **119** (1965), 125-149.

[M] C.B.Morrey, *Multiple integrals in the Calculus of Variations,* Springer, Berlin (1966).

[Sb] C.Sbordone, *Su alcune applicazioni di un tipo di convergenza variazionale,* Ann. Sc. Norm. Sup. Pisa **2** (1975), 617-638.

[S] J.Serrin, *On the definition and properties of certain variational integrals,* Trans. Am. Mat. Soc. **101** (1961), 135-167.

Dipartimento di Matematica e Applicazioni "R. Caccioppoli"

Università di Napoli

Via Mezzocannone 8

I-80134 NAPOLI

Istituto di Fisica, Matematica ed Informatica

Facoltà di Ingegneria

Università di Salerno

I-84100 FISCIANO (SA)

LIMITS OF OBSTACLE PROBLEMS
FOR THE AREA FUNCTIONAL

MICHELE CARRIERO GIANNI DAL MASO
ANTONIO LEACI EDUARDO PASCALI

Dedicated to Ennio De Giorgi on his sixtieth birthday

Summary. We study the general form of the limit, in the sense of Γ-convergence, of a sequence of variational problems for the area functional with one-side obstacles.

Introduction. Let Ω be a bounded open subset of R^n with Lipschitz continuous boundary $\partial\Omega$, let $\psi : \Omega \to \bar{R}$ be an arbitrary Borel function, and let $\phi \in L^1(\partial\Omega)$. Under these general assumptions we consider the problem of minimizing the area functional

$$(0.1) \qquad \int_\Omega \sqrt{1 + |Du|^2}$$

on $BV(\Omega)$ with the obstacle condition

$$(0.2) \qquad u_+ \geq \psi \quad H^{n-1} -\text{a.e. on } \Omega$$

and the boundary condition

$$(0.3) \qquad \gamma(u) = \phi \qquad H^{n-1} \text{ —a.e. on } \partial\Omega,$$

where u_+ denotes the approximate upper limit of u (see Section 1) and $\gamma : BV(\Omega) \to L^1(\partial\Omega)$ denotes the trace operator.

It is well known that this problem does not admit a solution for every ψ and ϕ. However, under a very mild condition relating ϕ with ψ, we have proved in a previous paper (see [4]) that the greatest lower bound of (0.1) under the constraints (0.2) and (0.3) coincides with the minimum value on $BV(\Omega)$ of the functional

$$
(0.4) \qquad
\begin{aligned}
F(u) &= \int_\Omega \sqrt{1 + |Du|^2} + \int_{\partial\Omega} |\gamma(u) - \phi| dH^{n-1} \\
&\quad + \int_\Omega [(\psi - u_+) \vee 0]\, d\sigma,
\end{aligned}
$$

where σ is the $(n-1)$-dimensional variational measure introduced by De Giorgi, Colombini, and Piccinini (see [12]). Moreover every minimizing sequence of (0.1) under the constraints (0.2) and (0.3) has a subsequence which converges in $L^1(\Omega)$ to a minimum point of (0.4) and, conversely, every minimum point of (0.4) is the limit in $L^1(\Omega)$ of a minimizing sequence of (0.1) under the constraints (0.2) and (0.3).

In this paper, given a sequence (ψ_h) of Borel functions from Ω into $\bar{R}$, we study the limit behaviour of the minimum points and of the minimum values of the functionals

$$
(0.5) \qquad
\begin{aligned}
F_h(u) &= \int_\Omega \sqrt{1 + |Du|^2} + \int_{\partial\Omega} |\gamma(u) - \phi| dH^{n-1} \\
&\quad + \int_\Omega [(\psi_h - u_+) \vee 0]\, d\sigma,
\end{aligned}
$$

related to the obstacles ψ_h.

We assume only that there exists $w \in W^{1,1}(\Omega)$ such that $\gamma(w) = \phi \, H^{n-1}$-a.e. on $\partial\Omega$ and $w_+ \geq \psi_h H^{n-1}$ -a.e. on Ω for every $h \in N$. We prove that, under this hypothesis, there exists a subsequence (ψ_{h_k}) such that the corresponding sequence of functionals (F_{h_k}) converges, in the sense of the Γ-convergence (see Section 1), to a functional F_∞ of the form

(0.6)
$$F_\infty(u) = \int_\Omega \sqrt{1 + |Du|^2} + \int_{\partial\Omega} |\gamma(u) - \phi| dH^{n-1} + \int_\Omega g(x, u_+) d\mu,$$

where μ is a non-negative Borel measure on Ω, absolutely continuous with respect to H^{n-1}, and $g : \Omega \times R \to [0, +\infty]$ is a Borel function, with $g(x, .)$ convex, decreasing, and lower semicontinuous on R for every $x \in \Omega$ (see Theorem 3.4).

This allows to obtain the desired results about the convergence of minimum points and minimum values by using a general property of Γ-convergence (see Theorem 4.3).

In the most common situations the integral term

(0.7)
$$\int_\Omega g(x, u_+) d\mu$$

has the form

(0.8)
$$\int_\Omega g(x, u_+) d\mu = \int_\Omega [(\psi - u_+) \vee 0] \, d\sigma$$

for a suitable function $\psi : \Omega \to \bar{R}$, so the functional (0.6) coincides with the functional (0.4) associated with an obstacle problem. But there are examples where (0.7) cannot be written in the form (0.8). For instance, an example can be given where

$$\int_\Omega g(x, u_+) d\mu = \int_\Omega |u \wedge 0| dx$$

(see [20], Theorem 4.1).

Our compactness theorem for sequences of the form (0.5) improves an analogous result obtained by C.Picard in [20], where an integral representation formula like (0.6) is proved only for $u \in W^{1,1}(\Omega)$.

The extention of (0.6) to $BV(\Omega)$ is necessary to apply to the case at hand the general theorems on convergence of minima related to the Γ-convergence. Our proof of (0.6) on $BV(\Omega)$ relies on a result concerning the approximation from above of a function of $BV(\Omega)$ by means of a sequence in $W^{1,1}(\Omega)$ (see [4], Theorem 3.3).

All results described here for the area functional are easily extended to more general integral functionals with linear growth in the gradient.

Similar problems for superlinear functionals have been extensively studied by many authors. We refer to [3], [13], [10], [5], [1] for the quadratic case and to [7], [8], [2] for the general case.

1. Notation and preliminaries.

1. By a *Borel measure* on R^n we mean a non-negative countably additive set function defined on the Borel σ-field of R^n.

We indicate by L^n the n-dimensional Lebesgue measure on R^n and by $H^m (0 \leq m \leq n)$ the m-dimensional Hausdorff measure (see [16], 2.10.2).

Let Ω be an open subset of R^n. For every $u \in L^1_{loc}(\Omega)$ we define (see [16], 2.9.12)

$$u_+(x) = \inf \left\{ t \in R : \limsup_{\rho \to 0} \rho^{-n} L^n(\{u > t\} \cap B_\rho(x)) = 0 \right\},$$

where

$$B_\rho(x) = \{y \in R^n : |x - y| < \rho\} \text{ and } \{u > t\} = \{y \in \Omega : u(y) > t\}.$$

2. By $BV(\Omega)$ we denote the space of all functions u of $L^1(\Omega)$ whose gradient Du (in the distributional sense) is a bounded vector measure on Ω. By $BV_{loc}(\Omega)$ we denote the space of all functions which belong to $BV(\Omega')$ for every open set $\Omega' \subset\subset \Omega$ (i.e. $\bar{\Omega}'$ compact and $\bar{\Omega}' \subset \Omega$).

If Ω is bounded, the *area* of u on Ω is defined as

(1.1)
$$A(u,\Omega) = \sup \Big\{ \int_\Omega \Big[\sum_{i=1}^n u D_i \phi_i + \phi_{n+1} \Big] dL^n :$$
$$\phi \in C_o^1(\Omega; R^{n+1}), |\phi| \le 1 \Big\}$$

for every $u \in L_{loc}^1(\Omega)$.

If $u_h, u \in BV(\Omega)$, we say that (u_h) *converges in area* to u on Ω if (u_h) converges to u in $L^1(\Omega)$ and $A(u_h, \Omega)$ converges to $A(u, \Omega)$.

For the main properties of the space $BV(\Omega)$ and of the area functional $A(u, \Omega)$ we refer to [18] and [19].

3. We denote by U the family of all bounded open subsets of R^n and by V a vector space of (L^n-equivalence class of) functions defined on R^n. In particular we shall consider $V = L_{loc}^1(R^n)$ and $V = BV_{loc}(R^n)$.

By a *local functional* on V we mean a functional

$$G : V \times U \to \bar{R}$$

such that $G(u, \Omega) = G(v, \Omega)$ for every $\Omega \in U$ and for every pair of functions $u, v \in V$ with $u = v$ L^n-a.e. on Ω.

Let G be a local functional on V. We say that G is *decreasing* if, for every $\Omega \in U$, the functional $G(., \Omega)$ is decreasing on V with respect to the order induced by L^n -a.e. inequalities.

We say that G is a *measure* if, for every $v \in V$, the set function $G(v, .)$ is the trace on U of a Borel measure.

Let G be a local functional on $L_{loc}^1(R^n)$ and let $\Omega \in U$. The functional $G(., \Omega)$, defined on $L_{loc}^1(R^n)$, can be extended in a natural way to $L^1(\Omega)$: for every $u \in L^1(\Omega)$ we set $G(u, \Omega) = G(v, \Omega)$, where v is an arbitrary function of $L_{loc}^1(R^n)$ which extends u. Since G is local, the definition of $G(u, \Omega)$ does not depend on the extension chosen.

Example 1.1. *The area functional* $A : L_{loc}^1(R^n) \times U \to [0, +\infty]$ *defined by (1.1) is a local functional on* $L_{loc}^1(R^n)$. *Moreover* A *is a measure.*

Example 1.2. *Let* $\psi : R^n \to \bar{R}$ *be a Borel function and let* $G_\psi : L^1_{loc}(R^n) \times U \to [0,+\infty]$ *be the functional defined by*

$$G_\psi(u,\Omega) = \begin{cases} 0 & \text{if } u_+ \geq \psi \ H^{n-1}-a.e. \text{ on } \Omega, \\ +\infty & \text{otherwise.} \end{cases}$$

Then G_ψ *is a decreasing local functional on* $L^1_{loc}(R^n)$. *Moreover* G_ψ *is a measure.*

Example 1.3. *Let* $g : R^n \times \bar{R} \to [0,+\infty]$ *be a Borel function such that* $g(x,.)$ *is decreasing on* $\bar{R}$ *for every* $x \in R^n$. *Let* μ *and* ν *be two non-negative Borel measures on* R^n. *Let* $G : L^1_{loc}(R^n) \times U \to [0,+\infty]$ *be the functional defined by*

$$G(u,\Omega) = \int_\Omega g(x,u_+)d\mu + \nu(\Omega).$$

Then G *is a decreasing local functional on* $L^1_{loc}(R^n)$. *Moreover* G *is a measure.*

4. To study the convergence properties mentioned in the previous section we need the notion of Γ-convergence introduced in [14]. This is a variational convergence for functionals defined on a topological space. It is called also epi-convergence (see [1]), because of the equivalence between Γ-convergence of functions and Kuratowski convergence of their epigraphs.

We say that a sequence (F_h) of functionals defined on a metric space X Γ-*converges* in X to a functional $F : X \to \bar{R}$ if the following conditions are satisfied:

(a)   for every $u \in X$ and for every sequence (u_h) converging to u in X

$$F(u) \leq \liminf_h F_h(u_h);$$

(b)   for every $u \in X$ there exists a sequence (u_h) converging to u in X such that

$$F(u) \geq \limsup_h F_h(u_h).$$

It is evident from the above definition that the Γ-convergence is preserved under the addition of a continuous term. The variational meaning of this convergence is given by the following property (see [14],Corollary 2.4).

Proposition 1.4. *Let* (F_h) *be a sequence of functionals which* Γ-*converges in* X *to a functional* F. *Suppose that for every* $t \in R$ *there exists a compact set* $K_t \subset X$ *such that*

$$\{v \in X : F_h(v) \leq t\} \subset K_t$$

for every $h \in N$. *Then* F *attains its minimum in* X *and*

$$\lim_{h} \inf_{v \in X} F_h(v) = \min_{v \in X} F(v).$$

Suppose, in addition, that each functional F_h *has a minimum point* u_h *in* X *and that* F *has a unique minimum point* u *in* X *such that* $F(u) < +\infty$. *Then* (u_h) *converges to* u *in* X.

2. A compactness theorem. In this section we study the compactness properties of a class $\mathcal{F}$ of local functionals on $L^1_{loc}(R^n)$ which includes the relaxed obstacle functionals

$$F(u,\Omega) = A(u,\Omega) + \int_\Omega [(\psi - u_+) \vee 0] \, d\sigma$$

considered in [4], section 6.

Definition 2.1. *We denote by* $\mathcal{F}$ *the set of all local functionals* F *on* $L^1_{loc}(R^n)$ *such that:*
(a) $F(u,\Omega) \geq A(u,\Omega)$ *for every* $u \in L^1(\Omega)$ *and for every* $\Omega \in U$;
(b) F *is a measure;*
(c) *for every* $\Omega \in U$ *the functional* $G(.,\Omega) : BV(\Omega) \rightarrow [0,+\infty]$, *defined by* $G(u,\Omega) = F(u,\Omega) - A(u,\Omega)$, *is decreasing on* $BV(\Omega)$.
We denote by $\mathcal{F}_s$ *the set of all functionals of the class* $\mathcal{F}$ *such that* $F(.,\Omega)$ *is lower semicontinuous on* $L^1(\Omega)$ *for every* $\Omega \in U$.

Example 2.2. *Let G be a non-negative decreasing local functional on $L^1_{loc}(R^n)$, and let F be the local functional on $L^1_{loc}(R^n)$ defined by*

$$F(u,\Omega) = A(u,\Omega) + G(u,\Omega).$$

If G is a measure, then F belongs to the class $\mathcal{F}$.

Example 2.3. *Let $\psi : R^n \to R$ be a Borel function and let $F_1, F_2 : L^1_{loc}(R^n) \times U \to [0, +\infty]$ be the local functionals defined by*

$$F_1(u,\Omega) = \begin{cases} A(u,\Omega) & \text{if } u_+ \geq \psi \ H^{n-1}-a.e. \text{ on } \Omega, \\ +\infty & \text{otherwise,} \end{cases}$$

$$F_2(u,\Omega) = A(u,\Omega) + \int_\Omega [(\psi - u_+) \vee 0] \, d\sigma,$$

where σ is the $(n-1)$-dimensional variational measure introduced by De Giorgi, Colombini and Piccinini (see [12]). From Example 2.2 it follows easily that F_1 and F_2 belong to the class $\mathcal{F}$ (see examples 1.2 and 1.3). Moreover for every $\Omega \in U$ the functional $F_2(.,\Omega)$ is lower semicontinuous on $L^1(\Omega)$, hence F_2 belongs to the class $\mathcal{F}_s$. To prove that $F_2(.,\Omega)$ is also convex, we consider the functional F_3 defined by

$$F_3(u,\Omega) = \begin{cases} A(u,\Omega) & \text{if } u \in W^{1,1}(\Omega) \text{ and} \\ & u_+ \geq \psi \ H^{n-1}-a.e. \text{ on } \Omega, \\ +\infty & \text{elsewhere on } L^1_{loc}(R^n) \times U. \end{cases}$$

Note that for every $\Omega \in U$ the functional $F_3(.,\Omega)$ is convex on $L^1(\Omega)$. Since $F_2(.,\Omega)$ is the greatest lower semicontinuous functional on $L^1(\Omega)$ majorized by $F_3(.,\Omega)$ (see [4]), we obtain easily that $F_2(.,\Omega)$ is convex. We remark that $F_1(.,\Omega)$ is neither convex nor lower semicontinuous on $L^1(\Omega)$, as one can see by simple examples even in dimension $n = 1$.

To state the compactness of the class $\mathcal{F}$ with respect to the Γ-convergence we need the following definition.

Definition 2.4. *A family $(\Omega_t)_{t \in R}$ in U is said to be a chain if $\Omega_s \subset\subset \Omega_t$ for every $s < t$. A subset $\mathcal{R}$ of U is said to be rich in U*

if, for every chain $(\Omega_t)_{t \in R}$ *in* U, *the set* $\{t \in R : \Omega_t \notin \mathcal{R}\}$ *is at most countable.*

We now prove the compactness of the class $\mathcal{F}$.

Theorem 2.5. *For every sequence* (F_h) *in* $\mathcal{F}$ *there exist a subsequence* F_{h_k}, *a functional* F *of the class* $\mathcal{F}_s$, *and a family* $\mathcal{R}$ *of open sets, rich in* U, *such that the functionals* $F_{h_k}(.,\Omega)$ Γ*-converge to* $F(.,\Omega)$ *in* $L^1(\Omega)$ *for every* $\Omega \in \mathcal{R}$.

PROOF. We give here a proof that can be extended easily to the more general case considered in Section 5. A different proof for the specific case of the area functional can be found in [20].

Let (F_h) be a sequence in $\mathcal{F}$. By a general compactness theorem with respect to Γ-convergence (see [11] Theorem 4.15 and Proposition 4.11), there exist a subsequence (F_{h_k}) of (F_h), a local functional $F : L^1_{loc}(R^n) \times U \to [0, +\infty]$, and a rich family $\mathcal{R}$ of bounded open sets such that

(i)   for every $\Omega \in \mathcal{R}$ the functionals $F_{h_k}(.,\Omega)$ Γ-converge to $F(.,\Omega)$ in $L^1_{loc}(R^n)$, (hence in $L^1(\Omega)$),

(ii)   for every $\Omega \in U$ the functional $F(.,\Omega)$ is lower semicontinuous in $L^1_{loc}(R^n)$, (hence in $L^1(\Omega)$),

(iii)   the set function $F(u,.)$ is increasing in U for every $u \in L^1_{loc}(R^n)$ and

$$F(u,\Omega) = \sup \{F(u,\Omega') : \ \Omega' \in U, \Omega' \subset\subset \Omega\}$$

for every $u \in L^1_{loc}(R^n)$ and for every $\Omega \in U$.

Since $A(.,\Omega)$ is lower semicontinuous on $L^1(\Omega)$, from the inequality $F_h(u,\Omega) \geq A(u,\Omega)$ it follows that $F(u,\Omega) \geq A(u,\Omega)$ for every $u \in L^1(\Omega)$ and for every $\Omega \in U$.

Let us define, for every $\Omega \in U$ and for every $u \in BV(\Omega)$,

(2.1)
$$G(u,\Omega) = F(u,\Omega) - A(u,\Omega)$$
$$G_h(u,\Omega) = F_h(u,\Omega) - A(u,\Omega) \qquad (h = 1,2,..)$$

Since $G_h(.,\Omega)$ is decreasing on $BV(\Omega)$, and

(2.2) $$A(u \vee v,\Omega) + A(u \wedge v,\Omega) \leq A(u,\Omega) + A(v,\Omega)$$

for every $u,v \in L^1(\Omega)$, we can adapt Lemma 3.3(3) of [2] to prove that $G(.,\Omega)$ is decreasing on $BV(\Omega)$ for every $\Omega \in U$.

To prove that $F \in \mathcal{F}$ it remains only to show that F is a measure. Since each F_h is a measure, it is easy to check, by using the definition of Γ-convergence, that the set function $F(u, .)$ is superadditive on U for every $u \in L^1_{loc}(R^n)$. Therefore, to prove that F is a measure, it remains to show that for every $u \in L^1_{loc}(R^n)$ the set function $F(u, .)$ is subadditive on U (see [15], Theorem 5.6).

To this aim, we fix $u \in L^1_{loc}(R^n)$ and $\Omega_1, \Omega_2 \in U$ and we put $\Omega = \Omega_1 \cup \Omega_2$. We have to prove that

$$(2.3) \qquad F(u, \Omega) \le F(u, \Omega_1) + F(u, \Omega_2).$$

We may assume that $F(u, \Omega_1) < +\infty$ and $F(u, \Omega_2) < +\infty$, hence $u \in BV(\Omega)$. By (iii), inequality (2.3) is equivalent to

$$(2.4) \qquad F(u, \Omega') \le F(u, \Omega_1) + F(u, \Omega_2)$$

for every $\Omega' \in \mathcal{R}$ with $\Omega' \subset\subset \Omega$.

Let us fix $\Omega' \in \mathcal{R}$ with $\Omega' \subset\subset \Omega$. Then there exist $\Omega_1', \Omega_2' \in U$ such that

$$\Omega' \subset\subset \Omega_1' \cup \Omega_2' \ , \ \Omega_1' \subset\subset \Omega_1 \ , \ \Omega_2' \subset\subset \Omega_2.$$

Let us choose Ω_1'' and Ω_2'' in $\mathcal{R}$ such that

$$\Omega_1' \subset\subset \Omega_1'' \subset\subset \Omega_1 \ , \ \Omega_2' \subset\subset \Omega_2'' \subset\subset \Omega_2.$$

To simplify the notation, we indicate by (F_h) the subsequence which satisfies (i).

By the definition of Γ-convergence, there exists two sequences (u_h^1) and (u_h^2) converging to u in $L^1_{loc}(R^n)$ such that

$$(2.5) \qquad \lim_h F_h(u_h^i, \Omega_i'') = F(u, \Omega_i'') \le F(u, \Omega_i) < +\infty$$

for $i = 1, 2$. We may assume that $u_h^i \in BV(\Omega_i'')$ for every $h \in N$ and for $i = 1, 2$.

We now want to modify u_h^i out of Ω_i', so that the modified function v_h^i is equal to u out of Ω_i'' and

$$A(v_h^i, \Omega') \le A(u_h^i, \Omega_i'') + A(u, \Omega' \setminus \overline{\Omega_i'}) + R_h^i,$$

with a careful estimate of the remainder R_h^i.

To this aim we use the J-property introduced in [11], Definition 2.2. Since the area functional $A(u, \Omega)$ satisfies the J-property (see [11], Theorem 6.4 and Corollary 6.7), for every $\epsilon > 0$ there exist a constant $M > 0$ and two sequences (v_h^1) and (v_h^2) converging to u in $L_{loc}^1(R^n)$ such that $v_h^i = u_h^i$ on Ω_i', $v_h^i = u$ on $R^n \backslash \Omega_i''$, and

$$
(2.6) \quad
\begin{aligned}
A(v_h^i, \Omega') \leq (1 + \epsilon) &\left[A(u_h^i, \Omega_i'') + A(u, \Omega' \backslash \overline{\Omega}_i') \right] + \\
&+ \epsilon \left[\|u_h^i\|_{L^1(\Omega_i'')} + \|u\|_{L^1(\Omega')} + 1 \right] + M \|u_h^i - u\|_{L^1(\Omega_i'')}
\end{aligned}
$$

for every $h \in N$ and for $i = 1, 2$. Since $u_h^i \in BV(\Omega_i'')$ and $u \in BV(\Omega')$, from (2.6) it follows that $v_h^i \in BV(\Omega')$ for every $h \in N$.

We define $v_h = v_h^1 \vee v_h^2$ for every $h \in N$. Then (v_h) converges to u in $L_{loc}^1(R^n)$ and $v_h \in BV(\Omega')$ for every $h \in N$. By property (i) and by the definition of Γ-convergence we have

$$
(2.7) \qquad F(u, \Omega') \leq \liminf_h F_h(v_h, \Omega').
$$

From (2.1) and (2.2) we obtain

$$
(2.8) \quad F_h(v_h, \Omega') \leq A(v_h^1, \Omega') + A(v_h^2, \Omega') - A(v_h^1 \wedge v_h^2, \Omega') + G_h(v_h, \Omega').
$$

Since $\Omega' \subset \Omega_1' \cup \Omega_2'$, from the inequalities $v_h \geq v_h^1$ and $v_h \geq v_h^2$ we get

$$
(2.9)
$$
$$
G_h(v_h, \Omega') \leq G_h(v_h^1, \Omega_1') + G_h(v_h^2, \Omega_2') = G_h(u_h^1, \Omega_1') + G_h(u_h^2, \Omega_2').
$$

By (2.6),(2.8),(2.9) we have

$$
F(v_h, \Omega') \leq
$$

$$
\leq (1 + \epsilon) \left[A(u_h^1, \Omega_1'') + A(u, \Omega' \backslash \overline{\Omega}_1') + A(u_h^2, \Omega_2'') + A(u, \Omega' \backslash \overline{\Omega}_2') \right]
$$

$$
+ G_h(u_h^1, \Omega_1') + G_h(u_h^2, \Omega_2')
$$

$$
+ \epsilon \left[\|u_h^1\|_{L^1(\Omega_1'')} + \|u_h^2\|_{L^1(\Omega_2'')} + 2\|u\|_{L^1(\Omega')} + 1 \right]
$$

$$
+ M \left[\|u_h^1 - u\|_{L^1(\Omega_1'')} + \|u_h^2 - u\|_{L^1(\Omega_2'')} \right] - A(v_h^1 \wedge v_h^2, \Omega') \leq
$$

$$\leq (1 + \epsilon) \left[F(u_h^1, \Omega_1'') + F(u_h^2, \Omega_2'') + A(u, \Omega') \right]$$

$$+ \epsilon \left[\|u_h^1\|_{L^1(\Omega_1'')} + \|u_h^2\|_{L^1(\Omega_2'')} + 2\|u\|_{L^1(\Omega')} + 1 \right]$$

$$+ M \left[\|u_h^1 - u\|_{L^1(\Omega_1'')} + \|u_h^2 - u\|_{L^1(\Omega_2'')} \right] - A(v_h^1 \wedge v_h^2, \Omega').$$

Since $A(., \Omega')$ is lower semicontinuous in $L^1_{loc}(R^n)$ and $(v_h^1 \wedge v_h^2)$ converges to u in $L^1_{loc}(R^n)$, by taking (2.5) and (2.7) into account we obtain

$$F(u, \Omega') \leq (1 + \epsilon) \left[F(u, \Omega_1) + F(u, \Omega_2) + A(u, \Omega') \right]$$

$$+ \epsilon \left[4\|u\|_{L^1(\Omega)} + 1 \right] - A(u, \Omega').$$

By taking the limit as $\epsilon \to 0$ we obtain (2.4), so F is a measure and the theorem is proved.

3. An integral representation theorem. The aim of this section is to prove the following integral representation theorem.

Theorem 3.1. *Let F be a functional of the class $\mathcal{F}_s$. Then there exist*

(a) *a Borel function $g : R^n \times R \to [0, +\infty]$, such that $g(x, .)$ is decreasing and lower semicontinuous on R for every $x \in R^n$,*

(b) *a non-negative Radon measure μ on R^n, absolutely continuous with respect to H^{n-1},*

(c) *a non-negative Borel measure ν on R^n, not necessarily finite on compact sets, such that*

(3.1) $$F(u, \Omega) = A(u, \Omega) + \int_\Omega g(x, u_+) d\mu + \nu(\Omega)$$

for every $\Omega \in U$ and for every $u \in L^1(\Omega)$.

If $F(., \Omega)$ is convex on $L^1(\Omega)$ for every $\Omega \in U$, then we can require in addition that $g(x, .)$ is convex on R for every $x \in R^n$.

To prove theorem 3.1 we need the following integral representation theorem for non-negative decreasing local functionals defined on $BV_{loc}(R^n)$.

Theorem 3.2. *Let G a non-negative decreasing local functional on $BV_{loc}(R^n)$. Assume that G is a measure and that for every $\Omega \in U$ the function $G(.,\Omega)$ is lower semicontinuous on $BV_{loc}(R^n)$ with respect to the convergence in area on Ω. Then there exist g, μ, ν as in Theorem 3.1 such that*

$$(3.2) \qquad G(u,\Omega) = \int_\Omega g(x,u_+)d\mu + \nu(\Omega)$$

for every $u \in BV_{loc}(R^n)$ and for every $\Omega \in U$.

PROOF. By the integral representation theorem for decreasing local functionals in $W^{1,1}(R^n)$ (see [9], Theorem 5.7 and Section 6), there exist g, μ, ν, satisfying conditions (a),(b),(c) of Theorem 3.1 such that

$$(3.3) \qquad G(u,\Omega) = \int_\Omega g(x,\tilde{u})d\mu + \nu(\Omega)$$

for every $u \in W^{1,1}(R^n)$ and for every $\Omega \in U$.

Let Ω be a bounded open set with a Lipschitz boundary and with $\nu(\Omega) < +\infty$. Since G is local and every function of $BV(\Omega)$ (resp. $W^{1,1}(\Omega)$) can be extended to a function of $BV(R^n)$ (resp. $W^{1,1}(R^n)$), the function $G(.,\Omega)$ is well defined on $BV(\Omega)$ and (3.3) holds for every $u \in W^{1,1}(\Omega)$. Therefore we can apply Lemma 6.3 of [4] to the function $G(.,\Omega) - \nu(\Omega)$. This proves (3.2) for every $u \in BV_{loc}(R^n)$ and for every $\Omega \in U$ with a Lipschitz boundary and with $\nu(\Omega) < +\infty$.

Suppose now that Ω is a bounded open set with a Lipschitz boundary and with $\nu(\Omega) = +\infty$. Then $G(u,\Omega) = +\infty$ for every $u \in W^{1,1}(\Omega)$. Since G is decreasing and every function of $BV(\Omega)$ is majorized by a function of $W^{1,1}(\Omega)$ (see, for instance, Theorem 3.3 of [4]) we have $G(u,\Omega) = +\infty$ for every $u \in BV(\Omega)$, hence (3.2) holds for every $u \in BV_{loc}(R^n)$ also in the case $\nu(\Omega) = +\infty$.

We can conclude that (3.2) holds for every $u \in BV_{loc}(R^n)$ and for every $\Omega \in U$ with a Lipschitz boundary. Since G is a measure, we can approximate every bounded open set by means of bounded open sets with a Lipschitz boundary, therefore (3.2) holds for every $\Omega \in U$ and the theorem is proved.

Remark 3.3. If, in Theorem 3.2, we assume also that the functions

$$s \to G(u + s, \Omega)$$

are convex on R for every $u \in BV_{loc}(R^n)$ and for every $\Omega \in U$, then we can require, in addition, that $g(x,.)$ is convex on R for every $x \in R^n$. In fact, a careful inspection in the proof shows that, in this case, the function $g(x, s)$ constructed in Theorem 5.7 of [9] is convex in s for every $x \in R^n$.

Proof of Theorem 3.1. Let F be a functional of the class $\mathcal{F}_s$ and let $G : BV_{loc}(R^n) \times U \to [0, +\infty]$ be the local functional defined by

$$G(u, \Omega) = F(u, \Omega) - A(u, \Omega).$$

Condition (c) of Definition 2.1 says that G is decreasing. Since F is a measure, G is a measure too. Since $F(., \Omega)$ is lower semicontinuous on $L^1_{loc}(R^n)$, $G(., \Omega)$ is lower semicontinuous on $BV_{loc}(R^n)$ with respect to the convergence in area on Ω.

Therefore there exist g, μ, ν satisfying conditions (a),(b) and (c) such that

$$G(u, \Omega) = \int_\Omega g(x, u_+) d\mu + \nu(\Omega)$$

for every $u \in BV_{loc}(R^n)$ and for every $\Omega \in U$.

This gives (3.1) for every $u \in BV_{loc}(R^n)$ and for every $\Omega \in U$. Since F and A are local measures, (3.1) holds also for $\Omega \in U$ and for every $u \in BV(\Omega)$. Since both sides of (3.1) are $+\infty$ if $u \notin BV(\Omega)$, equality (3.1) holds for every $\Omega \in U$ and for every $u \in L^1(\Omega)$.

If $F(., \Omega)$ is convex on $L^1(\Omega)$ for every $\Omega \in U$, then the function

$$s \to G(u + s, \Omega)$$

is convex on R for every $u \in BV_{loc}(R^n)$ and for every $\Omega \in U$. Therefore we can require that $g(x,.)$ is convex on R for every $x \in R^n$ (see remark 3.3.).

From Theorems 2.5 and 3.1 we obtain the following result concerning the Γ-limits of obstacle functionals.

Theorem 3.4. *Let* (ψ_h) *be a sequence of Borel functions from* Ω *into* $\bar{R}$ *and let* F_h *be the functionals of the class* $\mathcal{F}_s$ *defined by*

$$F_h(u, \Omega) = A(u, \Omega) + \int_\Omega [(\psi_h - u_+) \vee 0] \, d\sigma.$$

Then there exist a subsequence (F_{h_k}), *a functional* F *of the class* $\mathcal{F}_s$, *and a family* $\mathcal{R}$ *of open sets, rich in* U, *such that the functionals* $F_{h_k}(., \Omega)$ Γ-*converge to* $F(., \Omega)$ *in* $L^1(\Omega)$ *for every* $\Omega \in \mathcal{R}$. *Moreover there exist* g, μ, ν *as in Theorem 3.1, with* $g(x, .)$ *convex on* R *for every* $x \in R^n$, *such that*

$$F(u, \Omega) = A(u, \Omega) + \int_\Omega g(x, u_+) d\mu + \nu(\Omega)$$

for every $\Omega \in U$ *and for every* $u \in L^1(\Omega)$.

PROOF. The assertion concerning the Γ-convergence follows from Theorem 2.5. Since the functionals $F_h(., \Omega)$ are convex (Example 2.3), the functional $F(., \Omega)$ is convex on $L^1(\Omega)$ for every $\Omega \in U$, so the conclusion follows from Theorem 3.1.

4. Dirichlet boundary conditions. Throughout this section (F_h) is a fixed sequence of functionals of the class $\mathcal{F}$ and F is a functional of the class $\mathcal{F}_s$. We suppose that there exists a family $\mathcal{R}$, rich in U, such that

(4.1) $(F_h(., \omega))\Gamma-$converges to $F(., \omega)$ in $L^1_{loc}(R^n)$ for every $\omega \in \mathcal{R}$.

Let Ω be a bounded open subset of R^n with Lipschitz continuous boundary $\partial\Omega$, so that the trace operator $\gamma : BV(\Omega) \to L^1(\partial\Omega)$ is well-defined ([18], Theorem 2.10).

Our aim is to study the convergence of the solutions of the (relaxed) minimum problems for the functionals $F_h(., \Omega)$ with Dirichlet boundary conditions.

Let $\phi \in L^1(\partial\Omega)$ and let $w \in W^{1,1}_{loc}(R^n)$ such that $\gamma(w) = \phi$.

We suppose that the following condition holds:

(4.2) there exist a sequence (w_h) in $BV(\Omega)$ and a compact subset K
       of Ω such that

       (a) $\gamma(w_h) = \gamma(w) = \phi$,

       (b) (w_h) converges to w in $L^1(\Omega\backslash K)$,

       (c) $\limsup\limits_{h} F_h(w_h, \Omega\backslash K) \leq F(w, \Omega\backslash K) < +\infty$.

We consider the following functionals defined on $L^1(\Omega)$

(4.3)
$$F_h^\phi(u) = \begin{cases} F_h(u,\Omega) & \text{if } u \in BV(\Omega) \text{ and } \gamma(u) = \phi \ H^{n-1}-\text{a.e. on } \partial\Omega, \\ +\infty & \text{otherwise,} \end{cases}$$

(4.4)
$$\bar{F}_h^\phi(u) = F_h(u,\Omega) + \int_{\partial\Omega} |\gamma(u) - \phi| dH^{n-1},$$

(4.5)
$$\bar{F}^\phi(u) = F(u,\Omega) + \int_{\partial\Omega} |\gamma(u) - \phi| dH^{n-1}.$$

Theorem 4.1. *Assume that (4.1) and (4.2) are satisfied. Then the sequences (F_h^ϕ) and $(\bar{F}_h^\phi)$ Γ-converge to $\bar{F}^\phi$ in $L^1(\Omega)$.*

To prove Theorem 4.1, we need the following lemma.

Lemma 4.2. *If (4.1) and (4.2) are satisfied then for every $\Omega', \Omega'' \in U$ such that $K \subset\subset \Omega' \subset\subset \Omega'' \subset\subset \Omega$ we have*

$$\limsup\limits_{h} F_h(w_h, \Omega\backslash\bar{\Omega}'') \leq F(w, \Omega \backslash \bar{\Omega}').$$

PROOF. Let $\Omega^* \in \mathcal{R}$ such that $\Omega^* \subset\subset \Omega''\backslash K$. Then, by (4.1),

$$F(w,\Omega^*) \leq \liminf\limits_{h} F_h(w_h, \Omega^*) \leq \liminf\limits_{h} F_h(w_h, \Omega''\backslash K).$$

Hence

$$F(w, \Omega''\backslash K) \leq \liminf\limits_{h} F_h(w_h, \Omega''\backslash K).$$

By the previous inequality, we obtain

$$\limsup_h F_h(w_h, \Omega \setminus \overline{\Omega}'')$$

$$\leq \limsup_h F_h(w_h, \Omega \setminus K) - \liminf_h F_h(w_h, \Omega'' \setminus K)$$

$$\leq F(w, \Omega \setminus K) - F(w, \Omega'' \setminus K)$$

$$\leq F(w, \Omega \setminus \overline{\Omega}').$$

Before proving Theorem 4.1, we set

$$G_h(u, \omega) = F_h(u, \omega) - A(u, \omega)$$
$$G(u, \omega) = F(u, \omega) - A(u, \omega)$$

for every $\omega \in U$ and for every $u \in BV(\omega)$. Recall that the functionals $G_h(., \omega)$ and $G(., \omega)$ are decreasing on $BV(\omega)$ for every $\omega \in U$.

Proof of Theorem 4.1. For every $u \in L^1(\Omega)$ we define

$$F'(u) = \inf \left\{ \liminf_h \bar{F}_h^\phi(u_h) \; ; \; u_h \to u \text{ in } L^1(\Omega) \right\},$$

$$F''(u) = \inf \left\{ \limsup_h F_h^\phi(u_h) \; ; \; u_h \to u \text{ in } L^1(\Omega) \right\}.$$

Since $\bar{F}_h^\phi \leq F_h^\phi$, to prove the theorem it is enough to show that

(4.6) $$F''(u) \leq \bar{F}^\phi(u)$$

and

(4.7) $$\bar{F}^\phi(u) \leq F'(u)$$

for every $u \in L^1(\Omega)$. Since these inequalities are trivial for $u \notin BV(\Omega)$, it suffices to prove (4.6) and (4.7) for $u \in BV(\Omega)$.

Let us prove (4.6). We suppose first that $u \in BV(\Omega)$ and $\mathrm{spt}(u - w) \subset \Omega$.

Let $\epsilon > 0$; we pose $K_1 = K \cup \operatorname{spt}(u-w)$, and we consider $\Omega_1 \in U$ such that $K_1 \subset \Omega_1 \subset\subset \Omega$ and

$$(4.8) \qquad F(w, \Omega \backslash \bar{\Omega}_1) < \epsilon.$$

Let Ω_2 and Ω_3 be two open subsets of R^n such that $\Omega_1 \subset\subset \Omega_2 \subset\subset \Omega_3 \subset\subset \Omega$. By using the J-property for the area functional (see [11], Theorem 6.4 and Corollary 6.7), we obtain a sequence (w_h') in $BV(\Omega)$ converging to u in $L^1(\Omega)$ such that

$$w_h' = u \qquad \text{on } \Omega_2$$
$$w_h' = w_h \qquad \text{on } \Omega \backslash \Omega_3$$
$$\limsup_h A(w_h', \Omega) \le A(u, \Omega_3) + \limsup_h A(w_h, \Omega \backslash \overline{\Omega}_2),$$

so that, by Lemma 4.2 and by (4.8),

$$\limsup_h A(w_h', \Omega) \le A(u, \Omega) + F(u, \Omega \backslash \overline{\Omega}_1) \le A(u, \Omega) + \epsilon.$$

Let $\Omega_4 \in \mathcal{R}$ such that $\Omega_3 \subset\subset \Omega_4 \subset\subset \Omega$. By (4.1), there exists a sequence (u_h) in $BV(\Omega_4)$ converging to u in $L^1(\Omega_4)$ such that

$$F(u, \Omega_4) = \lim_h F_h(u_h, \Omega_4).$$

We define $u_h' = u_h \vee w_h'$. Since $u_h \wedge w_h' \to u$ in $L^1(\Omega_4)$, we have

$$A(u, \Omega_4) \le \liminf_h A(u_h \wedge w_h', \Omega_4)$$

and then

$$A(u, \Omega_4) + \limsup_h F_h(u_h', \Omega_4) \le$$

$$\le \limsup_h [A(u_h \wedge w_h', \Omega_4) + A(u_h \vee w_h', \Omega_4) + G_h(u_h \vee w_h', \Omega_4)]$$

$$\le \limsup_h A(w_h', \Omega_4) + \limsup_h [A(u_h, \Omega_4) + G_h(u_h, \Omega_4)]$$

$$\le A(u, \Omega) + \epsilon + A(u, \Omega_4) + G(u, \Omega_4).$$

Hence

$$(4.9) \qquad \limsup_h F_h(u_h', \Omega_4) \le F(u, \Omega) + \epsilon.$$

By using again the J-property for the area functional, we obtain a sequence (v_h) in $BV(\Omega)$ converging to u in $L^1(\Omega)$ such that

$$v_h = u'_h \quad \text{in an open neighbourhood of } \overline{\Omega}_3,$$
$$v_h = w_h \quad \text{on } \Omega\backslash\Omega_4,$$
$$v_h \geq u'_h \wedge w_h = w_h \quad \text{on } \Omega_4\backslash\Omega_3,$$
$$A(v_h,\Omega) \leq A(u'_h,\Omega_4) + A(w_h,\Omega\backslash\overline{\Omega}_3) + \epsilon_h$$

with $\lim_h \epsilon_h = 0$. Since $v_h \geq w_h$ on $\Omega\backslash\overline{\Omega}_3$ and $v_h = u'_h$ in a neighbourhood of $\overline{\Omega}_3$ we have

$$G_h(v_h,\Omega) \leq G_h(u'_h,\Omega_4) + G_h(v_h,\Omega\backslash\overline{\Omega}_3)$$
$$\leq G_h(u'_h,\Omega_4) + G_h(w_h,\Omega\backslash\overline{\Omega}_3).$$

Then by (4.9) and by Lemma 4.2 we have

$$F''(u,\Omega) \leq \limsup_h F_h^\phi(v_h) = \limsup_h F_h(v_h,\Omega)$$

$$\leq \limsup_h \left[A(u'_h,\Omega_4) + A(w_h,\Omega\backslash\overline{\Omega}_3) + G_h(u'_h,\Omega_4) + G_h(w_h,\Omega\backslash\overline{\Omega}_3) \right]$$

$$\leq \limsup_h F_h(u'_h,\Omega_4) + \limsup_h F_h(w_h,\Omega\backslash\overline{\Omega}_3)$$

$$\leq F(u,\Omega) + \epsilon + F(w,\Omega\backslash\overline{\Omega}_1) \leq F(u,\Omega) + 2\epsilon.$$

Since ϵ is arbitrary we obtain

$$(4.10) \qquad F''(u,\Omega) \leq F(u,\Omega) = \bar{F}^\phi(u)$$

for every $u \in BV(\Omega)$ with $\text{spt}(u - w) \subset\subset \Omega$.

We consider now an arbitrary $u \in BV(\Omega)$. By Lemma 7.4 of [4] there exists a sequence (u_h) in $BV(\Omega)$ converging to u in $L^1(\Omega)$ such that

$$(4.11) \qquad \text{spt}(u_h - w) \subset\subset \Omega,$$

$$(4.12) \qquad u_h \geq u \wedge w \quad L^n\text{-a.e in } \Omega,$$

(4.13) $u_h = u$ L^n—a.e. on $A_h = \{x \in \Omega : \text{dist}(x, \partial\Omega) \geq 1/h\}$,

(4.14) $\lim\limits_{h} A(u_h, \Omega) = A(u, \Omega) + \int_{\partial\Omega} |\gamma(u) - \phi| dH^{n-1}$.

By (4.10) we have

$$F''(u_h, \Omega) \leq F(u_h, \Omega)$$

for every $h \in N$.

Let H be a compact subset of Ω with $K \subset H \subset \Omega$. By (4.12) and (4.13) we have

$$G(u_h, \Omega) \leq G(u_h, A_h) + G(u_h, \Omega\backslash H) \leq G(u, \Omega) + G(u \wedge w, \Omega\backslash H)$$

for h sufficiently large.

Since $F''(., \Omega)$ is lower semicontinuous with respect to the $L^1(\Omega)$ topology, we have

$$F''(u, \Omega) \leq \liminf\limits_{h} F''(u_h, \Omega) \leq \liminf\limits_{h} [A(u_h, \Omega) + G(u_h, \Omega)]$$

$$\leq A(u, \Omega) + \int_{\partial\Omega} |\gamma(u) - \phi| dH^{n-1} + G(u, \Omega) + G(u \wedge w, \Omega\backslash H).$$

Since $w \in W^{1,1}_{loc}(R^n)$, the following inequality holds

$$(u \wedge w)_+ \geq u_+ \wedge w_+ \qquad H^{n-1}\text{—a.e. on } \Omega.$$

Then, by using the integral representation (3.2) of G, we have

$$G(u \wedge w, \Omega\backslash H) \leq G(u, \Omega\backslash H) + G(w, \Omega\backslash H),$$

and then

$$F''(u, \Omega) \leq A(u, \Omega) + \int_{\partial\Omega} |\gamma(u) - \phi| dH^{n-1} +$$

$$+ G(u, \Omega) + G(u, \Omega\backslash H) + G(w, \Omega\backslash H).$$

By (4.2), $G(w, .)$ is a finite measure on $\Omega\backslash H$, and we can assume also that $G(u, .)$ is a finite measure (otherwise the inequality (4.6) to be proved is trivial). Then, as $H \uparrow \Omega$, we obtain

(4.15) $F''(u, \Omega) \leq \bar{F}^\phi(u, \Omega)$

for every $u \in BV(\Omega)$.

We prove now (4.7) for every $u \in BV(\Omega)$. Let $\Omega' \in \mathcal{R}$, such that $\Omega' \subset\subset \Omega$ and $\int_{\partial\Omega'} |Du| = 0$. Let (u_h) be an arbitrary sequence in $BV(\Omega)$ converging to u in $L^1(\Omega)$. Since the functional $A(u, \Omega\backslash\bar{\Omega}') + \int_{\partial\Omega} |\gamma(u) - \phi| dH^{n-1}$ is lower semicontinuous on $L^1(\Omega)$ (see [18], Theorem 14.5 or [17], Theorem 1.3), we obtain

$$A(u, \Omega) + \int_{\partial\Omega} |\gamma(u) - \phi| dH^{n-1} + G(u, \Omega') =$$

$$= A(u, \Omega \setminus \bar{\Omega}') + \int_{\partial\Omega} |\gamma(u) - \phi| dH^{n-1} + A(u, \Omega') + G(u, \Omega')$$

$$\leq \liminf_h \left[A(u_h, \Omega\backslash\bar{\Omega}') + \int_{\partial\Omega} |\gamma(u_h) - \phi| dH^{n-1} \right]$$

$$+ \liminf_h [A(u_h, \Omega') + G_h(u_h, \Omega')]$$

$$\leq \liminf_h \left[\bar{F}_h^\phi(u_h, \Omega) + G_h(u_h, \Omega) \right].$$

As $\Omega' \uparrow \Omega$, we obtain

$$\bar{F}^\phi(u, \Omega) \leq F'(u, \Omega),$$

which proves (4.7) for every $u \in BV(\Omega)$ and concludes the proof of the theorem.

We now prove the convergence of minima and minimizers of the functionals F_h^ϕ and $\bar{F}_h^\phi$.

Theorem 4.3 *Suppose that (4.1) and (4.2) are satisfied. Then*

$$\lim_h \inf_{u \in BV(\Omega)} F_h^\phi(u) = \lim_h \inf_{u \in BV(\Omega)} \bar{F}_h^\phi(u) = \min_{u \in BV(\Omega)} \bar{F}^\phi(u).$$

Suppose, in addition, that each functional F_h^ϕ (resp. $\bar{F}_h^\phi$) has a minimum point u_h in $BV(\Omega)$ and that $\bar{F}^\phi$ has a unique minimum point u in $BV(\Omega)$ such that $\bar{F}^\phi(u) < +\infty$. Then (u_h) converges to u in $L^1(\Omega)$.

PROOF. Since

$$F_h^\phi(u) \geq A^\phi(u) = A(u,\Omega) + \int_{\partial\Omega} |\gamma(u) - \phi| dH^{n-1},$$

and A^ϕ is coercive in $L^1(\Omega)$ (see [17]), the sequence (F_h^ϕ) is equico-ercive, so the thesis of the theorem follows from Proposition 1.4.

5. Further results. In the previous sections we have considered the limit behaviour of sequences of minimum problems related to the area functional $A(u,\Omega)$. In the present section we extend these results to more general integral functionals with linear growth in the gradient.

Let $f : R^n \times R^n \to R$ be a continuous function with the following properties:

(a)   there exist two constants $c_2 \geq c_1 > 0$ such that

$$c_1|p| \leq f(x,p) \leq c_2(1 + |p|)$$

for every $(x,p) \in R^n \times R^n$;

(b) $f(x,.)$ is convex on R^n for every $x \in R^n$;

(c)   for every $x_o \in R^n$ and for every $\epsilon > 0$ there exists $\delta = \delta(\epsilon, x_o) > 0$ such that

$$|f(x,p) - f(x_o,p)| \leq \epsilon(1 + |p|)$$

for every $(x,p) \in R^n \times R^n$ with $|x - x_o| < \delta$.

Let $f_\infty : R^n \times R^n \to R$ be the recession of f with respect to p, defined by

$$f_\infty(x,p) = \lim_{t \to +\infty} \frac{1}{t} f(x, tp).$$

Note that hypothesis (c) guarantees that f_∞ is continuous.

For every open set $\Omega \subset R^n$ and for every $u \in BV(\Omega)$ we denote by $D_a u$ and $D_s u$ the regular and the singular part in the Lebesgue decomposition of the vector measure Du, i.e.

$$Du = D_a u \, dL^n + D_s u,$$

where $D_a u$ is a function of $L^1(\Omega)$ and the measure $D_s u$ is singular with respect to the Lebesgue measure. We denote by $\frac{dD_s u}{d|D_s u|}$ the

Radon-Nikodym derivative of the measure $D_s u$ with respect to its variation $|D_s u|$.

We are now in a position to introduce the local functional

$$\Phi : L^1_{loc}(R^n) \times U \to [0, +\infty]$$

defined by

$$\Phi(u, \Omega) = \begin{cases} \int_\Omega f(x, D_a u) dL^n + \int_\Omega f_\infty(x, \frac{dD_s u}{d|D_s u|}) d|D_s u| \\ \qquad \text{if } u \in BV(\Omega), \\ +\infty \qquad \text{otherwise.} \end{cases}$$

For every $\Omega \in U$ the functional $\Phi(., \Omega)$ is lower semicontinuous in $L^1_{loc}(R^n)$ (see [6], Theorem 3.1). More precisely, it is the greatest lower semicontinuous functional on $L^1_{loc}(R^n)$ which is less than or equal to

$$\int_\Omega f(x, Du) dL^n$$

for every $u \in W^{1,1}(\Omega)$ (see [6], Theorem 3.2). This implies that

$$\Phi(u \wedge v, \Omega) + \Phi(u \vee v, \Omega) \le \Phi(u, \Omega) + \Phi(v, \Omega)$$

for every $\Omega \in U$ and for every $u, v \in L^1_{loc}(R^n)$.

Moreover, for every $\Omega \in U$ the functional $\Phi(., \Omega)$ is continuous on $BV(\Omega)$ with respect to the convergence in area on Ω. This can be proved by writing $\Phi(u, \Omega)$ in terms of the hypograph $S(u)$ of u (see [6], lemma 2.2) and by using the characterization of the convergence in area in terms of hypographs (given, for instance, in Proposition 1.1 of [4]), which allows us to apply to the case at hand the continuity theorem of Reshetnyak (see [21], Theorem 3), as done in the third part of the proof of Theorem 3.2 in [6].

Using all these properties, we obtain the following result by simply replacing $A(u, \Omega)$ with $\Phi(u, \Omega)$ in the proofs of the theorems.

Theorem 5.1. *Theorems 2.5, 3.1, 4.1, and 4.3 remain valid if we replace $A(u, \Omega)$ with $\Phi(u, \Omega)$ in their statements and in Definition 2.1, and $\int_{\partial\Omega} |\gamma(u) - \phi| dH^{n-1}$ with $\int_{\partial\Omega} f_\infty(x, (\phi - \gamma(u))\nu) dH^{n-1}$ in (4.4) and (4.5), where ν denotes the outward unit normal to $\partial\Omega$.*

References

[1] H.Attouch, *Variational Convergence for Functions and Operators,* Research Notes in Mathematics, Pitman, London, 1984.

[2] H.Attouch, C.Picard, *Variational inequalities with varying obstacles: the general form of the limit problem,* J. Funct. Anal. **50** (1983), 329-386.

[3] L.Carbone, F.Colombini, *On convergence of functionals with unilateral constraints,* J. Math. Pures Appl. **59** (1980), 465-500.

[4] M.Carriero, G.Dal Maso, A.Leaci, E.Pascali, *Relaxation of the nonparametric Plateau problem with an obstacle,* J. Math. Pures Appl. **67** (1988).

[5] D.Cioranescu, F.Murat, *Un terme étrange venu d'ailleurs,* in Non-linear Partial Differential Equations and Their Applications, College de France Seminar, Vol II and III (H.Brezis and J.L.Lions, eds), Research Notes in Mathematics **60**, 98-138, and **70**, 154-178, Pitman, London, 1982.

[6] G.Dal Maso, *Integral representation on $BV(\Omega)$ of Γ-limits of variational integrals,* Manuscripta Mat. **30** (1980), 387-413.

[7] G.Dal Maso, *Limiti di problemi di minimo per funzionali convessi con ostacoli unilaterali* Atti Accad. Naz. Lincei, Rend. Cl. Sci. Fis. Mat. Natur. **73** (1982), 15-20.

[8] G.Dal Maso, *Limits of minimum problems for general integral functionals with unilateral obstacles,* Atti Accad. Naz. Lincei, Rend. Cl. Sci. Fis. Mat. Natur. **74** (1983), 55-61.

[9] G.Dal Maso, *On the integral representation of certain local functionals,* Ricerche Mat. **32** (1983), 85-114.

[10] G.Dal Maso, P.Longo, *Γ-limits of obstacles,* Ann. Mat. Pura Appl. **128** (1980), 1-50.

[11] G.Dal Maso, L.Modica, *A general theory of variational functionals,* in Topics in Functionals Analysis 1980-81, Quaderni della Scuola Normale Superiore di Pisa, Pisa, 1982.

[12] E.De Giorgi, F.Colombini, L.C.Piccinini, *Frontiere Orientate di Misura Minima e Questioni Collegate.* Quaderni della Scuola Normale Superiore di Pisa, Editrice Tecnico Scientifica, Pisa, 1972.

[13] E.De Giorgi, G.Dal Maso, P.Longo, *Γ-limiti di ostacoli.* Atti

Accad. Naz. Lincei, Rend. Cl. Sci. Fis. Mat. Natur. **68** (1980), 481-487.

[14] E.De Giorgi, T.Franzoni, *Su un tipo di convergenza variazionale*, Rend. Sem. Mat. Brescia **3** (1979), 63-101.

[15] E.De Giorgi, G.Letta, *Une notion générale de convergence faible pour des fonctions croissantes d'ensemble*, Ann. Sc. Norm. Sup. Pisa **4** (1977), 61-99.

[16] H.Federer, *Geometric Measure Theory*, Springer-Verlag, Berlin 1969.

[17] M.Giaquinta, G.Modica, J.Soucek, *Functionals with linear growth in the calculus of variations* I,II. Comment. Math. Univ. Carolin. **20** (1979), 143-156, 157-172.

[18] E.Giusti, *Minimal Surfaces and Functions of Bounded Variation*. Birkhaüser-Verlag, Basel, 1984.

[19] U.Massari, M.Miranda, *Minimal Surfaces of Codimension One*, Notas de Matematica. North-Holland, Amsterdam, 1984.

[20] C.Picard, *Surfaces minima soumises a une suite d'obstacles*, Thèse, Université de Paris-Sud, 1984.

[21] Yu.G.Reshetnyak, *Weak convergence of completely additive vector functions on a set*. Siberian Math. J. **9** (1968), 1039-1045 (translation of: Sibirsk Mat.Z. **9** (1968), 1386-1394).

M.C., A.L., E.P.

Dipartimento di Matematica

Università di Lecce

Via Arnesano

I-73100 LECCE

G.D.M.

SISSA

Strada Costiera, 11

I-34014 TRIESTE

SOME REMARKS ON THE WELL-POSEDNESS
OF THE CAUCHY PROBLEM IN GEVREY SPACES

LAMBERTO CATTABRIGA

Dedicated to Ennio De Giorgi on his sixtieth birthday

1. In [2] [1] F.Colombini, E.De Giorgi, S.Spagnolo proved that the Cauchy problem

$$(1.1) \quad \begin{cases} \partial_{tt}^2 u - \sum_{i,j=1}^n c_{ij}(t)\partial_{x_i x_j}^2 u = 0 & (t,x) \in]0,T[\times \mathbf{R}^n \\ u(0,x) = u_o(x) & x \in \mathbf{R}^n \\ \partial_t u(0,x) = u_1(x) & x \in \mathbf{R}^n \end{cases}$$

where c_{ij} are real integrable functions on $[0,T]$ such that for some constant $a_o > 0$

$$\sum_{i,j=1}^n c_{ij}(t)\xi_i\xi_j \geq a_o|\xi|^2 \quad , \quad \xi \in \mathbf{R}^n,$$

is well posed in the Gevrey space $\mathcal{E}^{\{s\}}(\mathbf{R}_x^n)$, for $s \in [1, 1/(1-\chi)[$, $\chi \in]0,1[$, if there exists $c \geq 0$ such that

$$(1.2) \quad \int_o^T |c_{ij}(t+\tau) - c_{ij}(t)|dt < C\tau^\chi, \quad \tau > 0 \quad [2].$$

---

[1] See also [8] and the references quoted there.

[2] Here $c_{ij}(t)=c_{ij}(T)$ for $t>T$.

Here $\mathcal{E}^{\{s\}}(\mathbf{R}_x^n), s > 0$, is the space of all $\varphi \in C^\infty(\mathbf{R}_x^n)$ such that for every compact $K \subset \mathbf{R}_x^n$ there exist $A > 0$ such that

$$\sup_{\beta \in \mathbf{Z}_+^n} \sup_{x \in \mathbf{R}_x^n} A^{-|\beta|} \beta!^{-s} |\partial_x^\beta \varphi(x)| < +\infty.$$

The sharpness of this result is shown in [2] by exhibiting a function $c \in C^{0,\chi}([0,T])$, $\chi \in]0,1[$, $c(t) \geq 1$ for every $t \in [0,T]$, such that for $n = 1$, *problem* (1.1) *is not well posed in any* $\mathcal{E}^{\{s\}}(\mathbf{R}_x^n)$, *when* $s > 1/(1-\chi)$.

It can also be proved that when (1.2) is satisfied Problem (1.1) is equivalent to the Cauchy problem for a 2×2 system

$$(1.3) \quad \begin{cases} (\partial_t I + i\Lambda(t,D_x) + A(t,D_x))u = 0 & (t,x) \in]0,T[\times\mathbf{R}^n \\ u(0,x) = u_o(x) & x \in \mathbf{R}^n \; (3), \end{cases}$$

where I is the unit matrix, Λ is a real diagonal matrix and the elements $a_{h,k}(t,D_x)$, $h,k = 1,2$, $D_{x_j} = -i\partial_{x_j}$, of A are such that (1.4)

$$\sup_{\substack{(t,\xi) \in [0,T] \times \mathbf{R}^n \\ |\xi| \geq 1}} \sup_{\alpha \in \mathbf{Z}_+^n} \alpha!^{-1} C^{-|\alpha|} (1 + |\xi|)^{-1+\chi+|\alpha|} |\partial_\xi^\alpha a_{h,k}(t,\xi)| < +\infty$$

for some positive constant C. Note also that the elements of Λ satisfy (1.4) for $-1 + \chi + |\alpha|$ replaced by $-1 + |\alpha|$.

Thus we are led to the problem of finding necessary conditions and sufficient conditions in order for problem (1.3) is well posed in $\mathcal{E}^{\{s\}}(\mathbf{R}_x^n)$ when $s \geq 1/(1-\chi)$.

In the case of a scalar equation this problem has been considered by S. Mizohata. By his microlocal energy method he proves the following result:

Theorem 1.1. [5] *Let* $\lambda, a \in C([0,T]; C^\infty(\mathbf{R}_x^n \times \mathbf{R}_\xi^n))$ *be such that*
(1.5_1)

$$\sup_{(t,x,\xi) \in [0,T] \times \mathbf{R}^n \times \mathbf{R}^n} \sup_{(\alpha,\beta) \in \mathbf{Z}_+^n \times \mathbf{Z}_+^n} \alpha!^{-1} \beta!^{-s} C^{-|\alpha+\beta|} (1 + |\xi|)^{-1+|\alpha|}.$$

$$\cdot |\partial_\xi^\alpha \partial_x^\beta \lambda(t,x,\xi)| < +\infty$$

---

(3) This is true also when the coefficients c_{ij} depend on both t and x. See T. Nishitani [6], K. Taniguchi [9], M. Cicognani [1].

(1.5$_p$)
$$\sup_{(t,x,\xi)\in[0,T]\times\mathbf{R}^n\times\mathbf{R}^n}\sup_{(\alpha,\beta)\in\mathbf{Z}^n_+\times\mathbf{Z}^n_+} \alpha!^{-1}\beta!^{-s}C^{-|\alpha+\beta|}(1+|\xi|)^{-p+|\alpha|}.$$

$$\cdot|\partial^\alpha_\xi\partial^\beta_x a(t,x,\xi)| < +\infty$$

for some constants $C > 0$, $p \in]0,1[$.

Suppose that λ is real valued and that

$$(1.6) \qquad a(t,x,\xi) = \overset{\circ}{a}(t,x,\xi) + a'(t,x,\xi)$$

where $\overset{\circ}{a}$ is homogeneous of order p in ξ, for large ξ, and a' satisfies (1.5$_{p'}$) with $p' < p$. Then in order for the Cauchy problem

$$(1.7) \qquad \begin{cases} (\partial_t + i\lambda(t,x,D_x) + a(t,x,D_x))u = 0 & (t,x) \in]0,T[\times\mathbf{R}^n \\ u(0,x) = u_o(x) & x \in \mathbf{R}^n \end{cases}$$

is well posed in

$$\gamma^s_{L^2} = \{\varphi \in C^\infty(\mathbf{R}^n_x); \exists A > 0 : \sup_{\beta\in\mathbf{Z}^n_+} \beta!^{-s}A^{-|\beta|}\|\partial^\beta_x\varphi\|_{L^2(\mathbf{R}^n_x)} < +\infty\}$$

for $s > 1/p$ it is necessary that

$$Re\ \overset{\circ}{a}(0,x,\xi) \geq 0 \quad \forall(x,\xi) \in \mathbf{R}^n \times \mathbf{R}^n.$$

In what follows we prove necessary conditions and sufficient conditions for the well posedness of Problem (1.7) when λ and a do not depend on x, without assuming (1.6).

Some results of the same type can be easily proved also in the case of problem (1.3). In this case however the necessary conditions we obtain appear to be not sharp enough to be violated by the counter-example exhibited in [2].

2. Instead of $\gamma^s_{L^2}$ we find it convenient to consider some spaces of functions and ultradistributions in $\mathbf{R}^n$ already introduced by V.P.Palamodov [7].

Let s,μ,A,h be positive given numbers and let

$$S^{s,A}_{\mu,h} = \{\varphi \in C^\infty(\mathbf{R}^n)\ ;$$

$$\sup_{x,\alpha} \alpha!^{-s}A^{-|\alpha|}\exp(h|x|^{1/\mu})|\partial^\alpha_x\varphi(x)| = \|\varphi\|^{s,A}_{\mu,h} < +\infty\}.$$

With the norm $\| \quad \|_{\mu,h}^{s,A}$ the Banach spaces $S_{\mu,h}^{s,A}$ are such that

$$S_{\mu,h}^{s,A} \subset S_{\mu,h}^{s,A'} \subset S_{\mu,h'}^{s,A'} \quad \text{for} \quad A < A', \ h' < h$$

with continuous embeddings. Define

$$S_{\{\mu\}}^{\{s\}} = \text{ind} \lim_{h \to 0+} \ \text{ind} \lim_{A \to +\infty} S_{\mu,h}^{s,A} \quad , \quad S_{(\mu)}^{\{s\}} = \text{proj} \lim_{h \to +\infty} \ \text{ind} \lim_{A \to +\infty} S_{\mu,h}^{s,A}$$

$$S_{\{\mu\}}^{(s)} = \text{ind} \lim_{h \to 0+} \ \text{proj} \lim_{A \to 0+} S_{\mu,h}^{s,A} \quad , \quad S_{(\mu)}^{(s)} = \text{proj} \lim_{h \to +\infty} \ \text{proj} \lim_{A \to 0+} S_{\mu,h}^{s,A}.$$

Denoting by $\mathcal{F}$ the Fourier transformation defined by

$$(\mathcal{F}\varphi)(\xi) = \tilde{\varphi}(\xi) = \int e^{-i<x,\xi>} \varphi(x)dx,$$

then, as proved in [7],

$$\mathcal{F}\left(S_{\{\mu\}}^{\{s\}}\right) = S_{\{s\}}^{\{\mu\}} \quad , \quad \mathcal{F}\left(S_{(\mu)}^{\{s\}}\right) = S_{\{s\}}^{(\mu)},$$

$$\mathcal{F}\left(S_{\{\mu\}}^{(s)}\right) = S_{(s)}^{\{\mu\}} \quad , \quad \mathcal{F}\left(S_{(\mu)}^{(s)}\right) = S_{(s)}^{(\mu)}.$$

Moreover the same relations hold when the indicated spaces are replaced by their duals $S_{\{\mu\}}^{\{s\}'}$, $S_{(\mu)}^{\{s\}'}$, $S_{\{\mu\}}^{(s)'}$, $S_{(\mu)}^{(s)'}$ respectively.

Let us then suppose that *the functions λ and a in Problem (1.7) do not depend on x and for every $\xi \in \mathbf{R}^n$ are integrable functions of t in $[0,T]$, satisfying (1.5_1) and (1.5_p) respectively. Furthermore assume that λ is real valued.*

If $u_o \in S_{\{1\}}^{(1/p)}$, or $u_o \in S_{\{1\}}^{(1/p)'}$, then $u(t,x)$ is a solution of (1.7) and $u(t,\cdot) \in S_{\{1\}}^{(1/p)}$, or $u(t,\cdot) \in S_{\{1\}}^{(1/p)'}$ for every $t \in [0,T]$ respectively if and only if

$$\begin{cases} (\partial_t + i\lambda(t,\xi) + a(t,\xi))\tilde{u}(t,\xi) = 0 & (t,\xi) \in [0,T] \times \mathbf{R}^n \\ \tilde{u}(o,\xi) = \tilde{u}_o(\xi) & \xi \in \mathbf{R}^n. \end{cases}$$

Thus

$$(2.1) \qquad \tilde{u}(t,\xi) = \tilde{u}_o(\xi) \exp\left(-i \int_o^t \lambda(\tau,\xi)d\tau - \int_o^t a(\tau,\xi)d\tau\right)$$

and (1.7) *is well posed in* $S_{\{1\}}^{(1/p)}$ *and in* $S_{\{1\}}^{(1/p)'}$.

We now prove the following

Theorem 2.1. *In order for Problem (1.7) where λ and a do not depend on x, be well posed in $S_{\{1\}}^{\{s\}}$ or in $S_{\{1\}}^{\{s\}'}$ for $s \geq 1/p$ it is necessary that*

$$(2.2) \qquad \liminf_{\rho \to +\infty} \rho^{-1/\sigma} \int_o^t Re \ a(\tau, \rho\eta) d\tau \geq 0$$

for every $\sigma \in [1/p, s]$, $t \in]0, T]$, $\eta \in S_{n-1} = \{\eta \in \mathbf{R}^n; |\eta| = 1\}$.

PROOF. First note that for $s \geq 1/p$ $S_{\{1\}}^{\{s\}} \subset S_{\{1\}}^{(1/p)'}$ and that from (2.1) it follows that $\tilde{u}(t, \xi)$ is analytic in ξ if $u_o \in S_{\{1\}}^{\{s\}}$. Moreover if we suppose that for a fixed $t \in]0, T]$, $u(t, \cdot) \in S_{\{1\}}^{\{s\}'}$, then $\tilde{u}(t, \cdot) \in S_{\{s\}}^{\{1\}'}$. It follows that for every $\epsilon > 0$ there exists $c_\epsilon > 0$ such that

$$(2.3) \qquad |\tilde{u}(t, \xi)| \leq c_\epsilon \exp(\epsilon|\xi|^{1/s}) \quad, \quad \xi \in \mathbf{R}^n.$$

On the other hand if (2.2) does not hold for a $\sigma \in [1/p, s]$, there exist $t^o \in]0, T]$, $\eta^o \in S_{n-1}$, $c > 0$ and a sequence $\rho_\nu \to +\infty$ such that

$$(2.4) \qquad \rho_\nu^{-1/\sigma} \int_o^{t^o} Re \ a(\tau, \rho_\nu \eta^o) d\tau \leq -c, \quad \nu = 1, \dots \ .$$

Thus from (2.1), (2.3), (2.4) it follows that for every $\epsilon > 0$

$$(2.5) \qquad |\tilde{u}_o(\rho_\nu \eta^o)| \leq c_\epsilon \exp(-(c - \epsilon\rho_\nu^{1/s - 1/\sigma})\rho_\nu^{1/\sigma}) \quad, \nu = 1, \dots \ .$$

Let now

$$u_o(x) = \prod_{j=1}^n g_h(x_j),$$

where

$$(2.6) \quad g_h(y) = \begin{cases} \exp(-h[(y - r)(r - y)]^{-1/(s-1)}), & \text{for } -r < y < r \\ 0 & \text{for } |y| \geq r, \end{cases}$$

$h > 0$.

As is well known $u_o \in S_{\{1\}}^{\{s\}} \backslash S_{\{1\}}^{(s)}$, moreover there exist positive constants d_o, d_1 independent of h and η^o such that for $\rho \to +\infty$

$$|\tilde{u}_o(\rho\eta^o)| = |\tilde{g}_h(0)|^{n-l}(d_o/h)^l(\rho/h)^{(-1+1/2s)l}\prod_j{}'|\eta_j^o|^{-1+1/2s}.$$

$$\cdot \exp\left(-d_1(\rho/h)^{1/s}\sum_j{}'|\eta_j^o|\right)(1+O((\rho/h)^{-1/s}))^{(4)}$$

where l is the number of j's such that $\eta_j^o \neq 0$ and $\prod_j'$ and $\sum_j'$ run over such j's. Thus (2.5) cannot hold when $\sigma < s$ and when $\sigma = s$, for $\epsilon \in]0,c[$ if h is chosen sufficiently large in (2.6).

Corollary 2.2. *Under the same hypotheses of Theorem 2.1, assume that*

$$a(t,\xi) = \sum_{h=o}^{r} a_h(t,\xi)$$

where a_h, $h = 0,...,r$, are supposed to be integrable functions of t in $[0,T]$ for every $\xi \in \mathbf{R}^n$ and to satisfy (1.5_{p_h}), $0 = p_0 < p_1 < ... < ... < p_r < 1$ and for $h = 1,...,r$ to be homogeneous of degree p_h in ξ, for large $\xi \in \mathbf{R}^n$. Then for the well posedness of problem (1.7) in $S_{\{1\}}^{\{s\}}$, $S_{\{1\}}^{\{s\}'}$, $s \geq 1/p_r$, it is necessary that for every $t \in [0,T]$, $\eta \in S_{n-1}$

(2.7)
$$\int_o^t Re\ a_r(\tau,\eta)d\tau \geq 0$$

and that

$$\int_o^t Re\ a_h(\tau,\eta) \geq 0 \quad for \quad h < r \quad such\ that \quad p_h \geq 1/s,$$

when

$$\int_o^t Re\ a_j(\tau,\eta)\ d\tau = 0 \quad , \quad j = h+1,...,r.$$

If a_r is continuous as a function of t, (2.7) gives the result of Theorem 1.1., when λ and a do not depend on x.

---

(4) See for example [3] and [4].

Remark 2.3. *From (2.1) it follows easily that if (2.2) holds for $\sigma = s$ uniformly w.r. to $\eta \in S_{n-1}$ then Problem (1.7) when λ and a do not depend on x is well posed in $S_{\{1\}}^{\{s\}}$ and $S_{\{1\}}^{\{s\}'}$. This always occurs for $s < 1/p$. Note also that, as a consequence of (1.5_p), condition (2.2) for $\sigma = 1/p$ holds in fact uniformly w.r. to $\eta \in S_{n-1}$ if it holds for every $\eta \in S_{n-1}$. Thus in this case condition (2.2) is both necessary and sufficient for the well posedness of (1.7) in $S_{\{1\}}^{\{1/p\}}$ and $S_{\{1\}}^{\{1/p\}'}$.*

3. Consider now the problem

$$(3.1) \quad \begin{cases} (\partial_t + i\Lambda(t, D_x) + A(t, D_x))u = 0 & (t, x) \in]0, T[\times \mathbf{R}^n \\ u(0, x) = u_o(x) & x \in \mathbf{R}^n, \end{cases}$$

where Λ and A are $m \times m$ matrices and Λ is hermitian. Furthermore assume that the elements of $\Lambda(t, \xi)$ and $A(t, \xi)$ are integrable functions of t in $[0, T]$ for every $\xi \in \mathbf{R}^n$ and satisfy the estimates (1.5_1) and (1.5_p) respectively.

As in the scalar case, when $u_o \in \left(S_{\{1\}}^{(1/p)}\right)^m$ or $u_o \in \left(S_{\{1\}}^{(1/p)'}\right)^m$ and $u(t, \cdot)$ is a solution of (3.1) in $\left(S_{\{1\}}^{(1/p)}\right)^m$ or in $\left(S_{\{1\}}^{(1/p)'}\right)^m$ for every $t \in]0, T[$, then

$$(3.2) \quad \begin{cases} (\partial_t + i\Lambda(t, \xi)) + A(t, \xi))\tilde{u}(t, \xi) = 0 & (t, \xi) \in]0, T[\times \mathbf{R}^n \\ \tilde{u}(0, \xi) = \tilde{u}_o(\xi) & \xi \in \mathbf{R}^n. \end{cases}$$

On the other hand if we let

$$\mu_o(t, \xi) = \inf_{\substack{v \in C^m \\ |v| = 1}} Re(A(t, \xi)v, v)$$

and

$$\mu^o(t, \xi) = \sup_{\substack{v \in C^m \\ |v| = 1}} Re(A(t, \xi)v, v)$$

the following estimates hold for the solution of (3.2)

$$(3.3) \quad |\tilde{u}(t, \xi)| \leq |\tilde{u}_o(\xi)| \exp\left(-\int_o^t \mu_o(\tau, \xi)d\tau\right)$$

(3.4) $$|\tilde{u}(t,\xi)| \geq |\tilde{u}_o(\xi)| \exp\left(-\int_o^t \mu^o(\tau,\xi)d\tau \right).$$

Since

$$|\mu_o(t,\xi)|, \; |\mu^o(t,\xi)| \leq M(1+|\xi|)^p, \; (t,\xi) \in [0,T] \times \mathbf{R}^n$$

for a positive constant M, from (3.3) it follows that Problem (3.1) is well posed in $\left(S_{\{1\}}^{(1/p)} \right)^m$ and in $\left(S_{\{1\}}^{(1/p)'} \right)^m$. Moreover by using (3.4) and (3.3) and arguing as in section 2 it can be proved

Proposition 3.1. *In order for Problem (3.1) is well posed in* $\left(S_{\{1\}}^{\{s\}} \right)^m$ *or in* $\left(S_{\{1\}}^{\{s\}'} \right)^m$ *for* $s \geq 1/p$, *it is necessary that*

$$\liminf_{\rho \to +\infty} \rho^{-1/\sigma} \int_o^t \mu^o(\tau,\rho\eta)d\tau \geq 0$$

$\forall \sigma \in [1/p, s]$, $t \in]0,T]$, $\eta \in S_{n-1}$, *and it is sufficient that* $\forall t \in]0,T]$,

$$\liminf_{\rho \to +\infty} \rho^{-1/s} \int_o^t \mu_o(\tau,\rho\eta)d\tau \geq 0$$

uniformly w.r. to $\eta \in S_{n-1}$.

References

[1] M.Cicognani, *The propagation of Gevrey singularities for some hyperbolic operators with coefficients Hölder continuous with respect to time*, in "Recent developments in hyperbolic equations", Pitman Research Notes in Math. **183** (1988), 38-58.

[2] F.Colombini, E.De Giorgi, S.Spagnolo, *Sur les équations hyperboliques avec des coefficients qui ne dépendent que du temps*, Ann. Scuola Norm. Sup. Pisa **6** (1979), 511-559.

[3] M.V.Fedoryuk, *Metod Perevala*, Nauka, Mosca, 1977.

[4] H.Komatsu, *Irregularity of hyperbolic operators*, Taniguchi Symp. HERT, Katata 1984, 155-179.

[5] S.Mizokata, *Microlocal energy method*, Taniguchi Symp. HERT, Katata 1984, 193-233.

[6] T.Nishitani, *Sur les équations hyperboliques à coefficients höldériens en t et de classe de Gevrey en x*, Bull. Sc. Math. **107** (1983), 113-138.

[7] V.P.Palamodov, *Fourier transform of strongly increasing infinitely differentiable functions*, (in russian) Trudy Mosk. Mat. Obsc. **11** (1962), 309-350.

[8] S.Spagnolo, *Analytic and Gevrey well-posedness of the Cauchy problem for second order weakly hyperbolic equations with coefficients irregular in time*, Taniguchi Symp. HERT, Katata 1984,363-380.

[9] K.Taniguchi, *Fourier integral operators in Gevrey class on* $\mathbf{R}^n$ *and the fundamental solution for a hyperbolic operator*, Publ. RIMS Kyoto Univ. **20** (1984), 491-542.

Dipartimento di Matematica

Università di Bologna

Piazza di Porta S.Donato,5

I-40127 BOLOGNA

APPROXIMATING MEASURES AND RECTIFIABLE CURVES

ANTONIO CHIFFI

Dedicated to Ennio De Giorgi on his sixtieth birthday

I began in [CZ] the study of the geometric meaning of discontinuities of approximating Hausdorff measures (cf. [F], §2. 10.1). Later those methods were developed and applied to questions regarding sets of constant width ([SZ], [St 1], [St 2]). In this paper I study some aspects regarding a plane rectifiable curve; precisely theor. 1.4 gives a sufficient continuity condition, while in §2 I pick out, on the rectifiable curve, subsets of the boundary of a set of constant width, which are responsible of the possible discontinuity. Symbols are explained at the beginning of each paragraph.

1. Let $\nu_t(I)$ be the *approximating measure* obtained with closed covers of the one dimensional Hausdorff measure μ^1 of a set $I \subset R^2$ (cf. [F]); let μ^1 be also the Lebesgue measure in R^1; let $\nu_{\delta-}(I)$ be the limit of $\nu_t(I)$ as $t \uparrow \delta$; let $s_\delta(I)$ be the difference $\nu_{\delta-}(I) - \nu_\delta(I)$; let $d(I)$ be the diameter of the set I. If Γ is a plane simple rectifiable curve, with diameter δ, let $\mathbf{f} : [0, L] \to R^2$ be its parametrization by arc lenght; for sake of simplicity we denote by Γ also the set $\mathbf{f}([0, L])$.

We pose:

$$g(s,t) = |\mathbf{f}(s) - \mathbf{f}(t)| \quad , \quad (s,t) \in [0,L] \times [0,L]$$

$$\gamma_\delta = \{(s,t) \in [0,L] \times [0,L] : g(s,t) = \delta\}$$

$$I_\delta = \{t \in [0,L]; \exists s \in I, \ (s,t) \in \gamma_\delta\}.$$

Let $\mathcal{A}$ be the family of subsets $A \subseteq [0,L]$, open in $[0,L]$, such that:

(1) $$(A \times [0,L]) \cup ([0,L] \times A) \supset \gamma_\delta.$$

Lemma 1.1. *Let $C \subseteq [0,L]$ be a closed set; the following statements are equivalent:*

$$d[\mathbf{f}(C)] < \delta$$
$$(C \times C) \cap \gamma_\delta = \emptyset$$
$$C \cap C_\delta = \emptyset$$

The proof is easy.

Lemma 1.2. *If $A \in \mathcal{A}$, then $d(\mathbf{f}([0,L]\backslash A)) < \delta$.*

PROOF. It follows from Lemma 1.1 and (1), which is equivalent to:

$$\{([0,L]\backslash A) \times ([0,L]\backslash A)\} \cap \gamma_\delta = \emptyset.$$

Lemma 1.3. *Let $I \subseteq [0,L]$ a μ^1 measurable set; if k is a number such that $\mu^1(I) \leq k$, then $\nu_t[\mathbf{f}(I)] \leq k$ for every $t > 0$.*

PROOF. It is a consequence of:

$$\nu_t[\mathbf{f}(I)] \leq \mu^1[\mathbf{f}(I)] = \mu^1(I).$$

Theorem 1.4. *Let Γ be a plane simple rectifiable curve, with diameter $\delta > 0$. If for every $\epsilon > 0$ a set $A \in \mathcal{A}$ exists, such that $\mu^1(A) < \epsilon$, then the function $t \to \nu_t(\Gamma)$ is continuous for $t = \delta$.*

PROOF. If t is such that $d(\mathbf{f}([0,L]\backslash A)) < t < \delta$ (lemma 1.2), then lemma 1.3 implies:

$$\nu_t(\Gamma) \leq \nu_t(\mathbf{f}([0,L])\backslash A) + \nu_t(\mathbf{f}(A)) < t + \epsilon < \delta + \epsilon$$

and the result follows recalling that:

$$\nu_\delta(\Gamma) = \delta.$$

Lemma 1.5. *If $I \subseteq [0,L]$ is a Borel set, then I_δ is μ^1 measurable.*

PROOF. The set I is the second projection of the set intersection between the Borel set $I \times [0,L]$ and the closed set γ_δ; the projection of a Borel set is μ^1 measurable (cf., e.g., [S],§88).

Lemma 1.6. *Let $I \subseteq [0,L]$ be a Borel set and let $\mu^1(I \cap I_\delta) \leq k$. Then:*

$$(2) \qquad \nu_{\delta-}[\mathbf{f}(I)] \leq \nu_\delta[\mathbf{f}(I)] + k \quad.$$

PROOF. If $\nu_\delta[\mathbf{f}(I)] < \delta$, then the function $t \to \nu_t[\mathbf{f}(I)]$ is continuous in $t = \delta$ ([CZ], corollary 5). Assuming now:

$$(3) \qquad \nu_\delta[\mathbf{f}(I)] = \delta$$

according to lemma 1.5 for every $\epsilon > 0$ a set A exists, open in $[0,L]$, such that $A \supset I \cap I_\delta, \mu^1(A) < \mu^1(I \cap I_\delta) + \epsilon$ and a closed set $C \subseteq I \backslash A$ exists, such that $\mu^1[(I \backslash A) \backslash C] < \epsilon$. Hence:

$$(4) \qquad \mu^1(I \backslash C) \leq \mu^1[(I \backslash A) \backslash C] + \mu^1(A) < k + 2\epsilon$$

Noting that $C \cap C_\delta \subset C \cap A = \emptyset$ and, by lemma 1.1: $d[\mathbf{f}(C)] < \delta$, we have, for every t, $0 < t < \delta$:

$$\nu_t[\mathbf{f}(I)] \leq \nu_t[\mathbf{f}(C)] + \nu_t[\mathbf{f}(I \backslash C)];$$

by (4):

$$\nu_t[\mathbf{f}(I)] < \delta + k + 2\epsilon;$$

and, by (3), (2) follows.

Corollary 1.7. *Let $I \subseteq [0, L]$ be a μ^1 measurable set, with $\mu^1(I \cap I_\delta) = 0$; then the function $t \to \nu_t[\mathbf{f}(I)]$ is continuous for $t = \delta$.* The proof follows from lemma 1.6, noting that it is not necessary to apply lemma 1.5 and to suppose I to be a Borel set.

Lemma 1.8. *Let $D = \{x \in [0, L] : \exists y \in [0, L], |\mathbf{f}(x) - \mathbf{f}(y)| = \delta\}$; there exists at least one set G of constant width δ, containing Γ, and $\partial G \supseteq \mathbf{f}(D)$.*

PROOF. A set G of costant width δ containing Γ exists (cf. [BF], §64). If some point of $\mathbf{f}(D)$ were interior to G, then G would have width greater then δ in some direction.

Lemma 1.9. *Let D the set defined in lemma 1.8. The set $P \subseteq D$ of points $p \in [0, L]$, such that $\{p\}_\delta$ consists of more then one point, is empty or countable.*

PROOF. Let $p \in [0, L]$ and let $\{p\}_\delta$ contains at least two distinct points; the point p is a vertex of ∂G and the set of vertex of a set of constant width is at most countable.

Lemma 1.10. *Let $A \in \mathcal{A}$; for every $\epsilon > 0$ a set $B \in \mathcal{A}$ exists such that $\mu^1(B \cap B_\delta) < \epsilon$ and*

$$\nu_\delta[\mathbf{f}(B)] \leq \nu_\delta[\mathbf{f}(A)] + \epsilon.$$

PROOF. Let $\mu^1(A \cap A_\delta) > 0$ and let P be the set as defined in lemma 1.9. If $(P \times [0, L]) \cup ([0, L] \times P) \supseteq \gamma_\delta$, put $A_1 = P$; otherwise consider a point $(p, q) \in \gamma_\delta$ such that $\{p\} = \{q\}_\delta$ and $\{q\} = \{p\}_\delta$ (this point exists by lemma 1.9). Let H be one of the two closed half planes determined by the straight line $\mathbf{f}(p)\mathbf{f}(q)$. Put:

$$A_1 = \{p\} \cup P \cup \{A \backslash (P_\delta \cup [A_\delta \cap \mathbf{f}^{-1}(H)])\}.$$

The set A_1 is μ^1 measurable by lemma 1.5 and $A_1 \cap (A_1)_\delta \subseteq P$ (the set $A_1 \cap (A_1)_\delta$ is empty if P is empty); therefore $\mu^1(A_1 \cap (A_1)_\delta) = 0$.

Now we prove $(A_1 \times [0, L]) \cup ([0, L] \times A_1) \supseteq \gamma_\delta$. Let $(x, y) \in \gamma_\delta$; either x or y belong to A; assume $x \in A$. If x does not belong to A_1, then $x \in A \cap A_\delta \cap \mathbf{f}^{-1}(H)$, and $\{y\} = \{x\}_\delta \subseteq (A \cap A_\delta) \backslash \mathbf{f}^{-1}(H)$; therefore $y \in A_1$, as we have said. Let B_1 and C_1 be subsets of $[0, L]$; B_1 open in $[0, L]$, C_1 closed, with $C_1 \subseteq A_1 \subseteq B_1$ and $\mu^1(B_1 \backslash C_1) < \epsilon$. The set $B = B_1 \backslash (C_1)_\delta$ is open in $[0, L]$ and it is easy to see that $B \cap B_\delta \subseteq (B_1 \backslash C_1) \cup P$. Therefore we have $\mu^1(B \cap B_\delta) < \epsilon$ and, by lemma 1.3:

$$\nu_\delta[\mathbf{f}(B)] \leq \nu_\delta[\mathbf{f}(A_1)] + \nu_\delta[\mathbf{f}(B \backslash A_1)] \leq \nu_\delta[\mathbf{f}(A)] + \epsilon.$$

Lemma 1.11. *The equality:*

$$\inf\{\nu_{\delta-}[\mathbf{f}(A)], A \in \mathcal{A}\} = \inf\{\nu_\delta[\mathbf{f}(A)], A \in \mathcal{A}\}$$

holds.

PROOF. By lemma 1.10 for every $\epsilon > 0$ a set $B \in \mathcal{A}$ exists, such that $\mu^1(B \cap B_\delta) < \epsilon$ and:

$$\nu_\delta[\mathbf{f}(B)] < \inf\{\nu_\delta[\mathbf{f}(A)], A \in \mathcal{A}\} + 2\epsilon.$$

By lemma 1.6:

$$\nu_{\delta-}[\mathbf{f}(B)] \leq \nu_\delta[\mathbf{f}(B)] + \epsilon \leq \inf\{\nu_\delta[\mathbf{f}(A)], A \in \mathcal{A}\} + 3\epsilon.$$

Since ϵ is arbitrary and the opposite inequality is obvious, the lemma follows.

Theorem 1.12. *Let Γ be a plane simple rectifiable curve, having diameter δ; then:*

$$s_\delta(\Gamma) \leq \inf\{\nu_\delta[\mathbf{f}(A)], A \in \mathcal{A}\}.$$

PROOF. For every $\epsilon > 0$ a set $B \in \mathcal{A}$ and a cover of $\mathbf{f}(B)$ by closed sets F_i exists such that:

$$\Sigma_i d(F_i) < \inf\{\nu_\delta[\mathbf{f}(A)], A \in \mathcal{A}\} + \epsilon.$$

It may be supposed, by Lemma 1.11, that $d(F_i) < \delta$ for every i, and, by the convergence of the serie $\Sigma d(F_i)$, that

$$\sup\{d(F_i) : i = 1, 2, ...\} = \Theta < \delta.$$

By the lemma 1.2 it follows $d(\mathbf{f}([0, L]\backslash B)) < \delta$; then for every $t < \delta$, greater than Θ and $d(\mathbf{f}([0, L]\backslash B))$ we have:

$$\nu_t(\Gamma) \leq \nu_t[\mathbf{f}([0, L]\backslash B] + \nu_t[\mathbf{f}(B)] < \delta + \Sigma_i d(F_i)$$
$$< \nu_\delta(\Gamma) + \inf\{\nu_\delta[\mathbf{f}(A)] : A \in \mathcal{A}\} + \epsilon.$$

This inequality proves the theorem.

2. Let $E \subset R^2$ be a set of constant width δ; let $\varphi_E : [0, 2\pi) \to R^2$ be the mapping of the curve ∂E in the terms of the orientation of the support line at each point of ∂E. Let $R_E(u)$, $u \in [0, 2\pi)$ be the radius of curvature of ∂E, if it exists, at the point $\varphi_E(u)$; let N_E be the subset of $[0, 2\pi)$ where $R_E(u) = 0$ and equality $R_E(u) + R_E((u + \pi)\mathrm{mod}2\pi) = \delta$ holds. For every $K \subseteq [0, 2\pi)$ let $K^s = \{u \in [0, 2\pi] : ((u + \pi)\mathrm{mod}\ 2\pi) \in K\}$ and let $\mathcal{K}$ be the family of μ^1 measurable subsets of $[0, 2\pi)$ such that $K \supseteq N_E$, $[0, 2\pi) \setminus K = K^s$.

Lemma 2.1. *Let E be a plane set of constant width δ, let $I \subseteq [0, 2\pi)$ be a μ^1 measurable set and let k be such that $\mu^1(I) \leq k$. Then: $\nu_t[\varphi_E(I)] \leq k\delta$, for every $t > 0$.*

PROOF. The proof follows from:

$$\nu_t[\varphi_E(I)] \leq \mu^1[\varphi_E(I)] = \int_I R_E(u)\mu^1(du) \leq \delta\mu^1(I) \leq k\delta.$$

Theorem 2.2. *Let Γ be a plane simple rectifiable curve and let δ be its diameter; let D be the set defined in lemma 1.8. Then a set E of constant width exists, such that ∂E contains $\mathbf{f}(D)$ with the possible exclusion of a countable set, and the radius of curvature of ∂E assumes only the two values 0 and δ, μ^1 almost everywhere in $\partial E \backslash \mathbf{f}(D)$.*

PROOF. Let G be a set of constant width δ containing $\mathbf{f}(D)$ as stated in Lemma 1.8; the set $\partial G\backslash\mathbf{f}(D)$ is open in ∂G; thus $H = \varphi_G^{-1}[\partial G\backslash\mathbf{f}(D)]$ is also open in $[0, 2\pi)$ and therefore the union of a at most countable family of intervals. From elementary properties of sets of constant width it follows that if $u \in \varphi_G^{-1}[\mathbf{f}(D)]$, also $(u + \pi)(\mathrm{mod}\ 2\pi) \in \varphi_G^{-1}(\mathbf{f}(D))$, therefore we may suppose, by perhaps increasing the number of intervals and changing H in a subset of zero μ^1 measure, that the points 0, $\pi/2$, $3\pi/2$ do not belong to any of these intervals and that if the interval (a, b) is contained in H, also $(a, b) + \pi$ $(\mathrm{mod}\ 2\pi)$ is contained in H. If (a, b) is contained in $(0, \pi/2)$ or in $(\pi/2, \pi)$, we will prove that there exist two numbers x and $y, a \le x \le y \le b$, such that, defining $R(u) = \delta$ for $u \in (x, y)$ and $R(u) = 0$ for $u \in (a, x) \cup (y, b)$:

$$(5) \qquad \int_a^b R(u) \sin\ u\ du = \int_a^b R_G(u) \sin\ u\ du$$

$$(6) \qquad \int_a^b R(u) \cos\ u\ du = \int_a^b R_G(u) \cos\ u\ du.$$

It is equivalent to show that the system in the unknowns x and y :

$$\int_x^y \delta \sin\ u\ du = \int_a^b R_G(u) \sin\ u\ du$$

$$\int_x^y \delta \cos\ u\ du = \int_a^b R_G(u) \cos\ u\ du$$

has solution in that set; for this purpose it is sufficient to study how the first members map the triangle $\{(x, y) \in R^2 : a \le x \le y \le b$ (for a similar proof cf. [K], th. 6). We now define $R(u)$ equal to $R_G(u)$ when u belongs to $[0, \pi)$ and does not belong to any of intervals (a, b); in this way the function $R(u)$ is μ^1 measurable and we may extend it on $[\pi, 2\pi)$ by the equality:$R((u + \pi)\mathrm{mod}\ 2\pi) = \delta - R(u)$. From (5) and (6) it follows:

$$\int_0^\pi R(u) \sin\ u\ du = \int_0^\pi R_G(u) \sin\ u\ du$$

$$\int_0^\pi R(u) \cos\ u\ du = \int_0^\pi R_G(u) \cos\ u\ du.$$

Therefore a set E exists such that $R(u) = R_E(u)$ (cf. [K], theor. 4) with support function $H_E(u) = \int_0^u R_E(v)\sin(u - v)dv$. Taking in mind that $R(u) = R_E(u) = R_G(u)$ for $u \in \varphi_G^{-1}[\mathbf{f}(D)]$, from (5) and (6) it follows $H_E(u) = H_G(u)$ for $u \in \varphi_G^{-1}(\mathbf{f}(D))$. The functions $H_E(u)$ and $H_G(u)$ are continuous with their partial derivatives, which are equal. These derivatives are the coordinates of the points of ∂E and ∂G respectively (cf. [BF], §(16), because the support lines meet the sets ∂E and ∂G only in one point. Therefore the set $\mathbf{f}(D)$ is contained in $\partial G \cap \partial E$, with the possible exception of a countable set.

Lemma 2.3. *Let* Γ *be a plane rectifiable curve, with diameter* $\delta > 0$; *let* E *be a set of constant width as in theorem 2.2. Then:*

$$\inf\{\nu_\delta[\mathbf{f}(A)] : A \in \mathcal{A}\} = \inf\{\nu_\delta[\varphi_E(K)]; K \in \mathcal{K}\}.$$

PROOF. Let $A \in \mathcal{A}$ and put:

$$K = N_E \cup \{\varphi_E^{-1}[\mathbf{f}(A)]\backslash\{N_E^s \cup \{\varphi_E^{-1}\{\mathbf{f}(A \cap A_\delta)\} \cap [0,\pi)\}\}.$$

The set K is μ^1 measurable because: $\mathbf{f}$ is a homeomorphism of $[0, L]$ onto Γ and therefore $\mathbf{f}(A)$ is a Borel set; $A \cap A_\delta$ is a Borel set because it is the second projection of the σ-compact set $(A \times [0, L]) \cap \gamma_\delta$. By the properties of E (theor.2.2) we have $K \in \mathcal{K}$ and:

$$\nu_\delta[\varphi_E(K)] \le \nu_\delta[\mathbf{f}(A) \cup \varphi_E(N_E)] = \nu_\delta[\mathbf{f}(A)].$$

Conversely let $K \in \mathcal{K}$ and for every $\epsilon > 0$ let B be a set open in $[0, 2\pi)$ such that $B \supseteq K, \mu^1(B\backslash K) < \epsilon$. The set $\varphi_E(B)$ is the union of an at most countable family of (perhaps degenerate) arcs of the curve ∂E and is therefore a Borel set; the set $B_1 = \mathbf{f}^{-1}[\varphi_E(B)]$ is then μ^1 measurable. Let $A \subseteq [0, L]$ be open in $[0, L]$, $A \supseteq B_1$ and $\mu^1(A\backslash B_1) < \epsilon$. By lemma 1.3, 2.1 we have:

$$\begin{aligned}
\nu_\delta[\mathbf{f}(A)] &\le \nu_\delta[\mathbf{f}(B_1)] + \nu_\delta[\mathbf{f}(A\backslash B_1)] < \nu_\delta[\varphi_E(B)] + \epsilon \\
&\le \nu_\delta[\varphi_E(K) + \nu_\delta[\varphi_E(B\backslash K)] + \epsilon \\
&< \nu_\delta[\varphi_E(K)] + \epsilon\delta + \epsilon.
\end{aligned}$$

Theorem 2.4. *Let* Γ *be a plane simple rectifiable curve with diameter* δ; *let* E *be a set of constant width as in theor. 2.2; then:*

$$s_\delta(\Gamma) \leq s_\delta(E) \leq \inf\{\nu_\delta[\varphi_E(K) : K] \in \mathcal{K}\}$$
$$= \inf\{\nu_\delta[\mathbf{f}(A)] : A \in \mathcal{A}\}.$$

PROOF. The first inequality follows from the fact that every cover of the set E is also a cover of Γ; the second inequality follows from [SZ] prop. 11; the last equality follows from lemma 2.3.

Corollary 2.5. *In the hypothesis of theorem 2.4, if:*

$$\inf\{\nu_\delta[\varphi_E(K)] : K \in \mathcal{K}\} = 0$$

then $s_\delta(\Gamma) = 0$.

References

[BF] T.Bonnesen, W.Fenchel, *Theorie der konvexen Körper*, Springer, Berlin (1934).

[CZ] A.Chiffi, G.Zirello, *Misure di Hausdorff e misure approssimanti*, Rend. Sem. Univ. Padova **69** (1983), 233-241.

[F] H.Federer, *Geometric Measure Theory*, Springer, Berlin (1969).

[K] K.Kallay, *Reconstruction of a plane convex body from the curvature of its boundary*, Israel J. Math. **17** (1974) 149-161.

[S] W.Sierpinski, *General Topology*, University of Toronto Press, 1956.

[St. 1] O.Stefani, *Condizioni di continuita' in una misura approssimante*, Rend. Sem. Mat. Univ. Padova **72** (1984), 191-202.

[St. 2] O.Stefani, *Misure approssimanti e frontiere di insiemi ad ampiezza costante*, Rend. Circ. Mat. Palermo **12** (supplemento) (1986), 277-289.

[SZ] O.Stefani, G.Zirello, *Misure approssimanti ed insiemi ad ampiezza costante*, Rend. Sem. Mat. Univ. Padova **72** (1984), 191-202.

Istituto di Matematica Applicata

Università di Padova

Via Belzoni, 7

I-35131 PADOVA

A NON-UNIQUENESS RESULT FOR THE OPERATORS WITH PRINCIPAL PART $\partial_t^2 + a(t)\partial_x^2$

FERRUCCIO COLOMBINI   SERGIO SPAGNOLO

Dedicated to Ennio De Giorgi on his sixtieth birthday

1. Statement of the result. After the classical theorems of Carleman [3] and the counter-examples of Plis [12] and De Giorgi [6], the question of uniqueness for the non characteristic Cauchy problem has been widely investigated (see Alinhac [1] and Zuily [14] for an extensive bibliography).

Here we shall be concerned with Cauchy problems, with initial data at $\{t = 0\}$, for linear differential operators of the form

$$P = \partial_t^m + \sum_{\substack{|\alpha|+j \leq m \\ j < m}} a_{\alpha,j}(t,x)\partial_x^\alpha \partial_t^j$$

where m is a integer ≥ 1, $t \in \mathbf{R}$, $x = (x_1, \cdots, x_n) \in \mathbf{R}^n, \partial_t = \frac{\partial}{\partial t}$ and $\partial_x = (\frac{\partial}{\partial x_1}, \cdots, \frac{\partial}{\partial x_n})$.

Through out this paper, we shall assume that the coefficients of P are complex valued C^∞ functions on $\mathbf{R}_t \times \mathbf{R}_x^n$, while the coefficients in the principal part, i.e. the functions $a_{\alpha,j}$ for $|\alpha| + j = m$, will always be assumed real valued.

In order to investigate the uniqueness of solution and the possible stability of such uniqueness under lower order perturbations, we give the following definitions.

Definition. *The operator* P *is said to have the* uniqueness *property* (UP) *at* $t = 0$, *if every* C^∞ *solution* $u(t, x)$ *to*

$$\begin{cases} Pu = 0 & on \ \mathbf{R}_t \times \mathbf{R}_x^n \\ u \equiv 0 & on \ \{t \le 0\} \end{cases}$$

is identically zero in some neighborhood of $\{t = 0\}$.

When the UP holds for all the operators having the same principal part as that of P, we say that P has the *strong uniqueness property*.

The results of Calderon-Carleman ([2] and [3]) state that every homogeneous operator with *real* coefficients having *simple* characteristic roots [1] has the strong UP. On the other hand, the examples of Cohen [4] and Hörmander [7], where for all $m \ge 1$ a smooth function $b(t, x)$ is constructed such that the operator $\partial_t^m + b(t, x)\partial_x$ does not have the UP, show, in particular, that the operator ∂_t^m does not have the strong UP for $m \ge 2$.

Here we shall confine ourselves to the second order operator

$$P_a = \partial_t^2 + a(t)\partial_x^2 \qquad (a(t) \ real)$$

in the space dimension $n = 1$, and we shall prove that P_a cannot have the strong UP when $a(t)$ vanishes to *infinite order* at $t = 0$.

More precisely we prove:

Theorem 1. *Let* $a(t)$ *be a real* C^∞ *function vanishing to infinite order at* $t = 0$. *Then the operator* P_a *does not possess the*

---

[1] A complex number τ is said to be a characteristic root for P at the point (t, x) if

$$P(t, x; i\tau, i\xi) = 0 \qquad \text{for some} \qquad \xi \in \mathbf{R}^n \setminus \{0\}.$$

strong uniqueness property; indeed there exist three C^∞ functions $b(t)$, $c(t,x)$, $u(t,x)$, with $b(t)$ real valued, such that

$$(*) \qquad u_{tt} + a(t)u_{xx} - ib(t)u_x + c(t,x)u = 0 \quad on \ \mathbf{R}_t \times \mathbf{R}_x$$

and

$$\text{supp } u = \{(t,x) : t \geq 0\}.$$

In order to motivate the assumption of infinite flatness of $a(t)$, it is convenient to restrict ourselves to the special cases when P_a is a *degenerate elliptic* operator $(a(t) \geq 0)$ or a *weakly hyperbolic* operator $(a(t) \leq 0)$.

In the first case, we have the following result of strong uniqueness:

Theorem 2 (Watanabe [13]). *If the coefficient $a(t)$ is non-negative for $t \geq 0$ and has a zero of finite order at $t = 0$, then the operator P_a has the strong uniqueness property.*

As to the hyperbolic case $a(t) \leq 0$, we have a completely different situation. Indeed:

Theorem 3 (Nakane [8], [9]). *If the coefficient $a(t)$ is non-positive for $t \geq 0$ and has a zero of finite order $k \geq 3$ at $t = 0$, then the operator P_a does not have the strong uniqueness property.*

As a matter of fact, Nakane proved a more general result ensuring, in particular, that the operator

$$\partial_t^2 + t^k a_0(t,x)\partial_x^2 + t^h b(t,x)\partial_x + c(t,x)$$

with $a_0(0,x_0) < 0$ and $\Im m \, b(0,x_0) \neq 0$ does not have UP near the point $(0,x_0)$ whenever $0 \leq h < (k/2) - 1$. Moreover, he exhibited a certain class of functions $a(t) \leq 0$, vanishing to infinite order at $t = 0$, for which the operator P_a does not have the strong UP.

Going back to Theorem 3, we recall that the case in which $a(t)$ is non-positive and vanishes at $t = 0$ of order ≤ 2, is that of the so called *effective hyperbolicity* of the operator P_a; in such a case it is well known since the work of Oleinik [11] (see §3 of the present paper) that the Cauchy problem for $P_a + Q$ is well-posed for any first order term Q, thus in particular P_a has the strong uniqueness property.

Observing that, when $a(t)$ has a zero of finite order at $t = 0$, then $a(t) \geq 0$ or $a(t) \leq 0$ in a right neighborhood of $t = 0$, we can summarize our Theorem 1 and the quoted results of Watanabe, Nakane and Oleinik as follows:

Corollary. *The operator P_a has the strong uniqueness property if and only if*

$$a(t) \geq 0 \qquad and \qquad a^{(j)}(0) \neq 0 \qquad for\ some\ j \geq 0$$

or

$$a(t) \leq 0 \qquad and \qquad a^{(j)}(0) \neq 0 \qquad for\ some\ j \leq 2$$

in a right neighborhood of $t = 0$.

2. Outline of the proof

2.1. The model equation. The starting point of our construction is an ordinary differential problem of the form

$$(1) \qquad \begin{cases} w'' + \alpha(\tau)w = 0 \\ w(0) = 1, \quad w'(0) = 0 \end{cases}$$

with $\alpha(\tau) > 0$, whose solution $w(\tau)$ is exponentially decaying for $|\tau| \to \infty$.

Indeed, we can find a C^∞ even function $\alpha(\tau) > 0$, periodic on $\{\tau > 0\}$ with period equal to $1/2$ and satisfying, for some $\delta > 0$,

$$(2) \qquad \alpha(\tau) \equiv 1 \qquad for\ |\tau| \leq \delta,$$

in such a way that the solution to (1) satisfies

$$(3) \qquad |w(\tau)| \leq 1$$

and

(4) $$w(\tau) = w_0(\tau)e^{-\epsilon|\tau|}$$

for some constant $\epsilon > 0$ and some (even) function $w_0(\tau)$, $(1/2)$-periodic on $\{\tau > 0\}$.

The existence of such a function $\alpha(\tau)$ is proved in [5], p.501.

In the following, we shall use the constants

(5) $$\lambda_0 = \min_\tau \alpha(\tau), \quad \Lambda_0 = \max_\tau \alpha(\tau), \quad \Lambda_1 = \max_\tau |\alpha'(\tau)|.$$

2.2. The parameters. Throughout our construction, we shall use three sequences of positive numbers $\{\rho_k\}$, $\{\delta_k\}$, $\{\eta_k\}$, all decreasing to zero, and a sequence $\{\nu_k\}$ of integers increasing to infinity. These sequences will be choosen in a suitable way at the end of the proof (see (57), (58) and (64)).

Assuming, for the moment, that these sequences have been defined and that

(6) $$\sum_{k=1}^\infty \rho_k < \infty,$$

we define (see fig.1)

$$t_k = \frac{\rho_k}{2} + \sum_{j=k+1}^\infty \frac{3}{2}\rho_j$$

$$I_k = \left[t_k - \frac{\rho_k}{2}, \; t_k + \frac{\rho_k}{2}\right] \qquad Z_k = \left[t_k + \frac{\rho_k}{2}, \; t_{k-1} - \frac{\rho_{k-1}}{2}\right]$$

$$J_k = [t_{k+1}, t_{k-1}].$$

We note that $\{t_k\}$ is converging to zero and that the interval Z_k has a length equal to $\rho_k/2$.

Finally, we put

(7) $$h_k = \left(\frac{\nu_k}{\rho_k}\right)^2 \delta_k^{-1}.$$

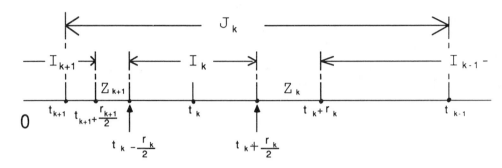

Figure 1

2.3. The solution $u(t,x)$. Assuming, for the moment, the coefficient $b(t)$ of equation $(*)$ as defined, we look for a null solution of the form

$$(8) \qquad u(t,x) = \sum_{k=1}^{\infty} \beta_k(t)\psi_k(t)e^{ih_k x},$$

where ψ_k is the solution to the ordinary differential problem

$$(9) \qquad \begin{cases} \psi_k'' - (h_k^2 a(t) - h_k b(t))\psi_k = 0 \\ \psi_k(t_k) = \eta_k \qquad \psi_k'(t_k) = 0, \end{cases}$$

while β_k is a *cut-function* of the form

$$(10) \qquad \beta_k(t) = \begin{cases} \beta(\frac{\nu_{k+1}}{\rho_{k+1}}(t - t_{k+1})) & \text{for } t \leq t_k \\ \beta(\frac{\nu_{k-1}}{\rho_{k-1}}(t_{k-1} - t)) & \text{for } t \geq t_k. \end{cases}$$

As $\beta(\tau)$ we take a non decreasing smooth function with $\beta(\tau)\equiv 0$ for $\tau \leq \delta/2$ and $\beta(\tau)\equiv 1$ for $\tau \geq \delta$, δ being given by (2). Let us observe that $\operatorname{supp}(\beta_k) \subset \overset{\circ}{J_k}$, so that the sum in (8) is *locally* given by only two terms.

2.4. Definition of the coefficient $b(t)$. Putting

$$(11) \qquad f_k(t) = -h_k a(t) + b(t),$$

we can re-write the problem (9) as

$$(12) \qquad \begin{cases} \psi_k'' + h_k f_k(t)\psi_k = 0 \\ \psi_k(t_k) = \eta_k, \qquad \psi_k'(t_k) = 0. \end{cases}$$

Let us then choose $b(t)$ in such a way that

$$(13) \qquad f_k(t) = \delta_k \cdot \alpha\left(\frac{\nu_k}{\rho_k}(t - t_k)\right) \qquad \text{on } I_k,$$

so that, by the change of variable

$$t \mapsto \tau = \frac{\nu_k}{\rho_k}(t - t_k),$$

the equation in (12) transforms, on I_k, to equation (1) and we have the following explicit expression of ψ_k :

$$(14) \qquad \psi_k(t) = \eta_k \cdot w\left(\frac{\nu_k}{\rho_k}(t - t_k)\right) \qquad \text{on } I_k.$$

Let us now complete the definition of $b(t)$; by (11) and (13) we have defined $b(t)$ on each interval I_k, thus we have to define $b(t)$ (or, equivalently by (11), to define $f_k(t)$) on the intervals

$$Z_k = \left[t_k + \frac{\rho_k}{2}, t_{k-1} - \frac{\rho_{k-1}}{2}\right].$$

To this end, we observe that $f_k = f_{k-1} + (h_{k-1} - h_k)a$ in virtue of (11) and hence from (13) (at the level $k - 1$) we derive

$$(15) \quad f_k(t) = \delta_{k-1}\alpha(\frac{\nu_{k-1}}{\rho_{k-1}}(t - t_{k-1})) + (h_{k-1} - h_k)a(t) \qquad \text{on } I_{k-1}.$$

Taking (2) into account, we see that (13) and (15) give, in particular, that

$$f_k(t) = \begin{cases} \delta_k, & \text{in a left nbd. of } t_k + \rho_k/2 \\ \delta_{k-1} + (h_{k-1} - h_k)a(t), & \text{in a right nbd. of } t_k + \rho_k. \end{cases}$$

Then, we take a non-increasing smooth function $\theta_k(t)$ such that $\theta_k \equiv 1$ on I_k, $\theta_k \equiv 0$ on I_{k-1} and

(16) $$|\theta_k^{(j)}(t)| \leq C_j \rho_k^{-j}$$

(note that the distance between I_k and I_{k-1} is equal to $\rho_k/2$), and we define

(17) $f_k(t) = \theta_k(t)\delta_k + (1 - \theta_k(t))(\delta_{k-1} + (h_{k-1} - h_k)a(t))$ on Z_k.

Summarizing, we have completely defined $b(t)$ (by (13),(17) and (11)) on $I_k \cup Z_k$ as

(18) $$b(t) = \begin{cases} \delta_k \cdot \alpha(\frac{\nu_k}{\rho_k}(t - t_k)) + h_k a(t), & \text{on } I_k \\ (\theta_k \delta_k + (1 - \theta_k)\delta_{k-1}) + (\theta_k h_k + (1 - \theta_k)h_{k-1})a(t), \\ \qquad\qquad\qquad\qquad\qquad\qquad\qquad \text{on } Z_k, \end{cases}$$

and hence $b(t)$ is defined on the whole interval $]0, t_1]$.

Clearly, it results that $b(t)$ is a C^∞ function on $]0, t_1]$. Later on, we shall see that $b(t)$ tends to zero, for $t \to 0^+$, together with all its derivates, thus it can be extended to a C^∞ function on $\mathbf{R}$ which is identically zero for $t \leq 0$.

2.5. Definition of the coefficient $c(t,x)$. From the definition of $b(t)$ it follows, in particular, that the solutions $\psi_k(t)$ of (12) are uniquely defined, and, consequently, also the function

$$u(t,x) = \sum_{k=1}^{\infty} \beta_k(t)\psi_k(t)e^{ih_k x}$$

is defined (see (10) for the definition of the β_k's).

Hence we are forced to define

$$(19) \qquad c(t,x) = \frac{u_{tt} + a(t)u_{xx} - ib(t)u_x}{u}$$

but we must preliminary show that the function

$$Lu(t,x) = u_{tt} + a(t)u_{xx} - ib(t)u_x =$$

$$= \sum_{k=1}^{\infty}(\beta_k''\psi_k + 2\beta_k'\psi_k')e^{ih_k x}$$

is identically vanishing on some neighborhood of the null-set of $u(t,x)$. To this regard, we observe that $Lu \equiv 0$ in each interval in which all the $\beta_j(t)$ are constant and that, by the definition (10), the β_j's are everywhere constant except for a small neighborhood of t_k, say the interval

$$(20) \qquad Y_k = \left[t_k - \frac{\rho_k}{\nu_k}\delta, \ t_k + \frac{\rho_k}{\nu_k}\delta\right] \qquad (k = 1,2,\ldots)$$

(δ being given by (2)) where β_{k-1} and β_{k+1} are non-constant.

Later on, we shall prove some estimates on $\psi_{k-1}, \psi_k, \psi_{k+1}$ ensuring, in particular, that ψ_{k-1} and ψ_{k+1} are dominated by ψ_k on Y_k: this will assure that u cannot vanish in $Y_k \times \mathbf{R}_x$ and, hence, that (19) makes sense.

2.6. Regularity of $b(t)$.
From the definition (18), using (16), we derive that [2]

$$(21)_k \qquad |b^{(j)}|_{I_k \cup Z_k} \le \tilde{C}_j \delta_{k-1}\left(\frac{\nu_k}{\rho_k}\right)^j \qquad \text{, with } \tilde{C}_0 = \Lambda_0 + 1,$$

---

[2] In the following, we shall use the notation

$$|g|_I = |g(t)|_I = \sup_{t \in I} |g(t)|.$$

for all $j \geq 0$, where $\Lambda_0 = \max \alpha(\tau)$, provided that for some M_j

$$(22)_k \qquad h_k |a^{(j)}|_{I_k \cup Z_k} \leq M_j \delta_{k-1} \left(\frac{\nu_k}{\rho_k}\right)^j \qquad \text{, with } M_o = 1.$$

Hence, to ensure that $b^{(j)}(t)$ tends to zero for $t \to 0^+$, it will be sufficient to take the parameters δ_k so large that (22) holds and so small that

$$(23) \qquad \delta_{k-1} \left(\frac{\nu_k}{\rho_k}\right)^j \to 0 \qquad \text{for } k \to \infty.$$

2.7. Estimates of the ψ_k's. In order to prove the regularity of $u(t, x)$ (see (8)) near $t = 0$, we estimate the behaviour of $|\psi_k^{(j)}|_{J_k}$ for $k \to \infty$, where $\psi_k(t)$ is the solution of (9), while to prove the regularity of $c(t, x)$ (see (19)) we need also an appropriate *lower* estimate of ψ_k near t_k.

Let us split the interval J_k as follows:

$$J_k = J_k^- \cup I_k \cup J_k^+$$

putting

$$J_k^- = [t_{k+1}, t_k - \frac{\rho_k}{2}] \, , \quad J_k^+ = [t_k + \frac{\rho_k}{2}, t_{k-1}].$$

i) *Estimates on I_k.*

Using (2), (3), (4) and the explicit form (14) of $\psi_k(t)$ on I_k, we get, for some constants K_j,

$$(24)_k \qquad |\psi_k^{(j)}|_{I_k} \leq K_j \eta_k \left(\frac{\nu_k}{\rho_k}\right)^j \qquad \text{with } K_0 = 1$$

$$(25) \qquad \psi_k(t_k \pm \frac{\rho_k}{2}) = \eta_k e^{-\epsilon \nu_k / 2}, \quad \psi_k'(t_k \pm \frac{\rho_k}{2}) = 0$$

$$(26)_k \qquad \psi_k(t) \geq \frac{\eta_k}{2} \qquad \text{on } Y_k = \left[t_k - \frac{\rho_k}{\nu_k} \delta, t_k + \frac{\rho_k}{\nu_k} \delta\right],$$

with δ given by (2).

ii) *Estimates on* J_k^-.

Let us go back to the equation (9), i.e. to

(27)
$$\psi_k'' + h_k f_k(t)\psi_k = 0$$

where $f_k(t)$ is given by (11).

In J_k^-, it is convenient to consider the *Kovalewskian* energy

(28)
$$E_k(t) = h_k \delta_k \psi_k^2(t) + \psi_k'^2(t)$$

thus, taking (25) as initial data, we derive from (27) that

(29)
$$\sqrt{E_k(t)} \leq \eta_k \sqrt{h_k}\sqrt{\delta_k}\exp\frac{1}{2}\left[-\epsilon\nu_k + \sqrt{h_k}\int_{J_k^-}\frac{|f_k(t) - \delta_k|}{\sqrt{\delta_k}}dt\right].$$

To estimate the integral in (29), let us observe that $J_k^- \subset I_{k+1}\cup Z_{k+1}$ so that by (11) and $(21)_{k+1}$ we have, for some constants L_j,

(30)
$$|f^{(j)}|_{J_k^-} \leq L_j\delta_k\left(\frac{\nu_{k+1}}{\rho_{k+1}}\right)^j \qquad \text{with } L_0 = \Lambda_0 + 2$$

(provided that $(22)_{k+1}$ is fulfilled).

Introducing (30) with $j = 0$ in (29) and remembering that $\sqrt{\delta_k h_k} = \nu_k\rho_k^{-1}$ (see (7)) and $|J_k^-| = \rho_{k+1}$, we find

$$\sqrt{E_k(t)} \leq \eta_k\frac{\nu_k}{\rho_k}\exp\left[\frac{\nu_k}{2}\left(-\epsilon + (\Lambda_0 + 3)\frac{\rho_{k+1}}{\rho_k}\right)\right]$$

and hence

(31)
$$\sqrt{E_k(t)} \leq \eta_k\frac{\nu_k}{\rho_k}e^{-\epsilon\nu_k/4} \qquad \text{in } J_k^-$$

provided that

(32)
$$(\Lambda_0 + 3)\frac{\rho_{k+1}}{\rho_k} \leq \frac{\epsilon}{2}.$$

Now, we estimate $\psi_k(t)$ and all its successive derivatives. From (28) and (31) we get (using that $\{\nu_k\rho_k^{-1}\}$ is increasing)

$$|\psi_k(t)| \le \eta_k e^{-\epsilon\nu_k/4}$$

$$|\psi_k'(t)| \le \eta_k \frac{\nu_k}{\rho_k} e^{-\epsilon\nu_k/4} \le \eta_k \frac{\nu_{k+1}}{\rho_{k+1}} e^{-\epsilon\nu_k/4}$$

for $t \in J_k^-$, while for $\psi_k^{(j)}(j \ge 2)$ it is convenient to go back to equation (27) observing that

$$h_k|f_k^{(j)}(t)| \le L_j\left(\frac{\nu_{k+1}}{\rho_{k+1}}\right)^{j+2} \qquad \text{in } J_k^-$$

as a consequence of (30) and (7).

Thus, [3] we get for some constants C_j

$$(33)_k \qquad\qquad |\psi_k^{(j)}(t)|_{J_k^-} \le C_j\eta_k\left(\frac{\nu_{k+1}}{\rho_{k+1}}\right)^j e^{-\epsilon\nu_k/4}.$$

iii) *Estimate on J_k^+.*

Let us go back to equation (27), considering now, instead of (28), the *hyperbolic* energy

$$(34) \qquad\qquad \tilde{E}_k(t) = h_k f_k(t)\psi_k^2(t) + \psi_k'^2(t).$$

Indeed, an estimate like (29) would be too weak for our purpose since $f_k(t)$ is very large in J_k^+. [We'll choose the parameters δ_k such that it results $f_k(t) > 0$ on $\bar{J}_k^+$].

---

[3] **Lemma:** *If*

$$\psi_k'' + g_k(t)\psi_k = 0$$

and

$$|g_k^{(j)}| \le L_j\lambda_k^{j+2}, \quad |\psi_k| \le \mu_k, \quad |\psi_k'| \le \mu_k\lambda_k,$$

then

$$|\psi_k^{(j)}| \le C_j\mu_k\lambda_k^j.$$

Taking (25) as initial data and observing that $f_k(t_k + \rho_k/2) = \delta_k$ by (13), we derive from (27)

$$(35) \qquad \sqrt{\tilde{E}_k(t)} \le \eta_k \sqrt{h_k} \sqrt{\delta_k} \exp\left[\frac{1}{2}\left(-\epsilon\nu_k + \int_{J_k^+} \frac{|f_k'|}{f_k} dt\right)\right].$$

Let us now estimate the integral in (35), observing that $J_k^+ \subset Z_k \cup I_{k-1}$.

In the interval Z_k, we have by (17)

$$f_k(t) = g_k(t) + (1 - \theta_k(t))(h_{k-1} - h_k)a(t)$$

having put

$$g_k(t) = \theta_k(t)\delta_k + (1 - \theta_k(t))\delta_{k-1}.$$

Now, $g_k(t)$ is a positive function which increases from the value δ_k to δ_{k-1}, hence (using (16)) we have, for $t \in Z_k$,

$$\frac{|f_k'(t)|}{f_k(t)} \le \frac{g_k'(t) + Ch_k(\rho_k^{-1}|a|_{Z_k} + |a'|_{Z_k})}{g_k(t) - h_k|a|_{Z_k}}$$

$$\le \frac{g_k'(t)}{g_k(t)/2} + \tilde{C}\rho_k^{-1},$$

provided that, for some M,

$$(36) \qquad h_k|a|_{Z_k} \le \frac{\delta_k}{2}, \quad h_k|a'|_{Z_k} \le M\frac{\delta_k}{\rho_k}.$$

Integrating on Z_k and remembering that $|Z_k| = \rho_k/2$, we then find

$$(37) \qquad \int_{Z_k} \frac{|f_k'|}{f_k} dt \le 2\log\frac{\delta_{k-1}}{\delta_k} + \frac{\tilde{C}}{2}.$$

On the other side, in the interval I_{k-1} we have (see (15))

$$\frac{|f_k'(t)|}{f_k(t)} \le \frac{\Lambda_1\delta_{k-1}\nu_{k-1}\rho_{k-1}^{-1} + h_k|a'|_{I_{k-1}}}{\lambda_0\delta_{k-1} - h_k|a|_{I_{k-1}}}$$

$$\le L_0\frac{\nu_{k-1}}{\rho_{k-1}} \qquad \text{with } L_0 = \frac{2}{\lambda_0}(\Lambda_1 + 1),$$

where $\Lambda_1 = \max |\alpha'(\tau)|$ and $\lambda_0 = \min \alpha(\tau)$, provided that

(38) $$h_k |a|_{I_{k-1}} \leq \frac{\lambda_0}{2} \delta_{k-1} , \quad h_k |a'|_{I_{k-1}} \leq \delta_{k-1} \frac{\nu_{k-1}}{\rho_{k-1}}.$$

Thus

(39) $$\int_{I_{k-1}} \frac{|f_k'|}{f_k} dt \leq L_0 \nu_{k-1}$$

and hence, introducing (37) and (39) in (35) and using again (7), we find, for $t \in J_k^+$,

$$\sqrt{\tilde{E}_k(t)} \leq C \eta_k \frac{\nu_k}{\rho_k} \frac{\delta_{k-1}}{\delta_k} \exp \left[\frac{\nu_k}{2} \left(-\epsilon + L_0 \frac{\nu_{k-1}}{\nu_k} \right) \right].$$

Since $L_0 = 2(\Lambda_1 + 1)/\lambda_0$, the last inequality gives

(40) $$\sqrt{\tilde{E}_k(t)} \leq C \eta_k \frac{\nu_k}{\rho_k} \frac{\delta_{k-1}}{\delta_k} e^{-\epsilon \nu_k / 4} \quad \text{in } J_k^+,$$

provided that

(41) $$\frac{2}{\lambda_0} (\Lambda_1 + 1) \cdot \frac{\nu_{k-1}}{\nu_k} \leq \frac{\epsilon}{2}.$$

Now, by $\{(17), (36),\}$ and $\{(15), (38)\}$ we have

(42) $$f_k(t) \geq \frac{\delta_k}{2} \quad \text{in } J_k^+$$

provided that

(43) $$\delta_k \leq \lambda_0 \cdot \delta_{k-1},$$

hence (40) and (34) imply, for $t \in J_k^+$,

(44) $$\begin{cases} |\psi_k(t)| \leq C \eta_k \frac{\delta_{k-1}}{\delta_k} e^{-\epsilon \nu_k / 4} \\ |\psi_k'(t)| \leq C \eta_k \frac{\nu_k}{\rho_k} \frac{\delta_{k-1}}{\delta_k} e^{-\epsilon \nu_k / 4}. \end{cases}$$

Let us now estimate $\psi_k^{(j)}$ for $j \geq 2$. From (15), (16) and (17) we derive that, for some constants C_j,

$$(45) \qquad |f_k^{(j)}|_{J_k^+} \leq C_j \delta_{k-1} \left(\frac{\nu_k}{\rho_k} \right)^j$$

provided that, for some M_j, the following conditions are fulfilled:

$$(46) \qquad h_k |a^{(j)}|_{Z_k} \leq M_j \delta_{k-1} \left(\frac{1}{\rho_k} \right)^j$$

$$(47) \qquad h_k |a^{(j)}|_{I_{k-1}} \leq M_j \delta_{k-1} \left(\frac{\nu_{k-1}}{\rho_{k-1}} \right)^j .$$

[Note that (46) is a consequence of $(22)_k$, while (47) is stronger that $(22)_{k-1}$].

Thus, using the Lemma in the footnote [3] with

$$g_k(t) = h_k f_k(t), \quad \mu_k = \frac{\delta_{k-1}}{\delta_k}, \quad \lambda_k = \sqrt{\frac{\delta_{k-1}}{\delta_k}} \frac{\nu_k}{\rho_k}$$

and observing that, by (45) and (7),

$$h_k |f_k^{(j)}|_{J_k^+} \leq C_j \left(\sqrt{\frac{\delta_{k-1}}{\delta_k}} \right)^{j+2} \left(\frac{\nu_k}{\rho_k} \right)^{j+2} ,$$

we get the estimate (for all $j \geq 0$)

$$(48)_k \qquad |\psi_k^{(j)}|_{J_k^+} \leq C_j \eta_k \left(\frac{\delta_{k-1}}{\delta_k} \right)^{1+j/2} \left(\frac{\nu_k}{\rho_k} \right)^j e^{-\epsilon \nu_k /4} .$$

Summarizing, we have proved the estimates (24), (26), (33) and (48) under the conditions $\{(32), (41), (43)\}$ on the parameters $\{\rho_k, \delta_k, \nu_k\}$ and the conditions $\{(22), (36), (38), (46), (47)\}$ which involve also the given function $a(t)$ and can be derived from the following one:

$$(49) \qquad h_{k+1} |a^{(j)}|_{I_k \cup Z_k} \leq M_j \delta_k \left(\frac{\nu_k}{\rho_k} \right)^j$$

$$\text{for some } M_j \text{ with } M_0 = \frac{\lambda_o}{2} .$$

2.8. Regularity of $u(t,x)$. From the definition (10) of $\beta_k(t)$ and the estimates (24), (33), (48) of $\psi_k(t)$, we derive

$$(50) \quad |\partial_t^j \partial_x^p \beta_k(t)\psi_k(t)e^{ih_k x}| \le C_j \eta_k \left(\frac{\nu_{k+1}}{\rho_{k+1}}\right)^j \left(\frac{\delta_{k-1}}{\delta_k}\right)^{1+j/2} h_k^p.$$

Thus, taking into account that $\operatorname{supp}(\beta_k) \subset J_k$ and $\{J_k\} \to 0$ for $k \to \infty$, we can conclude that the solution $u(t,x)$, defined by (8), is converging to zero for $t \to 0^+$ together with all its derivatives, provided that

$$(51) \quad \eta_k \left(\frac{\nu_{k+1}}{\rho_{k+1}}\right)^j \left(\frac{\delta_{k-1}}{\delta_k}\right)^{1+j/2} h_k^p \to 0 \quad \text{for} \quad k \to \infty$$

and that the conditions (6), (32), (41), (43), (49) are fulfilled.

2.9. Regularity of $c(t,x)$. By the definition (see (19)) we have

$$c(t,x) = 0 \quad \text{for } t \notin \bigcup_{k=1}^{\infty} Y_k$$

where

$$Y_k = \left[t_k - \frac{\rho_k}{\nu_k}\delta, \ t_k + \frac{\rho_k}{\nu_k}\delta\right],$$

thus we must only estimate $c(t,x)$ on each interval Y_k. To this end, let us split Y_k into the two sub-intervals

$$Y_k^- = Y_k \cap \{t \le t_k\} \quad , \quad Y_k^+ = Y_k \cap \{t \ge t_k\}.$$

In Y_k^+ we have $\beta_k \equiv 1$ and $\beta_j \equiv 0$ for $j \ne k-1$, hence

$$c(t,x) = \frac{(\beta_{k-1}''\psi_{k-1} + 2\beta_{k-1}'\psi_{k-1}')e^{ih_{k-1}x}}{\psi_k e^{ih_k x} + \beta_{k-1}\psi_{k-1}e^{ih_{k-1}x}},$$

i.e.

$$(52) \qquad c(t,x) = \sum_{n=1}^{\infty} \varphi_{k,n}^{+}(t) e^{-in(h_k - h_{k-1})x}$$

where

$$\varphi_{k,n}^{+} = (\beta''_{k-1}\psi_{k-1} + 2\beta'_{k-1}\psi'_{k-1})(-\beta_{k-1}\psi_{k-1})^{n-1}\psi_k^{-n}.$$

Now $Y_k^{+} \subseteq J_{k-1}$, hence from $(10)_{k-1}$ and $(33)_{k-1}$ we derive that, in Y_k^{+},

$$|\beta_{k-1}^{(j)}(t)| \leq C_j \left(\frac{\nu_k}{\rho_k}\right)^j$$

$$|\psi_{k-1}^{(j)}(t)| \leq C_j \eta_{k-1} e^{-\epsilon\nu_{k-1}/4} \left(\frac{\nu_k}{\rho_k}\right)^j ;$$

on the other hand, we have also $Y_k^{+} \subseteq I_k$ whence, by $(24)_k$,

$$|\psi_k^{(j)}(t)| \leq K_j \eta_k \left(\frac{\nu_k}{\rho_k}\right)^j \qquad (K_0 = 1).$$

Finally in Y_k we have (see $(26)_k$)

$$(53) \qquad \psi_k(t) \geq \frac{\eta_k}{2}.$$

Thus, in virtue of Lemma 3 of [5] (p.508), we see that, in Y_k^{+},

$$|\varphi_{k,n}^{+(j)}(t)| \leq B_j n^j \left(\frac{\nu_k}{\rho_k}\right)^2 \left(B \frac{\eta_{k-1}}{\eta_k} e^{-\epsilon\nu_{k-1}/4}\right)^n \left(\frac{\nu_k}{\rho_k}\right)^j$$

for some constants B_j, B, so that going back to (52) we conclude that

$$\operatorname*{Max}_{x} |\partial_t^j \partial_x^p c(t,x)|_{Y_k^{+}} \to 0 \qquad (k \to \infty)$$

if

$$(54) \qquad h_k^p \left(\frac{\nu_k}{\rho_k}\right)^{j+2} \frac{\eta_{k-1}}{\eta_k} e^{-\epsilon\nu_{k-1}/4} \to 0 \qquad (k \to \infty).$$

Passing to the right sub-interval Y_k^-, we see that in such interval

$$c(t,x) = \sum_{n=1}^{\infty} \varphi_{k,n}^-(t) e^{-in(h_k - h_{k+1})x}$$

where

$$\varphi_{k,n}^- = (\beta_{k+1}'' \psi_{k+1} + 2\beta_{k+1}' \psi_{k+1}')(-\beta_{k+1}\psi_{k+1})^{n-1}\psi_k^{-n}.$$

Now $Y_k^- \subseteq J_{k+1}^+$, hence in Y_k^- we have (thanks to $(10)_{k+1}$ and $(48)_{k+1}$)

$$|\beta_{k+1}^{(j)}(t)| \le C_j \left(\frac{\nu_k}{\rho_k}\right)^j \le C_j \lambda_k^j$$

and

$$|\psi_{k+1}^{(j)}(t)| \le C_j \eta_{k+1} \frac{\delta_k}{\delta_{k+1}} e^{-\epsilon\nu_{k+1}/4}\lambda_k^j$$

for some constants C_j, where we have put by brevity

(55) $$\lambda_k = \sqrt{\frac{\delta_k}{\delta_{k+1}} \cdot \frac{\nu_{k+1}}{\rho_{k+1}}}.$$

On the other side $Y_k^- \subseteq I_k$, so that

$$|\psi_k^{(j)}(t)| \le K_j \eta_k \left(\frac{\nu_k}{\rho_k}\right)^j \le K_j \eta_k \lambda_k^j \quad (K_0 = 1).$$

Using again (54) and Lemma 3 of [5], we find

$$|\varphi_{k,n}^-(t)| \le \tilde{B}_j n^j \lambda_k^2 \tilde{B}\left(\frac{\eta_{k+1}}{\eta_k}\frac{\delta_k}{\delta_{k+1}}e^{-\epsilon\nu_{k+1}/4}\right)^n \lambda_k^j$$

and we conclude (taking (55) into account) that

$$\underset{x}{\text{Max}}\, |\partial_t^j \partial_x^p c(t,x)|_{Y_k^-} \to 0 \qquad (k \to \infty)$$

if

(56) $$h_{k+1}^p \left(\frac{\nu_{k+1}}{\rho_{k+1}}\right)^{j+2} \frac{\eta_{k+1}}{\eta_k} \left(\frac{\delta_k}{\delta_{k+1}}\right)^{j/2+2} e^{-\epsilon\nu_{k+1}/4} \xrightarrow[k\to\infty]{} 0.$$

In conclusion, to ensure that $c(t, x) \to 0$ for $t \to 0^+$ together with all its derivatives, it will be sufficient to choose the parameters in such a way that, besides (6), (32), (41), (43) and (49), the conditions (54) and (56) be fulfilled.

2.10. Choice of the parameters. To complete the proof of Theorem 1, we show that there exist some parameters $\{\rho_k\}$, $\{\delta_k\}$, $\{\eta_k\}$ and $\{\nu_k\}$ such that all the conditions (6), (23), (32), (41), (43), (51), (54), (56) and (49) (which is the only one which involves also the given function $a(t)$) are fulfilled. [We recall that the constants λ_0, Λ_0, Λ_1 appearing in someones of the previous conditions are defined in (5), while the h_k's are related to the other parameters by (7)].

Let us firstly define $\{\rho_k\}$ and $\{\nu_k\}$ as

$$(57) \qquad \rho_k = A^{-k} \ , \ \nu_k = A^k$$

where A is a positive number sufficiently large (with respect to $\lambda_0, \Lambda_0, \Lambda_1, \epsilon$) that the conditions (6), (32) and (41) be satisfied; then let us define $\{\eta_k\}$ by taking

$$(58) \qquad \eta_1 = 1, \qquad \eta_k = \exp\left(-\frac{\epsilon}{8}(\nu_1 + \ldots \nu_{k-1})\right).$$

With these choices, condition (23) becomes

$$(59) \qquad \delta_k \cdot A^{kj} \to 0 \qquad (k \to \infty) \qquad \forall j \geq 0,$$

while conditions (51), (54) and (56) are fulfilled as soon as

$$(60) \qquad \delta_k^{-j} \cdot \exp\left(-\frac{\epsilon}{9} A^k\right) \to 0 \quad (k \to \infty) \quad \forall j \geq 0,$$

and condition (49) is fulfilled if

$$(61) \qquad |a^{(j)}|_{I_k \cup Z_k} \leq M_j \cdot \delta_{k+1}^2 A^{-2(k+1)} \qquad \forall j \geq 0$$

for some constants M_j with $M_0 = \lambda_0/2$.

Hence, we must only find $\{\delta_k\}$ such that (59), (60) and (61) hold. In view of (61), we observe that, thanks to the basic assumption that

$a(t)$ is vanishing to infinite order at $t = 0$, we can find a function $\varphi(t)$ such that, for all $j \geq 0$,

(62) $$0 \leq \varphi(t) \leq C_j' t^j$$

and

(63) $$|a^{(j)}(t)| \leq C_j'' \varphi(t) \quad \text{with} \quad C_0'' = 1,$$

in the interval $[0, t_1]$.

Thus, if we define $\{\delta_k\}$ as

(64) $$\delta_k = \max\left\{ \sqrt{\frac{2}{\lambda_0}} A^k |\varphi|_{I_{k-1} \cup Z_{k-1}}^{1/2}, \ \exp\left(-\frac{A^k}{k}\right)\right\},$$

we easily see that (60) and (61) hold (by (63)).

On the other hand, by (62) and the definition (57) of $\{\rho_k\}$, we have

$$\varphi(t) \leq \tilde{C}_i A^{-ki}, \quad \forall i \geq 0, \quad \text{in} \quad I_{k-1} \cup Z_{k-1},$$

and hence (59) follows.

This completes the proof.

Remark. Although the leading coefficient $a(t)$ in equation $(*)$ is not necessarily a negative (neither positive) function, an essential point in the proof has been to choose the lower order coefficient $b(t)$ in such a way that the O.D.E. $(9)_k$ be of *hyperbolic type* on a certain interval J_k^+ (see the step 7, (iii) of the proof).

3. Levi-type conditions. We now consider the inhomogeneous operator

$$P_{a,b} = \partial_t^2 + a(t)\partial_x^2 + b(t,x)\partial_x \qquad (a(t) \ real)$$

and we ask if the UP holds for all the operators of the form $P_{a,b} + c(t,x)$.

The results of non-uniqueness quoted in §1, namely the example of Cohen-Hörmander (where $a(t) = 0$), the Nakane's Theorem (where

$a(t) \leq 0$ and vanishes at $t = 0$ of a finite order $k \geq 3$) and our Theorem 1 (where $a(t)$ vanishes to infinite order), show the necessity to impose some conditions on $b(t, x)$ in order that $P_{a,b} + c(t, x)$ has the UP for any $c(t, x)$.

In this context, a classical Levi-type condition is that, for some $C \geq 0$,

$$(65) \qquad |b(t, x)|^2 \leq C|a(t)| \qquad (t \geq 0);$$

in fact, such a condition leads to some uniqueness results provided that $a(t)$ satisfies some preliminary condition of *quasi-monotonicity*, such as the following: [4]

$$(66) \qquad \exists C \geq 0 \text{ s.t. } a'(t) + Ca(t) \quad \text{does not change sign for } t \geq 0.$$

To make more precise the above mentioned uniqueness results, let us firstly observe that (except for the trivial case $a(0) \neq 0$, wherein the UP is well known) (66) implies, in particular, that also $a(t)$ does not change sign for $t \geq 0$, more precisely $a(t) \geq 0$ (resp. ≤ 0) if $a'(t) + Ca(t) \geq 0$ (resp ≤ 0) for some $C \geq 0$.

Hence, we can consider separately the case in which $P_{a,b}$ is (degenerate) elliptic and that in which $P_{a,b}$ is (weakly) hyperbolic, and we quote the following results.

Theorem (Oleinik [11]). *Assume that $a(t) \leq 0$ and that (66) holds. Then, (65) is a sufficient condition in order that the operator $P_{a,b} + c(t, x)$ has the uniqueness property for all $c(t, x)$.*

More generally, the same conclusion holds true if the assumptions $\{(65), (66)\}$ are replaced by the following weaker condition:

$$t|b(t, x)|^2 \leq C_1 a'(t) - C_2 a(t) \qquad (t \geq 0)$$

for some $C_1, C_2 \geq 0$.

---

[4] If $a(t)$ does not satisfy (66), it can happen that $P_{a,b}+c(t,x)$ does not have the UP even in the simple case $b=0$. This fact is proved in [5], where in addition $a(t)$ is non-positive.

Theorem (Nirenberg [10]). *Assume that $a(t) \geq 0$ and that (66) holds. Then, (65) is a sufficient condition in order that $P_{a,b}$ possess the compact uniqueness property.*[5]

Going back to the Theorems 1,2,3 of §1, we note that the condition (66) is always fulfilled in the case in which $a(t)$ has a zero of finite order at $t = 0$. The theorems of Watanabe and Nakane show that, in such a case, the Levi condition (65) is superfluous (in order to get the uniqueness) for the elliptic equations but not for the hyperbolic ones.

As to the complementary case (which can occur even if $a(t)$ satisfies (66)) wherein $a(t)$ vanishes to infinite order, our Theorem 1 shows that in such a case some kind of Levi condition of the type (65) must be imposed on the coefficient $b(t, x)$ in order to have the uniqueness for all the operators $P_{a,b} + c(t, x)$.

References

[1] S.Alinhac, *Uniqueness and non-uniqueness in the Cauchy problem*, Contemp. Math. **27** (1984), 1-22.

[2] A.P.Calderon, *Uniqueness in the Cauchy problem for P.D.E.*, Amer. J. Math. **80** (1958), 16-36.

[3] T.Carleman, *Sur un problème d'unicité pour les systèmes d'équations aux dérivées partielles à deux variables indépendantes*, Ark. Mat. Astr. Fys. **26** B, n.17 (1939), 1-9.

[4] P.Cohen, *The non-uniqueness of the Cauchy problem*, O.N.R. Techn. Report n.**93**, Stanford University 1960.

[5] F.Colombini, E.Jannelli and S.Spagnolo, *Non-uniqueness in hyperbolic Cauchy problems*, Ann. of Math. **126** (1987), 495-524.

[6] E.De Giorgi, *Un esempio di non unicità della soluzione del problema di Cauchy relativo ad una equazione differenziale lineare a derivate parziali di tipo parabolico*, Rend. di Mat. **14** (1955), 382-387.

---

[5] An operator P is said to have the *compact uniqueness property* at $t=0$ if all the solutions u of $Pu=0$ for which $supp(u) \cap \{t \leq 0\}$ is a compact set, vanish identically in a neighborhood of $t=0$.

[7] L.Hörmander, *Linear Partial Differential Operators*, Springer-Verlag, Berlin 1963.

[8] S.Nakane, *Non-uniqueness in the Cauchy problem for partial differential operators with multiple characteristic I*, Comm. in P.D.E. **9** (1984), 63-106.

[9] S.Nakane, *Uniqueness and non-uniqueness in the Cauchy problem for a class of operators of degenerate type-II*, Proc. Japan Acad. **59** (1983), 318-320.

[10] L.Nirenberg, *Uniqueness in the Cauchy problem for a degenerate elliptic second order equation*, in Differential Geometry and Complex Analysis, Springer-Verlag 1985, 213-218.

[11] O.A.Oleinik, *On the Cauchy problem for weakly hyperbolic equations*, Comm. on Pure and Appl. Math. **23** (1970), 569-586.

[12] A.Plis, *The problem of uniqueness of the solutions of a system of partial differential equations*, Bull. Acad. Pol. Sci. **2** (1954), 55-57.

[13] K.Watanabe, *L'unicité du prolongement des solutions des équations elliptiques dégénérées*, Tohoku Math. J. **34** (1982), 239-249.

[14] C.Zuily, *Uniqueness and non-uniqueness in the Cauchy problem*, Birkhäuser Boston 1983.

Dipartimento di Matematica

Università di Pisa

Via Buonarroti, 2

I-56127 PISA

A NOTE ON DUALITY AND
THE CALCULUS OF VARIATIONS

UBIRATAN D'AMBROSIO

Dedicated to Ennio De Giorgi on his sixtieth birthday

Duality has been one of the most powerful methods in Mathematics, with a wide range of applications. In the Calculus of Variations we have probably the most important use of this method in modern mathematics. Hamilton, when applying the duality of poles and polars to the Calculus of Variations, the much in use by geometers, was responsible for a decisive step towards the introduction of a separate dual space. Although poles and polars were of substantive importance in Apollonius work on the conics (III century B.C.) and reappeared in the "Brouillon project" of Gérard Desargues (1591-1661). Although analysed by Fermat and by Pascal, this "project" was forgotten for about two centuries and was revived by Chasles, making its way to Hamilton. Together with Legendre's transformation, Hamiltonian theory is related to Minkowski's duality of convex figures. This is not surprising, since Minkowski's dual convex figures is the key for Hahn-Banach Theorem. In fact, the existence of supporting hyperplane was important in Minkowski's work on convex bodies. But even before, the proof of Euler's condition relies on a lemma which says that if f is continuous in an interval $(0,1)$ and $\int_o^1 f\, g\ dt = 0$ for every sufficiently smooth g, then f is identically zero. This is closely related to the theory of distributions of L.Schwartz, as it is shown in [LY2]. This, together with an unpublished paper by L.C.Young on "The developemnt of duality", presented at a "Symposium on the History of Mathematical Methods" (UNICAMP, Campinas, 1979) are basic references for the role of duality in the calculus of Variations. L.C.Young himself was responsible for a major step when introducing in the early 30's the concept of generalized curves as elements of a dual space where he developed his analytical tools. This was done in close connection with the early developments of quantum mechanics, where we find the concept of duality as a major step towards enlarging our perception of space. Since the 60's much progress has ocurred in the Calculus of Variations, mainly in the classical Plateau'problem and Minimal Surfaces in general through a geometrical view of duality in functional analysis. We want to stress the fact that the succesful application of duality in this case is of the same nature of another kind of set theoretic duality which are the fuzzy.

A major feature of theories which relies on working within the framework of dual space is the possibility of looking together at what might be called the "sensorial" space, reached by our perception, and the dual space where our analytical tools are developed. This step, of going from classical curves, regarded as tracks, to generalized curves, or from Fréchet surfaces to generalized surfaces, find much of its roots in Cantor's theory. The concept of set itself depends on a track to which a precise, analytical, concept of pertinence is inherent. Thus, the concept of set is effectively expressed by the so called characteristic function of the set: $\chi_A : A \to \{0,1\}$. Recently, in dealing with problems of electric engineering, L.A.Zadeh introduced the notion of "fuzzy", which essentially enlarges the Cantorian concept of a set by allowing the two-valued characteristic function χ_A to take values in the interval $[0,1]$. Thus a *fuzzy* is a doublet $|X,\mu|$ where X is a set and μ is a map $\mu : X \to [0,1]$.

L.C.Young considered, in his basic work on generalized curves of the early 30's [LY1], non-negative linear functionals in the space $C(\sum)$ of continuous functions defined in $R^m \times S^m$, where S^m is the unit sphere in R^m. By working with *admissible* functions

$$f : R^m \times R^m \to R$$

which are continuous and homogeneous in the sense

$$f(x, kp) = k \ f(x, p)$$

for every $k \geq 0$, call *elementary curve* every functional L in $C(\sum)$ which can be represented as

$$L(f) = \int_a^b \tilde{f}(x(t), x'(t))dt$$

where $\tilde{f}$ is the extension of f to an admissible function and $x(t)$ is absolutely continuous in $[a, b]$ such that there exists a bound $K > 0$ such that $|x'(t)| = K$ almost everywhere. This definition expresses the duality inherent to the Calculus of Variations. Our objects are now functionals. A step further in L.C. Young's theory is given by introducing weak convergence in the space of elementary curves, i.e. $(L_n) \to L$ if $(L_n(f)) \to L(f)$ for every admissible f. Thus a *generalized curve* is a linear functional in $C(\sum)$ which is the weak limit of a sequence of elementary curves (L_n) all with support (in the sense of functionals) in $R^m \cap B(0, r)$, for a given $r > 0$.

We have shown in [DA2] that through a classical representation theorem as proposed by E.L. Mc Shane in the early 40's, it is possible to interpret Young's generalized curves in terms of fuzzies. For a generalized curve L has a representation by a pair $(x(t), \mu(t))$. Thus we have a fuzzy $|X, \mu|$, where

$$X = \{x(t) : t \in [0, 1], x(t) \quad lipschitzian\}$$

and the map $\mu : X \to [0, 1]$ is defined as the non-negative linear functional defined in $C(\sum)$ occurring in the representation theorem. This can be carried on to higher dimensions. Thus, through the representation theorem, fuzzies $|x(t), u(t)|$ can be interpreted as generalized geometric entities in the same category of generalization as L.C. Young's entities and this fuzzy duality has been decisive in the new methods introduced in the Calculus of Variations since the 60's.

In the early 50's E. De Giorgi has initiated a duality approach to higher dimensional measures which would be the geometric equivalent of the then emerging theory of distributions [DG1]. This was carried on to further applications to minimal surfaces and to the concept of *set of finite perimeter* in [DG2]. We recall that any Borel

E set whose closure belongs to an open bounded Ω of R^n is a *set of finite perimeter* if

$$P(E) = \int_\Omega |\text{grad } \chi_E|$$

$$= \sup \left\{ \int_E \text{div } g \, dx : g \in [C_o^1(\Omega)], \quad |g(x)| \le 1 \right\} < +\infty.$$

As it is established the equivalence of De Giorgi's sets of finite perimeter and L.C. Young's generalized objects, for example in [DA1], the fuzzy duality characteristics of De Giorgi's generalized geometric entities can be established.

References

[DA1] U.D'Ambrosio, *Semicontinuity Theorems for Multiple Integrals of the Calculus of Variations*, Anais Acad. Bras. Ciencias **38** (1966), 2.

[DA2] U.D'Ambrosio, *Sobre Conjuntos Nubilos e Processos Otimais Generalizados*, Bul. Inst. Polit. Iasi, Tomul XXIV (XXVIII), fasc. 1-2 (1978), 21-23.

[DG1] E.De Giorgi, *Su una teoria generale della misura $(n-1)$ dimensionale in uno spazio ad r dimensioni*, Ann. Mat. **36** (1954), 191-213.

[DG2] E.De Giorgi, *Frontiere orientate di misura minima*, Seminario di Matematica, Scuola Normale Superiore, Pisa, 1960-61.

[LY1] L.C.Young, *Generalized Curves and the Existence of an Absolute Minimum in the Calculus of Variations*, C.R. de la Société des Sciences et des Lettres de Varsovie **30** (1973), 212-234.

[LY2] L.C.Young, *Calculus of Variations and Optimal Control Theory*, W.B. Saunders Co., Philadelphia, 1969.

Institute of Mathematics
Universidad Estadual de Campinas
CP 6063
13100 CAMPINAS (Brasile)

SOME RESULTS ON PERIODIC SOLUTIONS
OF HAMILTON-JACOBI EQUATIONS
IN HILBERT SPACES

GIUSEPPE DA PRATO

Dedicated to Ennio De Giorgi on his sixtieth birthday

1. Introduction. We shall consider the problem of existence, uniqueness and stability of periodic solutions of Hamilton-Jacobi equations in Hilbert spaces.

Such a problem arises in optimal control with infinite horizon with a time dependent cost function. When the cost is time independent the problem reduces to the usual infinite horizon problem and the periodic solution to a stationary solution of the Hamilton-Jacobi equation.

2. Notations and some known results. Let H be a real Hilbert space (norm $||$, inner product $<,>$); for any $r > 0$ we set $B_r = \{x \in H; |x| \leq r\}$. We shall denote by K the set of all continuous mappings $\varphi : H \to \mathbf{R}$ which are convex and such that $0 \in \partial\varphi(0)$ where $\partial\varphi$ represents the sub-differential of φ. Moreover K_T will represent the set of all continuous mappings $\varphi : [0, T] \times H \to \mathbf{R}$ such that $\varphi(t,.) \in K$ for any $t \in [0,T]$. We set $K_\infty = \bigcap_{T \geq 0} K_T$.

We shall define now several function spaces:

1° $C^k(H), k = 0, 1,.$ is the set of all the mappings $\varphi : H \to \mathbf{R}$ which are continuous with their Frèchet derivatives up to k and such that:

$$(2.1) \qquad |\varphi|_{i,r} = \sup\{|\varphi^{(i)}(x)|; \quad x \in B_r\} < +\infty$$

for $i = 0, .., k$ and $r > 0$. $\varphi^{(i)}$ represents the ith derivative of φ.

2° $C^k_{Lip}(H)$ is the set of all $\varphi \in C^k(H)$ such that:

$$(2.2) \quad \|\varphi\|_{i,r} = \sup\left\{\frac{|\varphi^{(i)}(x) - \varphi^{(i)}(y)|}{|x - y|}; \ x, y \in B_r, x \neq y\right\} < +\infty.$$

3° $B([0,T]; C^k(H))$ is the set of all continuous mappings $\varphi : [0,T] \times H \to \mathbf{R}$ such that:

i) The mapping

$$[0,T] \times H \to \mathbf{R}, (t,x) \to \varphi_x^{(i)}(t,x)(z_1, z_2, .., z_i)$$

is continuous for any $i = 1, 2, .., k$ and any $z_1, z_2, .., z_i \in H$.

ii) For any $i = 1, 2, .., k$ we have:

$$\sup\{|\varphi(t,.)|_{i,r}; t \in [0,T]\} < +\infty.$$

4° $B([0,t]; C^k_{Lip}(H))$ is the set of all $\varphi \in B([0,T]; C^k(H))$ such that:
$$\sup\{|\varphi(t,.)|_{k,r}; t \in [0,T]\} < \infty$$

for any $r > 0$.

We are concerned with the following Hamilton-Jacobi equation:

$$(2.3) \quad \psi_t(t,x) - \frac{1}{2}|\psi_x(t,x)|^2 + <Ax, \psi_x(t,x)> +g(t,x) = 0$$
$$\psi(T,x) = \varphi_o(x)$$

under the following hypotheses:

$$(2.4)$$
i) A is the infinitesimal generator of a strongly continuous semi $-$ group e^{tA}. There exists $\omega \in \mathbf{R}$ such that $|e^{tA}| \le e^{t\omega}$

ii) $g \in K_\omega$

iii) $\varphi_o \in K$.

ψ is called a *strong solution* of problem (2.3) if:

$$(2.5)$$
i) $\psi \in K_T \cap B([0,T];C^1(H))$ and the set $\{\psi(t,.); t \in [0,T]\}$ is bounded in $C^1_{Lip}(H)$

ii) $\psi \in C^1([0,T] \times D(A); \mathbf{R})$

iii) equations (2.3) are satisfied for all $t \in [0,T]$ and $x \in D(A)$.

ψ is called a *weak solution* of Problem (2.3) if there exist three sequences, $\{\varphi_o^n\} \subset K, \{g^n\} \subset K_T$ and $\{\psi^n\} \subset B([0,T];C_1(H))$ such that:

$$(2.6)$$
i) $\varphi_o^n \to \varphi_o$ and $g^n \to h$ in $B([0,T];C(H))$

ii) $\psi^n \to \psi$ in $B([0,T];C(H))$

iii) The set $\{\psi^n(t,.); t \in [0,T], n \in \mathbf{N}\}$ is bounded in $C^1_{Lip}(H)$

iv) ψ^n is a strong solution of the equation :

$$\psi_t^n + \frac{1}{2}|\psi_x^n|^2 - <Ax, \psi_x^n> +g_n = 0; \psi^n(T,x) = \varphi_o^n(x).$$

The following result is proved in [1] (see Theorem 3, Remark 5 and Proposition 2 of Chapter 2).

Proposition 2.1. *Assume that hypotheses (2.4) hold. Then Problem (2.3) has a unique weak solution ψ. ψ is given by the formula:*

$$(2.7) \quad \psi(t,x) = inf\{ \int_t^T (g(s,y(s)) + \frac{1}{2}|u(s)|^2)ds + \varphi_o(y(T)); $$
$$u \in L^2(t,T;H), \; y' = Ay + u, \; y(t) = x\}$$

Moreover if $\varphi_o \leq \overline{\varphi}_0, g \leq \overline{g}, \overline{\varphi}_o \in K, \overline{g} \in K_T$ and if $\overline{\psi}$ is the weak solution of (2.3) corresponding to $\overline{\varphi}_o$ and $\overline{g}$, then we have $\psi \leq \overline{\psi}$.

Finally the following regularity result is proved in [2].

Proposition 2.2. *Assume (2.4) and let ψ be the strong solution of (2.3). Then:*

i) *for any $t \in [0,T]$, $\psi(t,.)$ is Gateaux differentiable*
ii) *for any $x \in D(A)$, $\psi(.,x)$ is Lipschitz continuous*
iii) *for any $x \in D(A)$ and almost every $t \in [0,T]$ we have :*

$$(2.8) \quad \psi_t(t,x) - \frac{1}{2}|\psi_x(t,x)|^2 + \; < Ax, \psi_x(t,x) > \; + g(t,x) = 0.$$

3. Existence of periodic solutions. We are here concerned with the equation:

$$(3.1) \quad \psi_t(t,x) - \frac{1}{2}|\psi_x(t,x)|^2 + \; < Ax, \psi_x(t,x) > \; + g(t,x) = 0$$

under the following hypotheses:

(3.2)
   i) Hypotheses (2.4) hold.
   ii) $g(t+p,x) = g(t,x), t \geq 0, x \in H$; such that $0 \leq M|x|^\alpha$
   iii) There exists $M > 0$ and $\alpha > 0$ such that $0 \leq g(t,x) \leq m|x|^\alpha$

We say that $\psi \in K_\infty$ is a *weak solution* to equation (3.1) if for any $T > 0$, ψ is the weak solution to (2.3) with $\varphi_0 = \psi(T, \cdot)$. If $\psi(t + p, x) = \psi(t, x)$ for any $t > 0$ and any $x \in H$ we say that ψ is a *weak periodic solution* to (3.1).

Theorem 3.1. *Assume (3.2). Then there exists a non negative weak periodic solution of equation (3.1) $\hat{\psi}$ such that $\hat{\psi}(t, 0) = 0$ for any $t \geq 0$. $\hat{\psi}$ is given by:*

$$
(3.3) \quad \hat{\psi}(t, x) = \inf\{ \int_t^\infty (g(s, y(s)) + \frac{1}{2}|u(s)|^2) ds \, |
$$
$$
u \in L^2(t, \infty; H), \ y'(s) = Ay(s) + u(s), y(t) = x\}.
$$

Moreover $\hat{\psi}$ is the minimal weak solution to (3.1) which is non negative.

PROOF. Let ψ^n be the weak solution to the problem:

$$
(3.4) \quad \psi_t^n - \frac{1}{2}|\psi_x^n|^2 + \langle Ax, \psi_x^n \rangle + g(t, x) = 0
$$
$$
\psi^n(np, x) = 0.
$$

Since $\psi^{n+1}(np, x) \geq 0 = \psi^n(np, x)$, by the last statement of Proposition 2.1 it follows that the sequence $\{\psi^n(t, x)\}$ is increasing for any $t \geq 0$ and any $x \in H$. We prove now that the set $\{\psi^n(t_o, x); x \in B_r\}$ is bounded for any $t_o \geq 0$ and $r > 0$. To this aim fix $u \in L^2(t, \infty; H)$ and let y be the mild solution of the linear problem:

$$
(3.5) \qquad y'(s) = Ay(s) + u(s) \quad , \quad y(t_0) = x; \quad s \geq t_o
$$

then by (2.7) it follows:

$$
(3.6) \qquad \psi^n(t_o, x) \leq \int_{t_o}^{np} (g(t, y(t)) + \frac{1}{2}|u(t)|^2) dt.
$$

Choose now:

$$
(3.7) \quad \begin{aligned} y(t) &= \exp((t - t_o)(A - \omega - 1))x, \\ u(t) &= -(\omega + 1)\exp((t - t_o)(A - \omega - 1))x \end{aligned}
$$

for any $t \geq t_o$. Then by (3.2)-iii) it follows

$$(3.8) \qquad \psi^n(t, x) \leq M|x|^\alpha + \frac{1}{4}|x|^2;$$

thus, there exists the limit:

$$(3.9) \qquad \psi(t, x) = \lim_{n \to \infty} \psi^n(t, x) \quad ; \quad t \geq 0, x \in X$$

and moreover $\hat{\psi}(t, \cdot) \in K$ for any $t \geq 0$.

Remark that $\hat{\psi}$ is periodic with period p as we see letting n tend to infinity in the equality:

$$\psi^n(t - p, x) = \psi^{n-1}(t, x) \quad ; \quad t \in [0, p(n-1)], \ x \in H.$$

We have to prove now that (3.3) holds and that $\hat{\psi}$ is a weak solution to (3.1). For this we start from the equality:

$$(3.10) \qquad \psi^n(t, x) = \int_t^{np} (g(s, \hat{y}_n(s)) + \frac{1}{2}|\hat{u}_n(s)|^2) ds$$

where $(\hat{y}_n, \hat{u}_n)$ is an optimal pair for the problem:
Minimize

$$J(u) = \int_t^{np} (g(s, y(s)) + \frac{1}{2}|u(s)|^2) ds$$

over all $u \in L^2(t, \infty; H)$ subject to $y' - Ay = u, y(t) = x$.
We have:

$$(3.11) \qquad \hat{y}_n'(s) = A\hat{y}_n(s) + \hat{u}_n(s) \quad ; \quad \hat{y}_n(t) = u.$$

Choose now two subsequences $\{y_{n_k}\}$, $\{u_{n_k}\}$ such that:

$$y_{n_k} \rightharpoonup \hat{y} \quad , \quad u_{n_k} \rightharpoonup \hat{u} \quad \text{weak in } L^2(t, \infty; H)$$

then it is easy to check that $\hat{y}$ is the mild solution of the problem:

$$(3.12) \qquad \hat{y}'(s) = A\hat{y}(s) + \hat{u}(s) \quad ; \quad \hat{y}(t) = x.$$

Since $g(s, \cdot)$ is convex, letting n tend to infinity in (3.10) we obtain:

$$(3.13) \qquad \hat{\psi}(t, x) \geq \int_t^\infty (g(s, \hat{y}(s))) + \frac{1}{2}|\hat{u}(s)^2)ds.$$

Moreover, letting n tend to infinity in (3.6), we see that the pair $(\hat{y}, \hat{u})$ is optimal, so that (3.3) holds and we have:

$$
\begin{aligned}
(3.14) \qquad \hat{\psi}(t, x) &= \int_t^\infty (g(s, \hat{y}(s)) + \frac{1}{2}|\hat{u}(s)|^2)ds \\
&= \int_t^T (g(s, \hat{y}(s)) + \frac{1}{2}|\hat{u}(s)|^2)ds + \hat{\psi}(T, \hat{y}(T)).
\end{aligned}
$$

By (3.14) and by Proposition 2.1 it follows that $\hat{\psi}$ is the weak solution of probelm (2.3) with $\varphi_o = \hat{\psi}(T, \cdot)$. Thus we have proved that $\hat{\psi}$ is a weak periodic solution of (3.1) and that (3.2) holds. It remains to prove the last statement.

Let ψ be any non negative weak solution of (3.1). Since:

$$\psi(np, x) \geq 0 = \psi^n(np, x)$$

by Proposition 2.1 it follows that $\psi(t, x) \geq \psi^n(t, x)$, so that $\psi \geq \hat{\psi}$.

Corollary 3.2.*Assume (3.2) and let $\hat{\psi}$ be the minimal periodic weak solution of (3.1). Then the following statements hold:*

   *i) for any $t \in$ **N**, $\hat{\psi}(t, \cdot)$ is Gateaux differentiable*

   ii) for any $x \in D(A), \hat{\psi}(\cdot, x)$ is Lipschitz continuous

   *iii) for any $x \in D(A)$ and almost every $t < $ **R** we have :*

$$(3.15) \qquad \hat{\psi}_t(t, x) - \frac{1}{2}|\hat{\psi}_x(t, x,)|^2 + <Ax, \hat{\psi}_x(t, x) > +g(t, x) = 0.$$

PROOF. The assertion follows from Proposition 2.2 since $\hat{\psi}$ is a weak solution of (2.3) with $\varphi_o = \hat{\psi}(T, \cdot)$ for any $T > 0$.

4. Uniqueness. We consider here equation (3.1) under hypotheses (3.2) and denote by Z_p the set of all non negative p-periodic weak solutions ψ such that $\psi(t,0) = 0$. By Theorem (3.1) we know that there exists $\hat{\psi} \in Z_p$ given by the formula (3.3); moreover $\hat{\psi}$ is minimal.

We remark that in general Z_p may consist of several elements. Consider in fact the particular case $g(t,x) = \frac{1}{2}|Rx|^2$ where R is a linear bounded operator in H; then, setting $\psi(t,x) = <Q(t)x, x>$ equation (3.1) reduces to Riccati equation:

$$(4.1) \qquad Q' + A^*Q + QA - Q^2 + R^*R = 0$$

and if (A^*, R) is not detectable equation (4.1) may have several stationary solutions (see for instance (5)).

In order to prove uniqueness we need the following additional assumptions:

i) There exists $N > 0$ and $\beta > 0$ such that :

$$(4.2) \qquad g(t,x) \geq N|x|^\beta \; ; \; x \in H, t \geq 0$$

ii)e^{tA} is compact for any $t > 0$.

We remark that hypothesis (4.2)-i) implies detectability in the linear quadratic control problem whereas (4.2)-ii) is made for technical reasons.

The main result of this section is the following:

Theorem 4.1.*Assume (3.2) and (4.2); then we have $Z_p = \{\hat{\psi}\}$, where $\hat{\psi}$ is given by (3.3).*

PROOF. Let $\psi_1, \psi_2 \in Z_p$ with $\psi_1 = \hat{\psi}$; then we have:

$$(4.3) \qquad \psi_2(t,x) \geq \psi_1(t,x) \; ; \; t \in \mathbf{N} \, , \, x \in H$$

since $\hat{\psi}$ is minimal. In order to prove the Theorem we have to show that:

$$(4.4) \qquad \psi_2(t,x) \leq \psi_1(t,x) \; ; \; t \in \mathbf{N} \, , \, x \in H.$$

Fix $T > 0$ and set $\varphi_{o,i} = \psi_i(T,\cdot), i = 1, 2$. Since ψ_i are weak solutions to (2.3) in $[0,T]$ there exist sequences:

$$\{\varphi^n_{o,i}\} \subset K, \quad \{g^n_i\} \subset K_T, \quad \{\psi^n_i\} \subset B([0,T]; C^1(H))$$

such that (see Definition 2.1)

$$(4.5) \qquad \varphi_{o,i}^n \to \varphi_{o,i} \quad \text{in} \quad C(H)$$

$$(4.6) \qquad g_i^n \to g, \quad \psi_i^n \to \psi_i \quad \text{in} \quad B([0,T]; C(H))$$

$$(4.7) \qquad \begin{array}{c} \text{the sets} \quad \{\psi_i^n(y,\cdot) \quad ; \quad t \in [0,T], n \in [0,T]\} \\ \text{are bounded in} \quad C_{Lip}^1(H) \end{array}$$

$$(4.8) \qquad \begin{array}{c} \psi_i^n \quad \text{is a weak solution to the equation} \\ \psi_{i,t}^n - \dfrac{1}{2}|\psi_{i,n}^n|^2 + <Ax, \psi_{i,x}^n> +g_i = 0. \end{array}$$

We set:

$$(4.9) \qquad L(t,x) = Ax - \psi_{1,x}(t,x) \quad ; \quad t \geq 0 \quad , \quad x \in D(A)$$

$$(4.10) \qquad L^n(t,x) = Ax - \psi_{1,x}^n(t,x) \quad ; \quad t \geq 0 \quad , \quad x \in D(A).$$

Since $\psi_{1,x}^n(t,\cdot)$ and $\psi_{1,x}(t,\cdot)$ are maximal monotone (see (4)) it is not difficult, taking into account (4.7), to see that the following problems:

$$(4.11) \qquad \begin{cases} v_t^n(t,x) = L^n(t, v^n(t,x)) \\ v^n(0,x) = x \end{cases}$$

and:

$$(4.12) \qquad \begin{cases} v_t(t,x) = L(t, v(t,x)) \\ v(0,x) = x \end{cases}$$

have unique solution v^n and v. Moreover the following estimates hold:

$$(4.13) \qquad |v(t,x)| \leq |x| \quad ; \quad |v^n(t,x)| \leq |x|$$

for any $x \in H$ and any $t \in [0,T]$.

We proceed now in three steps:

1st *step.* $\lim_{n\to\infty} v^n(t,x) = v(t,x)$. Due to (4.7) and (4.13) $\psi_{1,x}^n(t, v^n(t,x))$ is bounded in n, thus there exists a sub-sequence $\{n_k\}$ of **N** and $w(t,x) \in H$ such that:

$$(4.14) \qquad \psi_{1,x}^n(t, v^{n_k}(t,x)) \rightharpoonup w(t,x) \quad \text{weak.}$$

On the other hand we have:

$$(4.15) \qquad v^{n_k}(t,x) = e^{tA}x - \int_o^t e^{(t-s)A}\psi_{1,x}^{n_k}(s, v^{v_k}(s,x))ds$$

thus, by (4.14) and the compactness of e^{tA}, it follows that there exists $u(t,x)$ such that

$$(4.16) \qquad v^{n_k}(t,x) \to u(t,x) \quad \text{strong in} \quad H.$$

Now, for any $y \in H$ we have:
$$(4.17)$$
$$\psi_1^{n_k}(t,y) - \psi_1^{n_k}(t, v^{n_k}(t,x)) \geq < \psi_{1,x}^{n_k}(t, v^{n_k}(t,x)), y - v^{n_k}(t,x) >;$$

letting k tend to infinity we have:

$$(4.18) \qquad \psi_1(t,y) - \psi_1(t, u(t,x)) \geq < w(t,x), y - u(t,x) >$$

so that $w(t,x) = \psi_{1,x}(t, u(t,x))$. By (4.15) it follows:

$$(4.19) \qquad u(t,x) = e^{tA}x - \int_o^t e^{(t-s)A}\psi_{1,x}(s, u(s,x))ds$$

thus $v(t,x) = u(t,x)$ and the statement is proved.

2nd *step.* We have:

$$(4.20) \qquad \int_o^\infty |v(t,x)|^\beta dt \leq \psi_1(0,x).$$

In fact, recalling (4.8),

$$\frac{d}{dt}\psi_1^n(t, v^n(t,x)) = -\frac{1}{2}|\psi_1^n(t, v^n(t,x))|^2 - g^n(v^n(t,x))$$

it follows:

$$\psi_1^n(t, v^n(t, x)) + \frac{1}{2} \int_0^t (|\psi_{1,x}^n(s, x))|^2 + g^n(v^n(s, x)))ds = \psi_1^n(0, x)$$

which implies:

$$\int_0^T g^n(v^n(s, x))ds \le \psi_1^n(0, x).$$

Thus (4.20) follows since T is arbitrary.

3th *step.* Conclusion.

Setting $z = \psi_1 - \psi_2$, $z^n = \psi_1^n - \psi_2^n$, we obtain:

$$(4.21) \qquad \begin{aligned} z_t^n(t, x) + &< L(t, x), z_x^n(t, x) > + \frac{1}{2}|z_x^n(t, x)|^2 \\ &+ g_1^n(t, x) - g_2^n(t, x) = 0 \end{aligned}$$

it follows:

$$(4.22) \qquad \begin{aligned} \frac{d}{dt} z^n(t, v^n(t, x)) = &- \frac{1}{2}|z_x^n(t, v^n(t, x))|^2 \\ &- g_1^n(t, v^n(t, x)) - g_n^2(t, v^n(t, x)) \end{aligned}$$

so that, for $t \in [0, T]$,

$$(4.23) \qquad z^n(t, v^n(t, x)) \le z^n(0, x) \int_0^t (g_1^n - g_2^n)(s, v^n(s, x))ds.$$

As $n \to \infty$ we obtain:

$$(4.24) \qquad z(t, v(t, x)) \le z(0, x) \quad ; \quad x \in H \quad , \quad t \ge 0.$$

By (4.20) there exists $T_n \to \infty$ such that $v(T_n, x) \to 0$; thus by (4.24), we have:

$$(4.25) \qquad \psi_1(0, x) \ge \psi_2(0, x) \quad ; \quad x \in H;$$

it follows $\psi_1(p, x) \ge \psi_2(p, x)$ and then (recalling Proposition 2.1):

$$(4.26) \qquad \psi_1(t, x) \ge \psi_2(t, x) \quad ; \quad x \in H \quad , \quad t \in \mathbf{R}.$$

(4.26), along with (4.3), proves (4.4) and the proof is complete.

5. A generalization. Consider equation (3.1) under the following hypotheses:

i) hypotheses (2.4) hold.

(5.1)   ii) there exists $M > 0$ and $\alpha > 0$ such that

$$0 \leq g(t,x) \leq M|x|^{\alpha}.$$

Then the following theorem is proved as Theorem 3.1.

Theorem 5.1.*Assume (5.1). Then there exists a non negative weak solution $\hat{\psi}$ of (3.1) which is bounded in $[0,\infty[\times B_r$ for any $r > 0$ and such that $\hat{\psi}(t,0) = 0$, for any $x \in H$. Moreover we have:*

$$(5.2) \quad \hat{\psi}(t,x) = \inf\{ \int_t^{\infty} (g(s,y(s)) + \frac{1}{2}|u(s)|^2)ds;$$

$$u \in L^2(t,\infty;H) \ , \ y'(s) = Ay(s) + u(s), \ y(t) = x\}.$$

Finally $\hat{\psi}$ is the minimal weak solution to (3.1) which is non negative.

Remark 5.2 Proceeding as in the proof of Theorem 4.1 one can show that if Hypotheses (5.1) and (4.2) hold then $\hat{\psi}$ is the unique weak solution of (3.1) which is bounded on bounded sets of H and such that $\hat{\psi}(t,0) = 0$.

References

[1] V.Barbu, G.Da Prato, *Hamilton-Jacobi equations in Hilbert spaces*, Res. Notes in Math. **86**, Pitman, 1983.

[2] V.Barbu, G.Da Prato, *Hamilton-Jacobi equations in Hilbert spaces, variational and semigroup approach*, Ann. Mat. Pura e Appl. **142** (1985), 303-349.

[3] M.G.Crandall, A.Pazy, *Nonlinear evolution equations in Banach spaces*, Israel J. Math. **11** (1972), 57-94.

[4] W.H.Fleming, R.W.Rishel, *Deterministic and stochastic Optimal Control*, Springer-Verlag, New York, 1975.

[5] A.J.Pritchard, J.Zabczyk, *Stability and stabilizability of infinite dimensional systems*, SIAM Rev. **23** (1981), 25-52.

Scuola Normale Superiore

Piazza dei Cavalieri, 7

I-56126 PISA

GENERALIZED SOLUTIONS TO ORDINARY DIFFERENTIAL EQUATIONS WITH DISCONTINUOUS RIGHT-HAND SIDES VIA Γ-CONVERGENCE

ZOFIA DENKOWSKA   ZDZISLAW DENKOWSKI

Dedicated to Ennio De Giorgi on his sixtieth birthday

Summary. In the present paper we develop an idea of Prof.E. De Giorgi, which can be formulated as follows: "solutions to ODE may be obtained as minimizers of some functionals that are Γ-limits of appropriately chosen sequences of functionals defined on suitable functional spaces ".

In this way we obtain immediately the existence of Carathéodory solutions ($\mathcal{C}$-solutions) using functionals conneted to ODE with retarded argument. Next, using other functionals connected to ODE having a Borel function as the right-hand side, we get generalized solutions - so called K-solutions (Krasowskii solutions) and $\mathcal{D}$-solutions (we have named them $\mathcal{D}$-solutions in honour of Prof. De Giorgi). Both these classes of generalized solutions are different from the class of Filippov solutions ($\mathcal{F}$-solutions). In this part of the paper an important role is played by the projection theorem and the Aumann selection theorem for measurable multivalued functions with not necessarily closed values.

0. Introduction. The aim of this paper is to present a unified approach to the problem of finding a global solutions to the Cauchy problem

$$(0.1) \qquad x' = f(t,x) \quad , \quad t \in [0,1], \ x \in \mathbf{R}^n$$

$$(0.2) \qquad\qquad\qquad x(0) = a$$

without the continuity assumption on the function f.

When f is discontinuous only with respect to the time variable t (e.g. $f(\cdot,x)$ is measurable), the Carathéodory solutions ($\mathcal{C}$-solutions) are obtained. In the case where f is discontinuous with respect to the state variable x, (e.g. f is a Borel function of both variables), the problem is more complex. A natural way of solving it is to assign to every point (t,x) a whole set $F(t,x)$ of vectors (an orientor) rather than one vector $f(t,x)$, and then to consider the differential inclusion

$$(0.3) \qquad x' \in F(t,x) \quad t \in [0,1], \ x \in \mathbf{R}^n$$

$$(0.4) \qquad\qquad\qquad x(0) = a$$

regarding its solutions as generalized solutions of the starting Cauchy problem (0.1), (0.2). Of course, there are several resonable ways of defining the sets $F(t,x)$. We begin with the smallest sets taken as $F(t,x)$-this gives (see Section 2) the generalized solutions that we call (after Prof. E.De Giorgi's name) $\mathcal{D}$-solutions, then we consider $\mathcal{F}$-solutions (Filippov - see [14],[15],[16],[7],[2]) and $\mathcal{K}$-solutions (Krasovskii - see [15],[16]). Our approach to all of the mentioned above solutions is based on the Γ-convergence theory developed by E.De Giorgi and his collaborators ([12],[8]). The idea that Prof. E.De Giorgi suggested (explained in detail in section 3 - see Remark 3.6) is, roughly speaking, the following: first to find a variational formulation of the problem (0.1), (0.2) - so that the solutions of the problem become minimizers of an appropriately defined functional and then to find conditions which assure the existence of such minimizers. That may be done for instance by proving (this is what we actuallly do here) that this functional is the Γ-limit of an appropriately defined sequence of functionals, each of them attaining the minimum equal zero and then finding a convergent sequence of minimizers.

In the paper [1] L.Ambrosio proved the existence of the minimizer assuring the zero-minimum of the functional which appears in the variational formulation of the problem (0.1), (0.2) in a slightly different way and in the one-dimensional ($x \in \mathbf{R}$) case. This other way consist in relaxing the functional (which actually amounts to proving that it is the Γ-limit of a constant sequence). Also in the one dimensional case, P.Binding [4] obtained some existence results using completely different methods.

We would like to underline here that the existence of solutions to ODE with right-hand sides which are discontinuous with respect to the state variable x is of great importance in optimal control theory, especially in the problems of synthesis of optimal control (called also feedback or close loop control). Namely, (see Boltyanskii [5], Brunovsky [7], Hajek [16]) the synthesis (i.e. the optimal control u as a function of the state x) $u = v(x)$ is usually a discontinuous function, so passing from the equation with control $x' = f(t, x, u)$ to the ordinary differential equation $x' = f(t, x, v(x))$ we obtain ODE with a discontinuous right-hand side.

The organization of the present paper is the following: After introducing notations in Section 1, we give in Section 2 (Preliminaries) all the definitions of generalized solutions to ODE we will use together with the variational formulation of the Cauchy problem that corresponds to them. In Section 3 we quote some basic facts of the theory of Γ-convergence and in Remark 3.6 we explain in detail the idea of all the existence results we obtain. Section 4 contains an illustration of how the theory works. We give there an immediate proof of the existence of Carathéodory solutions. Section 5 contains the results from the theory of multivalued mappings, measurable, but with not necessarily closed values. We quote there the projection and the Aumann selection theorem and prove some lemmas we need for our proofs. Finally, in Section 6 we prove the existence of D- and K-solution for the Cauchy problem with the function $f(t, x)$ that is a Borel function satisfying some natural estimates.

We are preparing a sequel to this paper, concerning the existence of Filippov solutions, obtained in frame of the same unified approach. For the lack of room we couldn't include all of it in this work (see Final Remark).

Acknowledgments. The authors would like to express their deepest gratitude to Prof. E.De Giorgi for having suggested and discussing the problem, as well as, for his constant interest in our work, for helping us in having such good and fruitful contacts with the whole group of Italian mathematicians who work in G and Γ-convergence theory.

We thank very much Prof. G. Dal Maso for his suggestions concerning the use of "chattering lemma" and Prof. J.Traple for the discussion about the stating of the problem.

1. Notation. In the paper we use the standard notation:

$|x|=$ the Euclidean norm in $\mathbf{R}^n$ (in a part of Section 6 we use an equivalent norm $|x| = \max\{|x_j| : j = 1, ..., n\}$).

$C^1=$ the space of all continuously differentiable functions on $[0,1]$ (at the ends of the interval we intend the unilateral derivatives).

$W^{1,1}=$ the Sobolev space (see Brezis [6]) of functions summable with the first derivative on $[0,1]$. Each element of the space, which is actually an equivalence class, can be represented by an absolutely continuous function on $[0,1]$.

L^p, $(p < +\infty)=$ the space of all equivalence classes of real measurable functions such that $|x|^p$ is integrable on $[0,1]$, $L^\infty=$ the space of real measurable functions that are essentially bounded on $[0,1]$,

$\overline{co}A=$ the closed and convex hull of the set $A \subset \mathbf{R}^n$,

$\bar{A}=$ the closure of the set $A \subset \mathbf{R}^n$,

$B=$ the unit ball in $\mathbf{R}^n$,

$\rho(x, A)=$ the distance from the point x to the closed set $A \subset \mathbf{R}^n$,

$\|x\|_X=$ the norm of a vector x in a normed space X,

$a.e=$ almost everywhere.

2. Preliminaires. We consider the Problem (0.1),(0.2) of the Introduction where the function $f : [0,1] \times \mathbf{R}^n \to \mathbf{R}^n$, and the vector $a \in \mathbf{R}^n$ are given. If the function f is continuous, then the classical theory of differential equations is well set.

Definition 2.1 *By the classical solution to problem (0.1),(0. 2) we mean a C^1 function $x : [0, 1] \to \mathbf{R}^n$ which satisfies the conditions (0.2) and (0.1) in the following sense:*
(2.1) $x'(t) = f(t, x(t))$ for every $t \in [0, 1]$.

In this case the classical theorems of Cauchy-Peano, together with prolongation theorems give the existence of such (global) solutions, provided they are bounded on [0,1]. If, however, the right-hand side of (0.1) is discontinuous, we can distinguish two cases:
1) f is discontinuous with respect to t, but still continuous with respect to x; for example if f is the Carathéodory type, i.e. $f(\cdot, x)$ is measurable for every x and $f(t, \cdot)$ is continuous for every $t \in [0, 1]$
2) f is discontinuous also with respect to the state variable x (this is the case more important for applications to the optimal control theory).

In case 1), for Carathéodory functions f, there is a theory concerning the existence of so called Carathéodory solutions accordingly to the following:

Definition 2.2. *An absolutely continuous function $x : [0, 1] \to \mathbf{R}^n$ is called C-solution to the Cauchy-problem (0.1),(0.2) iff it satisfies the initial condition (0.2) and the equation (0.1) in the following sense:*

$$(2.2) \qquad x'(t) = f(t, x(t)) \quad a.e. \ in \ [0, 1].$$

In case 2), as we said in Introduction, a natural approach to the problem is to assign to every point (t, x) an orientor $F(t, x)$ and consider the differential inclusion (0.3), (0.4).

As there are many possibilities of constructing such sets $F(t, x)$ starting from $f(t, x)$, we get several "generalized" solutions to problem (0.1),(0.2). We quote below some of these definitions.
Putting:

$$(2.3) \qquad F(t, x) = \mathcal{D}f(t, x) = \bigcap_{\epsilon > 0} \overline{f(t, x + \epsilon B)}$$

$$(2.4) \qquad F(t, x) = \mathcal{K}f(t, x) = \bigcap_{\epsilon > 0} \overline{co}f(t, x + \epsilon B)$$

$$(2.5) \qquad F(t, x) = \mathcal{F}f(t, x) = \bigcap_{\epsilon > 0} \bigcap_{m(z) = 0} \overline{co}f(t, x + \epsilon B)$$

where Z ranges over all the subsets of $\mathbf{R}^n$ which have the Lebesgue measure zero, we adopt:

Definition 2.3. *An absolutely continuous function* $x : [0,1] \to$
$\mathbf{R}^n$ *satisfying the initial condition (0.2) is called*
the $\mathcal{D}$-*solution of (0.1), (0.2) iff* $x'(t) \in \mathcal{D}f(t, x(t))$ *a.e. in [0,1]*
the $\mathcal{K}$-*solution of (0.1), (0.2) iff* $x'(t) \in \mathcal{K}f(t, x(t))$ *a.e. in [0.1]*
the $\mathcal{F}$-*solution of (0.1), (0.2) iff* $x'(t) \in \mathcal{F}f(t, x(t))$ *a.e. in [0.1]*

Remark 2.4. The operator $\mathcal{D}$ given by (2.3), is related to Kuratowski property (K) in optimal control theory (see Chapter 8.5 in [10] and [18]) which in turn is closely related to the Γ-convergence theory (compare the so called (K)-convergence in [13]), while $\mathcal{K}$ and $\mathcal{F}$ operators given by (2.4) and (2.5), respectively, are known as the Krasovskii and Filippov operators (see [14], [15], [16], [7], [2]). All the above mentioned solutions to problem (0.1), (0.2) can be defined in a very natural variational way as the vectors at which appropriately defined non-negative functionals attain their minimum equal to zero. Namely, considering absolutely continuous functions as elements of Sobolev space $W^{1,1}$ we can formulate

Proposition 2.5. *If f is a continuous function, then for $x \in C^1$*
we have:
x is classical solution to (0.1), (0.2) iff

$$\int_0^1 |x'(t) - f(t, x(t))| dt = 0 \ , x(0) = a$$
(which means $0 = \|x'(t) - f(t, x(t))\|_{L^1} =$
$\min\{\|y'(t) - f(t, y(t))\|_{L^1} : y \in C^1, \ y(0) = a\}$),

similarly, if f is a Borel function, then for $x \in W^{1,1}$ we have:
x is a C-solution to (0.1), (0.2) iff

$$\int_0^1 |x'(t) - f(t, x(t))| dt = 0 \ , \ x(0) = a$$
(which means $0 = \|x'(t) - f(t, x(t))\|_{L^1} =$
$\min\{\|y'(t) - f(t, y(t))\|_{L^1} : y \in W^{1,1}, y(0) = a\}$),

analogously for $x \in W^{1,1}$ we have:
  x is a $\mathcal{D}$-solution to (0.1), (0.2) iff

$$\int_0^1 \rho(x(t), \mathcal{D}f(t, x(t)))dt = 0 \ , \ x(0) = a,$$

x is a $\mathcal{K}$-solution to (0.1), (0.2) iff

$$\int_0^1 \rho(x'(t), \mathcal{K}f(t, x(t)))dt = 0 \ , \ x(0) = a,$$

x is an $\mathcal{F}$-solution to (0.1),(0.2) iff

$$\int_0^1 \rho(x'(t), \mathcal{F}f(t, x(t)))dt = 0 \ , \ x(0) = a.$$

3. Basic facts form Γ-convergence theory. From Proposition 2.5 it follows that in order to get the existence results for generalized solutions we can use variational methods, especially the theory of Γ-convergence. Here we give the definition of the so called Γ sequential convergence (see [8]) and some simple properties of this convergence. For complete study of Γ-convergence we refer to [12] and the references in there.

Suppose we are given a sequence of function $f_k : X \to \bar{\mathbf{R}}$, $k = 1, 2, \ldots$ where X is a topological space and $\bar{\mathbf{R}} = \mathbf{R} \cup \{-\infty, +\infty\}$. We set

$$(3.1) \qquad S_x = \{\{x_k\} \subset X : x_k \to x, \text{ as } k \to \infty\}$$

$$(3.2) \qquad \Gamma(X^-)\liminf_{k\to\infty} f_k(x) = \inf_{S_x}(\liminf_{k\to\infty} f_k(x_k))$$

$$(3.3) \qquad \Gamma(X^-)\limsup_{k\to\infty} f_k(x) = \inf_{S_x}(\limsup_{k\to+\infty} f_k(x_k))$$

and we adopt

Definition 3.1. *The function* $f : X \to \bar{R}$ *is called the sequential* Γ-*limit of the sequence* $\{f_k\}_{k=1,2...}$ *iff for every* $x \in X$ *we have*

$$f(x) = \Gamma(X^-)\liminf_{k\to\infty} f_k(x) = \Gamma(X^-)\limsup_{k\to\infty} f_k(x).$$

We then write simply $f(x) = \Gamma(X)\lim_{k\to\infty} f_k(x)$ (omitting the sign -), or, in order to underline the topology τ in X with respect to which the convergence in (3.1) is meant, we will write

$$f(x) = \Gamma(\tau - X)\lim_{k\to\infty} f_k(x) .$$

Remark 3.1 *The above definition of* Γ-*sequential convergence coincides with the general definition of* Γ-*convergence given by De Giorgi and Franzoni in [12] if the topology in* X *satisfies the I axiom of countability.*

In the sequel, in order to verify the Γ-sequential convergence, we will use the following elementary observation:

Proposition 3.2. *Given a sequence* $\{f_k\}$ *of functions* $f_k : X \to$ $\bar{R}$, *on a topological space* (X, τ), *if the following two conditions hold:*

(i) for every $x \in X$ *and every sequence* $x_k \to x$,

$$\liminf_{k\to\infty} f_k(x_k) \geq f(x),$$

(ii) for every $x \in X$ *there exists a sequence* $x_k \to x$ *such that*

$$\liminf_{k\to\infty} f_k(x_k) = f(x),$$

then

$$f(x) = \Gamma(\tau - X)\lim_{k\to\infty} f_k(x).$$

Now we quote a theorem that is essential in the sequel:

Theorem 3.3. *Suppose* $\Gamma(\tau - X)\lim f_k = f$ *and there is a sequence* $\{\hat{x}_k\} \subset X$ *such that* $\hat{x}_k \to \hat{x}$ *in topology* τ *and*

$$(3.4), \qquad \liminf_{k \to +\infty} f_k(\hat{x}_k) = \liminf_{k \to +\infty}(\inf_X f_k(x))$$

then

$$f(\hat{x}) = \inf f(x) = \lim_{k \to \infty}(\inf_X f_k(x)).$$

Remark 3.4 Notice, that if $\hat{x}_k$ realizes the infimum of f_k i.e. $f(\hat{x}_k) = \inf_X f_k(x)$ for $k = 1, 2, \dots$ then (3.4) obviously holds. But (3.4) may hold also for other sequences which we call then sequences of quasiminimizers.

Remark 3.5 In the sequel we set $h = \frac{1}{k}$, so $h \to 0$. We then write $\Gamma(\tau - X)\lim_{h \to 0} f_h$ instead of $\Gamma(\tau - X)\lim_{k \to \infty} f_k$, and all the above mentioned results remain true with obvious reformulations.

We end this Section with a remark on which all the existence results of the paper are based.

Remark 3.6. In order to get an existence result for the problem (0.1), (0.2) we have to implement the following three steps:

1) Find a good variational formulation of the problem so that the existence of a solution be equivalent to the existence of a vector in a suitable vector space which realizes the minimum equal to 0 of some non-negative functional defined on this space.

2) Find a suitable sequence of appropriately defined functionals and then prove that the sequence is Γ convergent in some topology to the functional found in step 1). In general the proof of Γ-convergence is based on Proposition 3.2.

3) Construct a sequence of minimizers or quasiminimizers (in the sense of Remark 3.4) $\{\hat{x}_k\}$ and prove that it posseses a subsequence

which converges to a vector $\hat{x}$ in the topology in which we have proved the Γ-convergence in step 2).

All the three steps 1)-3) done, it is due to De Giorgi-Franzoni Theorem 3.3 that $\hat{x}$ is a solution of the problem (0.1), (0.2).

4. The Carathéodory solutions via Γ-convergence. We have picked up this case for its simplicity, as a "model case" for all results of this paper concerning the existence of generalized solutions of problem (0.1), (0.2) based on Remark 3.6.

As we will see below, the implementation of the three steps of Remark 3.6 in the case of Carathéodory solutions is extremely simple.

Suppose the Carathéodory function $f : [0,1] \times \mathbf{R}^n \to \mathbf{R}^n$ satisfies the conditions:

(4.1) $f(\cdot, x)$ is measurable for every fixed x

(4.2) $f(t, \cdot)$ is continuous for every fixed t

(4.3) $|f(t,x)| \leq \alpha(t) + \beta(t)|x|$, for some functions $\alpha, \beta \in L^1$.

Owing to Proposition 2.5 we easily find the form of the limit functional:

$$F_o(x) = \begin{cases} \int_o^1 |x'(t) - f(t,x(t))|dt & \text{if } x \in W^{1,1}, x(0)=a \\ +\infty & \text{otherwise} \end{cases}$$

and then we define (for any $h > 0$) the functionals:

$$F_h(x) = \begin{cases} \int_0^1 |x'(t) - f(t,x(t-h))|dt & \text{if } x \in W^{1,1} \text{ and } x(0) = a \\ +\infty & \text{otherwise.} \end{cases}$$

Above we adopted the convention that for any $h > 0$ we can extend to the left the functions satisfying the initial condition (0.2) setting $x(t) = a$ for $-h \leq t \leq 0$.

In Sobolev space $W^{1,1}$ we define topology τ by indicating the class of all convergent sequences as follows:

$$x_h \xrightarrow{\tau} x \quad \text{iff} \quad \begin{cases} x_h \to x & \text{uniformly in } [0,1] \\ \\ x'_h \to x' & \text{w-}L^1 \text{ (weakly in } L^1). \end{cases}$$

It is easy to check that this class of sequences introduces a topology, as it satisfy the axioms of Kisynski [17].

Theorem 4.1 *Under the above notations we have*

$$\Gamma(\tau - W^{1,1}) \lim_{h \to 0} F_h = F_o.$$

PROOF. Owing to Proposition 3.2. it suffices to prove the conditions (i) and (ii). To prove (i), suppose $x_h \xrightarrow{\tau} x$ and consider

$$F_h(x_h) = \|x'_h(t) - f(t, x_h(t - h))\|_{L^1}$$

in the case $x_h(0) = x(0) = a$ (other cases being trivial). As, by (4.4) $x'_h \xrightarrow{\tau} x'$ in $w - L^1$ and $x_h(t - h) \to x(t)$ as $h \to 0$ and owing to (4.2) (4.3) we have $f(t, x_h(t - h)) \to f(t, x(t))$ strongly in L^1 (hence, by force, weakly in L^1).

Thus, due to the lower semicontinuity of the norm in L^1 we get:
(i) $\liminf F_h(x_h) \geq F(x)$.

To prove (ii), let us observe that, given x, it is enough to put $x_h = x$ for every h. Then, owing to the continuity of the norm in L^1 we have
(ii) $\lim F_h(x_h) = F(x)$, which completes the proof of the proposition.

Passing to the construction of minimizers we set $h = \frac{1}{k}$, k being a fixed integer $t_j = j \cdot h$, $j = 0, ..., k$ and define $\hat{x}_h : [-h, 1] \to \mathbf{R}^n$ by the formula:

$$(4.5) \qquad \hat{x}_h(t) = \begin{cases} a & , \; t \in [-h, 0] \\ \hat{x}_j = \hat{x}_h(t_j) \, , \; t = t_j \, , \; j = 0, ..., k \\ \hat{x}_j + \int_{t_j}^t f(\tau, \hat{x}_h(\tau - h)) d\tau \, , \\ \qquad t \in [t_j, t_{j+1}] \, , \; j = 0, ..., k - 1. \end{cases}$$

Thus, for $t \in [0, 1]$ we have

$$(4.6) \qquad \hat{x}_h(t) = a + \int_0^t f(\tau, \hat{x}_h(\tau - h)) d\tau$$

so

(4.7) $\hat{x}_h'(t) = f(t, \hat{x}_h(t - h))$ a.e. in $[0, 1]$,

and in consequence

(4.8) $\min_{W^{1,1}} F_h(x) = F_h(\hat{x}_h) = 0.$

Now, it remains to show that $\{\hat{x}_h\}$ possesses a subsequence convergent in the topology τ.

To this end notice, that owing to assumption (4.3) and to Gronwall's Lemma, $\hat{x}_h$ are uniformly bounded by a costant, let's say $C > 0$.

Hence, due to (4.7) and (4.3) again, we get

$$|\hat{x}_h'(t)| \leq \alpha(t) + \beta(t) \cdot C =: \phi(t) \quad, \text{ and } \phi(t) \in L^1.$$

In consequence, owing to Theorem 4 of Chapter 0 in Aubin-Cellina [2] there is a function $\hat{x} \in W^{1,1}$, $\hat{x}(0) = a$ and a subsequence $\{\hat{x}_{h_\nu}\}$ such that $\hat{x}_{h_\nu} \to \hat{x}$ uniformly and $\hat{x}_{h_\nu}' \to \hat{x}'$ weakly in L^1, which means $\hat{x}_{h_\nu} \to \hat{x}$ in topology τ.

So, owing to Remark 3.6, we have proved the following

Theorem 4.2. *If the function f satisfies the assumptions (4.1)-(4.3), then the Cauchy problem (0.1), (0.2) posseses a global C-solution.*

5. Measurable multifunctions and selections. We start with the following

Definition 5.1 *We say the multifunction $K : [0, 1] \ni t \to K(t) \subset \mathbf{R}^n$ (with not necessarily closed values) is measurable iff its graph $G_k = \{(t, x) \in [0, 1] \times \mathbf{R}^n : x \in K(t)\}$ is measurable with respect to the product σ-algebra $\mathcal{L} \times \mathcal{B}$, where $\mathcal{L}$ and $\mathcal{B}$ denote, respectively, the σ-algebras of the Lebesgue sets in $[0,1]$ and the Borel sets in $\mathbf{R}^n$.*

The aim of this section is to prove the following two theorems

Theorem 5.2. *Suppose* $f : [0,1] \times \mathbf{R}^n \to \mathbf{R}^n$ *is a Borel function and* $x : [0,1] \to \mathbf{R}^n$ *is an absolutely continuous function. Then*
$1°$ *the multifunction* $[0,1] \ni t \to D(t) := Df(t, x(t)) \subset \mathbf{R}^n$, *where* $Df(t, x(t))$ *is given by (2.3), is measurable.*
$2°$ *there exists a measurable selection* $z(t) \in D(t)$ *a.e. in [0,1] such that* $|x'(t) - z(t)| = \rho(x'(t), Df(t, x(t)))$ *a.e. in [0,1].*

Theorem 5.3 *Assume* f *and* x *satisfy the assumptions of the Theorem 5.2 and let* $p \in L^1$ *be given. Then*
$1°$ *the multifunction* $[0,1] \ni t \to \mathcal{K}f(t, x(t)) \subset \mathbf{R}^n$, *where* $\mathcal{K}f(t, x(t))$ *is given by formula (2.4), is measurable,*
$2°$ *there exists the uniquely (up to a set of measure 0) defined measurable selection* $z(t) \in \mathcal{K}f(t, x(t)))$ *a.e. in [0,1] such that* $|x'(t) + p(t) - z(t)| = \rho(x'(t) + p(t), \mathcal{K}f(t, x(t)))$ *a.e. in [0,1].*

Here an essential role is played by the projection theorem and the Aumann selection theorem [9]. We quote below their simplified versions which are sufficient for us and can be found in [11] (see Proposition 8.4.4 and Corollary 8.5.4, respectively).

Theorem 5.4. *If a set* $H \subset [0,1] \times \mathbf{R}^n$ *is* $\mathcal{L} \times \mathcal{B}$-*measurable, then its projection on [0,1], i.e. the set*

$$p_1(H) = \{t \in [0,1] : \exists x \in \mathbf{R}^n \ (t, x) \in H\} \quad is \quad \mathcal{L} - measurable.$$

Theorem 5.5. *Every measurable (in the sense of Definition 5.1) multifunction* K *with values* $K(t) \neq \emptyset$ *for a.e.* $t \in [0,1]$ *admits a measurable selection (i.e. there exists a measurable function* $g : [0,1] \to \mathbf{R}^n$ *such that* $g(t) \in K(t)$ *a.e. in [0,1]).*

For the proof of Theorems 5.1 and 5.2 we need three Lemmae which we give below.

Lemma 5.6. *Given two measure spaces* (X, μ) *and* (Y, ν) , *if the function* $\alpha : X \to \mathbf{R}$ *if* μ-*measurable and the function* $\beta : Y \to \mathbf{R}$

is ν-measurable, then the tensor product $\alpha \otimes \beta : X \times Y \ni (x,y) \to \alpha(x).\beta(y) \in \mathbf{R}$ is measurable in the product measure $\mu \times \nu$. (the proof, being elementary, is omitted).

Lemma 5.7. Suppose $\phi : \mathbf{R}^\ell \times \mathbf{R}^s \ni (u,w) \to \phi(u,w) \in \mathbf{R}$ (ℓ,s are fixed positive integers) is a function such that:

(i)   for every fixed $w \in \mathbf{R}^s$ the function $\phi(\cdot,w)$ is measurable with respect to a (fixed) measure ν_ℓ in $\mathbf{R}^\ell$.
(ii) for every fixed $u \in \mathbf{R}^\ell$ the function $\phi(u,.)$ is continuous and let a (fixed) Borel measure β_s in $\mathbf{R}^s$ be given.
Then ϕ is measurable in the product measure $\nu_\ell \times \beta_s$.

PROOF. We use the induction argument with respect to s.
Step 1. In order to get the assertion for $s = 1$, we will construct a sequence of functions $\phi_n(u,w)$ which are $\nu_\ell \times \beta_s$-measurable and then we will show that

(5.1) $\phi_n(u,w) \to \phi(u,w)$ as $n \to \infty$, for every $(u,w) \in \mathbf{R}^\ell \times \mathbf{R}$.

To this end, let's fix n and decompose $\mathbf{R}$ into intervals $[a_{i-1}^{(n)}, a_i^{(n)}]$ of length $1/n$, then define:

$$(5.2) \quad \phi_n(u,w) = \frac{a_i^{(n)} - w}{a_i^{(n)} - a_{i-1}^{(n)}} \phi(u, a_{i-1}^{(n)}) + \frac{w - a_{i-1}^{(n)}}{a_i^{(n)} - a_{i-1}^{(n)}} \phi(u, a_i^{(n)})$$

if $w \in [a_{i-1}^{(n)}, a_i^{(n)}]$.

Owing to Lemma 5.6, these functions are $\nu_\ell \times \beta_1$-measurable. Now, to verify (5.1) let's fix arbitrarily (u_o, w_o) and $\epsilon > 0$. We have:

$$|\phi(u_o, w_o) - \phi_n(u_o, w_o)| \leq$$

$$\leq \frac{a_i^{(n)} - w_o}{a_i^{(n)} - a_{i-1}^{(n)}} |\phi(u_o, w_o) - \phi(u_o, a_{i-1}^{(n)})| +$$

$$+ \frac{w_o - a_{i-1}^{(n)}}{a_i^{(n)} - a_{i-1}^{(n)}} |\phi(u_o, w_o) - \phi(u_o, a_i^{(n)})| < \epsilon,$$

for n sufficiently large, by the definition on ϕ_n and the continuity of $\phi(u_o, \cdot)$ at w_o.

Step 2. Let's suppose the lemma is true for $s \leq k - 1(k \geq 2)$, we will prove it is true for $s = k$. To this end we will represent vectors $v \in \mathbf{R}^s$ in the form $v = (w, z) \in \mathbf{R} \times \mathbf{R}^{k-1}$ and we will regard ϕ as the function: $\tilde{\phi} : \mathbf{R}^{\ell+1} \times \mathbf{R}^{k-1} \ni ((u, w), z) \to \phi(\tilde{u}, z) \in \mathbf{R}$, with $\tilde{u} = (u, w)$, to which the induction assumption can be applied. This gives the $\nu_{\ell+1} \times \beta_{k-1}$-measurability of $\tilde{\phi}$, provided we know that for every fixed $z \in \mathbf{R}^{k-1}$ the function $\tilde{\phi}(., ., z)$ is $\nu_{\ell+1}$-measurable, where we denote by $\nu_{\ell+1}$ the product measure $\nu_\ell \times \beta_1$. We obtain this last fact by applying *Step 1* to the function $\hat{\phi} : \mathbf{R}^\ell \times \mathbf{R} \ni (u, w) \to \phi(u, w, z) \in \mathbf{R}$. The proof of Lemma 5.7 is completed.

Lemma 5.8. *Suppose* $f : [0, 1] \times \mathbf{R}^n \to \mathbf{R}^n$ *is a Borel function,* $x : [0, 1] \to \mathbf{R}^n$ *is a continuous function, and put* $g(t, y) := f(t, x(t) + y)$. *Then* $g : [0, 1] \times \mathbf{R}^n \to \mathbf{R}^n$ *is also a Borel function.*

(We omit the proof of this lemma for its simplicity).

Now we will pass to the

Proof of Theorem 5.2. For the assertion 1° of the theorem, let us observe that $\mathcal{D}(t) = Df(t, x(t)) = \cap_{\epsilon > 0} \overline{f(t, x(t) + \epsilon B)} = \cap_{\epsilon \in Q^+} \overline{f(t, x(t) + \epsilon B)}$, where Q^+ denotes the rational positive numbers. Thus, for the measurability of the multifunction $\mathcal{D}$ it suffices to prove that for every $\epsilon > 0$ the multifunction $\mathcal{D}_\epsilon$ defined as: $\mathcal{D}_\epsilon : [0, 1] \ni t \to \mathcal{D}_\epsilon(t) := \overline{f(t, x(t) + \epsilon B)} \subset \mathbf{R}^n$ is measurable i.e. its graph is $\mathcal{L} \times \mathcal{B}$-measurable (cf. Definition 5.1). This follows from 5.7 applied to the function $\phi(t, z) := \rho(z, f(t, x(t) + \epsilon B))$, as the graph of $\mathcal{D}_\epsilon$ is equal to $\phi^{-1}(\{0\})$ in these notations.

Thus it suffices to verify that the ϕ defined above satisfies the assumptions of Lemma 5.7. It is obvious that for every $t \in \mathbf{R}, \phi(t, \cdot)$ is continuous, so it remains to check whether for every $z \in \mathbf{R}^n$, the function $\phi(\cdot, z)$ is $\mathcal{L}$-measurable. This follows from the definition of ϕ and from the projection theorem, as we have:

$$\phi(\cdot, z)^{-1}((-\infty, \alpha)) = \{t, \phi(t, z) < \alpha\} =$$
$$= p_1 \left(\{(t, y) \in [0, 1] \times B : \rho(z, f(t, x(t) + \epsilon y)) < \alpha\} \right).$$

By the projection theorem (Th.5.4) the last set is $\mathcal{L}$-measurable, because the projected set is a Borel set, owing to Lemma 5.8. This proves 1°.

Assertion 2° will follow from the selection theorem 5.5 applied to the multifunction $\mathcal{Z}(t)$ defined as follows: $\mathcal{Z}(t) = \{z \in \mathcal{D}(t) : |x'(t) - z| = \psi(t)\}$, where $\psi(t) = \rho(x'(t), \mathcal{D}(t))$, provided we prove that $\mathcal{Z}(t)$ is $\mathcal{L} \times \mathcal{B}$-measurable.

Since $\mathcal{D}(t)$ is closed for every $t \in [0, 1]$, the function ψ is well defined and $\mathcal{Z}(t) \neq \emptyset$. Let's observe first that ψ is $\mathcal{L}$-measurable.

Indeed, given an $\alpha \in \mathbf{R}$, we have:

$$
\begin{aligned}
\psi^{-1}((-\infty, \alpha)) &= \{t : \psi(t) < \alpha\} \\
&= \{t : \exists z \in \mathcal{D}(t); \rho(x'(t), z) < \alpha\} \\
&= p_1(\mathcal{G}_\mathcal{D} \cap \{(t, z) : \rho(x'(t), z) < \alpha\}),
\end{aligned}
$$

where $\mathcal{G}_\mathcal{D}$ denotes the graph of the measurable, as we already know, multifunction $\mathcal{D}$ and the set $\{(t, z) : \rho(x'(t), z) < \alpha\}$ is also $\mathcal{L} \times \mathcal{B}$-measurable, since by Lemma 5.7. the function $(t, z) \to \rho(x'(t), z)$ is $\mathcal{L} \times \mathcal{B}$-measurable.

Now that we have the $\mathcal{L}$-measurability of ψ, in order to prove the measurability of the multifunction $\mathcal{Z}$ we represent the graph of $\mathcal{Z}$ as follows:

$$
\mathcal{G}_\mathcal{Z} = \{(t, z) \in \mathcal{G}_\mathcal{D} : \rho(x'(t), z) = \psi(t)\}
$$

$$
= \mathcal{G}_\mathcal{D} \cap \{(t, z) : \rho(x'(t), z) - \psi(t) = 0\}.
$$

Since the function $(t, z) \to \rho(x'(t), z)$ and $(t, z) \to \psi(t)$ are both $\mathcal{L} \times \mathcal{B}$-measurable, the set $\mathcal{G}_\mathcal{Z}$ is $\mathcal{L} \times \mathcal{B}$-measurable as well, which completes the proof of 2°. Now Theorem 5.2 is proved.

For the proof of Theorem 5.3 let us observe that in this case we deal with multivalued functions which have *closed and convex* values. The measurability of such functions is easier to get and we can either obtain it by some known theorems about multivalued functions with closed and convex values (see [9]) or repeat the proof of Theorem 5.2 with obvious modifications.

6. The existence of $\mathcal{D}$ - and $\mathcal{K}$ - solutions. To get $\mathcal{D}$ and $\mathcal{K}$-solutions for Cauchy problem (0.1), (0.2) with Borel right-hand side f we follow the general schema described in Remark 3.6. So, first we define the functionals:

$$(6.1) \qquad F_h(x,y) =$$

$$= \begin{cases} \int_o^1 |x'(t) - f(t, x(t) + y(t))| dt, \\ \qquad \text{if } x \in W^{1,1}, \ x(0) = a, \ y \in L^\infty, \ |y(t)| \le h \text{ a.e.} \\ +\infty \qquad\qquad\qquad\qquad \text{otherwise} \end{cases}$$

$$(6.2) \qquad F_0(x,y) =$$

$$= \begin{cases} \int_0^1 \rho\left(x'(t), \mathcal{D}f(t, x(t))\right) dt \\ \qquad \text{if } x \in W^{1,1}, \ x(0) = a, \ y \in L^\infty, \ y(t) = 0 \text{ a.e.,} \\ +\infty \qquad\qquad\qquad\qquad \text{otherwise} \end{cases}$$

in the case of $\mathcal{D}$-solution, and

$$(6.3) \qquad F_h(x,y,p) =$$

$$= \begin{cases} \int_0^1 |x'(t) - f\left(t, x(t) + y(t)\right) + p(t)| dt, \\ \qquad \text{if } x \in W^{1,1}, \ x(0) = a, \ y \in L^\infty, \ |y(t)| \le h \text{ a.e., } p(t) \in L^1 \\ +\infty \qquad\qquad\qquad\qquad \text{otherwise} \end{cases}$$

$$(6.4) \qquad F_0(x,y,p) =$$

$$= \begin{cases} \int_0^1 \rho\left(x'(t) + p(t), \mathcal{K}f(t, x(t))\right) dt, \\ \qquad \text{if } x \in W^{1,1}, \ x(0) = a, \ p(t) \in L^1, \ y \in L^\infty, \ y(t) = 0 \text{ a.e.} \\ +\infty \qquad\qquad\qquad\qquad \text{otherwise} \end{cases}$$

for $\mathcal{K}$-solutions.

Then we prove the theorems:

Theorem 6.1. *If f is a Borel function satisfying (4.3), F_h and F_0 are given by (6.1) and (6.2), respectively, then*

(6.5)
$$\Gamma(s - W^{1,1}, s - L^\infty) \lim_{h \to 0} F_h(x, y) = F_o(x, y)$$

(s – stands for the strong topology).

Theorem 6.2. *If f is a Borel function satisfying (4.3) with α, $\beta, \in L^\infty$, the functionals F_h and F_o are given by (6.3) and (6.4), respectively, then we have:*

(6.6) $$\Gamma(\tau - W^{1,1}, s - L^\infty, w - L^1) \lim_{h \to 0} F_h(x, y, p,) = F_0(x, y, p),$$

where the topology τ is defined by (4.4), s-stands for the strong and w- for the weak topology, respectively.

Before passing to the proofs of these theorems, we formulate two propositions, whose elementary proofs will be omitted.

Proposition 6.3. *Given $a \in \mathbf{R}^n$, a sequence $\{b_k\}_{k=1,2,\ldots}$ and a closed set $C \subset \mathbf{R}^n$, if every convergent subsequence of $\{b_k\}$ has its limit in C, then $\liminf_{k \to \infty} |a - b_k| \geq \rho(a, C)$*

Proposition 6.4. *Suppose $x_h(t) \to x(t)$ and $y_h(t) \to 0$ both a.e. in [0,1], as $h \to 0$, and denote by $g(t)$ the limit of a convergent subsequence of $\{f(t, x_h(t) + y_h(t))\}$. Then*

$$g(t) \in \bigcap_{\epsilon > 0} \overline{f(t, x(t) + \epsilon B)} = \mathcal{D}f(t, x(t)), \quad a.e. \ in[0,1].$$

while, if $f(t, x_h(t) + y_h(t)) \to z(t)$ in $w - L^1$, then

$$z(t) \in \bigcap_{\epsilon > 0} \overline{co}f(t, x(t) + \epsilon B) = \mathcal{K}f(t, x(t)), \quad a.e. \ in[0,1].$$

Proof of Theorem 6.1. Owing to the Propositon 3.2 it is enough to prove the conditions (i) and (ii). For (i), suppose $x_h \to x$ in $s - W^{1,1}, x_h(0) = a = x(0)$, and $y_h \to 0$ uniformly (other cases being trivial). Notice that $x_h \to x$ in $s - W^{1,1}$ implies $x_h(t) \to x(t)$ for every t and both $x_h \to x$, $x'_h \to x'$ in $s - L^1$. Hence, and from the inequality

$$\left| \int_0^1 |x'_h - f(t, x_h + y_h)| dt - \int_0^1 |x' - f(t, x_h + y_h)| dt \right| \le$$

$$\le \int_0^1 |x'_h - x'| dt$$

it follows

(6.7)
$$\liminf_{h \to 0} \int_0^1 |x'_h(t) - f(t, x_h(t) + y_h(t))| dt =$$

$$= \liminf_{h \to 0} \int_0^1 |x'(t) - f(t, x_h(t) + y_h(t))| dt.$$

Thus, using (6.7) and the Fatou lemma we get

(6.8) $$\liminf_{h \to 0} F_h(x_h) \ge \int_0^1 \liminf_{h \to 0} |x'(t) - f(t, x_h(t) + y_h(t))| dt,$$

which in view of Proposition 6.4 and Propositon 6.3 yields

(6.9) $$\liminf_{h \to 0} F_h(x_h, y_h) \ge \int_0^1 \rho(x'(t), \mathcal{D}f(t, x(t))) dt = F_0(x, 0).$$

which proves the condition (i).

Passing to the condition (ii) we fix $x \in W^{1,1}$, $x(0) = a$ (other cases being trivial) and we define $x_h = x$ for every $h > 0$.
Owing to Theorem 5.2 the multifunction $t \to \mathcal{D}(t) = \mathcal{D}f(t, x(t))$ is measurable and we can find a measurable selection $z(t) \in \mathcal{D}(t)$ a.e. such that

$$(6.10) \qquad |x'(t) - z(t)| = \rho\left(x'(t), \mathcal{D}f(t, x(t))\right).$$

Now, for any fixed $h > 0$ we define another multifunction by the formula

$$(6.11) \quad Y_h(t) = \{y \in \mathbf{R}^n : |y| \leq h, \ | f(t, x(t) + y) - z(t)| < h\}$$

with non-empty (it is due to the relation $z(t) \in \mathcal{D}f(t, x(t))$) but not necessarily closed values. In (6.11) it is more convenient to use the norm defined by the formula $|a| = \max_{1 \leq i \leq n} |a_i|$ for $a \in \mathbf{R}^n$.

Let us observe that the graph of multifunctions Y_h can be represented as follows

$$(6.12) \qquad \mathcal{G}_{Y_h} = \{(t, y) \in [0, 1] \times \mathbf{R}^n : y \in Y_h(t)\} =$$

$$= ([0, 1] \times hB) \cap \{(t, y) : |f_i(t, x(t) + y) - z_i(t)| < h, \quad i = 1, ..., n\}$$

$$= ([0, 1] \times hB) \cap \bigcap_{i=1}^{n} \{(t, y) : f_i(t, x(t) + y) < z_i(t) + h\} \cap$$

$$\cap \bigcap_{i=1}^{n} \{(t, y) : z_i(t) - h < f_i(t, x(t) + y)\}$$

and all the three sets on the right-hand side are $\mathcal{L} \times \mathcal{B}$-measurable. To see this for the second set (the case of the third set is analogous) notice that

$$\{(t,y) : f_i(t, x(t) + y) < z_i(t) + h\} =$$

$$\bigcup_{w \in \mathbf{Q}} (\{(t,y) : f_i(t, x(t) + y) < w\} \cap \{(t,y) : w < z_i(t) + h\}),$$

where $\mathbf{Q}$ denotes the rational numbers.

Since the functions $(t, y) \to f_i(t, x(t) + y)$ are Borel owing to Lemma 5.8, and the functions $(t, y) \to \tilde{z}_i(t, y) := z_i(t)$ are $\mathcal{L} \times \mathcal{B}$-measurable by Lemma 5.6, all the above sets are $\mathcal{L} \times \mathcal{B}$-measurable. In consequence, $\mathcal{G}_{Y_k}$ is $\mathcal{L} \times \mathcal{B}$-measurable, which prove the measurability of the multifunction Y_h. Hence by the Theorem 5.5 there is a measurable selection

$$(6.13) \qquad y_h(t) \in Y_h(t) \text{ a.e. in } [0.1].$$

Thus, by the definition of x_h, y_h and Y_h, we have $(x_h, y_h) \to (x, 0)$ in the product topology $s - W^{1,1} \times s - L^\infty$,

$$F_h(x_h, y_h) = \int_o^1 |x_h'(t) - f(t, x_h(t) + y_h(t))| dt =$$

$$= \int_0^1 |x'(t) - f(t, x(t) + y_h(t))| dt$$

and by (6.11), (6.10) this integrals converge to

$$F_0(x, 0) = \int_0^1 |x'(t) - z(t)| dt = \int_0^1 \rho\left(x'(t), \mathcal{D}f(t, x(t))\right) dt,$$

provided $|f(t, x(t) + y_h(t))|$ are bounded by a locally summable function uniformly with respect to h (for instance if f satisfies (4.3)). This proves (ii) and the proof of the theorem is completed.

Proof of Theorem 6.2. It goes similarly to that of Theorem 6.1 (with the use of Theorem 3.2). We begin with the condition (i). Assume we are given a sequence of functions $\{(x_h, y_h, p_h)\}$ such that

(6.14) $(x_h, y_h, p_h) \to (x, 0, p)$, as $h \to 0$,

in the $\tau - W^{1,1} \times (s - L^\infty) \times (w - L^1)$ topology, with $x(0) = a$ (all other cases to check are trivial). We can now pass to a subsequence (we always keep the same indices for subsequences) which realizes the limes inferior:

(6.15)
$$\liminf_{h \to 0} F_h(x_h, y_h, p_h) = \lim_{h \to 0} \|x'_h(t) - f(x_h(t) + y_h(t)) + p_h(t)\|_{L^1}.$$

Now observe that $x'_h \to x'$ in $w-L^1$ by (4.4), $p_h \to p$ in $w-L^1$ by (6.14) and $\|f(t, x_h(t) + y_h(t)\|_{L^1} \leq \alpha(t) + C\beta(t)$ by (4.3), where C is a constant such that $|x_h(t) + y_h(t)| \leq C$ in [0,1]. Since $\alpha(t) + C\beta(t) \in L^1$, we can infer (by a reasoning similar to that in the proof of Th.4 of Chapter 0 in [2]) there is a subsequence $\{f(t, x_h(t) + y_h(t))\}$ coverging weakly in L^1 to a function $z(t) \in L^1$. So by Proposition 6.4 we have

$$z(t) \in \bigcap_{\epsilon > 0} \overline{co} f(t, x(t) + \epsilon B) = \mathcal{K}f(t, x(t)) \quad \text{a.e. in } [0,1].$$

Hence, from (6.15) and from the lower semicontinuity of the norm in the weak topology we obtain

$$\liminf_{h \to 0} F_h(x_h, y_h, p_h) \geq \|x'(t) + p(t) - z(t)\|_{L^1}$$
$$\geq \int_0^1 \rho(x'(t) + p(t), \mathcal{K}f(t, x(t)))dt = F_o(x, 0, p),$$

which completes the proof of the condition (i).

For the condition (ii) let us assume that we are given a point $(x, 0, p) \in W^{1,1} \times L^\infty \times L^1$, $x(0) = a$ (other cases being trivial).

As we proved in the Theorem 5.3, the multifunction $\mathcal{K}(t) = \mathcal{K}f(t, x(t))$ is measurable and according to the assertion 2° of its thesis, there is a measurable selection $z(t) \in K(t)$ such that

$$(6.16) \qquad |x'(t) + p(t) - z(t)| = \rho(x'(t) + p(t), \mathcal{K}(t)).$$

In a way analogous to that of the proof of (ii) of Theorem 6.1 one can show that the multifunctions $Y_h(t)$ defined now by the formula

$$Y_h(t) := \left\{ \begin{array}{c} (\lambda_o, ..., \lambda_n, y_o, ..., y_n) : \lambda_i \in \mathbf{R}^+, \Sigma\lambda_i = 1, \ y \in hB \\ \text{and } |\Sigma\lambda_i f(t, x(t) + y_i) - z(t)| < h \end{array} \right\}$$

(having non-empty values for a.e. $t \in [0,1]$ since $z(t) \in \mathcal{K}(t)$ a.e.) is measurable (in the sense of Def.5.1). So by Theorem 5.5 there are measurable selections $\lambda_i^h(t)$, $y_i^h(t)$, $i = 1, ..., n$, such that

$$(6.17) \qquad (\lambda_1^h(t), ..., \lambda_n^h(t), y_0^h(t), ..., y_n^h(t)) \in Y_h(t) \text{ a.e. in } [0,1].$$

Now we use the "Chattering lemma" (see [3]), (with $p_i(t) = \lambda_i^h(t)$ and $f_i(t) = f(t, x(t) + y_i^h))$. In this way we get a sequence $\{\tilde{y}_h(t)\}$ such that

$$(6.18) \qquad \tilde{y}_h(t) = y_i^h(t), \text{ if } t \in I_{ij}, \ ([0,1] = \cup I_j, \ I_j = \bigcup_0^n I_{ij}),$$

$$(6.19)$$

$$\left| \int_{t'}^{t''} \left(\sum_0^n \lambda_i^h(t) f(t, x(t) + y_i^h(t)) - f(t, x(t) + \tilde{y}_h(t)) \right) dt \right| < h$$

for every $t', t'' \in [0,1]$.

Thus, in view of (6.18) and the definition of $Y_h(t)$ we obtain, for every $t', t'' \in [0,1]$:

$$(6.20) \qquad \left| \int_{t'}^{t''} f(t, x(t) + \tilde{y}_h(t)) - z(t) dt \right| \leq$$

$$\leq \left| \int_{t'}^{t''} \left(f(t, x(t) + \tilde{y}_h(t)) - \sum_{i=0}^{n} \lambda_i^h(t) f(t, x(t) + y_i^h(t)) \right) dt \right| +$$

$$+ \left| \int_{t'}^{t''} \left(\sum_{i=0}^{n} \lambda_i^h(t) f(t, x(t) + y_i^h(t)) - z(t) \right) dt \right| \leq 2h.$$

Now, from (6.20) we can infer, using Lemma 6.5 below (with $w_n(t) = f(x(t) + y_{1/n}(t)) - z(t))$ that

$$(6.21) \qquad f(t, x(t) + \tilde{y}_h(t)) \to z(t) \text{ in } w - L^1 \text{ as } h \to 0.$$

Thus, setting $x_h = x$ for every h, $y_h = \tilde{y}_h$ and

$$p_h = f(t, x(t) + \tilde{y}_h(t)) - z(t) + p(t),$$

we get, by (6.18), (6.20) and (6.16):

$$(x_h, y_h, p_h) \to (x, 0, p) \text{ in the } \tau - W^{1,1} \times s - L^\infty \times w - L^1 \text{ topology,}$$
and

$$F_h(x_h, y_h, p_h) = \|x_h'(t) - f(t, x_h(t) + y_h(t)) + p_h(t)\|_{L^1}$$

$$= \|x'(t) + p(t) - z(t)\|_{L^1} = \int_0^1 \rho(x'(t) + p(t), \mathcal{K}f(t, x(t))) dt$$

$$= F_o(x, 0, p),$$

which proves the condition (ii) and thus completes the proof of the theorem.

We will sketch here an elementary proof of the simple lemma used in the proof above.

Lemma 6.5. *Suppose $\{w_n\} \subset L^1$ is such that for every $t', t'' \in [0,1]| \int_{t'}^{t''} w_n(t)dt| < 1/n$ and $\|w_n\|_{L^\infty} \le C$. Then $w_n \to 0$ in $w - L^1$.*

PROOF. For any piecewise constant function X, equal to c_i on the intervals $[t_i, t_{i+1}]$, $i = 1, ..., k$, we have:

$$\left| \int_0^1 X w_n dt \right| = \left| \Sigma \int_{t_i}^{t_{i+1}} c_i w_n dt \right| \le (\max c_i)k/n.$$

Now, given $\phi \in L^\infty$ and $\epsilon > 0$, we can find a piecewise constant function X such that

$$\left| \int_0^1 (\phi - X) w_n dt \right| < \epsilon$$

(this is due to the fact that $\phi \in L^1$ as well and piecewise functions are dense in the L^1 norm). Hence we obtain

$$\left| \int_0^1 \phi w_n dt \right| \le \left| \int_0^1 (\phi - X) w_n dt \right| + \left| \int_0^1 X w_n dt \right|$$

$$\le \|\phi - X\|_L^1 \cdot \|w_n\|_{L^\infty} + C \cdot 1/n < \epsilon,$$

for n sufficiently large.

Now we are in a position to prove the existence result for $\mathcal{K}$-solutions.

Theorem 6.6. *Suppose f is a Borel function satisfying (4.3) with $\alpha, \beta \in L^\infty$. Then there exists a $\mathcal{K}$-solution to the Problem $\{(0.1), (0.2)\}$.*

PROOF. Owing to Remark 3.6 and Theorem 6.2 it is enough to construct a sequence of minimizers. To this end, we take $h = 1/k$, $k \in \mathbf{N}, t_j = j \cdot h, j = 0, ..., k$, and we define (see [16]):

$$\hat{x}_h(t) = \begin{cases} a & \text{if } t = t_o = 0 \\ \hat{x}_{hj} = \hat{x}_h(t_j) & \text{if } t = t_j, \ j = 1, ..., k \\ \hat{x}_{hj} + \int_{t_j}^t f(\tau, \hat{x}_{hj})d\tau & \text{if } t \in [t_j, t_{j+1}], \ j = 1, ..., k. \end{cases}$$

So, we have

$$(6.22) \qquad \hat{x}_h(t) = a + \sum_{i=0}^{j-1} \int_{t_j}^{t_{j+1}} f(\tau, \hat{x}_{hj}) d\tau + \int_{t_j}^{t} f(\tau, \hat{x}_{hj}) d\tau,$$

$$\text{for } t \in [t_j, t_{j+1}],$$

and

$$(6.23) \qquad \hat{x}_h'(t) = f(t, \hat{x}_{hj}) \qquad \text{a.e. in } [0,1]$$

Hence, setting

$$(6.24) \qquad y_h(t) = \hat{x}_h(t_j) - \hat{x}_h(t) \qquad \text{if } t \in [t_j, t_{j+1}]$$

we get

$$(6.25) \qquad \hat{x}_h'(t) = f(t, \hat{x}_h(t) + y_h(t)) \qquad \text{a.e. in } [0,1].$$

From (6.22) and (4.3), owing to Gronwall's lemma, we obtain that $|\hat{x}_h|$ are uniformly bounded by a constant C. Hence, and from (6.23), we get $|\hat{x}_h'(t)| \leq \alpha(t) + \beta(t)C$, which implies, by Th.4 in Chapter 0 of [2] that there exists a subsequence $\{\hat{x}_{h_\nu}\}$ of $\{\hat{x}_h\}$ and an absolutely continuous function $\hat{x}(t)$ such that $\hat{x}_{h_\nu} \to \hat{x}$ uniformly in [0,1] and $\hat{x}_{h_\nu}' \to \hat{x}'$ in $w - L^1$. This means that $\hat{x}_{h_\nu} \to \hat{x}$ in topology τ given by (4.4). As from (6.25) it follows that $F_{h_\nu}(\hat{x}_{h_\nu}, y_{h_\nu}, 0) = 0$ and $(\hat{x}_{h_\nu}, y_{h_\nu}, 0) \to (\hat{x}, 0, 0)$ in $\tau - W^{1,1} \times s - L^\infty \times w - L^1$ topology, we can apply Remark 3.6 and infer that $\hat{x}$ is a $\mathcal{K}$-solution of problem $\{(0.1),(0.2)\}$, which completes the proof.

Concerning the $\mathcal{D}$-solutions, here we confine ourselves to the following:

Remark 6.7. In the case of $\mathcal{D}$-solutions, it is more difficult to find a suitable sequence of minimizers or the conditions on f

sufficient for the existence of $\mathcal{D}$-solutions. As the following simple example shows:

$$f : \mathbf{R}^2 \to \mathbf{R}^2 \ , \ f(x_1, x_2) = \begin{cases} (1, 1) & \text{for } x_2 \leq 0 \\ (1, -1) & \text{for } x_2 > 0, \end{cases}$$

$\mathcal{D}$-solutions to $\{(0.1), (0.2)\}$ need not exist even for an apparently regular f.

Here, we see two possible approaches to the existence problem. If a concrete function f is given, one can try to repeat the construction of the sequence $\{\hat{x}_h\}$ we have just given for $\mathcal{K}$-solutions and then check the convergence, or else one can try to construct a sequence of quasi-minimizers (see Remark 3.6) for F_h in this concrete situation. In a general setting, we believe that the following conjecture is true:

Conjecture 6.8. *Suppose that a Borel function $f : [0, 1[\times \mathbf{R}^n \to \mathbf{R}^n$ satisfies the estimate $|f(t, x)| \leq \alpha(t) + \beta(t)|x|$ with some $\alpha, \beta \in L^1$ and, additionally, satisfies the following geometrical conditions: there exist N multivalued functions $M_1, ... M_N$, $M_i : [0, 1] \to \mathcal{P}(\mathbf{R}^n)$, each of them measurable (in the sense of definition 5.1) and such that*

(a) $\forall t \in [0, 1]$ *the family $\{M_i(t)\}$ is a C^1-stratification of $\mathbf{R}^n$,*

(b) *the multifunctions A_i : graph $M_i \ni (t, x) \to A_i(t, x) := \mathcal{D}f(t, x) \cap T_x M_i \subset \mathbf{R}^n$ have nonempty values (here $T_x M$ denotes the tangent space at x),*

(c) $\forall i \in \{1, ..., N\}$, *the differential inclusion $x' \in A_i(t, x)$ has a local solution for every initial value $x_0 \in M_i$.*

Then for any initial condition a there exists a $\mathcal{D}$-solution to the problem $\{(0.1),(0.2)\}$.

However, now we have only a geometrical idea of the proof of this conjecture and besides, we would like to find an analytic rather than geometrical condition on f that would ensure the existence of $\mathcal{D}$-solutions to $\{(0.1), (0.2)\}$.

Final Remark: As concerns Filippov solutions, it sufficies to put

$$f_h(t,x) = \frac{1}{\mathrm{vol}(hB)} \int_{|y| \le h} f(t,x,y)\, dy,$$

$$F_h(x,y,p) =$$

$$\begin{cases} \int_0^1 |x'(t) - f_h(t,x(t)+y(t)) + p(t)|dt, \\ \qquad \text{if } x \in W^{1,1},\ x(0)=a,\ y \in L^\infty, |y(t)| \le h \text{ a.e., } p \in L^1, \\ \\ +\infty \qquad\qquad\qquad\qquad\qquad\qquad\quad \text{otherwise.} \end{cases}$$

$$F_0(x,y,p) =$$

$$\begin{cases} \int_0^1 \rho(x'(t)+p(t), \mathcal{F}f(t,x(t)))dt, \\ \qquad \text{if } x \in W^{1,1},\ x(0)=a,\ y \in L^\infty,\ y(t)=0 \text{ a.e., } p \in L^1, \\ \\ +\infty \qquad\qquad\qquad\qquad\qquad\qquad\quad \text{otherwise.} \end{cases}$$

Then, in a similar (though more subtle) manner as in the case of $\mathcal{K}$-solutions, one can show

$$F_0(x,y,p) = \Gamma(\tau - W^{1,1}, s - L^\infty, w - L^1) \lim_{h \to 0} F_h(x,y,p).$$

Thus, applying Remark 3.6 (with $\hat{x}_h$ =Carathéodory solutions to the problem $x' = f_h(t,x)$, $x(0) = a$, f_h being obviously of the Carathéodory type, and $\hat{y}_h = 0$, $\hat{p}_h = 0$) we get the existence of $\mathcal{F}$-solutions.

Namely we have

Theorem 6.9. *Suppose f is a Borel function satisfying (4.3) with $\alpha, \beta \in L^\infty$, then there exists an $\mathcal{F}$-solution to the problem (0.1), (0.2).*

References

[1] L.Ambrosio, *Relaxation of autonomous functionals with discontinuous integrands*, to appear.

[2] J.P.Aubin, A.Cellina, *Differential inclusions*, Springer-Verlag, Berlin Heidelberg, New York-Tokyo (1984).

[3] L.D.Berkovitz, *Optimal control theory*, Springer-Verlag, New York-Heidelberg-Berlin (1974).

[4] P.Binding, *The differential equation $x' = f \circ x$*, J. Diff. Eq. **31** (1979), 183-199.

[5] V.G.Boltyanskii, *Sufficient conditions for optimality and the justification of the dinamic programming method*, SIAM J. Control **4** (1966), 326-361.

[6] H.Brezis, *Analisi funzionale. Teoria e applicazioni*, Liguori Editore, Napoli (1986).

[7] P.Brunovsky, *The closed-loop time-optimal control I: Optimality*, SIAM J. Control **12** (1974), 624-634.

[8] G.Buttazzo, G.Dal Maso, *Γ-convergence and optimal control problems*, JOTA **38** (1982), 385-401.

[9] C.Castaing, M.Valadier, *Convex analysis and measurable multifunctions*, Springer Lectures Notes in Mathematics **580** (1977).

[10] L.Cesari, *Optimization-theory and applications*, Springer-Verlag, New York-Heidelberg-Berlin (1983).

[11] D.L.Cohn, *Measure theory*, Birkhäuser, Boston (1980).

[12] E.De Giorgi, T.Franzoni, *Su un tipo di convergenza variazionale*, Rend.Sem.Mat. Brescia **3** (1979), 63-101.

[13] Z.Denkowski, *On Mosco type caractherization of (K)-convergence and lower semicontinuity of integral functionals*, Universitatis Jagellonicae Acta Math. **24** (1984), 189-201.

[14] A.F.Filippov, *Differential equations with discontinuous right-hand sides*, Mat. Sb. **51** (1960), 99-128.

[15] A.F.Filippov, *Differenzialnyie uravnenia s razryvnoj pravoj chastju* (in Russian), Nauka, Moskva (1985).

[16] O.Hajek, *Discontinuous differential equations* I, J. Diff. Eq. **32** (1979), 149-170.

[17] J.Kisynski, *Convergence du type L*, Coll. Math. **7** (1960), 205-211.

[18] C.Olech, *Existence theory in optimal control problems-the underlying ideas*, Int.Conf.Diff. Equations, Academic Press, New York (1975), 612-635.

Institute for Information Sciences
Jagellonian University
KRAKOW

Institute of Mathematics
Jagellonian University
KRAKOW

STATIONARY SOLUTIONS OF NONLINEAR SCHRÖDINGER EQUATIONS WITH AN EXTERNAL MAGNETIC FIELD

MARIA J.ESTEBAN   PIERRE-LOUIS LIONS

Dedicated to Ennio De Giorgi on his sixtieth birthday

Abstract. In this paper we study the existence of stationary solutions of some Schrödinger equations with an external magnetic field. We obtain these solutions by solving appropriate minimization problems for the corresponding energy-functional. These problems which are a priori not compact are solved by the use of the so-called concentration-compactness method. We also prove the existence of solutions for the generalized Hartree-Fock equations which model the interaction of electrons and static nucleii through a coulombic potential and under the action of an external magnetic field.

Key words: Schrödinger operators, Hartree-Fock equations, nonlinear field equations, external magnetic field, minimization problem, concentration-compactness.

Résumé. Dans ce travail on étudie l'existence de solutions stationnaires pour des équations de Schrödinger non linéaires avec un champ magnétique externe. Nous obtenons ces solutions en résolvant certains problémes de minimisation pour la fonctionnelle d'énergie correspondante, et nous faisons ceci en utilisant la méthode de concentration-compacité. Nous démontrons aussi l'existence de solutions pour les équations de Hartree-Fock qui modélisent l'interaction d'électrons avec des noyaux à travers un potentiel coulombien et sous l'action d'un champ magnétique externe.

Mots-clés: opérateurs de Schrödinger, équations de champs non linéaires, équations de Hartree-Fock, champ magnétique externe, problème de minimisation, concentration-compacité.

1. Introduction. In this paper, we study some semilinear, second-order, elliptic equations in $\mathbf{R}^N$ with $N = 2$ or $N = 3$ (even if the techniques and methods used below extend to higher dimensions without any change). The linear part of the equations (some examples are given below) is a Schrödinger operator with a so-called vector potential for instance an operator of the form

$$(1.1) \qquad L_A = -\Delta - 2iA \cdot \nabla + |A|^2 - i \operatorname{div} A$$

where A - the vector potential - is a given function from R^N into R^N. Assumptions upon A will be made later on. Here and below, Δ is the Laplacian, ∇ is the gradient. Terms involving A model the presence in some quantum model of an external magnetic field: actually, the magnetic field is nothing but $B = \operatorname{curl} A$ (if $N = 3$).

Linear operators like L_A-with possibly the addition of a Schrödinger potential V-have been studied in J.Avron, I.Herbst and B.Simon [3,4,5]; J.Combes, R.Schrader and R.Seider [15]; T.Kato [21]; M.Reed and B.Simon [30]; B.Simon [33].

A typical example of the nonlinear equations we are interested in is the following

$$(1.2) \qquad L_A u + \lambda u = |u|^{p-2}u \ , \quad u \not\equiv 0 \ , \quad u \in L^2(R^N) \quad (*)$$

$(*) \ L^2(\mathbf{R}^N) = \{\varphi : \mathbf{R}^N \to C, \quad \int_{R^N} |\varphi|^2 dx < \infty\}$

where $p \in (2, \infty), \lambda \in \mathbf{R}$. Next, we remark that L_A is (at least formally) self adjoint and the associated quadratic form is

$$(1.3) \qquad (L_A u, u) = \int_{\mathbf{R}^N} |\nabla u + iAu|^2 dx$$

(here and below we denote by (f, g) the scalar product $\int_{\mathbf{R}^N} fg dx$). We will use this fact in a fundamental way to obtain solutions of (1.2) (for some A, p, λ) via simple variational problems. Two typical examples that we will consider are

$$(I_{A,V}) \quad I_{A,V} = \inf \left\{ \int_{\mathbf{R}^N} |\nabla u + iAu|^2 + V|u|^2 dx \Big/ \int_{\mathbf{R}^N} |u|^p dx = 1 \right\}$$

where V is a given potential i.e. a given function from $\mathbf{R}^N$ into $\mathbf{R}$, or

$$(J_A) \qquad J_A = \inf \left\{ \frac{1}{2} \int_{\mathbf{R}^N} |\nabla u + iAu|^2 dx - \frac{1}{p} \frac{\displaystyle\int_{\mathbf{R}^N} |u|^p dx}{\displaystyle\int_{\mathbf{R}^N} |u|^2 dx} = 1 \right\}$$

We will also denote by ∇_A the operator $\nabla + iA$.

In fact, in order to be rigorous, the minimizing class has to be described a bit more carefully. In particular, it is natural to understand the space of functions u in $L^2(\mathbf{R}^N)$ such that $\nabla u + iAu \in L^2$. We briefly recall in Section 2 below what is known about such questions.

Next, in Section 3 we study the important particular case of a constant magnetic field say $B = (0, 0, b)$ (if $N = 3$) or $B = b$ (if $N = 2$) with $b \in \mathbf{R}$ and (for instance) $A = b/2(-x_2, x_1, 0)$ (if $N = 3$) or $A = b/2(-x_2, x_1)$ (if $N = 2$). To give an example of our results, we show that if $\lambda > -|b|$ then (1.2) admits infinitely many distinct solutions (a "double infinity" if $N = 2$) and that $I_{A,\lambda}$ (still for $\lambda > -|b|$) admits a minimum provided $p < 6$ if $N = 3$.

In Section 4, we study the general case and give various existence results. Finally, Section 5 is devoted to more general nonlinear equations of the form (1.2) and the so-called Hartree-Fock equations corresponding to an atom, an ion or a molecule with an external magnetic field acting on it.

At this stage, we want to mention that equations like (1.2) have been studied extensively in the case when $A = 0$ (i.e. no vector potential) and we refer to Z.Nehari [28], G.Ryder [31], C.V.Coffman [13], M.Berger [11], W.Strauss [35], H.Berestycki and P.L.Lions [9], P.L.Lions [25], H.Berestycki, T.Gallouët and O.Kavian [10], H.Brezis and E.H.Lieb [12], F.V. Atkinson and L.A.Peletier [1]... In particular, when studying minimization problems like $(I_{A,V})$ or (J_A) we will use the concentration-compactness method developed in [25], [26] -see also M.J.Esteban and P.L.Lions [17] for a relevant technical improvement. And when studying the existence of critical points, we will use some critical point theorems derived in [29].

We wish to conclude this introduction by a fundamental property enjoyed by the operators L_A, ∇_A, the equation (1.2) or the minimization problems $(I_{A,V}), (I_A)$ namely the gauge invariance. Indeed, since only the magnetic field B really matters, one should expect some flexibility in the choice of the potential A such that $B = curl\ A$. More precisely, if $A' = A + \nabla V$ where V is a real-valued function on $\mathbf{R}^N$, then denoting by $u'(x) = e^{-iV(x)}u(x)$ we have

$$(1.4) \qquad \nabla_{A'}u'(x) = e^{-iV(x)}\nabla_A u(x)$$

$$(1.5) \qquad L_{A'}u'(x) = e^{-iV(x)}L_A u(x).$$

Notation. In all that follows, we denote by

$$B_R = \{x \in \mathbf{R}^N / |x| < R\} \quad , \quad B(y, R) = \{x \in \mathbf{R}^N / |x - y| < R\}$$

and the energy functional in $(I_{A,V})$ by $\mathcal{E}_{A,V} = \int_{\mathbf{R}^N} |\nabla_A u|^2 + V|u|^2 dx$. Finally for any $z \in C$, we denote by $\bar{z}$ the conjugate of z.

Acknowledgement. The second author would like to thank E.H.Lieb for mentioning to him the questions studied here.

Summary

1. Introduction
2. Some functional results about Schrödinger operators with magnetic fields.
3. Constant magnetic fields.
3.1. Existence of a minimum in the subcritical case.
3.2. Existence of symmetric solutions in the case $N = 3$.
3.3. Existence of symmetric solutions in the case $N = 2$.
3.4. The limit case.
4. General magnetic fields.
4.1. Existence of a minimum without compactness.
4.2. Some compact or partially compact cases.
4.3. Symmetric solutions.
4.4. The limit case.
5. Hartree-Fock equations for Coulomb systems with magnetic fields.

2. Some functional results about Schrödinger operators with magnetic fields. Let $A \in L^\alpha_{loc}(\mathbf{R}^N, \mathbf{R}^N)$ with $\alpha > 2$ if $N = 2$, $\alpha = 3$ if $N = 3$. A typical Schrödinger operator with a vector potential (or a magnetic field deriving from this potential) is given by

$$(2.1) \qquad L_A = \Delta - 2iA \cdot \nabla + |A|^2 - i \operatorname{div} A.$$

Of course, this is the linear P.D.E. operator arising in the Euler-Lagrange equation associated with $I_{A,0}$. Formally, the quadratic form associated with the self-adjoint operator L_A is

$$\|u\|_A = \left(\int_{\mathbf{R}^N} |\nabla u + iAu|^2 dx \right)^{1/2}.$$

It is therefore natural to try to understand the relations between spaces of functions u on $\mathbf{R}^N$ such that $\|u\|_A < \infty$ and more usual spaces like for instance Sobolev spaces. For instance, we denote by

$$(2.2) \qquad H^1_A(\mathbf{R}^N) = \{u \in L^2(\mathbf{R}^N) / \nabla u + iAu \in L^2(\mathbf{R}^N)\}$$

and we endow this space with the norm

$$(|u|_{L^2}^2 + \|u\|_A^2)^{1/2}.$$

When $N = 3$, we will need to consider another space whose definition is slightly more complex (alternative definitions will be provided below): we denote by $X_A = \mathcal{D}_A^{1,2}(\mathbf{R}^3)$ the completion of $\mathcal{D}(\mathbf{R}^3)$ for the norm $\|\cdot\|_A$. However, to be rigorous we need to check that $\|\cdot\|_A$ is a norm, say on $\mathcal{D}(\mathbf{R}^3)$. This, and nearly everything that follows, is easily deduced from the fundamental relation (see for example A.Jaffe and C.Taubes [20])

$$|\nabla|\varphi|| = |Re\left(\nabla\varphi\frac{\bar{\varphi}}{|\varphi|}\right)| = |Re\left\{(\nabla\varphi + iA\varphi)\frac{\bar{\varphi}}{|\varphi|}\right\}|$$

since A is real valued; this relation holds a.e. . Hence

(2.3) $|\nabla|\varphi|| \leq |\nabla\varphi + iA\varphi|$ a.e.

for all $\varphi \in \mathcal{D}(\mathbf{R}^3)$ or more generally for all $\varphi \in H_{loc}^1(\mathbf{R}^3)$.

We may now check that $\|\cdot\|_A$ is a norm on $\mathcal{D}(\mathbf{R}^3)$: indeed, we just have to check that if $\|\varphi\|_A = 0, \varphi \in \mathcal{D}(\mathbf{R}^3)$, then $\varphi = 0$. In view of (2.3), if $\|\varphi\|_A = 0, \varphi \in \mathcal{D}(\mathbf{R}^3)$ we deduce that $|\varphi|$ is constant and since φ has compact support, $\varphi \equiv 0$.

A few properties of H_A^1 and X_A are given in the

Proposition 2.1.

i) $\mathcal{D}(\mathbf{R}^N)$ is dense in $H_A^1(\dot{\mathbf{R}}^N)$.

ii) $H_A^1(\mathbf{R}^2)$ is continuously embedded in $L^p(\mathbf{R}^2)$ for all $p \in [2, \infty)$ and $H_A^1(\mathbf{R}^2) \subset H_{loc}^1(\mathbf{R}^2)$.

iii) $X_A = \{u \in L^6(\mathbf{R}^3)/\nabla u + iAu \in L^2(\mathbf{R}^3)\}$ and $X_A \subset H_{loc}^1(\mathbf{R}^3)$.

iv) $H_A^1(\mathbf{R}^3)$ is continuously embedded in X_A.

PROOF. In view of (2.2) and (2.3), $H^1_A(\mathbf{R}^2)$ is continuously embedded in $L^p(\mathbf{R}^2)$ and by Sobolev inequalities ii) follows and we also deduce that X_A is continuously embedded in $L^6(\mathbf{R}^3)$. Since iv) follows easily from i), there just remains to show that if u belongs to $H^1_A(\mathbf{R}^N)$ or to $\{u \in L^6(\mathbf{R}^3)/\nabla u + iAu \in L^2(\mathbf{R}^3)\}$ then we may find φ in $\mathcal{D}(\mathbf{R}^N)$ arbitrarily close to u in the H^1_A norm or the $\|\cdot\|_A$ norm.

This is obtained by the standard truncation and mollification method: we first show that it is enough to consider the case when u has compact support. To this end, we consider φ in $\mathcal{D}(\mathbf{R}^n)$ such that $0 \leq \varphi \leq 1$, $\varphi \equiv 1$ if $|x| \leq 1$, $\varphi \equiv 0$, if $|x| \geq 2$ and we set

$$u_n(\cdot) = u\varphi\left(\frac{\cdot}{n}\right) \quad , \quad \text{for} \quad n \geq 1.$$

Obviously, u_n converges to u in L^2 (if $u \in H^1_A$) or in L^6 and

$$\nabla(u - u_n) + iA(u - u_n) = (1 - \varphi_n)(\nabla u + iAu) - \nabla\varphi_n u.$$

Therefore, u_n converges to u in H^1_A or in the $\|\cdot\|_A$ norm since

$$(1 - \varphi_n)(\nabla u + iAu)\underset{n}{\longrightarrow}0 \quad \text{in} \quad L^2$$

while $\||\nabla\varphi_n|u\|^2_{L^2} \leq C \int_{n \leq |x| \leq 2n} \frac{1}{n^2}|u|^2 dx \to 0$ if $u \in L^2$ or if $u \in L^6(\mathbf{R}^3)$ (apply Hölder inequality).

This argument shows that we may assume that u has compact support. Then, to conclude we use a standard mollification procedure that we skip observing only that $Au \in L^2(\mathbf{R}^N)$ (this is the only place where we use the assumption on A namely $A \in L^\alpha_{loc}$).

The following result is a generalized version of a theorem due to Avron, Herbst and Simon (theorem 2.9 in [3]) which will be used to obtain L^2-bounds for functions in X_A.

Proposition 2.2. *Let us denote by* $P_j = \partial/\partial x_j + iA_j$ *where* $A = (A_1, \ldots, A_N) \in W^{1,\infty}_{loc}(\mathbf{R}^N, \mathbf{R}^N)$. *Let* $j, k \in \{1, \ldots, N\}$. *Then we have for any* $u \in \mathcal{D}(\mathbf{R}^N)$

$$(2.4) \qquad |((\partial_j A_k - \partial_k A_j)u, u)| \leq \|P_j u\|^2_{L^2} + \|P_k u\|^2_{L^2}.$$

And if A *satisfies*

$$(2.5) \qquad \begin{array}{l} \epsilon\{\partial_j A_k - \partial_k A_j\} \geq w \quad a.e \ in \quad \mathbf{R}^N \quad , \quad w \in L^\beta(\mathbf{R}^N), \\ for \quad \epsilon = +1 \quad or \quad \epsilon = -1 \end{array}$$

then (2.4) holds for $u \in H_A^1(\mathbf{R}^N)$ if $\beta \in (1, +\infty]$ and $N = 2$ or $\beta \in [3/2, +\infty]$ and $N = 3$. Similarly, (2.4) holds for $u \in X_A$ if $\beta = 3/2$.

Remark 2.3. If (2.5) holds then $((\partial_j A_k - \partial_k A_j)u, u)$ makes sense in $\mathbf{R} \cup \{+\infty\}$ if $\epsilon = 1$ (in $\mathbf{R} \cup \{-\infty\}$ if $\epsilon = -1$) for any $u \in H_A^1$ or $u \in X_A$.

Remark 2.4. If the potential A is such that for some j, k in $\{1, \ldots, N\}$ the function $\partial_j A_k - \partial_k A_j$ is bounded from below away from 0, (2.4) immediately implies that an L^2 bound on u is deduced from an L^2 bound on $\nabla_A u$. Actually, we only need $P_j u$ and $P_k u$ to be in $L^2(\mathbf{R}^N)$.

We will come back on such consequences after the

Proof of Proposition 2.2. Observe that for any j, k the commutator $[P_j, P_k]$ is equal to $i(\partial_j A_k - \partial_k A_j)$. Moreover, it can be easily verified that for any smooth functions f, g, the following identity holds

$$(P_j f, g) = -(f, P_j g).$$

Therefore, if $u \in \mathcal{D}(\mathbf{R}^N)$ we have

$$|([P_j, P_k]u, u)| = |((P_j P_k - P_k P_j)u, u)|$$
$$= |(P_k u, P_j u) - (P_j u, P_k u)| \leq 2\|P_j u\|_{L^2(\mathbf{R}^3)} \|P_k u\|_{L^2(\mathbf{R}^3)}$$
$$\leq \|P_j u\|_{L^2(\mathbf{R}^3)}^2 + \|P_k u\|_{L^2(\mathbf{R}^3)}^2.$$

To complete the proof, we use the density of $\mathcal{D}$ proved in Proposition 2.1.

Remark 2.5. As we will see in the next section, an important example of the vector potential A is

$$A = \frac{1}{2}(-x_2, x_1) \quad \text{in} \quad \mathbf{R}^2 \quad \text{or} \quad A = \frac{1}{2}(-x_2, x_1, 0) \quad \text{in} \quad \mathbf{R}^3.$$

In this case $\partial_1 A_2 - \partial_2 A_1 \equiv 1$ and we deduce from (2.4) the following inequality

(2.6) $\|u\|_{L^2} \leq \|u\|_A \quad \text{for all} \quad u \in H_A^1 \quad \text{or} \quad u \in X_A.$

We claim that this inequality is sharp i.e. 1 is the best constant for the embedding $H_A^1 \subset L^2$ or $X_A \subset L^2$ (see [5]). Indeed, if $N = 2$ we consider

$$u(x) = \exp(-|x|^2/4)$$

and we compute $\|u\|_{L^2} = \|u\|_A = (2\pi)^{1/2}$. While if $N = 3$, we take any $\varphi \in \mathcal{D}(\mathbf{R})$ with $\int_\mathbf{R} |\varphi|^2 dx = 1$ and we consider

$$u_n(x) = \exp(-(x_1^2 + x_2^2)/4) \cdot n^{-1/2} \varphi(\frac{x_3}{n}).$$

Then, $\|u_n\|_{L^2} = (2\pi)^{1/2}$ while $\|u_n\|_A \xrightarrow[n]{} (2\pi)^{1/2}$.

Remark 2.6. The proof of (2.4) only requires $A \in L_{loc}^1(\mathbf{R}^N, \mathbf{R}^N)$ by an easy density argument if $u \in \mathcal{D}(\mathbf{R}^N)$ while if $u \in H_A^1(\mathbf{R}^N)$ (resp. $u \in X_A$) it is enough to assume that $A \in W_{loc}^{1,p}(\mathbf{R}^N, \mathbf{R}^N), p \in [N/2, +\infty]$ (resp. $p = N/2$).

We now list a few consequences of Proposition 2.2.

Proposition 2.7. *Let* $A \in W_{loc}^{1,\infty}(\mathbf{R}^N, \mathbf{R}^N)$. *We assume there exist* $j, k \in \{1, \ldots, N\}$ *such that*

(2.7)
$$\begin{aligned} &\exists \, \nu > 0, \ \exists \, R > 0, \ \exists \, \epsilon \in \{-1, +1\}, \\ &\epsilon(\partial_j A_k - \partial_k A_j) \geq \nu \quad a.e. \ for \quad |x| \geq R \end{aligned}$$

Then $\| \cdot \|_A$ *is an equivalent norm on* $H_A^1(\mathbf{R}^N)$ *and* X_A. *In addition, if we assume*

(2.8)
$$\begin{aligned} &\inf_{|x| \geq R} ess \ \epsilon(\partial_j A_k - \partial_k A_j) \to +\infty \\ &as \quad R \to +\infty, \quad with \quad \epsilon = \pm 1 \end{aligned}$$

then the embedding of H_A^1 *into* L^2 *is compact.*

PROOF. The first part may be proved by contradiction assuming there exists $(u^n)_n$ in H_A^1 such that $\|u^n\|_A \xrightarrow[n]{} 0$, while $\|u^n\|_{L^2} = 1$. In view of (2.3) this implies that $|u^n|$ converges weakly in H^1 to 0 and strongly in H^1_{loc} by Rellich-Kondrakov theorem. On the other hand we deduce from proposition 2.2 that

$$\int (\partial_j A_k - \partial_k A_j)|u_n|^2 dx \xrightarrow[n]{} 0$$

and since $u_n \xrightarrow[n]{} 0$ in L^2_{loc}, (2.7) implies that $u_n \xrightarrow[n]{} 0$ in L^2. The contradiction proves the first part of the Proposition 2.3.

The second part is easily deduced from Rellich-Kondrakov theorem and the following observations

$$\int_{|x|\geq R} |u|^2 dx \leq C_R \int_{|x|<R} |u|^2 dx + \delta_R \int \epsilon(\partial_j A_k - \partial_k A_j)|u|^2 dx$$

$$\leq C_R \int_{|x|<R} |u|^2 dx + \delta_R \|u\|^2_A$$

for some $\delta_R \to 0$ as $R \to \infty$, where $C_R < \infty$.

3. Constant magnetic fields. All throughout this section we will consider magnetic fields B which are constant in $\mathbf{R}^N$, i.e. for $N = 3$, $B = (b_1, b_2, b_3)$, $b_i \in \mathbf{R}$, $i = 1, 2, 3$. Actually when $N = 3$ we will assume, without loss of generality, that $b_1 = b_2 = 0$ and $b_3 \neq 0$. Indeed notice that if this is not the case we can perform a rotation to obtain it. Hence, in what follows we will consider that B is equal to $(0, 0, b)$ if $N = 3$, $B = b$ if $N = 2$ for some $b \in \mathbf{R} - \{0\}$. For such a magnetic field a particular choice for the corresponding potential A is the following

$$(3.1) \quad \begin{aligned} A &= \frac{1}{2}(B \times x) = \frac{b}{2}(-x_2, x_1, 0) \quad \text{if} \quad N = 3, \\ A &= \frac{b}{2}(-x_2, x_1) \quad \text{if} \quad N = 2. \end{aligned}$$

As we mentioned in the introduction, the choice of a particular potential A is not important since a simple change of gauge allows us to exchange two different potentials corresponding to the same magnetic field. In the Physics litterature the potential A defined by (3.1) corresponds to the so-called Lorentz gauge.

3.1. Existence of a minimum in the subcritical case: $N=3$, $2<p<6$; $N=2$, $2<p<+\infty$.

This section will be devoted to the existence of solutions for the minimization problems we described briefly in the introduction, i.e.

(3.2)

$$I_{A,V} = \inf\{\int_{\mathbf{R}^N} |\nabla_A u|^2 + V|u|^2 dx / \; u \in H^1_A, \int_{\mathbf{R}^N} K|u|^2 dx = 1\}$$

(3.3) $\;\; J_A = \inf\{\int_{\mathbf{R}^N} \frac{1}{2}|\nabla_A u|^2 - \frac{1}{p}|u|^p dx / \; u \in H^1_A, \int_{\mathbf{R}^N} |u|^2 dx = 1\}$

where A is defined by (3.1) and $N = 2$ or 3.

Remember that solutions of (3.2), (3.3) are particular solutions of equation (1.2).

We will concentrate here on the case

(3.4) $$V \equiv \omega \in \mathbf{R} \quad , \quad K \equiv 1 \quad \text{in} \quad \mathbf{R}^N$$

and we indicate how to generalize the results obtained in this case to more general situations.

We state now our main result.

Theorem 3.1. *Assume (3.4) with $\omega > -|b|$. Then any minimizing sequence $\{u_n\}_n$ of (3.2) is relatively compact in $L^q(\mathbf{R}^N)$ ($2 \leq q < 6$ if $N = 3$, $2 \leq q < +\infty$ if $N = 2$) up to a translation and change of gauge. More precisely, there exists $\{x_n\}_n \subset \mathbf{R}^3$ such that $\{e^{i\bar{A}_n} u_n(\cdot + x_n)\}_n$ is relatively compact in $L^q(\mathbf{R}^N)$, where $\bar{A}_n(x) \equiv A(x_n) \cdot x$.*

In particular, there is a minimum of problem (3.2).

Let us now extend the above theorem to the case when the function V is not constant.

Theorem 3.2. *Let $K \equiv 1$ and $V = V_o + \omega$, where $\omega \in \mathbf{R}$, $\omega > -|b|$ and $V_o \in L^{r_1}(\mathbf{R}^N) + L^{r_2}(\mathbf{R}^N)$, with $N/2 < r_1$, $r_2 < +\infty$, $N = 2, 3$. Let $\{u_n\}$ be a minimizing sequence for (3.2). Then:*

(i) There exists $\{x_n\} \subset \mathbf{R}^N$ such that $\{e^{i\bar{A}_n} u_n(\cdot + x_n)\}$ is relatively compact in $L^q(\mathbf{R}^N)$ for $2 \leq q < 6$ if $N = 3$ and $2 \leq q < +\infty$ if $N = 2$, where $\bar{A}_n$ is defined as in the preceding theorem.

(ii) If the sequence $\{x_n\}$ is bounded, then $\{u_n\}$ is relatively compact in H_A^1 and each converging subsequence converges to a minimum of (3.2).

(iii) On the other hand, if $\{x_n\}$ is unbounded, then for any subsequence $\{x_{n'}\}$ of $\{x_n\}$ such that $\{e^{i\bar{A}_{n'}}u_{n'}(\cdot + x_{n'})\}$ converges weakly in H_A^1 and strongly in $L^q(\mathbf{R}^N)$, it converges to a minimum of $I_{A,\omega}$.

(iv) Finally, any minimizing sequence of (3.2) is relatively compact in $H_A^1(\mathbf{R}^3)$ if and only if the following holds:

(3.5) $$I_{A,V_o+\omega} < I_{A,\omega}.$$

In particular, if (3.5) holds, there is a minimum of (3.2).

Remarks.
(i)   If $V_o \leq 0, V_o \neq 0$ in $\mathbf{R}^N$, (3.5) holds.
(ii) If $V_o \geq 0, V_o \neq 0$, then $I_{A,V_o+\omega} = I_{A,\omega}$ and there is no minimum for $I_{A,V_o+\omega}$.
(iii) We may also be be interested in studying problem (3.2) when K is not a constant function. Actually we may assume that K is any function in $C(\mathbf{R}^N)$ such that $K(x) \underset{|x|\to+\infty}{\longrightarrow} K^\infty > 0$.

To see how theorems 3.1 and 3.2 can be generalized to treat this case and more general ones, see [25] and [17].

(iv) We may also consider functions V which are more general than $V_o + \omega$, $V_o \in L^{r_1} + L^{r_2}$, $r_1, r_2 > N/2$, by using the same kind of arguments as in [17] and above.

Proof of theorem 3.1. This theorem may be proved by following and adapting the concentration-compactness method introduced in [25]. For any $\lambda \in \mathbf{R}^+$ let us define $I_{A,V}(\lambda)$ as we defined $I_{A,V}$, but replacing now the constraint

$$\int_{\mathbf{R}^N} |u|^2 dx = 1 \qquad \text{by} \qquad \int_{\mathbf{R}^N} |u|^2 dx = \lambda.$$

Therefore $I_{A,V}(1) = I_{A,V}$. We observe immediately that

(3.6) $$I_{A,V}(\lambda) = \lambda^{2/p} I_{A,V},$$

and hence, if $p > 2$, for all $\alpha \in (0, 1)$, we have

$$(3.7) \qquad I_{A,V} < I_{A,V}(\alpha) + I_{A,V}(1 - \alpha).$$

Now we may recall briefly the arguments of [25]. For each n, let us define the bounded non negative measure $\mu_n = |\nabla_A u_n|^2 + |u_n|^2$. Since $\{u_n\}$ is a minimizing sequence of (3.2), there exists $C > 0$ such that $\mu_n(\mathbf{R}^N) \leq C$ for all $n \in N$. Moreover, since $\omega > -|b|$ Proposition 2.2 implies that $\{u_n\}$ is bounded in $H_A^1(\mathbf{R}^N)$.

Then the first concentration-compactness lemma (see [25]) states that there are three possible behaviours for the sequence $\{\mu_n\}$: vanishing, dichotomy or tightness up to a translation. The first case corresponds to the following situation:

$$(3.8) \qquad \lim_{n \to +\infty} \sup_{y \in \mathbf{R}^N} \mu_n(B(y, R)) = 0 \quad \text{for all} \quad R > 0.$$

This situation is easily avoided in our case by using Lemma 1.1 in [25, part 2] and (2.3). Indeed, from (3.8) we deduce that for all $R > 0$

$$\sup_{y \in \mathbf{R}^N} \int_{B(y,R)} |\nabla_A u_n|^2 + |u_n|^2 dx \xrightarrow[n \to +\infty]{} 0$$

and by (2.3) we obtain also that

$$\sup_{y \in \mathbf{R}^N} \int_{B(y,\mathbf{R})} |\nabla |u_n||^2 dx \xrightarrow[n \to +\infty]{} 0.$$

We apply then Lemma 1.1 in [25, part 2] and infer that $u_n \xrightarrow[n \to +\infty]{} 0$ in $L^q(\mathbf{R}^N)$ for $2 < q < 6$ if $N = 3$ and $2 < q < +\infty$ if $N = 2$, which contradicts the constraint. Hence vanishing cannot occur.

Dichotomy is easily avoided by using inequality (3.7). Roughly speaking dichotomy means that if $\lambda = \lim_{n \to +\infty} \mu_n(\mathbf{R}^N)$, there exists $\beta \in (0, \lambda)$ such that for all $\epsilon > 0$, $R_1 < +\infty$, there exist $R_o \geq R_1$, $R_n \xrightarrow[n \to +\infty]{} +\infty$, $\{y_n\} \subset \mathbf{R}^N$ such that

$$(3.9) \quad |\mu_n(B(y_n, R_1)) - \beta| \leq \epsilon, \quad |\mu_n(B(y_n, R_n)^c) - (\lambda - \beta))| \leq \epsilon.$$

Then, as in [25], we may find u_n^1, u_n^2 such that

$$(3.10) \qquad \text{dist}(\text{supp } u_n^1, \text{supp } u_n^2) \xrightarrow[n \to +\infty]{} +\infty \qquad \text{and}$$

$$(3.11) \quad \left| \|u_n^1\|_{H_A^1}^2 - \beta \right| \leq C\epsilon \quad , \quad \left| \|u_n^2\|_{H_A^1}^2 - (\lambda - \beta) \right| \leq C\epsilon,$$

$$\|u_n - u_n^1 - u_n^2\|_{H_A^1} \leq C\epsilon$$

for some $C > 0$ independent of n and ϵ. Therefore,

$$(3.12) \quad |\mathcal{E}_{A,\omega}(u_n) - \mathcal{E}_{A,\omega}(u_n^1) - \mathcal{E}_{A,\omega}(u_n^2)| \leq C\epsilon$$

$$(3.13) \quad \left| \int |u_n|^p - |u_n^1|^p - |u_n^2|^p dx \right| \leq C\epsilon.$$

Moreover (3.11) implies also the existence of a positive constant δ such that

$$(3.14) \quad \mathcal{E}_{A,\omega}(u_n^1), \quad \mathcal{E}_{A,\omega}(u_n^2) \geq \delta \quad \text{for all} \quad n.$$

Hence, from (3.12)-(3.13),

$$\lim_{n \to +\infty} \mathcal{E}_{A,\omega}(u_n^1) \geq I_{A,\omega}(\theta) \quad , \quad \lim_{n \to +\infty} \mathcal{E}_{A,\omega}(u_n^2) \geq I_{A,\omega}(\rho)$$

with $\theta + \rho = 1$. Then, we easily reach a contradiction with (3.7) by using (3.6).

Then, since vanishing and dichotomy have be avoided, the only remaining possibility is the tightness of μ_n up to a translation. That is, there must exist $\{x_n\} \subset \mathbf{R}^N$ such that

$$(3.15) \quad \text{for all} \quad \epsilon > 0 \quad \text{there exists} \quad R > 0 \text{ such that, for all } n,$$

$$\mu_n(\mathbf{R}^N / B(x_n, R)) \leq \epsilon.$$

Since $\{u_n\}$ is bounded in H_A^1, (3.15) implies that $\{u_n(\cdot + x_n)\}$ is relatively compact in $L^2(\mathbf{R}^N)$ by Proposition 2.1. The relative compactness in $L^p(\mathbf{R}^N)$, $2 \leq q < 6$ if $N = 3$, $2 \leq q < +\infty$ if $N = 2$, then follows by Hölder inequalities and Proposition 2.1. Next we observe that for $\hat{u}_n(\cdot) = u_n(\cdot + x_n)$ we have:

$$\int_{\mathbf{R}^N} |\nabla_A u_n|^2 dx = \int_{\mathbf{R}^N} |\nabla \hat{u}_n + i(A + A(x_n))\hat{u}_n|^2 dx,$$

because A is linear. Now we define the function $\bar{A}_n$ in $\mathbf{R}^N$ by $\bar{A}_n(x) = A(x_n) \cdot x$ and we observe that $\nabla \bar{A}_n \equiv A(x_n)$. Hence, by using the observations made in the introduction about the changes of gauge we deduce that

$$(3.16) \qquad \int_{\mathbf{R}^N} |\nabla_A u_n|^2 dx = \int_{\mathbf{R}^N} |\nabla_A \underline{u}_n|^2 dx$$

where $\underline{u}_n(\cdot) = e^{i\bar{A}_n} u_n(\cdot + x_n)$.

Next since $|\underline{u}_n(\cdot)| = |u_n(\cdot + x_n)|$ a.e. in $\mathbf{R}^N$, we may use (3.15) again, this time for the sequence $\{\underline{u}_n\}$ to conclude the proof. Indeed, by (3.15), extracting if necessary a subsequence, $\{\underline{u}_n\}$ converges in $L^q(\mathbf{R}^N)$ $(2 \le q < 6$ if $N = 3$, $2 \le q < +\infty$ if $N = 2)$ to some $\underline{u}$ and

$$\int_{\mathbf{R}^N} |\nabla_A \underline{u}|^2 dx \le \lim_{n \to +\infty} \inf \int_{\mathbf{R}^N} |\nabla_A u_n|^2 dx,$$

because $A \in L^\alpha_{loc}(\mathbf{R}^N)$.

Remark. Observe that the basic property of A which allows us to conclude when V is constant is that for all $y \in \mathbf{R}^N$, $A(\cdot + y) = A(\cdot) + A(y)$, i.e. A is linear. Actually what we really use is the existence for $|y|$ large of $T_y : \mathbf{R}^N \to \mathbf{R}$ such that $A(\cdot + y) = A(\cdot) + \nabla T_y(\cdot)$. In this case $T_y(x) = A(y) \cdot x$.

Proof of Theorem 3.2. The first part of this proof can be done exactly as above. The only difference which appears when $V = V_o + \omega$, $V_o \ne 0$, is that even if (3.16) still holds, in this case we have

$$(3.17) \qquad \int_{\mathbf{R}^N} V |u_n|^2 dx \ne \int_{\mathbf{R}^N} V |\underline{u}_n|^2 dx.$$

Actually, when $V_o \ne 0$ we have

$$(3.18) \qquad \int_{\mathbf{R}^N} V |u_n|^2 dx = \int_{\mathbf{R}^N} V_n |\underline{u}_n|^2 dx,$$

with $V_n(\cdot) = V(\cdot + x_n)$. Then Theorem 3.2 follows since $\{\underline{u}_n\}$ is relatively compact in $L^q(\mathbf{R}^N)$, $2 \le q < 6$ if $N = 3$, $2 \le q < +\infty$ if $N = 2$, and $\{V_n\}$ is bounded in $L^{r_1}(\mathbf{R}^N) + L^{r_2}(\mathbf{R}^N)$. Hence, up to the extraction of some subsequence, we have

$$(3.19) \qquad \int_{\mathbf{R}^N} V_n |\underline{u}_n|^2 dx \xrightarrow[n \to +\infty]{} \omega \int_{\mathbf{R}^N} |\underline{u}|^2 dx, \quad \text{if } |x_n| \to +\infty.$$

Let us finally sketch what happens when we study problem (3.3) instead of (3.2). In this case we can prove the following:

Theorem 3.3. *Assume that $2 < p < 2 + 4/N$. Then any minimizing sequence $\{u_n\}$ of (3.3) is relatively compact in $L^q(\mathbf{R}^N)$, $2 \leq q < 6$ if $N = 3$, $2 \leq q < +\infty$ if $N = 2$, up to a translation and a change of gauge. More precisely, there exists a sequence of points of $\mathbf{R}^N$, $\{x_n\}$, such that $\{e^{i\bar{A}_n}u_n(\cdot + x_n)\}$ is relatively compact in $L^q(\mathbf{R}^N)$, where $\bar{A}_n$ is defined as above. Moreover, there is a minimum of problem (3.3).*

The proof of this theorem follows easily from the arguments used in the proofs above and those in [25]. Therefore we will skip it.

Remark. We may generalize Theorem 3.3 to treat problem (3.3) with the term $\frac{1}{p}\int_{\mathbf{R}^N} |u|^p dx$ replaced by $\int_{\mathbf{R}^N} F(u)dx$, where F is a function satisfying:

$$(3.20) \qquad \lim_{|t|\to+\infty} \frac{F(t)}{|t|^{(2+4/N)}} = 0 \quad , \quad \lim_{|t|\to 0} \frac{F(t)}{|t|^2} = 0.$$

In this case Theorem 3.3 still holds for $N = 3$ while in the case $N = 2$ we also have to assume that F satisfies

$$(3.21) \qquad F(t^{1/2}u) > tF(u) \quad \text{for all} \quad u > 0 \quad , \quad t > 1.$$

The only thing that we have to prove in order to extend the proof of Theorem 3.3 to this case is the analogue of inequality (3.7). Since F is not a power function, we cannot use (3.6) to prove the equivalent of (3.7) any more. Here we can prove it as follows. Let us consider the case $N = 3$. We observe that if $\int_{\mathbf{R}^N} |u|^2 dx = \alpha > 0$ and if we define u_λ by $u_\lambda(x', z) = u(x', \frac{z}{\lambda})$, $\lambda > 1$, where $x' \in \mathbf{R}^{N-1}$, $z = x_n$, we have

$$(3.22) \qquad \int_{\mathbf{R}^N} |u_\lambda|^2 dx = \lambda\alpha \qquad \text{and}$$
$$\int_{\mathbf{R}^N} \frac{1}{2}|\nabla_A u_\lambda|^2 - \frac{1}{p}|u_\lambda|^p dx < \lambda \int_{\mathbf{R}^N} \frac{1}{2}|\nabla_A u|^2 - \frac{1}{p}|u|^p dx$$

and from this and using the inequality $J_A(1) < |b|$ one shows that for all $\alpha > 0$, $\lambda > 1$, $J_A(\alpha\lambda) < \lambda J_A(\alpha)$ which implies $J_A(\lambda) < J_A(\alpha) + J_A(\lambda - \alpha)$, $\forall \lambda > 0$, $\forall \alpha \in (0, \lambda)$.

When $N = 2$ we use inequality (3.21) to derive the desired result.

3.2. Existence of symmetric solutions in the case $n=3$.

When B is a constant magnetic field and A is defined by (3.1), we may prove the existence of (many) solutions of equation (1.2) with particular symmetry properties.

The symmetry property that we will consider is defined by the set

$$(3.23) \quad \begin{aligned} S = \{u \in L^6(\mathbf{R}^3) | u \quad &\text{is spherically symmetric in } x', \\ &\text{where } x' = (x_1, x_2)\}, \end{aligned}$$

i.e. $u \in S$ if and only if $u \in L^6(\mathbf{R}^3)$ and for all $x \in \mathbf{R}^3$, $u(x) = u(s, z)$, with $s^2 = x_1^2 + x_2^2, z = x_3$.

Let $\lambda_k \in \mathbf{R}^+$ be defined by

$$\lambda_k = \inf \left\{ \int_{\mathbf{R}^3} |\nabla v|^2 + \left(\frac{bs}{2} - \frac{k}{s}\right)^2 |v|^2 dx \ / \ v \in S, \int_{\mathbf{R}^3} |v|^2 dx = 1 \right\}.$$

Then we can prove the

Lemma 3.4. $\lambda_o = |b|$ *and for all* $k \in \mathbf{Z}$, $\lambda_k \geq \max\{|b|, -bk\}$. *In particular,* $\lim \lambda_k = +\infty$ *as* k *goes to -(sign b)*∞.

Before proving this lemma let us state our main result about the existence of solutions of equation (1.2) with particular symmetry properties.

Theorem 3.5. *Let* p *be in (2.6). Then for every* $\lambda \in \mathbf{R}$ *and every* $k \in \mathbf{Z}$ *such that* $\lambda > -\lambda_k$, *equation (1.2) has a solution* u_k *of the form*

$$(3.24) \quad u_k(x) = C_k \left(\frac{x_2 + ix_1}{s}\right)^k v_k(x)$$

for some $C_k \in \mathbf{R}\backslash\{0\}$, $v_k \in S$.

Remark. Lemma 3.4 shows that for all $\lambda \in \mathbf{R}$ there exists an infinity of integers k such that (1.2) has a solution of the form (3.24). In particular if $\lambda > -|b|$ there is such a solution for all $k \in \mathbf{Z}$.

Remark. For all k such that $\lambda > -\lambda_k$, v_k actually is a solution of the following minimization problem

$$(N_k) \qquad N_k = \inf\left\{ \int_{\mathbf{R}^3} |\nabla v|^2 + \left(\frac{bs}{2} - \frac{k}{s}\right)^2 |v|^2 + \lambda|v|^2 dx \;\; \middle/ \right.$$

$$\left. v \in S, \quad \int_{\mathbf{R}^3} |v|^p dx = 1 \right\}.$$

Remarks.
(i) If v_k is a minimum of (N_k), then u_k (given by (3.24)) is a critical point of $\mathcal{E}_{A,\lambda}$ in the set $\{u \in H_A^1(\mathbf{R}^3)/ \int_{\mathbf{R}^3} |u|^p dx = 1\}$. However in general u_k is not a global minimum of $\mathcal{E}_{A,\lambda}$ in this set.
(ii) Under convenient assumptions, the above result may be extended to treat the case $V \neq \lambda$, V being spherically symmetric in $x' = (x_1, x_2)$.

Proof of Lemma 3.4. Let h_k be defined in $\mathbf{R}^3/\{x_1 = x_2 = 0\}$ by $h_k(x) = (x_2 + ix_1/s)^k$ for any $k \in \mathbf{Z}$. One can easily prove that for any $k \in \mathbf{Z}$ and any $r \in (1,2)$ $h_k \in W_{loc}^{1,r}(\mathbf{R}^3)$ and $\nabla h_k = (-ikh_k/s^2)$ $(-x_2, x_1, 0) = -(2ik/bs^2) Ah_k$ in $\mathcal{D}'(\mathbf{R}^3)$. Therefore for all $v \in \mathcal{D}^{1,2}(\mathbf{R}^3)$, $w = h_k v$ belongs to $W_{loc}^{1,s}(\mathbf{R}^3)$ for $s > 1$ sufficiently small and

$$\nabla_A w = \nabla h_k v + h_k(\nabla v + iAv) = h_k\left(\nabla v + i\left(1 - \frac{2k}{bs^2}\right)Av\right) \;\; \text{in } \mathcal{D}'(\mathbf{R}^3)$$

so, for $v \in \mathcal{D}^{1,2}(\mathbf{R}^3)$

$$(3.25) \qquad |\nabla_A w|^2 = |\nabla v|^2 + \left(\frac{bs}{2} - \frac{k}{s}\right)^2 |v|^2$$

which together with Proposition 2.2 and Remark 2.5 easily proves the lemma.

Proof of Theorem 3.5. Let v_k be a solution of Problem (N_k). Then if we define w_k by

$$(3.26) \qquad w_k(x) = \left(\frac{x_2 + ix_1}{s}\right)^k v_k(x),$$

we use the computations done in the proof of the above lemma, and specially (3.25), to deduce that w_k is a minimum of $\mathcal{E}_{A,\lambda}$ in the set

$$\{u \in H_A^1(\mathbf{R}^3)/u = h_k v, \ v \in S, \ \int_{\mathbf{R}^3} |v|^p dx = 1\}.$$

Therefore w_k is a critical point of $\mathcal{E}_{A,\lambda}$ in the set

$$\{u \in H_A^1(\mathbf{R}^3)/ \int_{\mathbf{R}^3} |u|^p dx = 1\}.$$

Furthermore, v_k being a minimum of (N_k), it satisfies the following equation

$$(3.27) \qquad -\Delta v_k + \left(\frac{bs}{2} - \frac{k}{s}\right)^2 v_k + \lambda v_k = \alpha_k |v_k|^{p-2} v_k \quad \text{in } \mathbf{R}^3$$

for some $\alpha_k \in \mathbf{R}\backslash\{0\}$. Next we use the fact that $w_k \in W_{loc}^{1,s}(\mathbf{R}^3)$ for some $s > 1$ to conclude.

Hence, the proof will be over as soon as we prove the existence of v_k. We may do it by using again standard concentration-compactness arguments in the case of partial translation invariance (see [25]). Actually if $\{v_n\}$ is a minimizing sequence for N_k and if $\lambda - bk > 0$ then the sequence $\{v_n\}$ is bounded in $H^1(\mathbf{R}^3)$. And this will be enough to prove that up to a translation in z, $\{v_n\}$ is relatively compact in $L^q(\mathbf{R}^3)$ for $2 \leq q < 6$. Since the set $\{k \in \mathbf{Z}/\lambda - bk > 0\}$ is not finite, this concludes the proof. Indeed, if we now fix $k \in \mathbf{Z}$, it is clear that when $\lambda > -\lambda_k$ all minimizing sequences for (N_k) are bounded in $H^1(\mathbf{R}^3)$ and then we go on as above.

Once we have proved the existence of infinitely many solutions of equation (1.2) via the problem $(I_{A,\lambda})$, let us say now what can be done about symmetric solutions of problem (J_A). We will state the corresponding result without proving it, since it easily follows from the same arguments as above.

Theorem 3.6. *Let* $2 < p < 10/3$. *Then every* $k \in \mathbf{Z}$ *there exists* $v_k \in S$ *such that* $w_k = (x_2 + ix_1/s)^k v_k$ *is a solution of equation (1.2). Actually* v_k *is a solution of the following minimization problem*

$$M_k = \inf \left\{ \int_{\mathbf{R}^3} \frac{1}{2} |\nabla v|^2 + \frac{1}{2} \left(\frac{bs}{2} - \frac{k}{s} \right)^2 |v|^2 - \frac{1}{p} |v|^p dx \middle/ \right.$$

$$v \in S, \quad \int_{\mathbf{R}^3} |v|^2 dx = 1 \Big\}.$$

3.3.   Existence of symmetric solutions in the case
N=2. In this section we will consider $A = b/2(-x_2, x_1)$. As in the preceding section we will look here for solutions of equation (1.2) in $\mathbf{R}^2$ which have particular symmetry properties. As we show below, in this case we will be able to obtain "more" solutions than in the case $N = 3$.

Let us define the space $H_r^1(\mathbf{R}^2)$ as the space of spherically symmetric functions of $H^1(\mathbf{R}^2)$. Then we may prove the following

Theorem 3.7. *Assume* $2 < p < +\infty$. *Then for all* $\lambda \in \mathbf{R}$ *and for all* $k \in \mathbf{Z}$ *such that* $\lambda > -\lambda_k$ *there is a sequence of distinct functions* $\{v_{k,n}\}_n \subset H_r^1(\mathbf{R}^2)$ *such that* $u_{k,n} = c_{k,n}(x_2 + ix_1/s)^k v_{k,n}$ *is a solution of equation (1.2) in* $\mathbf{R}^2$ *for some* $c_{k,n} \in \mathbf{R} \backslash \{0\}$. *Moreover the functions* $v_{k,n}$ *are obtained as critical points of the functional* $M(v) = -\int_{\mathbf{R}^2} |v|^p dx$ *restrained to* S_1, *the unit sphere of* $H_r^1(\mathbf{R}^2)$ *when we endow it with the norm*

$$|||v||| = \left(\int_{\mathbf{R}^2} |\nabla v|^2 + \left(\frac{bs}{2} - \frac{k}{s} \right)^2 |v|^2 + \lambda |v|^2 dx \right)^{1/2}.$$

Remark. Note that what we say is that $||| \cdot |||$ will be a norm on $H_r^1(\mathbf{R}^2)$ only when k is such that $\lambda > -\lambda_k$. Actually if $\lambda > -|b|$, $||| \cdot |||$ will be a norm on $H_r^1(\mathbf{R}^2)$ for all $k \in \mathbf{Z}$ and for any $\lambda \in \mathbf{R}$ there is an infinity of k such that $\lambda > -\lambda_k$ and the theorem holds.

Proof of Theorem 3.7. Let $k \in \mathbf{Z}$ be such that $\lambda - bk > 0$ and define $W : H_r^1(\mathbf{R}^2) \to \mathbf{R}^+$ by $W(v) = \||v|\|^2$. We can easily prove that W and M are meaningful and of class C^1 in $H_r^1(\mathbf{R}^2)$. Moreover M' maps weakly into strongly convergent sequences of $H_r^1(\mathbf{R}^2)$ since $H_r^1(\mathbf{R}^2)$ is compactly imbedded in $L^q(\mathbf{R}^2)$ for all $q > 2$ (see Lemma 1 and 2 in [35]). Thus we may apply a classical result in critical point theory (see by instance Theorem 2.10 or Lemma 2.11 in [29]) to infer the existence of a sequence of distinct critical points of $M_{|S_1}$. This proves the existence of $v_{k,n}$.

The rest of the proof follows by using the same arguments as above, i.e. a "singular" change of gauge of the form (3.26).

Remark. The above result may be extended as follows. Let V be a spherically symmetric function on $\mathbf{R}^2$ such that the quadratic form

$$(3.28) \qquad \int_{\mathbf{R}^2} |\nabla_A u|^2 + (\lambda + V)|u|^2 dx$$

is positive definite in $H_A^1(\mathbf{R}^2)$. Then for all $k \in \mathbf{Z}$, for all $n \in N$, there exist distinct functions $v_{k,n} \in S$ and constants $c_{k,n} \in \mathbf{R}\backslash\{0\}$ such that $u_{k,n} = c_{k,n}(x_2 + ix_1/s)^k v_{k,n}$ is a solution in $H_A^1(\mathbf{R}^2)$ of

$$L_A u + \lambda u + V u = |u|^{p-2} u \quad \text{in} \quad \mathbf{R}^2 \quad , \quad u \neq 0.$$

3.4. The limit case. In this section we will assume that $N = 3, p = 6$ and we will look for solutions of $(I_{A,0})$ and of equation (1.2) in this case. When $p = 6$, problem $(I_{A,0})$ is the following:

$$(3.29) \qquad \inf \left\{ \int_{\mathbf{R}^3} |\nabla_A u|^2 dx | u \in X_A \quad , \quad \int_{\mathbf{R}^3} |u|^6 dx = 1 \right\}.$$

We say that this a limit case for problem $(I_{A,0})$ because 6 is the limit exponent for the embeddings $H^1 \to L^p$ and $X_A \to L^p$.

When $A \equiv 0$ in $\mathbf{R}^3$ there exists a function u which minimizes $(I_{A,0})$ (see, by instance, [36], [26] or [2]). Thus, by taking into account the invariance of $(I_{A,0})$ by change of gauge, we may say that the same holds (i.e. there exists a minimum) when the vector potential A is the gradient of a function $T \in W_{loc}^{1,3}(\mathbf{R}^3)$. Actually we prove below that this is the only case in which $I_{A,0}$ is achieved. Therefore, if A is of the form (3.1) with $b \neq 0$, $I_{A,0}$ will not be achieved. However, by using the symmetry properties introduced in section 3.2, we will show that there may exist solutions of equation (1.2) for $p = 6$. Of course such solutions will be critical points of $\mathcal{E}_{A,0}$, but not minima. We start with the proof of the first statement above.

Theorem 3.7. *Let $A \in L_{loc}^3(\mathbf{R}^3, \mathbf{R}^3)$. Then $I_{A,0}$ is achieved if and only if there exists a function T in $W_{loc}^{1,3}(\mathbf{R}^3)$ such that $A \equiv \nabla T$ a.e. in $\mathbf{R}^3$.*

The proof of this theorem uses in a fundamental way the following result.

Proposition 3.8. *For any A in $L_{loc}^3(\mathbf{R}^3, \mathbf{R}^3), I_{A,0} = I_{0,0}$.*

Remark. In the terminology of the concentration-compactness method $(I_{0,0})$ is actually the problem at infinity of $(I_{A,0})$.

Proof of Proposition 3.8. By using inequality (2.3) we immediately see that $I_{0,0} \leq I_{A,0}$ for all A in $L_{loc}^3(\mathbf{R}^3, \mathbf{R}^3)$. Next, we show the reversed inequality.

Let $u \in \mathcal{D}(\mathbf{R}^3)$ be such that

$$(3.30) \qquad \int_{\mathbf{R}^3} |u|^6 dx = 1 \quad , \quad \int_{\mathbf{R}^3} |\nabla u|^2 dx \leq I_{0,0} + \epsilon$$

where ϵ is an arbitrarily small positive constant. Then, we define u_σ by

$$u_\sigma(\cdot) = \sigma^{-1/2} u(\frac{\cdot}{\sigma}) \quad , \quad \sigma > 0.$$

We compute $\|\nabla_A u_\sigma\|_{L^2(\mathbf{R}^3)}$ and obtain that

$$(3.31) \qquad \int_{\mathbf{R}^3} |\nabla u_\sigma + iAu_\sigma|^2 dx = \int_{\mathbf{R}^3} |\nabla u + i\sigma A^\sigma u|^2 dx,$$

where $A^\sigma(\cdot) = A(\sigma\cdot)$. Moreover,

$$\int_{\mathbf{R}^3} |\nabla u + i\sigma A^\sigma u|^2 dx \leq \int_{\mathbf{R}^3} |\nabla u|^2 dx +$$
$$+ \sigma^2 \int_{\mathbf{R}^3} |A^\sigma|^2 |u|^2 dx + 2\sigma \int_{\mathbf{R}^3} |A^\sigma| |u| |\nabla u| dx.$$

But $\sigma^2 \int_{\mathbf{R}^3} |A^\sigma|^2 |u|^2 dx \leq \left(\int_{B_R} \sigma^3 |A^\sigma|^3 dx \right)^{2/3} \left(\int_{\mathbf{R}^3} |u|^6 dx \right)^{1/3} = \left(\int_{B_{R\sigma}} |A|^3 dx \right)^{2/3}$, where $R > 0$ is such that supp $u \subset B_R$. Moreover

$$\sigma \int_{\mathbf{R}^3} |A^\sigma| |u| |\nabla u| dx \leq \left(\int_{\mathbf{R}^3} |\nabla u|^2 dx \right)^{1/2} \left(\int_{B_{R\sigma}} |A|^3 dx \right)^{1/3},$$

and hence, since $A \in L^3_{loc}(\mathbf{R}^3)$, $\int_{\mathbf{R}^3} |u_\sigma|^6 dx = 1$,

$$\varlimsup_{\sigma \to 0} \int_{\mathbf{R}^3} |\nabla u_\sigma + iA u_\sigma|^2 dx \leq I_{0,0} + \epsilon.$$

Finally, since ϵ is an arbitrary positive constant, the conclusion follows immediately.

Proof of Theorem 3.7. Assume that there is a minimum of $I_{A,0}$. Then there exists $u \in H^1_{loc}(\mathbf{R}^3) \cap L^6(\mathbf{R}^3)$ such that $\|u\|_{L^6(\mathbf{R}^3)} = 1$ and
(3.32)
$$L_A u = -\Delta u - 2i\, A \cdot \nabla u - i\, \text{div}\, Au + |A|^2 u = I_{A,0} |u|^4 u \quad \text{in} \quad \mathbf{R}^3.$$

Let v be a solution of

(3.33) $$-\Delta v = I_{A,0} |u|^5 \quad \text{in} \quad \mathbf{R}^3.$$

We may then use a result of Kato (see [21] or [3]) which states that if

(3.34) $$L_A u = f, \quad -\Delta v = |f| \quad \text{in} \quad \mathbf{R}^3 \quad, \text{ then}$$

(3.35) $$|u| \leq v \quad \text{a.e in} \quad \mathbf{R}^3.$$

Thus, by Proposition 3.15 we obtain

$$I_{0,0}\|v\|^2_{L^6(\mathbf{R}^3)} \leq \int_{\mathbf{R}^3} |\nabla v|^2 dx = I_{A,0} \int_{\mathbf{R}^3} |u|^5 v dx =$$

$$= I_{0,0} \int_{\mathbf{R}^3} |u|^5 v dx \leq$$

$$\leq I_{0,0} \left(\int_{\mathbf{R}^3} |u|^6 dx \right)^{5/6} \left(\int_{\mathbf{R}^3} |v|^6 dx \right)^{1/6}.$$

Moreover from (3.35) we deduce that $\int_{\mathbf{R}^3} |v|^6 dx \leq 1$ and hence

$$\int_{\mathbf{R}^3} |u|^5 v dx = \int_{\mathbf{R}^3} |v|^6 dx.$$

This together with the inequality $|u| \leq v$ shows that $|u|$ is actually equal to v almost everywhere in $\mathbf{R}^3$ and by the maximum principle, $|u| > 0$ a.e. in $\mathbf{R}^3$.

Let us now define $u_1 = Re\ u$, $u_2 = Im\ u$. Since $|u| = v$ a.e. in $\mathbf{R}^3$,

(3.36) $$\qquad -\Delta|u| = -\Delta v = I_{A,0}|u|^5 = I_{A,0}v^5$$

and

$$-\Delta|u| = \frac{1}{|u|^3}(u_1^2|\nabla u_1|^2 + u_2^2|\nabla u_2|^2 + 2u_1 u_2 \nabla u_1 \cdot \nabla u_2)$$

$$- \frac{1}{|u|}(|\nabla u_1|^2 + |\nabla u_2|^2 + u_1\Delta u_1 + u_2\Delta u_2)$$

$$= -\frac{1}{|u|^3}(u_2\nabla u_1 - u_1\nabla u_2)^2 + \frac{1}{|u|}Re(-\nabla u\ \bar{u})$$

$$= -\frac{1}{|u|^3}(Im\nabla u\ \bar{u})^2 +$$

$$+ \frac{1}{|u|}\ Re((L_A u)u + 2iA \cdot \nabla u\ u - |A|^2|u|^2 + i(\mathrm{div} A)|u|^2)$$

$$= -\frac{1}{|u|^3}(Im\nabla u\ \bar{u})^2 + I_{A,0}|u|^5 - \frac{2}{|u|}\ Im(A \cdot \nabla u\bar{u}) - |A|^2|u|.$$

Therefore, this and (3.36) imply that

$$\left(A|u|^{1/2} + \frac{Im(\nabla u\bar{u})}{|u|^{3/2}} \right)^2 = 0 \quad \text{a.e., i.e.}$$

(3.37)
$$A = -\frac{1}{|u|^2} Im \nabla u \bar{u} \quad \text{a.e. in} \quad \mathbf{R}^3.$$

Next, since $|u| = v > 0$ a.e., locally there exists some real valued regular function θ such that $u = e^{i\theta} v$.

Thus, locally, $\nabla u \bar{u} = \nabla v v + i v^2 \nabla \theta$ and by using (3.37) we see that for all $x \in \mathbf{R}^3$, there exists a neighbourhood of x and a function θ^x such that

(3.38)
$$A = -\frac{1}{|u|^2} (\nabla \theta^x) v^2 = \nabla \theta^x$$

in that neighbourhood. Finally, by the definition of θ^x, we see that we may define a function T in $\mathbf{R}^3$ such that $T \in W_{loc}^{1,3}(\mathbf{R}^3)$ and $A = \nabla T$ a.e. in $\mathbf{R}^3$.

Theorem 3.7 shows that we cannot hope to find non trivial ground states of equation (1.2) when $p = 6$ and $A = b/2(-x_2, x_1, 0)$, $b \neq 0$. However, as we prove below, there may exist solutions of (1.2). These solutions will be found, by performing a change of gauge, as critical points (which are not minima) of a certain variational problem. By following the arguments already used in the preceding sections, we know that solutions of equation (1.2) exist as soon as we are able to solve the following problem

(3.39)
$$N_k = \inf \left\{ \int_{\mathbf{R}^3} |\nabla v|^2 + (\frac{bs}{2} - \frac{k}{s})^2 + \lambda)|v|^2 dx / \right.$$
$$v \text{ spherically symmetric in } x' = (x_1, x_2),$$
$$\left. v \in \mathcal{D}^{1,2}(\mathbf{R}^3), \int_{\mathbf{R}^3} |v|^6 dx = 1 \right\}$$

As it is explained in [26], to study problem (3.39) we need to define another minimization problem. This new problem will be the so-called problem at infinity associated with (3.39). It is defined as follows:

(3.40)
$$N_k^\infty = \inf \left\{ \int_{\mathbf{R}^3} |\nabla v|^2 + \frac{k^2}{s^2}|v|^2 dx / \right.$$
$$v \quad \text{spherically symmetric in } x' = (x_1, x_2),$$
$$\left. v \in \mathcal{D}^{1,2}(\mathbf{R}^3), \quad \int_{\mathbf{R}^3} |v|^6 dx = 1 \right\}$$

By using the concentration-compactness method (see [26]), we prove the following result

Theorem 3.9. *Let* $\lambda > -|b|$ *and* $k \in \mathbf{Z}$*. Then all the minimizing sequences of (3.39) are relatively compact in* $D^{1,2}(\mathbf{R}^3)$ *(up to a translation in* z*) if and only if*

$$(3.41) \qquad\qquad N_k < N_k^\infty.$$

In particular, if (3.41) holds, there exists a solution of (3.39), v_k*, and some* $C_k \in \mathbf{R}\backslash\{0\}$*, such that* $u_k = c_k(x_2 + ix_1/s)^k v_k$ *is a solution of equation (1.2) with* $p = 6$*.*

Remark. To know whether inequality (3.41) holds or not we should have very precise information about the shape of all solutions of (N_k^∞), i.e., of the solutions of

$$-\Delta v + \frac{k^2}{s^2}v = v^5 \quad \text{in} \quad \mathbf{R}^3 \quad, \quad v > 0 \quad, \quad v \in S \cap \mathcal{D}^{1,2}(\mathbf{R}^3).$$

We will skip the proof of Theorem 3.9 since it is made by using [26] and the arguments already used in the preceding sections.

4. General magnetic fields. In this section we will study problems $(I_{A,V})$ and (J_A) for general vector potentials A or equivalently for general magnetic fields B.

4.1. The subcritical case. As in the case of constant magnetic fields, let us begin our study by considering the subcritical case, i.e. $2 < p < 6$ $(N = 3)$ or $2 < p < +\infty$ $(N = 2)$ in $(I_{A,V})$ and $2 < p < 2+4/N$ in (J_A). Under these assumptions we will generalize the results obtained in Section 3. Actually, as we show below, this can be done for the potentials A which satisfy the following property

Definition 4.1. *Let* A *be in* $L_{loc}^\alpha(\mathbf{R}^N, \mathbf{R}^N)$ *with* $\alpha > 2$ *if* $N = 2$*,* $\alpha = 3$ *if* $N = 3$*. We say that* A *satisfies the condition* (P) *if and only if for all sequence* $\{x_n\} \subset \mathbf{R}^N$ *with* $|x_n| \underset{n\to+\infty}{\longrightarrow} +\infty$*, the sequence* $\{\,\text{rot}(A(\cdot + x_n) - A(\cdot))\}_n$*, i.e.* $\{B(\cdot + x_n) - B(\cdot)\}_n$*, is bounded in*

$(\mathcal{D}_{\text{loc}}^{-1,\alpha}(\mathbf{R}^N))^\beta = \{f \in W_{\text{loc}}^{-1,\alpha}(\mathbf{R}^N)|\exists g \in (L_{\text{loc}}^\alpha(\mathbf{R}^N))^N, f = \text{div } g\}^\beta,$
with $\beta = 1$ *if* $N = 2, \beta = 3$ *if* $N = 3$.

Remark 4.2. Notice that if A satisfies the condition (P), then

(4.1) For any unbounded sequence $\{x_n\}_n$ of points of $\mathbf{R}^N$, there exists a sequence of functions $\{H_n\}$ in $W_{\text{loc}}^{1,\alpha}(\mathbf{R}^N)$ and the functions $A_n(\cdot) \equiv A(\cdot + x_n) - \nabla H_n$ are uniformly bounded in $L_{loc}^\alpha(\mathbf{R}^N, \mathbf{R}^N)$.

The condition (4.1) is what we really need to prove our results below.

Remark 4.3. The magnetic potentials considered in Section 3 trivially satisfy (P) and (4.1). Indeed since they are all linear it is enough to define H_n by $H_n(x) = A(x_n) \cdot x$.

If A satisfies (4.1), then for all $X = \{x_n\} \subset \mathbf{R}^N$ such that $|x_n| \to +\infty$ as $n \to +\infty$, there exists a subsequence $X' = \{x_{n'}\}$ such that $\{A_{n'}\}$ converges weakly to some $A^{X'}$ in $L_{loc}^\alpha(\mathbf{R}^3, \mathbf{R}^3)$. Then, let us call χ the following set

$$\chi = \{X = \{x_n\} \subset \mathbf{R}^N | |x_n| \to +\infty \text{ and } (*) \text{ holds}\}$$

where

$$(*) \qquad A_n \underset{n \to +\infty}{\longrightarrow} A^X \in L_{loc}^\alpha(\mathbf{R}^N, \mathbf{R}^N) - \text{ weak}$$

When A is of the form (3.1), we may take $A_n \equiv A$ for all n and therefore, for all $X \in \chi, A^X \equiv A$.

Next we begin by the study of problem $(I_{A,V})$. In all that follows we will assume that $A \in L_{loc}^\alpha(\mathbf{R}^N, \mathbf{R}^N)$ satisfies (4.1), i.e. the property (P). The basic assumption we make on V is the following

(4.2) For any $C > 0$, the set $\{u \in X_A; \mathcal{E}_{A,V}(u) \leq C\}$ is bounded in $H_A^1(\mathbf{R}^N)$.

Theorem 4.4. *If A satisfies (4.1), $K \equiv 1$ and A is such that for some $\lambda \in \mathbf{R}$, (4.2) holds for $V \equiv \lambda$ then*

(i) All minimization sequences of $(I_{A,\lambda})$, ((3.2)), are relatively compact in $L^q(\mathbf{R}^N), 2 \leq q < 6$ if $N = 3$, $2 \leq q < +\infty$ if $N = 2$ up to a translation and a change of gauge. That is, there exists $\{x_n\} \subset \mathbf{R}^N$ such that $\{e^{iH_n}u_n(\cdot + x_n)\}$ is relatively compact in $L^q(\mathbf{R}^N)$, and H_n is given by (4.1).

(ii) The strict inequality

$$(4.3) \qquad\qquad I_{A,\lambda} < \inf_{X \in \chi} I_{A^X,\lambda}$$

is necessary and sufficient for the relative compactness of all minimizing sequences of $(I_{A,\lambda})$. In particular, if (4.3) holds, there is a minimum of $(I_{A,\lambda})$.

Remark. As it is not difficult to prove, the large inequality always holds in (4.3), i.e.

$$I_{A,\lambda} \leq \inf_{X \in \chi} I_{A^X,\lambda}$$

When V is not a constant we have to be careful about the assumptions made on it. Let us start with a simple particular case: $V = \lambda + V_0, V_o \in L^{r_1}(\mathbf{R}^N) + L^{r_2}(\mathbf{R}^N)$ for some $r_1, r_2 > N/2$.

Theorem 4.5. *Let V be equal to $\lambda + V_o$ with $V_o \in L^{r_1}(\mathbf{R}^N) + L^{r_2}(\mathbf{R}^N)$ for some $r_1, r_2 > N/2$ and assume also that (4.2) is satisfied. Then, if A satisfies (4.1) and $K \equiv 1$,*

(i) All minimization sequences $\{u_n\}$ of $(I_{A,V}$ are relatively compact in $L^q(\mathbf{R}^N)$, $2 \leq q < 6$ if $N = 3$, $2 \leq q < +\infty$ if $N = 2$ up to a translation and a change of gauge; i.e., there exists $\{x_n\} \subset \mathbf{R}^N$ such that $\{e^{iH_n}u_n(\cdot + x_n)\}$ is relatively compact in $L^q(\mathbf{R}^N)$, where H_n is given by (4.1).

(ii) The strict inequality

$$(4.4) \qquad\qquad I_{A,V} < \inf_{X \in \chi} I_{A^X,\lambda}$$

is necessary and sufficient for the relative compactness of all mini-mization sequences of $(I_{A,V})$. *In particular, if (4.4) holds, there is a minimum of* $(I_{A,V})$ *and consequently, a solution of*

$$L_A + Vu = |u|^{p-2}u \quad in \quad \mathbf{R}^N.$$

Remark. The above result may be extended to consider more general functions V as follows. Assume that A and V satisfy (4.2). Then if for all $X = \{x_n\}_n \subset \mathbf{R}^N$ with $|x_n| \underset{n \to +\infty}{\longrightarrow} +\infty$, there exists a subsequence $X' = \{x'_n\}$ such that $\{V(\cdot + x_{n'})\}_{n'}$ converges weakly to some $V^{X'}$ in $L^r_{loc}(\mathbf{R}^N)$ for some $r > N/2$, then Theorem 4.5 still holds with (4.3) replaced by (4.4')

$$(4.4') \qquad I_{A,V} < \inf\{I_{A^{X'},V^{X'}} / X' \subset X \in \chi\}$$

where X' is any subsequence of X such that $\{V(\cdot + x_{n'})\}_{n'}$ converges in the above sense.

Proof of Theorem 4.4. We begin with (i). If we report the arguments used in the proof of Theorem 3.1, we see that for some $\{x_n\} \subset \mathbf{R}^N$, $\{u_n(\cdot + x_{n'})\}$ is relatively compact in $L^q(\mathbf{R}^N)$, $2 \le q < 6$ if $N = 3$, $2 \le q < +\infty$ if $N = 2$. Then, we use (4.1) to deduce

$$(4.5) \qquad \int_{\mathbf{R}^N} |\nabla_A u_n|^2 dx = \int_{\mathbf{R}^N} |\nabla_{A_n} \underline{u}_n|^2 dx,$$

where $\underline{u}_n(\cdot) = e^{iH_n} u_n(\cdot + x_n)$. Note that now (4.5) replaces (3.16).

Since we may assume without loss of generality that for all $n, u_n \in D(\mathbf{R}^N)$, we may use (2.3) and (4.5) to end the proof of (i).

In order to prove (ii) let us first see what happens when the sequence $\{x_n\}$ is unbounded. If $|x_n| \underset{n \to +\infty}{\longrightarrow} +\infty$ and if we suppose that $\{\underline{u}_n\}$ converges (up to the extraction of a subsequence) to $\underline{u}$ in $L^q(\mathbf{R}^N)$, $2 \le q < 6$ if $N = 3$, $2 \le q \le +\infty$ if $N = 2$ then by (4.2) and (4.5) we deduce that for all $R > 0$,

$$(4.6) \qquad \begin{aligned} \int_{B_R} |\nabla_{A^x} \underline{u}|^2 dx &\le \lim_{n \to +\infty} \inf \int_{B_R} |\nabla_{A_n} \underline{u}_n|^2 dx \\ &\le \lim_{n \to +\infty} \inf \int_{\mathbf{R}^3} |\nabla_{A_n} \underline{u}_n|^2 dx \end{aligned}$$

and therefore

$$(4.7) \qquad I_{A^x,\lambda} \leq \int_{\mathbf{R}^N} |\nabla_{A^x}\underline{u}|^2 + \lambda|\underline{u}|^2 dx \leq I_{A,\lambda}$$

for $\{u_n\}$ was a minimizing sequence for $(I_{A,\lambda})$.

Hence if (4.3) holds, the sequence $\{x_n\}$ cannot be unbounded. Moreover, if $\{x_n\}$ is bounded, then it is straightforward to see that $\{u_n\}$ is relatively compact in $L^q(\mathbf{R}^N)$. Next, if (4.3) does not hold, then there exists $\{X_m\}_m \subset \chi$ such that

$$(4.8) \qquad I_{A,\lambda} = \lim_{m\to+\infty} I_{A^{X_m},\lambda}.$$

Therefore, since for all $X \in \chi, I_{A^x,\lambda} \geq I_{A,\lambda}$, (4.8) implies the existence of sequences $\{x_{m,n}\}_n \subset \mathbf{R}^N$ such that $|x_{m,n}| \xrightarrow[n\to+\infty]{} +\infty$ for all $m \in \mathbf{N}$ and

$$(4.9) \qquad \mathcal{E}_{A_{m,n},\lambda}\left(e^{iH_{m,n}}u_{m,n}(\cdot + x_{m,n})\right) \xrightarrow[n\to+\infty]{} I_{A^{X_m},\lambda}$$

where $H_{m,n}$ is defined by $A(\cdot + x_{m,n}) = A_{m,n}(\cdot) + \nabla H_{m,n}$ and $\{A_{m,n}\}_n$ converges to A^{X_m} in $L^\alpha_{loc}(\mathbf{R}^N)$-weak for all m, as n goes to $+\infty$.

By taking a diagonal sequence we may find sequences $\{u_p\}_p \subset X_A, \{x_p\}_p \subset \mathbf{R}^N$ such that $|x_p| \xrightarrow[p\to+\infty]{} +\infty$ and

$$\mathcal{E}_{A_p,\lambda}\left(e^{iH_p}u_p(\cdot + x_p)\right) \xrightarrow[p\to+\infty]{} I_{A,\lambda}.$$

This implies that $\{u_p\}$ cannot be relatively compact, since $|x_p|$ is unbounded. This concludes the proof.

Proof of Theorem 4.5. Since V satisfies (4.2), the first part of this proof can be done exactly as above. Next, as far as (ii) is concerned, the only point which has to be modified is the following. If $\{x_n\}$ is unbounded we cannot use directly an inequality like (4.7). Indeed we observe that a new term appears here in the expression of $\mathcal{E}_{A,V}(u_n)$ namely

$$\mathcal{E}_{A,V}(u_n) = \int_{\mathbf{R}^N} |\nabla_{A_n}\underline{u}_n|^2 + V_n|\underline{u}_n|^2 dx$$

where $V_n(\cdot) = V(\cdot + x_n)$. Then, we apply our assumptions to deduce that

$$(4.10) \quad \int_{\mathbf{R}^N} V_n |\underline{u}_n|^2 dx \underset{n \to +\infty}{\longrightarrow} \lambda \int_{\mathbf{R}^N} |\underline{u}|^2 dx, \quad \text{when } |x_n| \underset{n}{\longrightarrow} +\infty,$$

since $\underline{u}_n \underset{n \to +\infty}{\longrightarrow} \underline{u}$ in $L^q(\mathbf{R}^N), 2 \le q < 6$ if $N = 3, 2 < q < +\infty$ if $N = 2$ and $V_o \in L^{r_1}(\mathbf{R}^N) + L^{r_2}(\mathbf{R}^N)$ with $r_1, r_2 > N/2$ and this allows us to conclude as above.

Finally let us see without proofs the analogous of the above results for the problem (J_A).

Theorem 4.6. *Let p be in $(2, 2 + 4/N)$ and let $A \in L^\alpha_{loc}(\mathbf{R}^N, \mathbf{R}^N)$ satisfy (4.1). Then*

(i) All minimizing sequences for $\{J_A\}$ are relatively compact in $L^q(\mathbf{R}^N), 2 \le q < 6$ if $N = 3, 2 < q < +\infty$ if $N = 2$, up to a translation and a change of gauge.

(ii) The strict inequality

$$J_A < \inf_{X \in \chi} J_{A^X}$$

is necessary and sufficient for the relative compactness of all minimizing sequences of (J_A).

4.2. Some compact or partially compact cases. In the previous section, we gave necessary and sufficient conditions for the relative compactness of all minimizing sequences for $(I_{A,\lambda})$ when A is a general vector potential with no particular structure. In some cases the properties satisfied by A allow us to give more precise results. In this section we will show that the behaviour of A or B at infinity is fundamental when trying to decide about the a priori relative compactness of all minimizing sequences. We begin with a general result and then give some examples of compactness or partial compactness induced by the behaviour of A and B at infinity.

Theorem 4.7. *Assume that $A \in W^{1,\infty}_{loc}(\mathbf{R}^N, \mathbf{R}^N), K \equiv 1$ and* *that there are three constants,* $\alpha_1, \alpha_2, \alpha_3 \in \mathbf{R}$ *such that*

(4.11) $$\inf_{|x| \geq r} \mathrm{ess} \left(\sum_{i=1}^{3} \alpha_i B_i(x) \right) \xrightarrow[r \to +\infty]{} + \infty,$$

where $B = (B_1, B_2, B_3) = \mathrm{curl}\ A.$ *Then if*

(4.12)

$$\lambda > \lambda_B = -\max \left\{ \inf_{x \in \mathbf{R}^3} \mathrm{ess} \left(\sum_{i=1}^{3} \underline{\alpha}_i B_i(x) \right) \Big/ \underline{\alpha}_i \in \mathbf{R}, \sum_{i=1}^{3} |\underline{\alpha}_i| = 1 \right\}$$

all minimizing sequences of $(I_{A,\lambda})$ *are relatively compact in* $L^q(\mathbf{R}^N)$ *for* $2 \leq q < 6$ *if* $N = 3, 2 \leq q < +\infty$ *if* $N = 2.$ *In particular, if* *(4.12) holds there is a minimum of* $(I_{A,\lambda}).$

PROOF. Let $\{u_n\}$ be any minimizing sequence for $(I_{A,\lambda})$. By using Propositions 2.2 and 2.7 and (4.11)-(4.12) we see that $\{u_n\}$ is bounded in $H_A^1(\mathbf{R}^N)$ and relatively compact in $L^2(\mathbf{R}^N)$. Moreover, since $H_A^1(\mathbf{R}^N) \subset L^{p'}(\mathbf{R}^N)$, for some $p' > p$, this implies that $\{u_n\}$ is relatively compact in $L^p(\mathbf{R}^N)$ and this is enough to conclude.

Remark. The point which is important in the above proof is that when A (or B) satisfies (4.11), $H_A^1(\mathbf{R}^N)$ is actually compactly embedded in $L^2(\mathbf{R}^N)$. Analogous compactness arguments allow us to prove the following results. Let us assume in all that follows that $\lambda \in \mathbf{R}^+$ is such that, in every case, the quadratic form $\mathcal{E}_{A,\lambda}(u)$ is positive definite in $H_A^1(\mathbf{R}^N)$. Then, using the same kind of arguments as above we may treat the following partially compact examples:

Example 1. Let $N = 3$ and A be equal to $(\alpha(x_2, x_3), 0, 0)$ with $\alpha \in W_{loc}^{1,\infty}(\mathbf{R}^3)$. Then the problem $(I_{A,\lambda})$ is invariant under translation in x_1. Moreover, by using again Proposition 2.7 we may prove that if for some $(a, b) \in \mathbf{R}^2 \backslash \{(0,0)\}$.

$$\inf_{|x_2| + |x_3| \geq r} \mathrm{ess}\ a \frac{\partial \alpha}{\partial x_2} + b \frac{\partial \alpha}{\partial x_3} \xrightarrow[r \to +\infty]{} + \infty$$

then all minimization sequences for $I_{A,\lambda}$ are relatively compact in $L^q(\mathbf{R}^3), 2 \leq q < 6$ up to a translation in x_1. Moreover $I_{A,\lambda}$ is achieved.

Example 2. If $A \in W_{loc}^{1,\infty}(\mathbf{R}^3)$ and depends only on x_3, we may assume without loss of generality that $A = (\alpha_1(x_3), \alpha_2(x_3), 0)$. In this case if there exist $a, b \in \mathbf{R}$ such that

$$\inf_{|x_3| \to +\infty} \mathrm{ess} \; (a\alpha_1'(x_3) + b\alpha_2'(x_3)) \xrightarrow[r \to +\infty]{} +\infty$$

then all minimizing sequences for $I_{A,\lambda}$ are relatively compact in $L^q(\mathbf{R}^3)$, $2 \le q < 6$ up to a translation in x_1, x_2 and $I_{A,\lambda}$ is achieved.

Another example that we may consider is that of a vector potential A which is "almost" periodic, i.e. which is periodic up to a gradient. Let us suppose that $A \in L_{loc}^3(\mathbf{R}^3)$ depends only on x_3. As before we may assume that $A = (\alpha_1(x_3), \alpha_2(x_3), 0)$. Then assume that there exist $t \in \mathbf{R}^+ \backslash \{0\}$ and $H \in W_{loc}^{1,3}(\mathbf{R}^3)$ such that for all $x \in \mathbf{R}^3, A(x + T) = A(x) + \nabla H$, when T denotes the vector $(0, 0, t)$. In this case we can prove the following

Proposition 4.8. *If A satisfies the above conditions and A and $\lambda \in \mathbf{R}$ are such that $\mathcal{E}_{A,\lambda}(\cdot)$ is a definite positive form in $H_A^1(\mathbf{R}^3)$, then all minimizing sequences for $(I_{A,\lambda})$ are relatively compact in $L^q(\mathbf{R}^3), 2 \le q < 6$, up to a translation in x_1, x_2 and a change of gauge.*
Moreover there is a minimum for $I_{A,\lambda}$.

Remark. In the particular case when $\nabla H \equiv 0$, i.e. A is periodic in x_3, then the result stated in the above proposition can be still more precise: in this case the change of gauge is not necessary to get the relative compactness of minimizing sequences.

Proof of Proposition 4.8. Let $\{u_n\}$ be a minimizing sequence for $I_{A,\lambda}$. As it is already standard, under our assumptions we may immediately prove that $\{u_n(\cdot + x_n)\}$ is relatively compact in $L^q(\mathbf{R}^3)$, for $2 \le q \le 6$ and for some $\{x_n\} \subset \mathbf{R}^3$. By hypothesis there exist $c_n \in \mathbf{Z}$ such that $x_n = c_n T + y_n$, with $|y_n^3| \le T$. Therefore if we define $y_n = (x_n^1, x_n^2, y_n^3), H_n = c_n H$ and $\underline{u}_n(\cdot) = e^{iH_n} u_n(\cdot + c_n T)$, we can see that the sequence $\{\underline{u}_n\}$ is relatively compact in $L^q(\mathbf{R}^3)$, $2 \le q < 6$, and

$$\mathcal{E}_{A,\lambda}(\underline{u}_n) = \mathcal{E}_{A,\lambda}(u_n).$$

The conclusion follows immediately.

Remark. Obviously all the compactness results that we have obtained for $(I_{A,V})$ in this section can be easily extended to study also the problem (J_A). We will not do it here since the ideas used above can be directly applied to this case.

4.3. Existence of solutions with symmetry properties for general potentials A.

Let A be any function in $L^\alpha_{loc}(\mathbf{R}^N, \mathbf{R}^N)$, with $\alpha > 2$ if $N = 2, \alpha = 3$ if $N = 3$. We have already shown in the preceding sections that for such a potential A solutions of the equation (1.2) can be obtained via the resolution of the minimization problem $(I_{A,\lambda})$. Moreover in sections 3.2-3, we proved that in the case of constant magnetic fields many solutions of (1.2) could be obtained by symmetry arguments. We will show here that this kind of arguments can be extended to treat the case of general potentials A. Assume that we look for solutions u of the form $u = hv$, where $v \in S$ if $N = 3$, and $v \in \mathcal{D}^{1,2}_r(\mathbf{R}^2)$ if $N = 2$. If h is smooth enough we compute $|\nabla_A u|^2$ to obtain:

$$|\nabla_A u|^2 = |\nabla v|^2 |h|^2 + |\nabla h + iAh|^2 |v|^2 + 2Re(\nabla v \cdot (\nabla \bar{h} - iA\bar{h})h\bar{v}).$$

Let us call this expression $F_A(v)$. Then if we denote by Y the space S if $N = 3$ and $\mathcal{D}^{1,2}_r(\mathbf{R}^2)$ if $N = 2$, we may consider the following minimization problem
(4.13)
$$\min\left\{\int_{\mathbf{R}^N} F_A(v)dx + \lambda \int_{\mathbf{R}^N} |h|^2 |v|^2 dx / v \in Y, \int_{\mathbf{R}^N} |hv|^p dx = 1\right\}.$$

It can be proved that if $h \in W^{1,q}_{loc}(\mathbf{R}^N)$ for some $q > 1$ and $|h|, |h|^{-1} \in L^q(\mathbf{R}^N), |h|$ being a function of s and z, then for any solution v of (4.13), hv is a solution of equation (1.2). Of course, solving (4.13) can be as difficult as solving $(I_{A,\lambda})$. Nevertheless a good choice of the function h can simplify considerably (4.13). For instance let us assume that there is an h satisfying the above conditions and

(4.14) $$\nabla h + iAh = \beta W \qquad \text{in} \quad \mathbf{R}^N,$$

with $W = (-x_2, x_1, 0)$ if $N = 3$, $W = (-x_2, x_1)$ if $N = 2$ and β is a function such that $|\beta|$ depends only on s and z. Then (4.13) can be rewritten as follows:

$$(4.15) \quad \min\left\{ \int_{\mathbf{R}^N} |h|^2 |\nabla v|^2 + (\lambda |h|^2 + |\beta|^2 s^4)|v|^2 dx \,\Big/ \right.$$
$$\left. v \in Y, \int_{\mathbf{R}^N} |hv|^p dx = 1 \right\}$$

and this new problem can be easily studied by using the concentration-compactness method. Then given A in $L^\alpha_{loc}(\mathbf{R}^N, \mathbf{R}^N)$ the difficult lies on finding a function h satisfying all the conditions above, and in particular (4.14). Of course from (4.14) we may obtain necessary conditions on A for the existence of β and h. We will not do it here in general. However we will consider a particular case which generalizes what we did in section 3.2. In that case we will analyse in a precise way what we can prove for problem (4.15).

Let us assume now that $A(x) = \gamma(s)W$ for some function $\gamma \in L^\alpha_{loc}(\mathbf{R}^+, \mathbf{R})$. Then the result we proved in the case $\gamma \equiv$ constant, i.e. Theorem 3.5, can be generalized here to prove

Theorem 4.9. *Let* $2 < p < 6$ *if* $N = 3, 2 < p < +\infty$ *if* $N = 2$, λ *be any real positive number and* $A = \gamma(s)W$, *with* $\gamma \in L^\alpha_{loc}(\mathbf{R}^+, \mathbf{R})$. *Then for every* $k \in \mathbf{Z}$ *there exist* $C_k \in \mathbf{R}\backslash\{0\}$ *and* $v_k \in Y$ *such that*

$$u_k(x) = C_k \left(\frac{x_2 + ix_1}{s} \right)^k v_k$$

is a solution of (1.2). Moreover if $N = 2$, *for every* k *there is an infinity of solutions of (1.2) of the form*

$$u_{k,n}(x) = C_{k,n} \left(\frac{x_2 + ix_1}{s} \right)^k v_{k,n} \quad,$$
$$C_{k,n} \in \mathbf{R}\backslash\{0\} \quad, \quad v_{k,n} \in \mathcal{D}^{1,2}(\mathbf{R}^2).$$

Finally exactly as in Theorem 3.5, we can characterize v_k *(resp.* $v_{k,n}$) *as a solution of a variational problem (see section 3.2).*

When $N = 3$ we can also be interested in the critical case $p = 6$. Let us consider A and γ as above and assume by instance that $\lambda = 0$.

Then the variational problem which gives us v_k in the above theorem is

$$(4.16) \quad \inf \left\{ \int_{\mathbf{R}^3} |\nabla v|^2 + |\gamma(s)s - \frac{k}{s}|^2 |v|^2 dx / v \in S, \int_{\mathbf{R}^3} |v|^6 dx = 1 \right\}$$

and there will be a solution of it if and only if the infimum value in (4.16) is strictly less than

$$\min_{i=1,2} \inf \left\{ \int_{\mathbf{R}^3} |\nabla v|^2 + \left(\bar{\gamma}_i(s) + \frac{k^2}{s^2} \right) |v|^2 dx / v \in S, \int_{\mathbf{R}^3} |v|^6 dx = 1 \right\}$$

where the functions $\bar{\gamma}_i$ are defined by

$$\bar{\gamma}_1(s) = \lim_{\sigma \to +\infty} \gamma^2(\sigma s)\sigma^4 s^2 - 2\gamma(\sigma s)\sigma^2 k$$

$$\bar{\gamma}_2(s) = \lim_{\sigma \to 0^+} \gamma^2(\sigma s)\sigma^4 s^2 - 2\gamma(\sigma s)\sigma^2 k$$

when these limits exist.

As we have already indicated, these results can be proved by using the concentration-compactness method in the subcritical and critical cases and for partially symmetric problems (see [25, 26] and the sections above).

Remark. There are easy cases in which we can be sure of the existence of solutions of equation (1.2) via symmetry arguments. Indeed assume that $A = \gamma W$, with $\bar{\gamma}_1(s) = \bar{\gamma}_2(s) = +\infty$ for all s, then the sufficient condition for the existence of a solution of (4.16) is automatically satisfied. This provides us with a potential A such that (1.2) has no ground state solution (see Section 3.4) but nevertheless it has an infinity of solutions, i.e. the ones obtained via the resolution of (4.16).

5. Hartree-Fock equations for Coulomb systems with magnetic fields.

In the quantum mechanical description of electrons interacting with static nucleii in the presence of an external magnetic field the energy has the the following form

$$\overline{\mathcal{E}}_A(\phi) =$$

$$(5.1) \quad \int_{\mathbf{R}^{3N}} |\nabla_A \phi|^2 dx + \int_{\mathbf{R}^{3N}} \left\{ \sum_{i=1}^N V(x_i) + \sum_{i<j} \frac{1}{|x_i - x_j|} \right\} |\phi|^2 dx,$$

where $V(x) = -\sum_{j=1}^{m} z_j(x - \bar{x}_j)^{-1}, m \geq 1$ being the number of nucleii, z_j is the charge of the nucleii located at $\bar{x}_j$ (fixed point of $\mathbf{R}^3$) and A is the vector potential associated with the magnetic field B (curl $A = B$). The ground state energy is then defined by

(5.2)
$$\inf \left\{ \overline{\mathcal{E}}_A(\phi) / \ \phi \in H_A^1(\mathbf{R}^{3N}), \ \phi \text{ antisymmetric in } x_1, \ldots, x_n, \right.$$
$$\left. \int_{\mathbf{R}^{3N}} |\phi|^2 dx = 1 \right\}.$$

The first term in the energy $\overline{\mathcal{E}}_A$ corresponds to the kinetic energy of the electrons, the second term is the 1-body attractive interaction between the electrons and the nucleii and the last term is the 2-body repulsive interaction between the electrons.

Problem (5.1) is very hard to deal with, because of the dimension $3N$. V.Fock [18] and J.C. Slater [34] proposed to simplify the problem by considering only functions ϕ having the form

(5.3)
$$\phi(x_1, \ldots, x_n) = \frac{1}{\sqrt{N!}} \det(\varphi_i(x_j)),$$

with $\varphi_i \in H_A^1(\mathbf{R}^3), i = 1, \ldots, N, \int_{\mathbf{R}^3} \varphi_i \bar{\varphi}_j dx = \delta_{ij} \ \forall i, j$. This simplification takes into account the Pauli principle and reduces the search of a function in dimension $3N$ to the search of N functions in dimension 3. The functions ϕ of the form (5.3) are called Slater determinants. By considering functions ϕ of the form (5.3) problem (5.1) becomes

(5.4) $I = \inf \left\{ \mathcal{E}_A(\varphi_1, ..., \varphi_N) | \varphi_i \in H_A^1(\mathbf{R}^3) \ \forall i, \ (\varphi_1, \ldots, \varphi_N) \in K \right\}$

where

$$K = \{ (\varphi_1, \ldots, \varphi_N) \in (L^2(\mathbf{R}^3))^N / \int_{\mathbf{R}^3} \varphi_i \bar{\varphi}_j dx = \delta_{ij} \ i, j = 1, \ldots, N \}$$

and

$$\mathcal{E}_A(\varphi_1, \ldots, \varphi_N) = \sum_{i=1}^{N} \int_{\mathbf{R}^3} |\nabla_A \varphi_i|^2 + V |\varphi_i|^2 dx +$$
$$+ \frac{1}{2} \int \int_{\mathbf{R}^3 \times \mathbf{R}^3} \rho(x) \frac{1}{|x - y|} \rho(y) dx \, dy$$
$$- \frac{1}{2} \int \int_{\mathbf{R}^3 \times \mathbf{R}^3} \frac{1}{|x - y|} |\rho(x, y)|^2 dx \, dy$$

with $\rho(x) = \sum_{i=1}^{N} |\varphi_i(x)|^2$, $\rho(x,y) = \sum_{i=1}^{N} \varphi_i(x)\bar{\varphi}_i(y)$.

For a more complete physical interpretation of this problem see [27], [19], [18], [34], [24]. In [27] problem (5.4) is studied in the case $A \equiv 0$. Here, we will consider $A = 1/2(-x_2, x_1, 0)$. However, the type of arguments we give below may be applied to a large class of magnetic potentials A.

The Euler-Lagrange equations corresponding to problem (5.4) can be written up to a unitary transform on $(\varphi, \ldots, \varphi_N)$ as follows

(5.5)
$$L_A \varphi_i + V \varphi_i + (\rho * \frac{1}{|x|}) \varphi_i - \int_{\mathbf{R}^3} \rho(x,y) \frac{1}{|x-y|} \varphi_i(y) dy + \epsilon_i \varphi_i = 0$$
$$\text{in} \quad \mathbf{R}^3, \quad i = 1, \ldots, N$$

where $-\epsilon_i$ is the Lagrange multiplier corresponding to φ_i and $(\varphi_1, \ldots, \varphi_N) \in K$.

Equations (5.5) are the so-called Hartree-Fock equations with magnetic field. There are other equations, as the Hartree equations, which arise when making in (5.1) simplifications which are different from (5.3). These other equations can be also studied in the case of a non null magnetic field. We will not do it here since it can be easily done by extending the results in [27] for the case $A \equiv 0$ in the same way as we do below for the Hartree-Fock equations. For further details about the Hartree equations or the Thomas-Fermi theory see [19], [22, 23], [8].

Remark. There are solutions of equations (5.5) which are not minima of problem (5.4). Actually the solutions which are not minima are critical point of $\mathcal{E}_A$ which correspond in a vague sense to excited states of the system and that can be found by using critical point methods which are a little bit more sophisticated than the ones used here.

By following very closely the proofs given in [27] in the case $A \equiv 0$, we can prove the following:

Theorem 5.1. *Let us assume that $Z = \sum_{i=1}^{m} z_i > N - 1$. Then, every minimizing sequence of problem (5.4) is relatively compact in $(H_A^1(\mathbf{R}^3))^N$. In particular there is a minimum for (5.4) and* *a solution for the system (5.5).*

Remark. When $Z \geq N$ there is actually a sequence of distinct solutions of the Hartree-Fock equations with external magnetic field (5.5), $(\varphi_1^k, \ldots, \varphi_N^k)$, $k \geq 1$, such that $\int_{\mathbf{R}^3} \varphi_i^k \bar{\varphi}_j^k dx = \delta_{ij}$ for all $1 \leq i, j \leq N$, $k \leq 1$. The proof of this result uses critical point theory together with some information on the index of the critical points: see A.Bahri [6], C.Viterbo [37], A.Bahri et P.L. Lions [7] and C.V.Coffman [14] for results showing the relation between critical point theory and critical point indices. Moreover see [27] for a proof close to ours. Indeed in [27] the case $A = 0$ is treated with this method. Our proof basically follows the steps of that in [27] for the case $A = 0$. The fact that $A \neq 0$ is dealt with as in the preceding sections and in the proof of Theorem 5.1 below.

Finally let us point out that the assumption $Z \geq N$ is not much more restrictive than that of Theorem 5.1, i.e. $Z > N - 1$, since from the Physics viewpoint Z is an integer.

PROOF.

First step. Some estimates. As we said before, this proof is very close to the one made in [27] corresponding to the case $A \equiv 0$. Actually the only real difference between these two proofs lies on a spectral result proved in Lemma 5.2 below.

Let $\{(\underline{\varphi}_1^n, \ldots, \underline{\varphi}_N^n)\}$ be any minimizing sequence. Then, by using general optimization results (see [16] by instance) we can find another sequence $\{(\varphi_1^n, \ldots, \varphi_N^n)\} \subset K$ such that

$$\sum_{i=1}^{N} \|\varphi_i^n - \underline{\varphi}_i^n\|_{L^2(\mathbf{R}^3)} \leq 1/n$$

and for all $n \in \mathbf{N}$, $(\varphi_1^n, \ldots, \varphi_N^n)$ is a minimum for a minimization problem which is obtained from (5.4) by adding a small perturbation in $L^2(\mathbf{R}^3)$. Moreover, the sequence $\{(\varphi_1^n, \ldots, \varphi_N^n)\}$ satisfies the following: there exist constants γ_n, $\epsilon_1^n, \ldots, \epsilon_N^n$ such that $\gamma_n \xrightarrow[n \to +\infty]{} 0$, $(\epsilon_1^n, \ldots, \epsilon_N^n) \in \mathbf{R}^N$ and

(5.6)
$$L_A \varphi_i^n + V \varphi_i^n + \left(\rho^n * \frac{1}{|x|} \right) \varphi_i^n +$$
$$- \left(\int_{\mathbf{R}^3} \rho^n(x, y) \varphi_i^n(y) \frac{1}{|x - y|} dy \right) + \epsilon_i^n \varphi_i^n \xrightarrow[n \to +\infty]{} 0$$

in the dual space of $H_A^1(\mathbf{R}^3)$ and

(5.7)
$$\int_{\mathbf{R}^3} |\nabla_A \Psi|^2 + V|\Psi|^2 + (\rho * \frac{1}{|x|})|\Psi|^2 + (\epsilon_i^n + \gamma^n)|\Psi|^2 -$$
$$- \int\int_{\mathbf{R}^3 \times \mathbf{R}^3} \rho(x, y)\frac{1}{|x - y|}\bar{\Psi}(x)\Psi(y)dydx \geq 0,$$

for all $\Psi \in L^2(\mathbf{R}^3)$ such that $\int_{\mathbf{R}^3} \Psi \bar{\varphi}_j^n dx = 0$, $\forall j$. i.e. the operator $\bar{H}_n = L_A + V + (\rho^n * \frac{1}{|x|}) - K$ has at most N negative eigenvalues strictly less than $(1 - \epsilon_i^n + \gamma^n)$, where the operator K is defined by

(5.8)
$$K\varphi = Re \int_{\mathbf{R}^3} \rho(x, y)\frac{1}{|x - y|}\varphi(y)dy.$$

Moreover, since the operator K is nonnegative (see [27]) the operator $H_n = \bar{H}_n + K$ has the same above spectral property (as $\bar{H}_n$). Now, if $Z > N$ we can use a spectral result proved in Lemma 5.2 below and the above information to insure the existence of $\bar{\epsilon} > 0$ such that

(5.9)
$$\epsilon_i^n \geq \bar{\epsilon} \quad \forall i \quad , \quad \forall n.$$

Second step: convergence to a minimum. First we prove that since $\{(\underline{\varphi}_1^n, \ldots, \underline{\varphi}_N^n)\}$ is a minimizing sequence for (5.4),$\{(\varphi_1^n, \ldots, \varphi_N^n)\}$ is bounded in $(H_A^1(\mathbf{R}^3))^N$. This can be done by using Cauchy-Schwarz inequality, the fact that $V \in L_{loc}^p(\mathbf{R}^3)$, $p > \frac{3}{2}$ and $V(x) \underset{|x| \to +\infty}{\longrightarrow} 0$ and a slight variation of the so-called uncertainty inequality:

(5.10)
$$\int_{\mathbf{R}^3} \frac{|\varphi(x)|^2}{|x - \bar{x}|}dx \leq C\|\varphi\|_{L^2(\mathbf{R}^3)}\|\nabla|\varphi|\|_{L^2(\mathbf{R}^3)} \leq C\|\nabla_A \varphi\|_{L^2(\mathbf{R}^3)}^2$$

which is obtained by using inequality (2.3).

Hence from this and from (5.3), we may assume, without loss of generality, that

(5.11)
$$\epsilon_i^n \underset{n \to +\infty}{\longrightarrow} \epsilon_i > 0, \quad i = 1, \ldots, N$$
$$\varphi_i^n \underset{n \to +\infty}{\longrightarrow} \varphi_i \quad \text{in} \quad H_A^1(\mathbf{R}^3) - \text{weak} \quad , \quad i = 1, \ldots N.$$

Then we use (5.5) to deduce that

(5.12)
$$L_A \varphi_i + V\varphi_i + (\rho * \frac{1}{|x|})\varphi_i - \int_{\mathbf{R}^{N3}} \rho(x,y)\frac{1}{|x-y|}\varphi_i(y)dy + \epsilon_i\varphi_i = 0$$

and therefore

$$\varlimsup_{n\to+\infty} \sum_i \epsilon_i^n \int_{\mathbf{R}^3} |\varphi_i^n|^2 dx =$$

$$= -\lim_{n\to\infty} \{\sum_i \int_{\mathbf{R}^3} |\nabla_A\varphi_i^n|^2 + V|\varphi_i^n|^2 dx +$$

$$+ \int\int_{\mathbf{R}^3\times\mathbf{R}^3} (\rho^n(x)\rho^n(y) - |\rho^n(x,y)|^2)\frac{1}{|x-y|}dxdy\}$$

and again by using Cauchy-Schwarz's inequality and (5.9) we obtain

(5.13) $$\varlimsup_{n\to+\infty} \sum_i \epsilon_i^n \int_{\mathbf{R}^3} |\varphi_i^n|^2 dx \le \sum_i \epsilon_i \int_{\mathbf{R}^3} |\varphi_i|^2 dx.$$

Obviously (5.13) implies the strong convergence of φ_i^n to φ_i in $L^2(\mathbf{R}^3)$. And this together with the weakly lower semicontinuity of $\mathcal{E}_A$ (easy to prove) allows us to conclude.

If $Z = N$ the above argument can be slightly modified to give the desired result (see [27]). The basic idea is that either $\int_{\mathbf{R}^3} \rho\, dx = N$ and then the compactness holds and there is a minimum or $\int_{\mathbf{R}^3} \rho\, dx < N$, and then we can apply Lemma 5.2 as above. We also refer to [27] for the case $N - 1 < Z < N$.

Lemma 5.2. *Assume that $Z = \sum_{i=1}^m z_i > \int_{\mathbf{R}^3} \rho\, dx$. Then the Schrödinger operator $H_A = L_A + V + \rho * \frac{1}{|x|}$ has infinitely many eigenvalues strictly below 1.*

Remark. Notice that the above result implies that there are also infinitely many eigenvalues of $H_A - K$ strictly below 1 if K is defined by (5.8) since K is nonnegative.

Proof of Lemma 5.2.
First step. Reduction to dimension 1. By means of the results obtained in Section 2 the proof of the lemma will be over as soon as we

prove that the operator

(5.14)
$$L = -\frac{d^2}{dz^2} + \frac{1}{2\pi}\int_{\mathbf{R}^2} V(x',z)e^{-|x'|^2/2}dx' +$$
$$+ \frac{1}{2\pi}\int_{\mathbf{R}^3}\int_{\mathbf{R}^2}\rho(y)\frac{1}{|x-y|}e^{-|x'|^2/2}dx'dy,$$

has infinitely many negative eigenvalues in $L^2(\mathbf{R})$. Indeed if this holds, we can show that for every $k \in \mathbf{N}$ there exists a subspace F_k of dimension k such that

(5.15)
$$\text{Max}\left\{ <H_Ag,g> |g \in F_K , g(x',z) = \right.$$
$$\left. = \frac{1}{\sqrt{2\pi}}e^{-|x'|^2/4}h(z), \int_{-\infty}^{+\infty}|h(z)|^2dz = 1\right\} < 0$$

and this implies the result stated in the lemma.

Second step. Spectral properties of L. In order to show that L has infinitely many negative eigenvalues let us define

(5.16)
$$V_1(z) = \frac{1}{2\pi}\int_{\mathbf{R}^2} V(x',z)e^{-|x'|^2/2}dx'$$

(5.17)
$$V_2(z) = \frac{1}{2\pi}\int_{\mathbf{R}^3}\int_{\mathbf{R}^2}\rho(y)\frac{1}{|z-y|}e^{-|x'|^2/2}dy\,dx'.$$

We will use a spectral result from [32] for one dimensional Schrödinger operators which says the following

Proposition 5.3. *Let H be a function in $L^2_{loc}(\mathbf{R})$ such that*

(i)
$$\int_a^{a+1}|H(z)|^2dz \xrightarrow[|a|\to+\infty]{} 0$$

and either

(ii)
$$\int_a^\infty H(z)dz = -\infty$$
for some a such that $H(z) \le 0$ for $z \ge a$

or

(iii)
$$\int_{-\infty}^{a} H(z)dz = -\infty$$

for some a such that $H(z) \le 0$ *for* $z \le a$.

Then, the operator $\underline{L} = -\frac{d^2}{dz^2} + H$ has infinitely many negative eigen-values.

In this case we will define $H = V_1 + V_2$ and then check that H satisfies all the assumptions of Proposition 5.3. This will allow us to conclude the proof of Lemma 5.2.

Third step. Behaviour of V_1. Let us define $v : \mathbf{R} \to \mathbf{R}^+$ by

$$v(z) = \int_{\mathbf{R}^2} \frac{e^{-|x'|^2/2}}{(|x'|^2 + |z|^2)^{1/2}} dx'.$$

It can be easily seen that $v \in C_o(\mathbf{R})$ and that $v(z)|z| \xrightarrow[|z| \to +\infty]{} 2\pi$. Then we observe that for all $\delta > 0$ there exists $C_\delta \ge 0$ such that, for all $z \ge 0$, $v(z)[(1 + \delta)z + C_\delta] \ge 2\pi$. Then

$$\frac{1}{\log R} \int_0^R v(z)dz \ge \frac{2\pi}{\log R} \int_o^R \frac{dz}{(1 + \delta)z + C_\delta} =$$
$$= \frac{2\pi}{\log R} \left(\log \left(\frac{C_\delta + (1 + \delta)R}{C_\delta} \right) \right) \frac{1}{1 + \delta} \xrightarrow[R \to +\infty]{} \frac{2\pi}{(1 + \delta)}$$

(5.18)
$$\lim_{R \to +\infty} \frac{1}{\log R} \int_o^R V_1(z)dz \le -Z,$$

and the same estimate holds when we integrate V_1 between $-R$ and 0, since $v(z)$ is even.

Finally, it is straightforward to check that $V_1 \in L^2_{loc}(\mathbf{R})$ and that

(5.19)
$$\lim_{|a| \to +\infty} \int_a^{a+1} |V_1(z)|^2 dz = 0.$$

Fourth step. Estimates on V_2. First we observe that since $\rho * \frac{1}{|x|} \in L^6(\mathbf{R}^3)$, $V_2(z) \in L^6(\mathbf{R})$ and therefore $V_2 \in L^2_{loc}(\mathbf{R})$ also. Moreover, this implies also that

$$\lim_{|a| \to +\infty} \int_a^{a+1} |V_2(z)|^2 dz = 0.$$

Hence the only thing that remains to be done in order to prove (ii) or (iii) is to compute $\int_o^R V_2(z)dz$ and $\int_{-R}^o V_2(z)dz$.

Let us then take $\chi \in \mathcal{D}^+(\mathbf{R}^N)$ such that $\chi \equiv 1$ in $[-1,1]$ and for all $R > 0$ define $\chi_R(\cdot)$ by $\chi(\frac{\cdot}{R})$. Then

$$(5.21) \qquad \int_{-R}^R V_2(z)dz \le \int_{-\infty}^{+\infty} V_2(x)\chi_R(z)dz.$$

Moreover by applying Plancherel equality we obtain

$$
\begin{aligned}
(5.22) \quad & \int_{-\infty}^{+\infty} V_2(z)\chi_{\mathbf{R}}(z)dz = \\
& = \frac{1}{2\pi} \int \int_{\mathbf{R}^3 \times \mathbf{R}^3} \rho(y)\frac{1}{|x-y|}e^{-|x'|^2/2}\chi_{\mathbf{R}}(z)dy\,dx'\,dz \\
& = \frac{1}{2\pi} \int_{\mathbf{R}^3} \hat{\rho}(\xi')\frac{\gamma}{|\xi'|^2}e^{-\alpha|\xi'|^2}\hat{\chi}(t)d\xi'\,dt = A_R
\end{aligned}
$$

where $\alpha, \gamma \in \mathbf{R}^+$, $\xi = (\xi', t) \in \mathbf{R}^2 \times \mathbf{R}$ and by $\hat{f}$ we denote the Fourier transform of f, defined by

$$\hat{f}(\xi) = \int_{\mathbf{R}^3} f(x)e^{-2\pi i x \cdot \xi}dx \quad , \quad \xi \in \mathbf{R}^3.$$

Then if we denote by $\phi_R(t)$ the function $|\hat{\chi}_R(t)|$ we see that

$$(5.23) \qquad \phi_R(t) = R\phi(tR) \quad , \quad \phi \equiv \phi_1 \quad \text{and}$$

$$(5.24) \qquad |A_R| \le C\|\hat{\rho}\|_{L^\infty(\mathbf{R}^3)} \int_0^{+\infty} \int_{-\infty}^{+\infty} \frac{re^{-\alpha r^2}}{r^2 + t^2} R\phi(tR)dr\,dt$$

where C is independent of ρ and of R.

We integrate by parts the integral in (5.24) to obtain

$$
\int_0^{+\infty} \int_{-\infty}^{+\infty} \frac{re^{-\alpha r^2}}{r^2 + t^2} R\phi(tR)dr \; dt =
$$

(5.25)
$$
= \alpha \int_0^{+\infty} \int_{-\infty}^{+\infty} \log(r^2 + t^2) re^{-\alpha r^2} R\phi(tR)dt \; dr
$$

$$
+ \int_{-\infty}^{+\infty} \log|t| R\phi(tR)dt.
$$

Moreover,

$$
\int_{-\infty}^{+\infty} \log|t| R\phi(tR)dt = \int_{-\infty}^{+\infty} \log\left|\frac{t}{R}\right|\phi(t)dt
$$

and thus

$$
\frac{1}{\log R} \int_{-\infty}^{+\infty} \log|t|\phi(tR)R \; dt =
$$

$$
= \int_{-\infty}^{+\infty} \frac{\log|t| - \log R}{\log R}\phi(t)dt \xrightarrow[R \to +\infty]{} -\|\phi\|_{L^1(\mathbf{R})}.
$$

On the other hand, the integral

$$
\int_0^{+\infty} \int_{-\infty}^{+\infty} \log(r^2 + t^2) re^{-\alpha r^2} \phi(tR)R \; dr \; dt
$$

can be studied separately in the sets $\{r^2 + t^2 \geq 1\}$ and $\{r^2 + t^2 \leq 1\}$:

(5.26)
$$
\left| \int\!\!\int_{r^2+t^2 \leq 1} \log(r^2 + t^2) re^{-\alpha r^2} \phi(tR)Rdt \; dr \right| \leq
$$

$$
\leq \int\!\!\int_{r^2+t^2 \leq 1} \log\left(\frac{1}{r^2}\right) e^{-\alpha r^2} r\phi(tR)R \; dt \; dr \leq C\|\phi\|_{L^1(\mathbf{R})}
$$

and

(5.27)
$$
\left| \int\!\!\int_{r^2+t^2 \geq 1} \log(r^2 + t^2) re^{-\alpha r^2} \phi(tR)R \; dt \; dr \right| \leq
$$

$$
\leq C\|\phi\|_{L^1(\mathbf{R})} + \frac{C}{R^2} \int_{-\infty}^{+\infty} \phi(t)t^2 dt.
$$

Finally, from (5.24), (5.25), (5.26), (5.27), (5.29) we obtain

(5.28)
$$\frac{1}{\log R}|A_R| \le$$

$$\le C\|\rho\|_{L^1(\mathbf{R})}\left(\left(1+\frac{1}{\log R}\right)\|\phi\|_{L^1(\mathbf{R})} + \frac{C}{R^2\log R}\int_{-\infty}^{+\infty}\phi(t)t^2dt\right)$$

and therefore, since C is independent of ρ and R, $A_R(\rho)/\log R$ is continuous in $L^1(\mathbf{R}^3)$. Hence, we may consider functions $\rho \in \mathcal{D}(R^3)$ such that $\|\rho - \underline{\rho}\|_{L^1(\mathbf{R}^3)}$ is small and study the limit of $A_R(\underline{\rho})/\log R$ as R goes to $+\infty$. For such a function $\underline{\rho}$ we have

$$\frac{1}{\log R}\left|\int_{\mathbf{R}^3}\underline{\hat{\rho}}(\xi')\widehat{\left(\frac{1}{|x|}\right)}e^{-\alpha|\xi'|^2}\hat{\chi}(tR)Rdt\,d\xi' + \right.$$

$$\left. - \int_{\mathbf{R}^3}\underline{\hat{\rho}}(0)\widehat{\left(\frac{1}{|x|}\right)}e^{-\alpha|\xi'|^2}\hat{\chi}(tR)R\,dt\,d\xi'\right| \le$$

$$\le \frac{C}{\log R}\int\int_{r^2+t^2\ge1}\frac{re^{-\alpha r^2}}{r^2+t^2}\phi(tR)R\,dt\,dr +$$

$$+ \frac{C}{\log R}\int\int_{r^2+t^2\le1}\frac{re^{-\alpha r^2}}{(r^2+t^2)^{1/2}}\phi(tR)R\,dt\,dr$$

$$\le \frac{C'}{\log R}\left(1+\frac{1}{R^2}\right)\xrightarrow[R\to+\infty]{}0.$$

Moreover,

$$\int_{\mathbf{R}^3}\underline{\hat{\rho}}(0)\widehat{\left(\frac{1}{|x|}\right)}e^{-\alpha|\xi'|^2}\hat{\chi}(tR)R\,dt\,d\xi' =$$

$$= \|\underline{\rho}\|_{L^1(\mathbf{R}^3)}\int_{\mathbf{R}^3}\frac{1}{|x|}e^{-|x'|^2/2}\chi_R(z)dz\,dx'.$$

Then we can use again the estimates obtained for v and V_1 obtained in the third step to prove that we can choose $\underline{\rho}$ and ξ to have

$$\lim_{R\to+\infty}\frac{1}{\log R}\int_{\mathbf{R}^3}\hat{\rho}(0)\widehat{\left(\frac{1}{|x|}\right)}e^{-\alpha|\xi'|^2}\hat{\chi}(tR)R\,dt\,d\xi'$$

as close to $4\pi\|\rho\|_{L^1(\mathbf{R}^3)}$ as we want. Therefore, either

$$\lim_{R\to+\infty}\left[\frac{1}{\log R}\int_{-R}^0 H(z)dz\right] \le (\|\rho\|_{L^1(\mathbf{R}^3)} - Z) < 0,$$

(and this proves (ii)) or we obtain (iii) in a similar way, since $H(z)$ is negative for $|z|$ large.

References

[1] F.V.Atkinson & L.A.Peletier, *Ground state of* $-\Delta u = f(u)$ *and the Emden-Fowler equation*, Arch. Rat. Mech. Anal. **93** (2) (1986), 103-127.

[2] T.Aubin, *Problèmes isopérimetriques et espaces de Sobolev*, J. Diff. Geom., **11** (1976), 575-598; announced in C.R. Acad. Sci. Paris **280** (1975), 279-282.

[3] J.Avron, I.Herbst, B.Simon, *Schrödinger operators with magnetic fields I. General interactions*, Duke Math. J. **45** (4) (1978), 847-883.

[4] J.Avron, I.Herbst, B.Simon, *Separation of center of mass in homogeneous magnetic fields*, Ann. Phys. **114** (1978), 431-451.

[5] J.Avron, I.Herbst, B.Simon, *Schrödinger operators with magnetic fields. 3. Atoms in homogeneous magnetic fields*, Comm. Math. Phys. **79** (1981), 529-572.

[6] A.Bahri, *Une méthode perturbative en théorie de Morse*, Thése d'Etat, Un4. P. et M. Curie, Paris, 1981.

[7] A.Bahri, P.L.Lions, *Remarques sur la théorie variationelle des points critiques et applications*. C.R. Acad. Sci. Paris **301** (1985), 145-147.

[8] R.Benguria, H.Brezis, E.H.Lieb, *The Thomas-Fermi-von Weizsacker theory of atoms and molecules*, Comm. Math. Phys. **79** (1981), 167-180.

[9] H.Berestycki, P.L.Lions, *Nonlinear scalar field equations*. Arch. Rat. Mech. Anal. **82** (1983), 313-345 and 347-375.

[10] H.Berestycki, T.Gallouët, O.Kavian, *Equations de champs scalaires euclidiens non linéaires dans le plan*, C.R.A.S. Paris **297**, série I (1983), 307-310.

[11] M.S.Berger, *On the existence and structure of stationary states for a nonlinear Klein-Gordon equation*, J.Funct. Anal. **9** (1978),

249-261.

[12] H.Brezis, E.H.Lieb, *Minimum action solutions of some vector field equations*, Comm. Math. Phys. **96** (1984), 97-113. See also E.H.Lieb, *Some vector field equations*, in Proc. Int. Conf. on Diff. Eq., Alabama, 1983.

[13] C.V.Coffman, *Uniqueness of the ground state solution for $\Delta u - u + u^3 = 0$ and a variational characterization of other solutions*, Arch. Rat. Mech. Anal. **46** (1972), 81-95.

[14] C.V.Coffman, *Ljusternik-Schnirelman theory: complementary principle and the Morse index*, preprint.

[15] J.M.Combes, R.Schrader, R.Seiler, *Classical bounds and limits for energy distributions of Hamilton operators in electromagnetic fields*, Ann. Phys. **111** (1978), 1-18.

[16] I.Ekeland, *Nonconvex minimization problems*, Bull. A.M.S. **1** *(3)* (1979), 443-474.

[17] M.J.Esteban, P.L.Lions, *Γ-convergence and the concentration-compactness method for some variational methods with lack of compactness*, Ricerche di Matem. **36** (1987),73-101.

[18] V.Fock, *Näherungsmethode zur lösing des quanten mechanischen Mehrkörperproblems*, Z.Phys. **61** (1930), 126-148.

[19] D.Hartree, *The wave mechanics of an atom with a non-coulomb central field. Part I. Theory and methods*, Proc. Cam. Phil. Soc. **24** (1928), 89-312.

[20] A.Jaffe, C.Taubes, *Vortices and Monopoles*, Birkhäuser.

[21] T.Kato, *Perturbation Theory for Linear Operators*, Springer, 1966.

[22] E.H.Lieb, *Thomas-Fermi and related theories af atoms and molecules*, Rev. Mod. Phys. **53** (1981), 603-641.

[23] E.H.Lieb, *Thomas-Fermi and Hartree-Fock theory*, in Proceedings of the International Congress of Mathematicians, Vancouver, Vol.2, 383-386.

[24] E.H.Lieb, B.Simon, *The Hartree-Fock theory for Coulomb systems*, Comm. Math. Phys. **53** (1977), 185-194.

[25] P.L.Lions, *The concentration-compactness principle in the calculus of variations: Part I*, Ann. IHP. Anal. Non Lin. **1** (1984), 109-145; Part 2, Ann. IHP. Anal. Non Lin. **1** (1984), 223-283.

[26] P.L.Lions, *The concentration-compactness principle in the calculus of variations. The limit case. Parts I and II: Part I. Rev. Mat. Iber. **I, 1** (1985), 145-200; Part 2. Rev Mat. Iber. **I, 2**

(1985), 45-121.

[27] P.L.Lions, *Solutions of Hartree-Fock equations for Coulomb systems*, Comm. Math. Phys. **109** (1987), 33-97.

[28] Z.Nehari, *On a nonlinear differential equation arising in nuclear physics*, Proc. R. Irish Acad. **62** (1963), 117-135.

[29] P.H.Rabinowitz, *Variational methods for nonlinear eigenvalue problems*, in Eigenvalues of Nonlinear Problems, Ed. Cremonese, Roma, 1974, 141-195.

[30] M.Reed, J.Simon, *Methods of Modern Mathematical Physics*. Academic Press, London-New York.

[31] G.H.Ryder, *Boundary value problems for a class of nonlinear differential equations*, Proc. J. Math. **22** (1967), 477-503.

[32] M.Schechter, *Operator Methods in Quantum Mechanics*, North-Holland, Amsterdam, 1981.

[33] B.Simon, *Functional Integration and Quantum Physics*, Academic Press, London-New York, 1979.

[34] J.C.Slater, *A note on Hartee's method*, Phys. Rev. **35** (1930), 210-211.

[35] W.A.Strauss, *Existence of solitary waves in higher dimensions* Comm. Math. Phys. **55** (1977), 149-162.

[36] G.Talenti, *Best constant in Sobolev inequality*, Ann. Mat. Pura Appl. **110** (1976), 353-372.

[37] C.Viterbo, *Indice de Morse des points critiques obtenus par minimax*, C.R. Acad. Sci. Paris, 1985.

Laboratoire d'Analyse Numérique

Université Pierre et Marie Curie

4, place Jussieu

F-75252 PARIS cedex 05

Ceremade

Université de Paris IX

Place de Lattre de Tassigny

F-75775 PARIS cedex 16

ON THE TOUCHING PRINCIPLE

ROBERT FINN   ENRICO GIUSTI*

Dedicated to Ennio De Giorgi on his sixtieth birthday

Abstract. The touching principle, used previously to obtain a-priori bounds depending only on domain of definition for functions satisfying mean curvature inequalities, is extended in two ways to classes of nonlinear elliptic operators.

1. Introduction. In 1974, Concus and Finn [1] proved the theorem:

Let $u(x)$ define a surface S over a ball $B_\delta(x_o) \subset \mathbf{R}^n$, and suppose that the mean curvature H of S satisfies an inequality $H \geq f(u)$, with $\lim_{u \to \infty} f(u) = \infty$. Then $u(x)$ is bounded above in B_δ, depending only on δ and on $f(u)$.

---

* The results of this paper were obtained while the authors were visiting the Max-Planck-Institut in Bonn.The first author was supported in part by a Fulbright award, and in part by a grant from the Nat. Sc. Found. Partial support for the second author came from the M.P.I.(40%)

The proof can be obtained by a procedure, described in [2] as the *touching principle*. Chose $\delta' < \delta$ and let $v(x)$ denote a lower hemisphere $S_{\delta'}$ of radius δ' lying over $B_{\delta'} \subset B_\delta$. Raise $S_{\delta'}$ until it lies entirely above S, then lower it until a first point of contact $(q, u(q))$ occurs. Since $v(x)$ is vertical on $\partial B_{\delta'}$, we have $q \notin \partial B_{\delta'}$. Thus $q \in B_{\delta'}$, $u(q) = v(q)$, and since $u(x) \leq v(x)$ in $B_{\delta'}$, we must have also $Du(q) = Dv(q)$. It follows that the curvature of any normal section of S at q cannot exceed the corresponding curvature of $S_{\delta'}$. Thus, $f(u_q) \leq H_q \leq 1/\delta'$, which is the mean curvature of $S_{\delta'}$. We conclude that $u(q) = v(q) \leq M_{\delta'} = \max\{u : f(u) \leq 1/\delta'\}$. Therefore $v(x_o) \leq v(q) \leq M_{\delta'}$, and there follows that for all $x \in B_{\delta'}$ there holds

$$u(x) \leq v(x) = v(x_o) + \delta' - \sqrt{\delta'^2 - x^2} \leq M_{\delta'} + \delta' - \sqrt{\delta'^2 - x^2}.$$

The stated result is now obtained by letting $\delta' \to \delta$.

The above proof is geometrical, and depends strongly on the particular properties of the mean curvature operator. We intend here to show that an analogue theorem for two fairly general classes of nonlinear differential operators. Central to our considerations is the following observation.

Proposition 1. *Let*

$$E[u] \equiv a_{ij}(x, u, Du)u_{x_i x_j} + P(x, u, Du)$$

be an operator with non negative principal part. Suppose $u(x) \leq v(x)$ in a ball B, and suppose there exists $q \in B$ at which $u(q) = v(q)$. Then $E[u] \leq E[v]$ at q.

PROOF. Since $u(x) \leq v(x)$ in B, and $u(q) = v(q)$, there follows $Du(q) = Dv(q)$, and thus the coefficients of $E[u]$ and of $E[v]$, and also the function P, coincide at q. We may thus chose the coordinates so that for both operators, $a_{ij} = \lambda_i \delta_{ij}$ at q, with $\lambda_i \geq 0$ by nonnegativity. For each fixed i we must have $u_{x_i x_i}(q) \leq v_{x_i x_i}(q) \leq v_{x_i x_i}(q)$, and the statement follows.

Our principal results take somewhat different forms, depending on the structure of the operator considered.

2. Divergence structure. We suppose the operator derivable from a symmetric variational principle; that is, we set

$$w = \sqrt{p_1^2 + p_2^2 + ... + p_n^2} \,,$$

and suppose that there exists $F(u, w)$ such that $a_{ij} = F_{p_i p_j}$.

Lemma 1. *Suppose* $F_{ww} > 0$, $F_w(u, 0) = 0$, $F_{uw} \leq 0$ *and* $|F_w| < M < \infty$ *for all* w *and all* $u \geq 0$. *Then for any ball* $B_\delta(x_o)$: $\{x : r \equiv |x - x_0| < \delta\}$ *and any* $v_o \geq 0$ *there exists* $v(r; v_o, \delta)$ *with*

$$v(0; v_o, \delta) = v_o, \quad v'(0; v_o, \delta) = 0,$$
$$v'(\delta; v_o, \delta) = \infty, \quad v''(r; v_o, \delta) > 0,$$

and such that

(1) $$\frac{1}{r^{n-1}} \frac{d}{dr} \{r^{n-1} F_w(v, v_r)\} \leq \frac{n}{\delta} F_w(0, \infty)$$

in B_δ.

PROOF. We seek to choose C such that

(2) $$\frac{1}{r^{n-1}} \frac{d}{dr} \{r^{n-1} F_w(v, v_r)\} \equiv C.$$

We thus require, for given $C > 0$,

(3) $$F_w(v, v_r) = \frac{1}{n} r C \; ; \; v_o = v(x_o).$$

Since $F_{ww} > 0$ and $F_w(v_o, 0) = 0$, we can regard (3) as a differential equation for v, which can be solved locally with initial data $v(0) = v_o$. For this solution we will have $v'(0) = 0$.

From (3) we obtain

(4) $$F_{ww} v'' + F_{uw} v' = \frac{1}{n} C$$

which yields that $v'' > 0$ in some initial interval, and hence that $v' > 0$, $v > 0$ in that interval. Therefore $v'' > C/n F_{ww}$, so that

v' increases as long as v can be continued as a solution. We assert that this interval is finite, and thus that $v'(R) = \infty$ at its end point $R < \infty$. For, by hypothesis, $F_w(v, v_r) < F_w(0, \infty)$, and thus if the solution can be continued to the value r, then by (3)

$$(5) \qquad\qquad r < \frac{n}{C} F_w(0, \infty).$$

From (5) it is immediate that $R \to 0$ as $C \to \infty$. On the other hand the solutions of (3) depend smoothly on C, and if $C = 0$ the unique solution with the given initial conditions is $v \equiv v_o$, for which $R = \infty$. We conclude that for given δ there exists $C > 0$ such that $R = \delta$. We then have, from (5):

$$(6) \qquad\qquad C < \frac{n}{\delta} F_w(0, \infty)$$

as it was to be shown.

Theorem 1. *Let $F(u, w)$ be as in Lemma 1, let $u(x)$ satisfy in B_δ an inequality*

$$(7) \qquad\qquad E[u] \equiv \frac{\partial}{\partial x_i} F_{p_i} \geq f(u) > 0$$

with $\lim_{u \to \infty} f(u) = \infty$. Let

$$u_o = \sup\{u : f(u) < \frac{n}{\delta} F_w(0, \infty)\},$$

and let

$$\hat{v}(r; \delta) = \sup_{0 \leq v_o \leq u_o} v(r; v_o, \delta).$$

Then

$$u(x) < \hat{v}(r; \delta)$$

in B_δ.

PROOF. Let $\delta' < \delta$. Then for sufficiently large v_o there will hold $v(r; v_o, \delta') > u(x)$ in $B_{\delta'}(x_o)$. Let v_o decrease until either $v_o = 0$ or a point of contact appears, whichever comes first. In the former case there is nothing further to prove. Suppose there exists $v_o > 0$ and $q \in$

$\bar{B}_{\delta'}$, such that $v(r; v_o, \delta') \geq u(x)$ in the ball $\bar{B}_{\delta'}$, and $v(r_q; v_o, \delta') = u(q)$ at q. We first observe that $q \notin \partial B_{\delta'}$, as $v_r = \infty$ there. Thus $q \in B_{\delta'}$, $v(q) = u(q)$, and $Dv(q) = Du(q)$. By Proposition 1 we have $E[v] \geq E[v]$ at q, while by Lemma 1

$$E[v] \equiv \frac{1}{r^{n-1}} \frac{d}{dr} \left(r^{n-1} F_w \right) \leq \frac{n}{\delta} F_w(0, \infty).$$

Thus

$$f(u) < \frac{n}{\delta} F_w(0, \infty)$$

so that $v(q) = u(q) < u_o$. Therefore $v_o = v(x_o) < u_o$, from which $u(x) < \hat{v}(r; \delta')$ in $B_{\delta'}$. The result now follows by letting $\delta' \to \delta$.

In the special case of the mean curvature operator $\left(F = \sqrt{1 + w^2}\right)$ the surface $\hat{v}$ becomes a lower hemisphere, and we retrieve the result described in the introduction.

3. General operators. We consider here an operator of the form

(8)
$$E[u] = a_{ij}(x, u, p) u_{x_i x_j} + P(x, u, p) ;$$
$$p = (p_1, p_2, ..., p_n), \quad p_i = u_{x_i}$$

subject to the ellipticity condition

(9)
$$a_{ij} \xi_i \xi_j > 0 , \quad |\xi|^2 = 1 ,$$

and we seek conditions under which a suitable majorant $v(r)$ can be constructed.

Lemma 2. *Let* $A = \{a_{ij}\}$, *set* $E_0[u] = a_{ij}(x, u, p) u_{x_i x_j}$, *and*

$$\mathcal{E}(p) = a_{ij} p_i p_j.$$

Suppose

(10)
$$\frac{\mathcal{E}(p)}{|p|^2} \leq \Phi(|p|)$$

with

(11)
$$\int_o^\infty \Phi(t)dt < \infty$$

and suppose

(12)
$$|p|tr(A) < M < \infty.$$

Then for every x_o and $\delta > 0$ there exist a function $v(r;\delta)$, $0 \leq r \equiv |x-x_o| \leq \delta$, with $v(0;\delta) = 0$, $v'(0;\delta) = 0$, $v''(r;\delta) > 0$, $v'(\delta;\delta) = \infty$, and a constant $C(\delta) < \infty$, such that $E_o[v] < C$ in $B_\delta(x_o)$. If in addition

(13)
$$\int_o^\infty t\Phi(t)dt < \infty,$$

then $v(\delta;\delta) < \infty$.

PROOF. For any function $v(r)$ we find

(14)
$$E_o[v] \equiv \frac{1}{r}\left(tr(A) - \frac{\mathcal{E}(p)}{|p|^2}\right)|p| + \frac{\mathcal{E}(p)}{|p|^2}|p|'$$

and setting $w = |p| = |v_x|$,

(15)
$$E_o[v] < \frac{1}{r}w\,tr(A) + \Phi(w)w'(r).$$

We choose for $w(r)$ the solution of the differential equation

(16)
$$\Phi(w)w'(r) = \frac{1}{\delta}\int_0^\infty \Phi(t)dt$$

with initial condition $w(0) = 0$. The solution can clearly be continued for all $r < \delta$, while $w(\delta) = \infty$. For this solution, we find $w(r) < Cr$ in $0 < r < \delta/2$. Using (12) we then find

(17)
$$\frac{1}{r}w\,tr(A) < C_1(\delta) < \infty$$

throughout B_δ. Thus from (15), (16),

$$(18) \qquad E_o[v] < C_1(\delta) + \frac{1}{\delta} \int_o^\infty \Phi(t)dt,$$

which provides the desired bound.

We have further

$$(19) \qquad v(r) = \int_o^r v_r dr =$$

$$= \frac{\delta}{\int_o^\infty \Phi(t)dt} \int_o^{v'(r)} t\Phi(t)dt$$

by (16), which provides the final statement of the lemma.

We note that for any given v_0, the function $v + v_o$ yields the identical properties and assumes the initial datum v_o.

Theorem 2. *Let $E[u] \equiv E_o[u] + P(u, Du)$, with $P(u, p) \le C < \infty$. Let $u(x)$ satisfy an inequality*

$$(20) \qquad E[u] \ge f(u)$$

in $B_\delta(x_o)$, with $f(u)$ as in Theorem 1, and suppose A satisfies (11) and (12).

Then for $0 \le r < \delta$ there exists $M(r; \delta) < \infty$ such that $u(x) < M(r; \delta)$ in $B_r(x_o)$. If in addition (13) holds, then there exists $M(\delta)$ such that $u(x) < M(\delta)$ throughout B_δ.

PROOF. We proceed as in the last section, using the function $v(r; \delta)$ of Lemma 2 as comparison function, and noting as above that any constant v_o can be added to v. We thus obtain from (18) and from the hypotheses on P,

$$(21) \qquad f(u) \le C_1(\delta) + \frac{1}{\delta} \int_o^\infty \Phi(t)dt + C + v(r) - v_o$$

which completes the proof.

An estimate for the right side of (21) can be obtained explicitly from the method in any particular case. We note, however, that for the mean curvature operator the procedure does not yields as good a result as that of the preceding section (even though $v(r)$ again appears as lower hemisphere). The reason is that the negative term in (14), which was neglected, makes a significant contribution. We improve the result in the next section by observing a remarkable simplification that can arise from the nonlinearity and that occurs - in particular - with the mean curvature operator.

4. The case of symmetric $E[u]$. It can happen that $tr(A)$ and $\mathcal{E}(p)$ depend only on $|p|$; further, cancellation can occur because the variables p_j in $\mathcal{E}(p)$ are identical to the p_i occurring in the coefficients. For the minimal surface operator we obtain, with $W^2 = 1 + |p|^2$,

$$a_{ij} = \frac{\delta_{ij}}{W} - \frac{p_i p_j}{W^3}$$

so that

$$\frac{E(p)}{|p|^2} = \frac{1}{W^3}, \quad v_r = \frac{r}{\sqrt{\delta^2 - r^2}},$$

$$\int_o^\infty \Phi(t)dt = 1,$$

$$(v - v_o - \delta)^2 + r^2 = \delta^2, \quad tr(A) = \frac{n}{W};$$

thus

$$E[v] = \frac{1}{r}\left(\frac{nW^2 - 1}{W^3}\right)v_r + \frac{1}{\delta} = \frac{n}{\delta},$$

which leads to a result identical to that of the introduction.

5. Solutions with Neumann type data. We consider solutions of (20) defined in a domain Ω, with normal derivative $\nu \cdot Du = \varphi(\tau)$ defined on $\Sigma = \partial\Omega$. We suppose that the operator $E[u]$ satisfies the assumptions of section 2 [section 3]. For $\delta > 0$, we denote by $v(r)$ the function defined in Lemma 1 [Lemma 2]. We have

Theorem 3. *If a ball B_δ can be situated so that $\Omega \cap B_\delta \neq \emptyset$ and $\nu \cdot Dv \geq \varphi(\tau)$ on $\sum \cap B_\delta$, then the bound of theorem 1 [theorem 2] holds throughout $\Omega \cap B_\delta$.*

PROOF. This follows by observing that B_δ can be moved so that $\nu \cdot Dv > \varphi(\tau)$, in which case the first contact point cannot occur on $\sum$. The bound in the shifted B_δ then follows as in the text, and B_δ can be returned to its original position without affecting the bound.

A case of particular interest occurs when there is an isolated conical point, with interior angle $2\alpha < \pi$. In this case Neumann data cannot be prescribed at the vertex; nevertheless the bounds of the theorems hold up to the vertex, without growth hypotheses on the solution $u(x)$. We see that by translating B_δ slightly into Ω so that the vertex is excluded, and if necessary decreasing its radius to maintain the boundary condition, then moving it continuously back to the original configuration.

In capillarity theory, in which $E[u]$ is the mean curvature operator, the boundary condition

$$\frac{1}{W}\nu \cdot Du = \varphi(\tau)$$

appears. The natural analogue for a variational problem with integrand $F(u, |p|)$ is the expression

$$\nu_i F_{p_i} = \varphi(\tau).$$

Under the ellipticity condition $F_{pp} > 0$ this case yield again the same result, with an analogue discussion. In fact, by using the condition $|F_p| < M$, the bounds can be obtained under very weak hypotheses on boundary regularity, as in [3], Chapters 6 and 7.

We wish to thank M. Meier for helpful comments.

References

[1] P.Concus and R.Finn, *On capillarity free surfaces in a gravitational field*, Acta Math.**132** (1974), 207-223.

[2] R.Finn, *Comparison principles in capillarity*, Abschluss Band des Sonderforschnungsbereiches 72, Univ. Bonn, to appear.

[3] R.Finn, *Equilibrium Capillary Surfaces*, Springer Verlag, 1986.

Department of Mathematics

Stanford University

California

Istituto Matematico

Università di Firenze

Viale Morgagni 67-A

I-50134 FIRENZE

GENERALIZED SOLUTIONS AND CONVEX DUALITY IN OPTIMAL CONTROL

WENDELL H. FLEMING

Dedicated to Ennio De Giorgi on his sixtieth birthday

1. Introduction. In the 1930's L. C. Young introduced the concept of generalized curve. His purpose was to obtain a class of objects in which calculus of variations problems have solutions, without convexity assumptions on the corresponding variational integrands. Young's ideas were subsequently adapted in optimal control theory, in terms of what are called "relaxed controls." See [10], [12]. About 1950 Young [11] introduced a concept of generalized surface, which was not a simple multidimensional generalization of that of generalized curve. A smooth surface of dimension k in Euclidean space R^n was identified with a particular kind of nonnegative measure on $R^n \times G_k^n$, where G_k^n is the associated Grassmann manifold. The generalized surfaces consisted of all nonnegative measures on $R^n \times G_k^n$ with compact support. This formulation allowed for the use of tools of linear functional analysis in calculus of variations, and in particular use of duality between spaces of continuous functions and spaces of measures.

Later Vinter and Lewis [8] [9] used methods adapted from those of L. C. Young together with ideas from convex analysis. In this work, which concerns deterministic control problems, the "generalized solutions" are objects in a certain convex, w^*-compact set $\mathcal{M}$ of measures. The generalized control problem is to minimize a linear w^*-continuous function on $\mathcal{M}$. The minimum cost is equal to the supremum of smooth subsolutions to the Hamilton-Jacobi (or dynamic programming) equation. This result expresses the equality of the minimum in the generalized control problem and the maximum for its dual problem.

In this paper we shall outline a simpler method to obtain the Vinter-Lewis results. The method is slightly adapted from Fleming-Vermes [3][4], which are concerned with similar results for stochastic optimal control of diffusion processes. Then, in Section 6, we shall see that the same methods apply to a particular class of piecewise deterministic control problems, in which the state dynamics depend on parameters which vary randomly according to a Markov chain. Related work on control of piecewise deterministic processes appears in Vermes [7].

2. Deterministic Control Problem. We consider the following familiar optimal control problem, on a finite time interval $t \leq s \leq T$. The dynamics of the process x_s being controlled are governed by the differential equation

$$(2.1) \qquad \dot{x}_s = f(s, x_s, u_s), \quad t \leq s \leq T,$$

with initial data $x_t = x$. The control process u_s takes values in a control space Y. The objective is to minimize

$$(2.2) \qquad J^u(t, x) = \int_t^T l_0(s, x_s, u_s)\, ds.$$

It is assumed that the control space Y is compact. Moreover, f is bounded and Lipschitz continuous and l_0 is continuous.

Let us introduce the optimal cost function

$$(2.3) \qquad \psi(t, x) = \inf_u J^u(t, x).$$

If l_0 is Lipschitz continuous, then ψ is also Lipschitz continuous on $[T_0, T]$ for any $T_0 > -\infty$. By Rademacher's Theorem ψ is differentiable almost everywhere. Let

$$(2.4) \qquad A\psi(t,x,y) = \frac{\partial \psi}{\partial t}(t,x) + \nabla_x \psi(t,x) \cdot f(t,x,y).$$

At each (t,x) where ψ is differentiable

$$(2.5) \qquad A\psi(t,x,y) + l_0(t,x,y) \geq 0, \quad \forall y \in Y.$$

See [2, p.83]. In fact, the Hamilton-Jacobi (or dynamic programming) equation

$$(2.6) \qquad 0 = \min_{y \in \overline{Y}} [A\psi(t,x,y) + l_0(t,x,y)]$$

holds at each point (t,x) of differentiability. We shall use (2.5) but not (2.6). The optimal cost ψ is the unique viscosity solution of (2.6) with the data $\psi(T,x) = 0$. Moreover $\psi \geq \phi$ for any Lipschitz continuous ϕ which satisfies (2.4) almost everywhere with $\phi(T,x) \leq 0$. See Lions [6]. While we do not use this result, it is related to Theorem 2 stated in Section 4.

We shall use the following technical lemma. Fix $T_0 < T$ and let $\| \ \|$ denote the sup norm.

Lemma 1. *Assume that l_0 is bounded and Lipschitz continuous. Then there exist ψ_n of class C^∞ on $[T_0, T] \times R^n$ such that:*

$$\psi_n(T,x) = 0, \quad \|\psi_n\| + \|\psi_{nt}\| + \|\nabla_x \psi_n\| \leq C, \ \lim_{n \to \infty} \|\psi - \psi_n\| = 0,$$

$$A\psi_n + l_0 \geq -\delta_n \quad \text{if} \quad T - t \geq \frac{1}{n}, \quad \text{where} \quad \lim_{n \to \infty} \delta_n = 0.$$

Let us merely indicate the proof and refer to [3, Lemma 3.2] or [4, Lemma 5.1] for details. We extend ψ for $t > T$ by

$$\psi(T + s, x) = \psi(T - s, x), \quad s > 0.$$

Let $\rho_n(t,x)$ be an approximation to the identity, with sptρ_n contained in the ball of radius $\frac{1}{n}$ and center $(0,0)$. We take $\psi_n = \psi * \rho_n$, where $*$ denotes convolution. Since $\nabla_x \psi$ is bounded,

$$\|(A\psi) * \rho_n - A(\psi * \rho_n)\| \leq K\theta_n$$

where $\theta_n \to 0$ as $n \to \infty$. Lemma 1 then follows routinely.

In [5, Lemma 5.1] a similar regularization technique was used in a multidimensional calculus of variations setting.

3. The Generalized Control Problem. Let us introduce the following notation. We fix T_0 and T, with $-\infty < T_0 < T < \infty$. Let

$$\Sigma = [T_0, T] \times R^n, \quad S = \Sigma \times Y.$$

Let $C_b(S)$ denote the space of real-valued, bounded continuous functions $l(t, x, y)$ on S, and $C_0(S)$ the space of continuous l such that

$$\lim_{|x| \to \infty} l(t, x, y) = 0,$$

uniformly with respect to (t, y). Let $\mathcal{M}(S)$ denote the space of finite signed measures on S. The scalar product

$$\langle l, M \rangle = \int_S l \, dM$$

provides a duality between $C_0(S)$ with the uniform topology and $\mathcal{M}(S)$ with the w^*-topology.

We define

(3.1)
$$C_b^1(\Sigma) = \{ \phi : \phi, \phi_t, \phi_{x_i} \in C_b(\Sigma), \quad i = 1, \ldots, n \},$$
$$D = \{ \phi \in C_b^1(\Sigma) : \phi(T, x) = 0 \quad \forall x \in R^n \},$$

Let $\mathcal{A} = \mathcal{A}(t)$ denote the space of measurable Y-valued functions u_s on $[t, T]$. Given an initial state $x = x_t$, the solution to (2.1) defines for each $u \in \mathcal{A}$ a measure $M^u \in \mathcal{M}(S)$ as follows:

(3.2)
$$\langle l, M^u \rangle = \int_t^T l(s, x_s, u_s) \, ds, \quad \forall l \in C_b(S).$$

We have

(3.3)
$$M^u \geq 0, \quad \|M^u\| = T - t,$$

and by the fundamental theorem of calculus

(3.4)
$$\phi(t, x) = -\langle A\phi, M^u \rangle, \quad \forall \phi \in D.$$

Let
$$\mathcal{M}^s = \{M^u : u \in \mathcal{A}\}$$
$$\mathcal{M}^w = \{M \in \mathcal{M}(S) : (3.5)(a) \text{ and } (b) \text{ hold}\},$$

where

(3.5)
$$(a)\, M \geq 0, \quad \|M\| \leq T - t$$
$$(b)\, \phi(t, x) = -\langle A\phi, M \rangle, \quad \forall \phi \in D.$$

Note that the sets $\mathcal{M}^s, \mathcal{M}^w$ depend on the initial data (t, x). We think of a measure $M^u \in \mathcal{M}^s$ as a control in the *usual* (or *strong*) sense, identified with $u \in \mathcal{A}$. A measure $M \in \mathcal{M}^w$ is regarded as a control in a *generalized* (or *weak*) sense. By (3.3) and (3.4), $\mathcal{M}^s \subset \mathcal{M}^w$. For $M \in \mathcal{M}^w$ we actually have $\|M\| = T - t$ in (3.5)(a). In fact, by considering $\phi = \phi(t)$ it can be seen that the projection $(t, x, y) \to t$ maps M onto Lebesgue measure restricted to $[t, T]$. We also consider

(3.6)
$$\mathcal{M}^g = cl\mathcal{M}^s,$$

where cl denotes w^*-closure. It can be shown that $M \in \mathcal{M}^g$ if and only if M has the disintegration

$$M(ds, dx, dy) = \lambda(ds)\delta_{x_s}(dx)\nu_s(dy)$$

where λ is Lebesgue measure on $[t, T]$, δ_ξ is the Dirac measure at ξ, ν_s is a probability measure on Y and x_s is the solution of the corresponding relaxed control problem. Thus, we call $M \in \mathcal{M}^g$ a *relaxed* control. We will see (Corollary 1) that $\mathcal{M}^w$ is the w^*-convex closure of $\mathcal{M}^s$. Once this is established the distinctions we have made between functions in C_b and C_0 become irrelevant. In fact, each $M \in \mathcal{M}^w$ has support in $\Gamma(t, x) \times Y$, where $\Gamma(t, x)$ is the closure of the reachable set using all possible $u \in \mathcal{A}$.

Theorem 1. $\psi = \psi^w$, *where*

(3.7)
$$\psi^w = \min_{\mathcal{M}^w} \langle l_0, M \rangle.$$

PROOF. It suffices to consider Lipschitz $l_0 \in C_0(S)$. Since $\mathcal{M}^s \subset \mathcal{M}^w$, clearly $\psi \geq \psi^w$. To prove the opposite inequality, choose ψ_n as in Lemma 1, and let

$$\phi_n = \psi_n - \delta_n(T - t) - \|\psi_n - \psi\|.$$

Then $\phi_n \in D$ and

$$A\phi_n + l_0 = A\psi_n + \delta_n + l_0 \geq 0$$

for any (s, ξ, y) with $s \leq T - \frac{1}{n}$. Moreover

$$\langle l_0, M \rangle = -\langle A\phi_n, M \rangle + \langle l_0 + A\phi_n, M \rangle$$
$$\geq \phi_n(t, x) - \|l_0 + A\phi_n\| M \left(\left[T - \frac{1}{n}, T \right] \times R^n \right)$$
$$= \phi_n(t, x) - \frac{1}{n} \|l_0 + A\phi_n\|$$

Since this is true for each $n = 1, 2, \ldots,$

$$\langle l_0, M \rangle \geq \psi(t, x), \quad \forall M \in \mathcal{M}^w.$$

Thus $\psi^w \geq \psi$, which proves Theorem 1.

Corollary 1. $\mathcal{M}^w$ *is the w^*-convex closure of $\mathcal{M}^s$.*

PROOF. Suppose not. Since $\mathcal{M}^w$ is convex and w^*-compact, a separation theorem implies that $l_0 \in C_0(S)$ and a exist such that

$$\langle l_0, M^u \rangle \geq a \quad \forall u \in \mathcal{A}$$
$$\langle l_0, M^0 \rangle < a,$$

where $M_0 \in \mathcal{M}^W - cl(co\mathcal{M}^s)$. But then $\psi(t, x) \geq a$, $\psi^w(t, x) < a$, a contradiction.

Problems with state constraints. Let $Q \subset \Sigma$ be closed, and let

$$\mathcal{M}_Q^s = \{M^u : (s, x_s) \in Q, \quad \forall s \in [t, T]\}.$$

To avoid trivialities assume that $\mathcal{M}_Q^s$ is not empty. Let

$$\mathcal{M}_Q^g = \{M \in \mathcal{M}^g : \text{spt} M \subset Q \times Y\}$$
$$\mathcal{M}_Q^w = \{M \in \mathcal{M}^w : \text{spt} M \subset Q \times Y\}.$$

Let $\psi_Q, \psi_Q^g, \psi_Q^w$ denote the infimum of $\langle l_0, M \rangle$ over $\mathcal{M}_Q^s, \mathcal{M}_Q^g, \mathcal{M}_Q^w$ respectively. We claim that

(3.8)
$$\psi_Q^g = \psi_Q^w.$$

This can be seen as follows. By Corollary 1 and the fact that $\mathcal{M}^g = cl\mathcal{M}^s$, given $M \in \mathcal{M}^w$ there exists a probability measure α on $\mathcal{M}^g$ such that

$$\langle l, M \rangle = \int_{\mathcal{M}^g} \langle l, \mu \rangle \alpha(d\mu), \quad \forall l \in C_0(S).$$

In particular, we may take $l = d_Q$, where $d_Q(t, x) = \text{dist}\,[(t, x), Q]$. For $M \in \mathcal{M}_Q^w$, we have $\langle d_Q, M \rangle = 0$, which implies $\langle d_Q, \mu \rangle = 0$ for α-almost all μ. Thus, $\mu \in \mathcal{M}_Q^g$ for α-almost all μ. This implies (3.8). Whether $\psi_Q = \psi_Q^g$ is a question in the theory of relaxed controls. It holds under suitable additional assumptions.

4. Duality. Let us now consider the problem from the viewpoint of convex analysis. We call $\phi \in C_b^1(\Sigma)$ a smooth subsolution of the Hamilton-Jacobi equation if

(4.1)
$$A\phi + l_0 \geq 0 \quad \text{and} \quad \phi(T, t) \leq 0$$

for all $x \in R^n$. Using a duality theorem of Rockafellar the optimal cost function ψ in (2.3) is obtained as the supremum of smooth subsolutions ϕ subject to the additional restriction $\phi(T, x) = 0$ (rather than ≤ 0). In fact,

Theorem 2. *For each* $(t, x) \in \Sigma$

(4.2) $\quad \psi(t, x) = \sup \{\phi(t, x) : \phi \in D, \phi \text{ a smooth subsolution}\}$

For a proof we refer to [3, Theorem 4.1] and [4, Theorem 1]. Very similar duality arguments were used earlier in [7, Section 5]. Roughly speaking the idea of the proof is as follows. Let

$$h(M) = \begin{cases} \langle l_0, M \rangle, & M \in \mathcal{M}^w \\ +\infty, & M \in \mathcal{M}(S) - \mathcal{M}^w. \end{cases}$$

Then $\psi^w(t, x)$ is the minimum of the convex, lower semicontinuous function h. The essential step is to identify the right side of (4.2) with the supremum of a dual concave function on $C_0(S)$. One then uses the fact that $\psi = \psi^w$ by Theorem 1.

The proof that $\psi^w(t, x)$ is the supremum of smooth subsolutions does not depend on the particular structure of the deterministic control problem. A similar result holds for a wide class of controlled Markov stochastic processes, including piecewise deterministic processes [7] and diffusion processes [4]. We shall describe a particular case in Section 6.

5. Inclusion of Terminal Costs.

Let us now suppose that the objective is to minimize

$$(5.1) \qquad J^u(t, x) = \int_t^T l_0(s, x_s, u_s)\, ds + L(x_T).$$

In previous sections we took $L \equiv 0$ (see (2.2)). If L is smooth this problem can be reduced to the previous one by the device of replacing l_0 by $l_0 + \nabla L \cdot f$. However, the problem with terminal costs can also be considered directly, by a modification of the methods above which has some appeal conceptually. We merely indicate this modification, and refer to [4, Sec. 8] for details. The idea is to introduce a pair of measures (M, N), where $N \geq 0$ is a measure on R^n with $\|N\| = 1$. We require again (3.5)(a), but instead of (3.5)(b) we require

$$(5.2) \qquad \phi(t, x) = -\langle A\psi, M \rangle + \langle \phi(T, \cdot), N \rangle, \quad \forall \phi \in D.$$

For $u \in \mathcal{A}, M^u$ is defined again by (3.2) and $N^u = \delta_{x_T}$ is a Dirac measure concentrated at the final state x_T. One can think of the pair of measures $(-\delta_{x_t}, \delta_{x_T})$ as corresponding to the boundary of the curve $\{(s, x_s) : t \leq s \leq T\}$. Correspondingly, one could interpret $(-\delta_{x_t}, N)$ as a kind of boundary in the generalized formulation, in analogy with Young's theory of generalized curves and surfaces. Since $x_t = x$ is fixed as initial data, reference to δ_{x_t} can be omitted.

In the definition of smooth subsolution, it is now required that

$$(5.3) \qquad A\phi + l_0 \geq 0 \quad \text{and} \quad \phi(T, x) \leq L(x)$$

for all $x \in R^n$. The optimal cost $\psi(t, x)$ is again the supremum of smooth subsolutions. See [4, Theorem 5].

Problems with a terminal constraint $x_T \in C$, where $C \subset R^n$ is closed can be treated by the method described at the end of Section 3. One observes that $\operatorname{spt} N \subset C$ if and only if $\langle d_C, N \rangle = 0$, where $d_C(x) = \operatorname{dist}(x, C)$.

6. Control Problems with Markov Chain Parameters.

To adapt the generalized formulation described above for deterministic control problems to stochastic control, one proceeds roughly as follows. For each $y \in Y$, a family A^y of linear operators is given, such that A^y generates a Markov process for each constant control y. If y is replaced by a Y-valued stochastic process u_s belonging to a suitable admissible class, then x_s becomes the controlled state process. The objective is to minimize an expectation

$$(6.1) \qquad J^u(t, x) = E_{tx} \int_t^T l_0\left(s, x_s, u_s\right) ds,$$

where the subscript on the expectation indicates the initial data $x_t = x$. See [1]. For example, for controlled diffusions in R^n the dynamics of x_s are described by a stochastic differential equation [2].

We let

$$(6.2) \qquad A\psi(t, x, y) = \frac{\partial \psi}{\partial t} + A^y \psi(t, x).$$

The identification of u with the measure M^u is to be made taking the expectation of the right side of (3.2). The Dynkin formula

$$(6.3) \qquad \psi(t, x) = -E_{tx} \int_t^T A\psi\left(s, x_s, u_s\right) ds + E_{tx} \psi\left(T, x_T\right),$$

replaces the fundamental theorem of calculus used for deterministic control to obtain (3.4).

To carry out this program, two results are needed. The first is that $\psi = \psi^w$; to prove this, particular features of the stochastic control problem need to be used. The second is the analogue of the convex duality arguments mentioned in Section 4; this works under quite general hypotheses.

Let us consider here only the following particular stochastic control problem, for which the results for the deterministic case are quite easily modified. Instead of (2.1) let us consider the dynamics

$$(6.4) \qquad \dot{x}_s = f\left(s, x_s, z_s, u_s\right), \quad t \leq s \leq T,$$

where z_s is a continuous time, finite state Markov chain. Thus, the dynamics evolve according to one of finitely many possible deterministic regimes, with random jumps from one regime to another. The pair (x_s, z_s) must now be regarded as the state, and not x_s itself. We now take

$$\Sigma = [T_0, T] \times R^n \times Z, \quad S = \Sigma \times Y,$$

where Z is the state space of the Markov chain z_s. Let

$$(6.5) \qquad A\phi(t, x, z, y) = \phi_t + \nabla_x \cdot f + B_z \phi,$$

where B is the generator of z_s:

$$Bg(z) = \sum_{\zeta \neq z} q_{z\zeta}[g(\zeta) - g(z)], \quad z \in Z.$$

The numbers $q_{z\zeta} \geq 0$ represent transition rates from state z to state ζ. The dynamic programming equation (2.6) becomes a system of first-order Hamilton-Jacobi equations in the variables (t, x), coupled in the zeroth order terms through the operator B. The arguments used in Sections 2-5 can be repeated with the evident notational changes to account for the additional discrete variable z. As control processes, Y-valued processes are admitted which are progressively measurable with respect to the collection $\mathcal{F}_s^z$ of σ- algebras generated by the z_s process. The analogue of (2.4) is

$$(6.6) \qquad A\psi(t, x, z, y) + l_0(t, x, z, y) \geq 0, \quad \forall y \in Y,$$

if $\psi(\cdot, \cdot, z)$ is differentiable at (t, x) for each $z \in Z$. This follows from the dynamic programming inequality

(6.7)

$$\psi(t, x, z) \leq E_{txz}\left\{ \int_t^{t+h} l_0\left(s, x_s, z_s, y\right) ds + \psi\left(t + h, x_{s+h}, z_{s+h}\right) \right\},$$

taking $u_s \equiv y$ on $[t, t + h]$.

References

[1] W. H. Fleming, *Optimal Control of Markov Processes*, Proc. Internat. Congress of Math., Warsaw (1983).

[2] W. H. Fleming and R. W. Rishell, *Deterministic and Stochastic Optimal Control*, Springer-Verlag, 1975.

[3] W. H. Fleming and D. Vermes, *Generalized solutions in the optimal control of diffusions*, in "IMA Volumes in Math. and Applica.", eds. W. H. Fleming and P.-L. Lions, vol. 10, Springer-Verlag, 1987, 119-127.

[4] W. H. Fleming and D. Vermes, *Convex duality approach to the optimal control of diffusions*, submitted to SIAM J. on Control and Optimiz.

[5] W. H. Fleming and L. C. Young, *A generalized notion of boundary*, Trans. Amer. Math. Soc. **76** (1954), 457-484.

[6] P.-L. Lions, *Generalized solutions to Hamilton-Jacobi equations*, Pitman Research Notes in Math. **69** (1982).

[7] D. Vermes, *Optimal control of piecewise deterministic processes*, Stochastics **14** (1985), 165-208.

[8] R. B. Vinter and R. M. Lewis, *The equivalence of strong and weak formulations for certain problems in optimal control*, SIAM J. Control **16** (1978), 546-570.

[9] R. B. Vinter and R. M. Lewis, *A necessary and sufficient condition for optimality of dynamic programming type, making no a priori assumptions on the controls*, SIAM J. Control **16** (1978), 571-583.

[10] L. C. Young, *Generalized curves and the existence of an attained absolute minimum in the calculus of variations*, Comptes Rendus de la Société des Sciences et des Lettres de Varsovie **30** (1937), 212-234.

[11] L. C. Young, *Surfaces paramétriques généralisées*, Bull. Soc. Math. France **79** (1951), 59-85.

[12] L. C. Young, *Lectures on the Calculus of Variations and Optimal Control Theory*, Saunders, Philadelphia, 1969.

Division of Applied Mathematics

Brown University

PROVIDENCE (RI) 02912

MODELS OF SELF-DESCRIPTIVE SET THEORIES

Marco Forti   Furio Honsell

Dedicated to Ennio De Giorgi on his sixtieth birthday

Introduction. It is well known that Zermelo-Fraenkel set theory has a limited self-descriptive power. In fact most of the basic set-theoretic relations, operations and properties (e.g. membership, union, sethood) cannot be represented as sets since the classes which correspond to them are too large. Many attempts have been made to define set theories consistent relative to ZF, which allow as sets many interesting classes having the size of the universe. Apart from W.V.O.Quine's NF [16], whose consistency strength is still unknown, we can mention the theories (all equiconsistent with ZF) considered by A.Church [1], H.Friedman [11], E.Mitchell [14], and A.Oberschelp [15]. These, however, are in some sense unsatisfactory, since each of them is not closed under some basic construction.

A very interesting class of set-theoretical models, closed under many basic operations but still possessing a lot of large sets, was introduced by R.J.Malitz in his thesis [13]. Unfortunately, he considered only wellfounded universes, thus utterly weakening the actual power of his construction. In fact the most interesting properties of the models he defined depended on a conjecture which is now almost completely disproved (see section 2).

However, by simply performing Malitz's costruction inside a nonwellfounded universe verifying a suitable "Free Construction Principle", the first author [7] succeeded in proving the consistency, relative to ZF, of the axiom schema GPK. This is a general "Positive Comprehension Schema", which postulates the existence of the set $\{x|\Phi(x)\}$ for any non-negative formula Φ (for a precise definition of the generalized positive comprehension GPK, see [7] and section 3 below).

On a different ground, wider self-referential power can also be achieved by considering non purely set-theoretic foundational theories. In these theories basic objects such as *properties, relations and operations* are considered as primitive notions and are not identified with their usual set-theoretic reductions. We refer in particular to the work inspired by E.De Giorgi and developed since the late seventies by him and several researchers attending his Seminar on Logic and Foundations at the Scuola Normale Superiore, Pisa (see [2], [3], [4]).

In this paper we discuss in depth, from topological and set-theoretical viewpoints, the constructions of [7]. Using techniques from the theory of infinitary trees we provide a number of counter-examples to Malitz's conjecture. We also generalize the construction of [7] to universes with (universe-many) urelements. The models thus obtained provide a suitable environment for modelling theories for the Foundations of Mathematics like [2], [3], [4]. In section 3 we explore the possibility of modelling significant sublists of the strong axioms of [2,§VI]. Theorems 3.3 and 3.4 are first results in this direction; a more detailed account of this will be given in [10].

It is well known that comprehension principles entailing the existence of universe-sized sets are often inconsistent with principles of choice (see [6]). In the last section of this paper we discuss various classical choice principles in connection to our models. We obtain *inter alia* the relative consistency of the axiom of choice and of the well-ordering principle with respect to the generalized positive comprehension schema GPK plus an axiom of infinity.

Finally the authors would like to express how deeply they are indebted to Ennio De Giorgi for his constant help and encouragement throughout their set-theoretic and foundational research.

1. The Basic Construction. We work in a non-wellfounded Zermelo-Fraenkel like set theory with urelements. We assume the axiom of choice and, instead of the axiom of foundation, a suitable *free construction principle*.

The axioms of our set theory are the following [1]:

ZF_o^- *Pairing* Pair, *Union* Un, *Power-set* PS, *Replecement* Rpl, *and infinity* Inf *as in Zermelo-Fraenkel's theory* ZF.

AC   *Zermelo's axiom of choice.*

WE   *Weak Extensionality with respect to a (possibly empty) set U of atoms, i.e.*

$$(x \in U \to t \notin x) \ \& \ (\exists x \notin U \forall t \ t \notin x) \&$$
$$(x \notin U \ \& \ y \notin U \ \& \ \forall t(t \in x \leftrightarrow t \in y). \to .x = y).$$

FC   *Unique Free Construction with respect to a set U of atoms, i.e. Given a function* $f : X \to \mathcal{P}(X) \cup U$ *such that* $f(a) = a$ *for any* $a \in X \cap U$, *there is a unique function* $g : X \to T$ *verifying* [2]

$$g(x) = \begin{cases} f(x) & \text{if } f(x) \in U \\ \hat{g}(f(x)) & \text{otherwise.} \end{cases}$$

The axiom FC generalizes the free construction axiom X_1 of [8] to set theories with atoms. A straightforward modification of the argument in the proof of Theorem 3 of [9] yields:

Theorem 1.0. *Given any model* $\mathcal{N}$ *of* $ZF_o^- + WE$ *there is, up to isomorphism, exactly one (inner) model* $\mathcal{R}$ *of* $ZF_o^- + WE + FC$ *with the same atoms and the same well-founded sets of* $\mathcal{N}$.

Therefore, as far as relative consistency and mutual interpretability are concerned, our theory $ZF_0^- + WE + FC$ is equivalent to ZF. The same holds for any extension of both theories obtained by

---

[1] For definitions and standard results on set theory we usually refer to [12]; when we adhere instead to the notation of [2] or [8], we shall mention it explicitly.

[2] We denote by $\hat{g}(x)$ the image of x under the function g; more generally, we put throughout the paper $\hat{x}(y) = \{v | \exists u \in y (u,v) \in x\}$.

adding any large cardinal axiom or any choice principle, in particular
AC (cfr [9]).

An easy consequence of the axiom FC is *the absence of nontrivial
atom-preserving $\in$-homomorphism*. This property of *atomic rigidity*,
analogous to the rigidity property implied by the axiom of Foundation, will be of some importance in the sequel, so we formulate it
explicity:

AR *If T is transitive and $h : T \rightarrow S$ verifies $h(x) = x$ for $x \in T \cap U$
     and $h(x) = \hat{h}(x)$ for $x \in T \backslash U$, then h is the identity on T.*

In particular, AR implies the following axiom of *strong extensionality up to atoms:*

SextA- *If two transitive sets are $\in$-isomorphic under an isomorphism
        which leaves any atom fixed, then they are equal.*

In defining our models, we shall use topological notions. [3]
In fact, we need a *uniform topology* with a *nested uniformity basis*
made up by *equivalences*. To this aim, we fix a regular cardinal κ and
we assume that the set U of the atoms carries a κ-*hypermetric*, i.e. a
distance $d : U^2 \rightarrow {}^*R$, where ${}^*R$ is any nonstandard model of the real
numbers with cofinality κ, satisfying the following properties:

(i) $d(a, b) = 0$ iff $a = b$;

(ii) $d(a, b) = d(b, a) \geq 0$ for all $a, b \in U$;

(iii) $d(a, b) \leq \max\{d(a, c), d(b, c)\}$ for all $a, b, c \in U$.

We are interested only in the uniform structure induced by d. We
assume ${}^*R$ to be a model of the reals only for sake of suggestivity. Actually, all that is needed is simply an ordered set od type $1 + \eta$, with
cof $\eta^* = \kappa$.

Therefore we fix a *strictly decreasing* κ-sequence $< \epsilon_\alpha | \alpha < \kappa >$
with infimum 0, and we define for any ordinal $\alpha \leq \kappa$ the α-*equivalence*
$\approx_\alpha$ on U by

(1.1) $\alpha \approx_\alpha b$ iff $d(a, b) < \epsilon_\beta$ for any $\beta < \alpha$.

---

[3] We shall only sketch some of the topological arguments in this paper. All
properties we shall state and use are straightforward modifications of standard
results and methods of the theory of metric and of compact spaces. We refer to
[5], where also a detailed treatment of general uniform spaces can be found.

Thus $\approx_o$ and $\approx_\kappa$ are respectively the trivial equivalence U^2 and the equality. Moreover, the chain $< \approx_\alpha | \alpha \leq \kappa >$ *is weakly decreasing and continuous* (i.e. $\approx_\alpha \subseteq \approx_\beta$ whenever $\alpha > \beta$, and $\approx_\lambda = \cap\{\approx_\alpha | \alpha < \lambda\}$ for limit λ), and *generates the uniformity* $\mathcal{U}$ associated to d, which is therefore either *discrete* or *of weight* κ.

Note also that the above defined sequence is made up of equivalences by virtue of the hypermetric inequality (iii), which implies that the set of all balls of any fixed radius is a partition of U. However, this condition is restrictive only for $\kappa = \omega$, since in the uncountable case any κ-distance d verifying the usual triangular inequality can be replaced by a uniformly equivalent one satisfying the hypermetric inequality (iii). Actually, it is easy to see that it is possible to define such a κ-hypermetric for any uniform space having a *nested uniformity basis of uncountable cofinality* κ. Only when $\kappa = \omega$, i.e. when $\mathcal{U}$ is *metrizable*, one has to check the supplementary condition that *no pair of different points can be connected by a finite set of arbitrarily small non-disjoint balls* (see [18] for more details about κ-metric spaces).

Following [7], we extend inductively on α the equivalences $\approx_\alpha$ to the whole universe V by

(1.2)

$$x \approx_\alpha y \quad \text{iff} \quad \forall \beta < \alpha \; \forall s \in x \; \forall u \in y \; \exists t \in y \; \exists v \in x \; s \approx_\beta t \; \& \; u \approx_\beta v.$$

Note that the sequence $< \approx_\alpha >_{\alpha < \kappa}$ is now *strictly* decreasing and continuous. Moreover, $\approx_o$ is again the trivial relation V^2, but $\approx_\kappa$ is no more the equality (e.g. all ordinals greater then κ are κ-equivalent, see [13]). However, we can extend the distance d to V^2 by putting

(1.3) $\qquad d(x,y) = \begin{cases} \epsilon_\alpha & \text{if } \exists \alpha < \kappa \; x \approx_\alpha y \; \& \; x \not\approx_{\alpha+1} y \\ 0 & \text{otherwise} \end{cases}$

thus obtaining a *pseudo-κ-hypermetric*, which verifies only conditions (ii) and (iii) above.

In order to obtain the corresponding κ-metric space we need to single out just one point from each ball of radius zero, but we can neither invoke the axiom of choice nor simply take the quotient, since many 0-balls are proper classes. We cannot even apply Scott's trick as in [13], since we are working in a non-well-founded universe; we use here instead the method of [7], the axiom FC playing the role of X_1 in the presence of urelements.

First of all we introduce the κ-*membership* $\in_\kappa$ on V by

$$(1.4) \qquad x \in_\kappa y \quad \text{iff} \quad \begin{cases} \forall \alpha < \kappa \exists z \in y \; x \approx_\alpha z, \\ \text{or equivalently} \\ \exists x' \approx_\kappa x \exists y' \approx_\kappa y \; x' \in y'. \end{cases}$$

Both κ-membership and κ-equivalence on V have nice topological characterizations, namely

Lemma 1.1. *Two points $x, y \in V \backslash U$ are κ-equivalent iff they have the same closure (considered as subsets of V). The κ-members of any point $x \in V \backslash U$ are precisely the ordinary members of its closure.*

PROOF. By the first definition of κ-membership, any κ-member of x is the limit of a κ-sequence of members of x, hence belongs to the closure of x; conversely, any point of the closure of x is such a limit, and the second assertion of the lemma follows.

Moreover, by the second definition of κ-membership, κ-equivalent sets have the same κ-members, hence the same closure. On the other hand, although the closure $y = \bar{x}$ of a given set x is possibly a proper class, nevertheless it satisfies *in toto* the condition (1.2) for being κ-equivalent to x; the lemma is thus completely proved.

$$\text{Q.E.D.}$$

By the above lemma, if the closure of any set were itself a set, then the closed sets together with the atoms would be a complete set of representatives for the κ-equivalence classes. Apparently, this is not the case, but our goal can be achieved if we restrict ourselves to a suitable subspace.

Lemma 1.2. *Suppose that the set X meets all κ-equivalence classes, i.e. that for any $y \in V$ there is $x \in X$ such that $x \approx_\kappa y$,*

Then there are a unique transitive set N and a unique function $g : X \to N$ verifying the following conditions:

(i) $x \approx_\kappa g(x)$ *for any* $x \in X$;

(ii) $x \approx_\kappa y$ *iff* $g(x) = g(y)$ *for any* $x, y \in X$;

(iii) $x \in_\kappa y$ *iff* $g(x) \in g(y)$ *for any* $x, y \in X$.

Therefore N is a transitive set of representatives for the κ-equivalence classes, and κ-membership agrees on N with ordinary membership; thus N provides a sort of "transitive collapse" of the "quotient structure" $(V/_{\approx_\kappa}, \in_\kappa /_{\approx_\kappa})$.

PROOF. Note that $X \supseteq U$, since any atom is the only memeber of its κ-equivalence class; define the function $f : X \to \mathcal{P}(X) \cup U$ by $f(u) = u$ for $u \in U$ and $f(x) = X \cap \bar{x}$ for $x \in X \backslash U$.

Let $g : X \to N$ be the unique function, given by the axiom FC, which is the identity on U and equal to $\hat{g} \circ f$ on $X \backslash U$. By definition, g has transitive range and is the identity on U : therefore, in proving (i)-(iii), we can restrict ourselves to consider only sets.

We shall prove by induction on α that $x \approx_\alpha g(x)$ for any $\alpha < \kappa$: note that the assertion is trivial for $\alpha = 0$ and that the limit steps are true by definition.

Assume now $t \approx_\alpha g(t)$ for any $t \in X$ and pick $x \in X \backslash U$: then $g(x) = \hat{g}(X \cap \bar{x}) = \{g(t) | t \in X \ \& \ t \in_\kappa x\}$. First pick $s \in x$: by hypothesis there is some $t \in X$ with $t \approx_\kappa s$, and surely $t \in_\kappa x$, hence $g(t) \in g(x)$. Conversely, pick $g(t) \in g(x)$: since $t \in_\kappa x$, t is the limit of some κ-sequence in x, hence there is some $s \in x$ such that $s \approx_\alpha t$. In both cases we have $s \approx_\alpha g(t)$, whence $x \approx_\alpha g(x)$.

The implication $g(x) = g(y) \Rightarrow x \approx_\kappa y$ is an immediate consequence of (i), and the remaining part of (ii) follows from the fact that κ-equivalent sets, having the same closure, have the same image under f.

Finally, what we have shown above, namely that

$$g(y) = N \cap \{g(x) | x \in_\kappa y\},$$

is a mere rephrasing of (iii).

Up to now, we have only used the existential part of the axiom FC. The uniqueness of g and N follows from the rigidity property AR, which is a consequence of the uniqueness part of FC.

Q.E.D.

We obtained the set N starting from a set X where all κ-equivalence classes were represented. But the role of X in FC is merely that of a *parameter set for defining the real membership* on the set one is looking for. Therefore all that we need in order to get N is a set *which parametrizes all κ-equivalence classes*, i.e. a set Y together with a mapping $\tau : V \to Y$ inducing the identity on U and verifying $x\approx_\kappa y$ whenever $\tau(x) = \tau(y)$.

Then we can put

$$f(y) = \{z \in Y | \exists u,v \ (u \in_\kappa v \ \& \ \tau(u) = z \ \& \ \tau(v) = y)\}$$

for $y \in Y\backslash U$ and $f(y) = y$ for $y \in U$; taking the function g given by the axiom FC and putting $\sigma = g \circ \tau$ we obtain a function which satisfies conditions (i), (ii), and (iii) for all $x,y \in V$ and has therefore the same range N.

There are several ways of defining such a mapping τ, and we choose one that gives supplementary information about all "quotients" $V/\approx_\alpha$ for $\alpha \leq \kappa$.

We fix functions $\tau_\alpha : U \to U$ for $\alpha \leq \kappa$, in such a way that $x\approx_\alpha \tau_\alpha(x)$ for any $x \in U$ and any $\alpha \leq \kappa$; we extend them inductively to V by putting, for $x \in V\backslash U$

$$\tau_o(x) = \tau_o(u) \quad \text{for some } u \in U,$$
$$\tau_{\alpha+1}(x) = \hat\tau_\alpha(x) = \{\tau_\alpha(y)|y \in x\} \quad \text{for any } \alpha < \kappa, \text{ and}$$
$$\tau_\lambda(x) =< \tau_\alpha(x) >_{\alpha<\lambda}= \{(\alpha,\tau_\alpha(x))|\alpha < \lambda\} \quad \text{for limit } \lambda.$$

Note that we can impose to the original τ_α's the supplementary condition that $\tau_\alpha(x) = \tau_\alpha(y)$ whenevre $x\approx_\alpha y$, thus getting a sequence of *choice functions* for the α-*equivalence classes* of U; but we can as well take all τ_α 's to be the identity on U, in order to make our construction independent of the axiom of choice.

In any case we obtain

Lemma 1.3.

(i)   If $\tau_\alpha(x) = \tau_\alpha(y)$, then $x\approx_\alpha y$.

(ii) If $x\approx_\alpha y$ implies $\tau_\alpha(x) = \tau_\alpha(y)$ for $x,y \in U$, then the same holds for any $x,y \in V$.

(iii) $\hat\tau_\alpha(V)$ is a set and $|\hat\tau_\alpha(V)| \leq \exp_{\alpha+1}(|\hat\tau_\alpha(U)|)$ for any $\alpha \leq \kappa$.[4]

---

[4] We define inductively the iterated exponential $\exp_\alpha(\kappa)$ in the usual way: $\exp_o(\kappa)=\kappa$, $\exp_{\alpha+1}(\kappa)=2^{\exp_\alpha(\kappa)}$ and, for limit λ, $\exp_\lambda(\kappa)=\sup\{\exp_\alpha(\kappa)|\alpha<\lambda\}$. We also put $\beth_\alpha=\exp_\alpha(\aleph_o)=\exp_{\omega+\alpha}(0)$.

PROOF. (i) We proceed by induction on α, assuming $x, y \notin U$ since (i) is true by definition for $x, y \in U$, and the hypothesis is never true when $x \in U$ and $y \in V \backslash U$.

The case $\alpha = 0$ is trivial. For limit α it suffices to recall that the equivalences $\approx_\beta$ are a decreasing and continuous sequence, hence $x \approx_\alpha y$ iff $x \approx_\beta y$ for all $\beta < \alpha$, whereas, by definition, $\tau_\alpha(x) = \tau_\alpha(y)$ iff $\tau_\beta(x) = \tau_\beta(y)$ for all $\beta < \alpha$.

Finally, assume (i) true for α and $\tau_{\alpha+1}(x) = \tau_{\alpha+1}(y)$, i.e. $\hat{\tau}_\alpha(x) = \hat{\tau}_\alpha(y)$; then we have that for any $s \in x$ there is some $t \in y$ such that $\tau_\alpha(s) = \tau_\alpha(t)$ and symmetrically, hence $x \approx_{\alpha+1} y$.

(ii) We proceed inductively on α, and once more both the initial and the limit steps are straightforward.

Assuming (ii) true for α and $x \approx_{\alpha+1} y$ with $x, y \notin U$, we have for any $s \in x$ some $t \in y$ such that $s \approx_\alpha t$, hence $\tau_\alpha(s) = \tau_\alpha(t)$, and symmetrically starting from $t \in y$.

Hence $\tau_{\alpha+1}(x) = \hat{\tau}_\alpha(x) = \hat{\tau}_\alpha(y) = \tau_{\alpha+1}(y)$.

(iii) Again we proceed by induction on α, the assertion being trivial for $\alpha = 0$, and we put $\kappa_\alpha = |\hat{\tau}_\alpha(U)|$ and $\nu_\alpha = |\hat{\tau}_\alpha(V)|$.

Since $\tau_{\alpha+1} = \hat{\tau}_\alpha$ on $V \backslash U$, we have $\hat{\tau}_{\alpha+1}(V \backslash U) \subseteq \mathcal{P}(\hat{\tau}_\alpha(V))$, hence $\nu_{\alpha+1} \leq \kappa_{\alpha+1} + 2^{\nu_\alpha} \leq \exp_{\alpha+2}(\kappa_{\alpha+1})$ by induction hypothesis.

For limit α, we have $\hat{\tau}_\alpha(V \backslash U) \subseteq \Pi_{\beta < \alpha} \hat{\tau}_\beta(V)$, hence

$$\nu_\alpha \leq \kappa_\alpha + \Pi_{\beta < \alpha} \nu_\beta \leq \kappa_\alpha + (\exp_\alpha(\kappa_\alpha))^{|\alpha|} \leq \exp_{\alpha+1}(\kappa_\alpha),$$

since, by induction hypothesis, $\nu_\beta \leq \exp_\alpha(\kappa_\alpha)$ for any $\beta < \alpha$.

$$\text{Q.E.D.}$$

We can summarize the preceding results as follows:

Theorem 1.4. *Suppose that the set U of all urelements carries a κ-hypermetric structure, let $< \approx_\alpha >_{\alpha \leq \kappa}$ be the chain of equivalences on the universe V associated to it according to (1.1)-(1.2), and let $\mathcal{U}$ be the generated uniformity.*

Then there are a unique transitive set $N = N_\kappa(\mathcal{U})$ and a unique projection $\sigma : V \to N$ verifying the following conditions:
(i) $x \approx_\kappa \sigma(x)$ for any $x \in V$;
(ii) $\sigma(x) = \sigma(y)$ iff $x \approx_\kappa y$ for any $x, y \in V$;

(iii) $\sigma(x) \in \sigma(y)$ iff $x \in_\kappa y$ for any $x, y \in V$.

In particular κ-membership agrees on N with ordinary membership.

The equivalences $< \approx_\alpha \cap N^2 >_{\alpha<\kappa}$ are a nested basis for a uniformity on N, compatible with the κ-hypermetric obtained by restricting (1.3) to N. In the corresponding uniform topology, $\sigma(X)$ is the closure of X for any subset X of N. Therefore N is the disjoint union of its clopen subsets U and $N\backslash U$, the latter being exactly the set of all closed subsets of N.

PROOF. The assertions about σ are merely a restatement of the above lemmata. Moreover, since $\approx_\kappa$ is the equality on N, the distance defined by (1.3) verifies also condition (i), hence is a κ-hypermetric on N, whose induced uniformity admits the equivalences $\approx_\alpha$ as a basis.

It remains to prove that, whenever $X \subseteq N, \sigma(X)$ is the closure of X in N, for the remainig assertions are easy consequences of this fact.

Given any set $x \in V\backslash U$, denote by $\bar{\bar{x}}$ its closure in V, and if $x \subseteq N$ put $\bar{x} = \bar{\bar{x}} \cap N$, thus $\bar{x}$ is the closure of x in N.

By Lemma 1.1, $\bar{\bar{x}}$ is satured w.r.t. the equivalence $\approx_\kappa$, hence in particular $\bar{\bar{x}} \supseteq \hat{\sigma}(\bar{\bar{x}})$. Then, if $x \subseteq N$, we have

$$\sigma(x) = \{\sigma(y)|y \in_\kappa x\} = \{\sigma(y)|y \in \bar{\bar{x}}\} =$$
$$\hat{\sigma}(\bar{\bar{x}}) \subseteq \bar{\bar{x}} \cap N = \bar{x} = \hat{\sigma}(\bar{x}) \subseteq \hat{\sigma}(\bar{\bar{x}});$$

therefore both inclusions are equalities, whence $\sigma(x) = \bar{x}$.

$$\text{Q.E.D.}$$

Since $N\backslash U$ is precisely the set of all its closed subsets, it is natural to consider on it the *Hausdorff κ-metric*

$$h(x,y) = \{\sup_{s \in x} \inf_{t \in y} d(s,t), \sup_{t \in y} \inf_{s \in x} d(t,s)\}$$

By means of (1.1)-(1.3) it is easy to compare the distances d and h, and obtain $h(x,y)^+ \leq d(x,y) \leq h(x,y)$, where $\epsilon_\alpha^+ = \epsilon_{\alpha+1}$: hence d and h are uniformly equivalent.

The same conclusion is reached by comparing the product distance of pairs $d_2((x,y),(u,v)) = \max\{d(x,u), d(y,v)\}$ with the distance d between the same pairs (intended *à la Kuratowski*) considered as subsets of N; hence $N \times N$ is a *closed uniform subspace* of N, and the same is true for any power N^n.

If we consider general function spaces, the situation is not so nice. However, from Theorem 1.4 and the above remarks, we can conclude that a function (or relation) graph belongs to N iff it is a closed subset of the product space $N \times N$; in particular all *continuous functions with closed domains* belong to N.

Moreover the uniformity induced by N on any function space is given by the Hausdorff distance of the graphs; in particular, on spaces of functions with the same domain, it agrees with the *uniformity of uniform convergence*. Since the situation becomes neater when the space N is κ-*compact*, we shall give a topological characterization of the function spaces which are members of N (Lemma 3.1) only after dealing with κ-compactness in section 2.

We conclude this section with some useful remarks.

Remark 1.5. For $A \subseteq U$, define the *cumulative hierarchy* $\Pi(A) = \cup_{\alpha \in ord}\Pi_\alpha(A)$ of the *sets wellfounded over A* by putting $\Pi_o(A) = A$, $\Pi_{\alpha+1}(A) = \Pi_\alpha(A) \cup \mathcal{P}(\Pi_\alpha(A))$, and $\Pi_\lambda(A) = \cup_{\alpha<\lambda}\Pi_\alpha(A)$ for limit λ. In particular $\Pi_\alpha = \Pi_\alpha(\emptyset)$ is the set of all well founded sets of rank less than α.

Call x α-*isolated* if it is the only element of its $\approx_\alpha$-class, and let U_α be the set of all α-isolated points of U.

It is easily seen by induction on β that any $x \in \Pi_\beta(U_\alpha)$ is $(\alpha + \beta)$-isolated: in particular any wellfounded set of rank less that α is α-isolated (cfr. [7], [13]). It follows that $N \supseteq \cup_{\alpha<\kappa}\Pi_\kappa(U_\alpha)$.

On the other hand all ordinals greater that α are α-equivalent to each other, hence in particular $\Pi \cap N = \Pi_\kappa$.

More generally one has that the elements of $\Pi_\alpha(A)$ are pairwise $(\beta + \alpha)$-inequivalent whenever the elements of A are pairwise β-inequivalent. Hence, putting $\kappa_\alpha = |U/\approx_\alpha|$ and $\nu_\alpha = |N_\kappa(\mathcal{U})/\approx_\alpha|$, one obtains the inequality $\exp_\alpha(\kappa_\beta) \leq \nu_{\beta+\alpha}$ for any $\alpha, \beta < \kappa$.

Better estimates can be obtained by observing that if $< S_\alpha | \alpha < \lambda >$ (λ limit) is an increasing sequence of sets of representatives for the α-equivalence on N, then all elements of $\mathcal{P}(\cup_{\alpha<\lambda}S_\alpha)$ are pairwise λ-inequivalent, and $\mathcal{P}(S_\alpha)$ is a set of representatives for the $(\alpha + 1)$-equivalence on $N \backslash U$.

Therefore

$$\nu_{\alpha+1} = 2^{\nu_\alpha} + \kappa_{\alpha+1} \quad \text{and} \quad \nu_\lambda = 2^{\sup\{\nu_\alpha | \alpha < \lambda\}} (\lambda \text{ limit})$$

In particular we obtain that, if $\kappa_{\alpha+1} \leq 2^{\kappa_\alpha}$ for $\kappa > \alpha > \beta$, then

$$\nu_\alpha = \; \beth_{\alpha+1} \text{ for any } \alpha > \beta.$$

Remark 1.6. If there are no urelements, i.e. $U = \emptyset$, the model $N_\kappa = N_\kappa(\emptyset)$ is exactly the same as the one introduced in [7]. This is more comprehensive than the corresponding model M_κ of [13], which contains only the representatives of the κ-equivalence classes of wellfounded sets.

Restricting the construction of Theorem 1.4 to all wellfounded sets, we can obtain a transitive collapse N_κ^{wf} of the model M_κ with the κ-membership $\in_\kappa$. Similarly, we can perform our construction only for the class $\Pi(U)$ of all wellfounded sets over the atoms. However, in view of the κ-compactness results of the next section, the full model N_κ seems more interesting.

Remark 1.7. If we consider models $N = N_\alpha(\mathcal{U})$ for α any limit ordinal, as in [7] and [13], the uniformity of N has then weight $\nu = \text{cof } \alpha$ and will therefore never be ν-compact or even ν-bounded, when α is singular (see section 2).

Moreover, stopping the construction of $N_\kappa(\mathcal{U})$ at $\alpha < \kappa$ amounts to starting it with a set of atoms isometric to the quotient space $U/\approx_\alpha$ (which is either discrete or has weight $\nu = \text{cof } \alpha$). On the other hand, proceeding up to $\alpha > \kappa$ is equivalent to starting with a discrete set of atoms, which can even be assumed pairwise 1-inequivalent.

The only interesting possibility is therefore to take, instead of the basic κ-sequence $< \epsilon_\alpha >_{\alpha < \kappa}$, a new λ-sequence $< \epsilon'_\alpha >_{\alpha < \lambda}$, where λ is any limit ordinal of cofinality κ.

In this way the uniform structure of U is preserved, and all wellfounded sets of rank less than λ are now present in $N_\lambda(\mathcal{U})$. However, in view of the results of the next section (Theorem 2.7), Cauchy completeness is preserved only for $\kappa = \text{cof } \lambda = \omega$. Moreover, κ-compactness is lost for singular λ, and with it many interesting comprehension properties of the model (see section 3).

2. Cauchy completeness and κ-compactness. As pointed out before, most of the interesting features of our models depend on additional topological properties, which, for κ-hypermetric spaces, can be characterized as follows:

Definition 2.0. *Let N be a κ-hypermetric space:*
N is Cauchy complete iff any Cauchy κ-sequence of N converges in N;
N is κ-bounded iff there are less than κ balls of any fixed radius;
N is κ-compact iff any κ-sequence in N has a convergent κ-subsequence.

It is easily seen that the above definition of κ-compactness is equivalent to each of the following classical properties:
(i) any open cover of N has a subcover of cardinality less that κ;
(ii) any strictly descending κ-chain of closed sets has non-empty intersection.

Moreover, any κ-compact κ-metric space is both κ-bounded and Cauchy complete, but the converse implication can fail for uncountable κ, e.g. for the tree T' defined in the proof of Lemma 2.3 (see also [18]).

We shall see below that κ-boundedness, Cauchy completeness and κ-compactness of the space $N = N_\kappa(\mathcal{U})$ can be obtained by combining the same properties of the subspace U of all atoms with suitable combinatorial properties of the cardinal κ.

We begin by considering κ-boundedness:

Lemma 2.1. $N = N_\kappa(\mathcal{U})$ *is κ-bounded iff U is κ-bounded and κ is strongly inaccessible.*

PROOF. First of all, N is the disjoint union of its clopen subsets U and $N\backslash U$, and $N\backslash U$ includes the set Π_κ of all hereditarily wellfounded sets of rank less that κ, by Remark 1.5.

Since any point $x \in \Pi_\alpha$ is α-isolated, the κ-boundedness of N yields both that U is κ-bounded and that $\beth_\alpha < \kappa$ for any $\alpha < \kappa$; therefore the given conditions are necessary.

On the other hand, by Lemma 1.3, the number of distinct α-equivalence classes in the whole universe V does not exceed $\exp_{\alpha+1}(\kappa_\alpha)$, where κ_α is the number of α-equivalence classes in U, and $\kappa_\alpha < \kappa$ when U is κ-bounded.

Therefore, if κ is inaccessible, the number of ϵ_α-balls in the whole universe is strictly less than κ, for $\exp_\eta(\xi) < \kappa$ whenever both ξ and η are less that κ.

Q.E.D.

We shall now investigate the notion of Cauchy completeness.

Lemma 2.2. *Let $\kappa = \lambda^+$ be a successor cardinal. Then $N = N_\kappa(\mathcal{U})$ is not Cauchy complete.*

PROOF. We define a "universal" 2^λ-ary tree [5] T of subsets of $\kappa + 1$ in the following way: for $\alpha < \kappa$, put

$$I_\alpha = \{\lambda.\alpha + \gamma \mid 0 \le \gamma < \lambda\},$$
$$T_\alpha = \{x \cup \{\kappa\} \mid x \in \mathcal{P}(\lambda.\alpha) \ \& \ x \cap I_\beta \ne \emptyset \ \forall \beta < \alpha\},$$

and, for $x, y \in T = \cup_{\alpha < \kappa} T_\alpha$, put

$$x <_T y \quad \text{iff} \quad \exists \alpha < \kappa \ x\backslash\{\kappa\} = y \cap \lambda.\alpha.$$

Clearly $(T, <_T)$ is a tree of height κ, whose α^{th} level is T_α; it is a universal 2^λ-ary tree, since any of its nodes has exactly 2^λ immediate successors and any of its branches of limit length has exactly one immediate successor, hence any 2^λ-ary tree is (isomorphically) embeddable into T, and the embedding can be taken level-preserving.

Moreover, recalling that all ordinals $\ge \alpha$ are α-equivalent, whereas those $< \alpha$ are pairwise α-inequivalent, we get that, for any $x \in T_\alpha$ and any $y \in T, x <_T y$ holds iff $x \approx_{\lambda\alpha+1} y$. Hence there is a natural correspondence between κ-branches of T and Cauchy κ-sequences of elements of T.

Let S be the set of all *bounded strictly increasing α-sequences* (with $\alpha < \kappa$) of elements of the *lexicographically ordered* set

$$Q = \{s \in \lambda^\omega \mid \exists m \forall n > m \ s_n = 0\}$$

---

[5] Recall that a partially ordered set $(T, <_T)$ is a tree iff the predecessors of any element (node) $x \in T$ are wellordered by $<_T$, their order-type (length) being the level of x. T is κ-ary iff any node of T has at most κ immediate successors and any branch of limit length at most one.

of all *eventually* 0 *sequences* of ordinals less than λ, and arrange S in a tree by inclusion.

S is a classical λ-*ary tree of height* κ *without any* κ-*branch,* and it is homogeneous in the sense that it is isomorphic to each of its full subtrees obtained by taking *all* successors of any node. Let T' be a subtree of T isomorphic to S, and let x_α be the αth level of T', i.e. the set of all nodes of T' corresponding to α-sequences of S. The κ-sequence $< x_\alpha >_{\alpha < \kappa}$ is Cauchy, and in fact $x_\alpha \approx_{\lambda \alpha + 2} x_\beta$ for $\alpha < \beta$, as can be seen by considering, through any $s \in x_\alpha$, a branch of T' of length greater than β.

But the sequence $< x_\alpha >$ cannot have a limit, since otherwise, putting $x = \lim x_\alpha$ and picking some $s \in x$, we would find for any $\alpha < \kappa$ elements $s_\alpha \in x_\alpha$ verifying $s_\alpha \approx_{\lambda \alpha + 1} s$, and these would constitute a κ-branch of the tree T'.

Q.E.D.

Lemma 2.3. *Let* κ *be inaccessible and assume that* $\kappa \to (\kappa)_2^2$.[6] *Then the space* $N = N_\kappa(\mathcal{U})$ *is not Cauchy complete.*

PROOF. In order to reach our conclusion, we follow closely the argument used in the case of a successor cardinal.

We define a tree T of subsets of $\kappa + 1$ by putting, for any infinite cardinal $\lambda < \kappa$,

$$I_\lambda = \{\gamma | \lambda \leq \gamma < \lambda^+\},$$
$$T_\lambda = \{x \cup \{\kappa\} \mid x \in \mathcal{P}(\lambda) \ \& \ \forall \xi < \lambda \quad x \cap I_\xi \neq \emptyset\},$$

---

[6] Recall that the partition property $\kappa \to (\kappa)_2^2$ holds iff given any partition of all doubletons from κ into two parts, there is a κ-sized subset of κ all of whose doubletons belong to the same part. It is well known (cfr. [12]) that $\kappa \to (\kappa)_2^2$ is equivalent to the binary tree property, saying that any binary tree of cardinal κ has a κ-branch, and that it implies that κ is strongly inaccessible. For a strongly inaccessible cardinal κ, the property $\kappa \to (\kappa)_2^2$ is also equivalent to the tree property (which says that any tree of size κ all whose levels have sizes less than κ has a κ-branch), as well as to weakly compactness (which says that any κ-complete filter over a κ-complete field $\mathcal{F}$ of subsets of κ is included in a κ-complete ultrafilter on $\mathcal{F}$). We include ω among the strongly inaccessible cardinals.

and, for $x, y \in T = \cup_{\lambda < \kappa} T_\lambda$,

$$x <_T y \quad \text{iff} \quad \exists \lambda < \kappa \; x \setminus \{\kappa\} = y \cap \lambda.$$

Clearly $(T, <_T)$ is a tree of height κ, whose α^{th} level is $T_{\aleph_\alpha}$; any node of T_λ has exactly 2^{λ^+} immediate successors and any branch of limit length in T has exactly one immediate successor.

Since, as above, for any $x \in T_\lambda$ and any $y \in T$, $x <_T y$ holds iff $x \approx_{\lambda+1} y$, there is again a natural correspondence between κ-branches of T and Cauchy κ-sequences of elements of T.

Assume now that κ is weakly, but not strongly inaccessible. Then, for some $\lambda < \kappa$, κ is less than 2^λ, and one can embed isomorphically into T any κ-ary tree of height κ.

Define a tree S in the following way: put $Q = \{(\lambda, \alpha) | \alpha < \lambda < \kappa \ \& \ \lambda \text{ is an infinite cardinal}\}$, and let $S = Q^{<\omega}$ be the set of all finite sequences of elements of Q.

Given $\quad s = <(\lambda_o, \alpha_o), ..., (\lambda_m, \alpha_m)> \quad$ and $t = <(\mu_o, \beta_o), ..., (\mu_n, \beta_n)>$ put $s < t$ iff $m \leq n$, $\lambda_i = \mu_i$ for $i \leq m$, $\alpha_i = \beta_i$ for $i < m$ and $\alpha_m \leq \beta_m$.

Clearly S becomes a κ-ary tree of height κ without any κ-branch. In fact any node of S has exactly κ immediate successors and the whole tree S is isomorphically embeddable into the subtree of all successors of any of its nodes.

Let T' be a subtree of T isomorphic to S, and let x_α be the αth level of T': the very same argument of the previous lemma now works and proves that the κ-sequence $< x_\alpha >_{\alpha < \kappa}$ is Cauchy, but cannot have any limit.

Finally, if κ is strongly inaccessible, then by hypothesis there is a *binary* tree S of height κ without any κ-branch. Let T' be a subtree of T isomorphic to S, and put $x_\alpha = \{t \in T' \cap T_{\aleph_\alpha} | t \text{ has } \kappa \text{ successors}\}$.

Since κ is strongly inaccessible, all levels of T have size less than κ, hence any element of x_α has successors in each x_β with $\beta > \alpha$. Therefore $< x_\alpha >_{\alpha < \kappa}$ is again a non-convergent Cauchy κ-sequence.

Q.E.D.

A criterion for Cauchy completeness can now be given, namely:

Lemma 2.4. *Suppose that the atom space U is κ-bounded. Then $N = N_\kappa(\mathcal{U})$ is Cauchy complete iff both $\kappa \to (\kappa)^2_2$ (i.e. κ is strongly inaccessible and weakly compact) and U is Cauchy complete.*

PROOF. By the lemmata above, the given conditions are necessary for Cauchy completeness. They are also trivially sufficient when κ-sequences of urelements are considered.

Thus assume $\kappa \to (\kappa)_2^2$ and let $< x_\alpha >_{\alpha < \kappa}$ be any Cauchy κ-sequence in $N \backslash U$: we can suppose w.l.o.g. that $x_\alpha \approx_\alpha x_\beta$ whenever $\alpha < \beta < \kappa$.

Put $S = \{x \in N^\kappa | x_\alpha \approx_\alpha x_\beta \forall \alpha < \beta < \kappa\}$.

Define the function $f : S \to \mathcal{P}(S) \cup U$ by setting

$f(x) = \{y \in S | y_\alpha \in x_{\alpha+1} \forall \alpha < \kappa\}$ if x is eventually outside U,

$f(x) = \lim x_\alpha$ otherwise (the limit exists in U by hypothesis).

Let g be the unique function given by the axiom FC. We claim that $g(y) \approx_\alpha y_\alpha$ for all $\alpha < \kappa$ and all $y \in S$. Then, in particular, $g(x) = \lim x_\alpha$, and we are done.

Our claim is trivial for $\alpha = 0$, and easily verified for any limit $\lambda < \kappa$, provided it holds for all $\alpha < \lambda$. Moreover, it holds by definition if x is eventually atomic: so we only need to prove the induction step from α to $\alpha + 1$ when all y_α's are non-empty sets.

By definition, we have, for any such $y \in S$,

$$g(y) = \hat{g}(f(y)) = \{g(z) | z \in S \ \& \ \forall \gamma < \kappa \ z_\gamma \in y_{\gamma+1}\},$$

hence, by induction hypothesis,

$$g(z) \approx_\alpha z_\alpha \in y_{\alpha+1} \quad \text{for any} \quad g(z) \in g(y).$$

Conversely, given $t \in y_{\alpha+1}$, we need to find $z \in S$ such that $t \approx_\alpha z_\alpha$ and $z_\gamma \in y_{\gamma+1}$ for all $\gamma < \kappa$. Then $t \approx_\alpha z_\alpha \approx_\alpha g(z)$ by induction hypothesis, hence $y_{\alpha+1} \approx_{\alpha+1} g(y)$, and our goal is achieved.

In order to find the κ-sequence z, we define a tree $T = \cup_{\alpha < \kappa} T_\alpha$ as follows:

$$T_\alpha = \{\alpha\} \times \{B_\alpha(x) | x \in y_{\alpha+1}\}, \quad \text{where} \quad B_\alpha(x) = \{z \in N | z \approx_\alpha x\},$$

and

$$(\alpha, A) <_T (\beta, B) \quad \text{iff} \quad \alpha < \beta \ \& \ A \supseteq B.$$

Any node of T lies on a κ-branch, since its successors constitute a tree of height κ with levels of size less than κ, by our hypotheses.

Pick a κ-branch $< (\gamma, C_\gamma) | \gamma < \kappa >$ of T through $(\alpha, B_\alpha(t))$. Any κ-sequence z such that $z_\gamma \in y_{\gamma+1} \cap C_\gamma$ for all $\gamma < \kappa$ is now suitable for our purposes.

Q.E.D.

We are now able to state the main result of this section, which generalizes to the present context Theorem 4.4. of [7]:

Theorem 2.5. *The space* $N = N_\kappa(\mathcal{U})$ *is κ-compact iff both* $\kappa \to (\kappa)_2^2$ *(i.e. κ is strongly inaccessible and weakly compact) and U is κ-compact.*

PROOF. As in Lemma 2.4, the conditions are obviously necessary. Thus assume $\kappa \to (\kappa)_2^2$ and let $< \kappa_\alpha >_{\alpha < \kappa}$ be any κ-sequence in N. Arrange the pairs (α, x_α) in a tree T in the following way: suppose the levels T_δ of T are already defined for $\delta < \gamma < \kappa$, consider the γ-equivalence classes of all elements x_α such that (α, x_α) is not yet arranged at any level $\delta < \gamma$ and, for each class, put in T_γ the pair (β, x_β) having the least index β. Given (α, x_α) at level δ and (β, x_β) at level γ, put
$$(\alpha, x_\alpha) <_T (\beta, x_\beta) \text{ iff both } \delta < \gamma \text{ and } x_\alpha \approx_\delta x_\beta.$$
It is immediate to verify that T becomes a tree whose γ^{th} level is indeed T_γ. Since the elements of the αth level of T are pairwise α-inequivalent and N is κ-bounded by Lemma 2.1, T is a tree of size κ with all levels of size less than κ, hence of height κ. It has therefore a κ-branch, since by hypothesis κ has the tree property.

By definition, the second components of any κ-branch of T constitute a Cauchy κ-subsequence of the original sequence, which is convergent since N is Cauchy complete by Lemma 2.4: the proof is thus complete.

$$\text{Q.E.D.}$$

Clearly, when the atom space U is discrete, it is κ-compact iff its size is less than κ. Hence if $|U| < \kappa$, then $N_\kappa(\mathcal{U})$ is κ-compact iff κ has the partition property.

On the other hand, a κ-metric space is κ-compact iff it is uniformly isomorphic to a closed subspace of the "universal" space 2^κ of all κ-sequences of 0's and 1's, equipped with the *first difference* κ-hypermetric $d(x, y) = \epsilon_\alpha$ iff $\alpha = \min\{\beta | x_\beta \neq y_\beta\}$.

It follows that if $N_\kappa(\mathcal{U})$ is κ-compact, then $|U| \leq 2^\kappa$.

We conclude this section with some remarks about Cauchy completeness.

Remark 2.6. The space $N_\omega(\mathcal{U})$ is Cauchy complete whenever U

is a complete metric space, since the argument of the proof of Lemma 2.4 works without the κ-boundedness hypothesis for $\kappa = \omega$. In fact, using the notation of that proof, given $y \in S$ and $t \in y_{n+1}$ one can always pick a sequence $z \in S$ with $z_n = t$ verifying $z_m \in y_{m+1}$ for all $m \in \omega$.

On the contrary, κ-boundedness is a necessary condition for Cauchy completeness for any uncountable κ, as we shall show below.

Assume U κ-unbounded and let A be a set of κ pairwise α-inequivalent atoms for some $\alpha < \kappa$. We shall assume, for sake of simplicity, that the elements of A are already 1-inequivalent.

For $a \in A$, consider the "generalized ordinals" a_α defined by
$a_o = a$, $a_{\alpha+1} = a_\alpha \cup \{a_\alpha\}$ and $a_\lambda = \cup_{\alpha<\lambda} a_\alpha$ for limit λ.

It is easy to verify that generalized ordinals built up over different atoms from A are pairwise 2-inequivalent. Moreover

$$a_\alpha \approx_\alpha a_\beta \quad \text{and} \quad a_\alpha \not\approx_{\alpha+1} a_\beta \quad \text{for} \quad \alpha < \beta.$$

Put $A_\alpha = \cup_{a\in A} a_\alpha$, $I_\alpha = A_{\alpha+1} \backslash A_\alpha$ and define a tree $T = \cup_{\alpha<\kappa} T_\alpha$ by

$$T_\alpha = \{x \cup I_\kappa | x \subseteq A_\alpha \ \& \ x \cap I_\beta \neq \emptyset \ \forall \beta < \kappa\} \quad \text{and}$$
$$s <_T t \quad \text{iff} \quad \exists \alpha < \kappa \, (s \in T_\alpha \quad \text{and} \quad s \cap A_\alpha = t \cap A_\alpha).$$

Clearly T is a tree whose α^{th} level is T_α and any node of T has exactly 2^κ immediate successors. Therefore one can embed in T the tree S defined in the proof of Lemma 2.3. Since for $s, t \in T$ one has

$$s \approx_{\alpha+1} t \quad \text{iff} \quad s \cap A_\alpha = t \cap A_\alpha,$$

the argument of the proof applies and gives a non-convergent Cauchy κ-sequence.

Remark 2.7. Let κ be a strongly inaccessible weakly compact cardinal, and let λ be any limit ordinal of cofinality κ.

According to the last part of Remark 1.7, we can build up the model $N_\lambda(\mathcal{U})$ so as to include all wellfounded sets of rank less than λ. Then we can fix an increasing κ-sequence $< \gamma_\alpha >_{\alpha<\kappa}$ of ordinals cofinal in λ and use the γ_α-equivalence instead of $\approx_\alpha$.

If $\kappa = \mathrm{cof}\,\lambda = \omega$, we can easily modify the initial argument of Remark 2.6 so as to obtain that $N_\lambda(\mathcal{U})$ is Cauchy complete iff its atom space U is.

On the other hand, if $\lambda > \kappa = \mathrm{cof}\,\lambda > \omega$, we can argue as in the proofs of Lemma 2.2 and 2.3. Namely, the κ-sequence $< \gamma_\alpha >_{\alpha<\kappa}$ can replace the ordinals less than κ in defining suitable trees of parts of $\lambda + 1$, so as to provide inside $N_\lambda(\mathcal{U})$ counterexamples to Cauchy completeness.

Summing up all results on Cauchy completeness we obtain the following general criterion:

Theorem 2.7. $N_\lambda(\mathcal{U})$ *is a Cauchy complete metric space iff* λ *has countable cofinality and U is Cauchy complete.*

If λ *has uncountable cofinality κ , then the space $N_\lambda(\mathcal{U})$ is Cauchy complete iff* $\lambda \to (\lambda)_2^2$ *(hence $\lambda = \kappa$) and U is both Cauchy complete and κ-bounded.*

In particular the models N_α of [7] are complete iff either $\mathrm{cof}\,\alpha = \omega$ or $\alpha \to (\alpha)_2^2$.

Remark 2.8. In his thesis [13], R.J.Malitz calls *crowded* a κ-metric space where any κ-sequence has a Cauchy κ-subsequence. Clearly, crowdedness implies κ-boundedness, whereas κ-compactness is equivalent to the conjunction of crowdedness and Cauchy completeness.

Many of the most relevant properties of the models M_α of [13] depend on the existence of some ordinal α such that M_α is both crowded and Cauchy complete (such ordinals are called *Malitz ordinals* in [7]), and Malitz conjectured that *all regular uncountable cardinals* have this property. However, since the counterexamples employed for the negative parts of the above theorems make use only of wellfounded sets, they apply also to Malitz's models. Therefore, if κ is *Malitz* then $\kappa \to (\kappa)_2^2$. On the contary, a free construction principle (although not necessarily FC) plays the essential role in proving the positive parts of Theorems 2.5 and 2.7. Thus all that one obtains from the argument of Theorem 2.5 is that M_α is *crowed* if and only if $\alpha \to (\alpha)_2^2$. Malitz himself proved that M_ω, unlike our N_ω, is *not Cauchy complete*. His argument can easily be carried out for any ordinal of countable cofinality, following the pattern of the proof of the

Theorem 2.7. The question as to whether Malitz cardinals exist at all is still open. As a matter of fact, the opinions of the authors are split in conjecturing an answer to this question. A positive solution would yield that the corresponding M_κ shares many of the comprehension properties of our models $N_\kappa(\mathcal{U})$.

3. Comprehension properties of κ-compact models.

As we noticed in the first section, it is easier to study functions and function spaces in the model $N_k(\mathcal{U})$ when this is a κ-compact space. In fact most properties of compact metric spaces have perfect analogues for any uncountable κ. E.g. the graph of a function f is closed in the product topology if and only if f is continuous and dom f is closed, and in this case f is a closed uniformly continuous map; the κ-compact-open topology on the space of all continuous functions is induced by the uniformity of uniform convergence; a set of continuous functions with the same domain is closed in the κ-compact-open topology iff it is equicontinuous, etc..

We summarize the results which are relevant in determining the comprehension properties of our models in the following lemma, and we refer to Chapter 8 of [5] for detailed proofs and more information on this topic (see in particular [5,8.2.4-10]).

Lemma 3.1. *Let $N = N_\kappa(\mathcal{U})$ be κ-compact. Then*

(i) A function f belongs to N iff it is continuous and its domain is closed, and in this case f is a closed uniformly continuous map. More generally, if $A \subseteq N$, a function $g : A \to N$ is κ-equivalent to a function $f \in N$ iff it is uniformly continuous on A (the domain of f being then the closure of A in N).

(ii) For any $X \in N \backslash U$ and any $Y \subseteq N$, the space $Y^X \cap N$ with the induced uniformity is precisely the set $\mathcal{U}(X,Y)$ of all uniformly continuous functions from X into Y with the uniformity of uniform convergence (which induces the κ-compact-open topology).

(iii) A set $F \subseteq N^X$ belongs to N iff F is equicontinuous and X is closed. In particular, if $|Y| > 1$, then $\mathcal{U}(X,Y) = Y^X \cap N$ belongs to N iff X is closed and discrete, i.e. iff $|X| < \kappa$, and then $\mathcal{U}(X,Y) = Y^X$.

PROOF.

(i) Since N is κ-compact, the graph of f is closed in $N \times N$ iff f is continuous and dom f is closed; moreover any subset of N is closed iff it is κ-compact, hence any continuous function maps closed sets onto closed sets, and is uniformly continuous on any closed set.

On the other hand, if g is uniformly continuous on A, then it has a unique uniformly continuous extension to $\bar{A}$, whose graph is clearly the representative of g in N.

(ii) As in the ordinary compact case, it is easy to see that if the space X is κ-compact, then the Hausdorff distance of the graphs induces on the space $\mathcal{U}(X,Y)$ of all (necessarily uniformly) continuous functions from X into Y both the uniformity of uniform convergence and the κ-compact-open topology. Since $\mathcal{U}(X,Y) = Y^X \cap N$ by (i) above, (ii) follows.

(iii) By Ascoli's theorem extended to κ-compact κ-metric spaces, if X is κ-compact, then a closed set $F \subseteq Y^X$ is κ-compact iff F is equicontinuous and $\{f(x)|f \in F\}$ has κ-compact closure for any $x \in X$. Since in this context closed and κ-compact are synonyms, we conclude the first assertion.

As to the second one, the condition is obviously sufficient, for then the points of X are α-isolated for some α, hence the set of all functions on X is equicontinuous. To prove the converse, let x be a cluster point of X, pick two different points z_o, z_1 in Y, and, for any $y \in X$, put $f_\alpha(y) = z_o$ if $y \approx_\alpha x$, $f_\alpha(y) = z_1$ otherwise. Clearly, $< f_\alpha >_{\alpha < \kappa}$ is a κ-sequence of uniformly continuous functions on X whose pointwise limit is not continuous, hence it cannot have any uniformly convergent subsequence. Therefore $\mathcal{U}(X,Y)$ is not κ-compact.

$$\text{Q.E.D.}$$

We shall now illustrate the selfdescriptive power of a κ-compact model $N = N_\kappa(\mathcal{U})$. As pointed out before, these models are closed under many basic operations. Simultaneously many interesting large classes are closed subsets of N, hence they belong to N. We begin by stating a theorem which transposes to the present situation, where urelements are allowed, all the assertions of Theorem 4 of [7 §4.2].

Following [7], we define the class GPF of the *generalized positive formulae* as the least class which includes the *atomic* formulae and is closed under *conjunction* , *disjunction, existential* and *universal quantification* as well as under the following rules of *bounded quantification* (which, strictly speaking, are non-positive):

if ϕ is GPF, then both $\forall x(x \in y \rightarrow \phi)$ and $\forall x(\theta(x) \rightarrow \phi)$ are GPF, where θ is any formula with exactly one free variable.

The *Generalized Positive Comprehension Principle* GPK is the axiom schema (denoted by Comp (GPF) in [7]) which postulates the existence of the set $\{x|\phi\}$ for any generalized positive formula ϕ,

GPK - $\exists x \forall y(y \in x \longleftrightarrow \phi)$ where ϕ is GPF and x is not free in ϕ.

Since the formula $z = (x, y)$ is GPF, the generalized positive comprehension principle GPK yields both the existence of many *fundamental graphs* (e.g. membership, inclusion, identity, singleton and power-set maps, projections, permutations and all natural manipulations of n-tuples, etc.) and the stability under many *basic operations* (e.g. union, intersection, cartesian product, domain, range, inversion, composition and fibred product of graphs, etc.).

In the light of the above remark, the strength of the following theorem will now be evident.

Theorem 3.2. *Assume that $N = N_\kappa(\mathcal{U})$ is κ-compact. Then*
(i) any subset of N of size less than κ belongs to N together with its complement w.r.t. N. Both the product of less than κ elements of $N\backslash U$ and the intersection of arbitrarily many of them belong to N.

(ii) N satisfies the Generalized Positive Comprehension schema GPK.

(iii) The "cumulative cardinals" $c_{<\lambda} = \{x \in N|\ |x| < \lambda\}$ belong to N for any cardinal $\lambda < \kappa$, whereas no "Frege-Russell cardinal"

$$f_\lambda = \{x \in N|\ |x| = \lambda\}$$

belongs to N for $\lambda > 1$.

PROOF. First of all, any set $X \subseteq N$ of size less than κ is well-spaced, i.e. its points are all α-isolated for some $\alpha < \kappa$, hence X is both clopen and discrete.

Now the first part of (i) is immediate. The second one follows directly from Lemma 3.1, any function on X being uniformly continuous and any set of functions on X equicontinuous. The last assertion of (i) is obvious, since any intersection of closed sets is closed.

In order to prove (ii), we make use of an analogue of the classical Bernay's theorem for Gödel-Bernays class theory, proved in [7, §4.1]. Namely, GPK holds in N provided that the following sets belong to N :

(1) $I = \{(x,y) \in N^2 | \ x = y\}$ and $E = \{(x,y) \in N^2 | x \in y\};$

(2)
$X^{-1} = \{(x,y) \in N^2 | (y,x) \in X\},$

$Q(X) = \{(x,y) \in N^2 | \forall z \in x((x,y),z) \in X\}, \ \{X,Y\}, \ \hat{X}(Y),$

and $X^*Y = \{((x,y),z) \in N^2 | (x,z) \in X \& (y,z) \in Y\},$

for all X,Y in N;

(3)
$$\check{X}(Y) = \{z \in N | \forall y \in Y \ \ (y,z) \in X\}$$
for any $X \in N$ and any $Y \subseteq N.$

Since a κ-sequence of pairs (x_α, y_α) converges to (x,y) iff both $\lim x_\alpha = x$ and $\lim y_\alpha = y$, it is easy to check that the identity I is closed, as well as X^{-1} and X^*Y whenever X and Y are closed.

Moreover, any fiber $\tilde{X}(y) = \{z | (y,z) \in X\}$ is closed provided X is closed; this yields (3), for $\check{X}(Y) = \bigcap \{\tilde{X}(y) | y \in Y\}$.

On the other hand, if both X and Y are closed, hence κ-compact, then also $\hat{X}(Y) = \hat{P}_2(X \cap Y \times V)$, being the image of a κ-compact set under the continuous map P_2 (second projection), is κ-compact, hence closed.

In order to obtain (ii), it remains to prove that E and $Q(X)$ are closed.

Let (x_α, y_α) be any κ-sequence in E, and suppose that $(x,y) = \lim(x_\alpha, y_\alpha)$. We may assume w.l.o.g. that $x_\alpha \approx_{\alpha+1} x$ and $y_\alpha \approx_{\alpha+1} y$; since $x_\alpha \in y_\alpha$, there is for any $\alpha < \kappa$ some $s_\alpha \in y$ such that $x_\alpha \approx_\alpha s_\alpha$. Then $\lim s_\alpha = \lim x_\alpha = x$ belongs to y and so E is closed.

Finally, let (x,y) be the limit of a κ-sequence (x_α, y_α) in $Q(X)$, and assume again that $x_\alpha \approx_{\alpha+1} x$ and $y_\alpha \approx_{\alpha+1} y$; pick $z \in x$ and, for any α, $z_\alpha \in x_\alpha$ such that $z_\alpha \approx_\alpha z$: by definition $((x_\alpha, y_\alpha), z) \in X$ for any α, hence also $((x,y),z) \in X$, and we are done.

It remains to prove (iii): to this aim, assume that we are given a κ-sequence x_α such that $|x_\alpha| < \lambda$ and $x_\alpha \approx_{\alpha+1} x$ for any $\alpha < \kappa$.

By considering the quotients modulo α-equivalence, we get $|x/\approx_\alpha| = |x_\alpha/\approx_\alpha| < \lambda$ for any $\alpha < \kappa$. If $|x| \geq \lambda$, consider any subset y of x of size λ : since $\lambda < \kappa$, the elements of y are pairwise β-inequivalent for some $\beta < \kappa$, hence $|y/\approx_\beta| \geq \lambda$, contradiction.

Finally, for fixed $\lambda < \kappa$, let $x_\alpha = \{\alpha + \gamma | \gamma < \lambda\}$.

Clearly, $\lim x_\alpha = \lim\{\alpha\} = \{\bar\kappa\}$ (where $\bar\kappa = \lim \alpha = \kappa \cup \{\bar\kappa\}$). Hence no Frege-Russel cardinal greater than 1 is closed.

$$\text{Q.E.D.}$$

Although compact $N_\kappa(\mathcal{U})$'s are highly self-referential, nevertheless interesting open relations, like *non-identity*, and discontinuous operations, like *binary intersection*, fail to be elements of the model. The most serious self-descriptive deficiencies of these models are ultimately due to the fact that full function spaces in general are not elements. In fact the set $X^Y \cap N = \mathcal{U}(Y, X)$ *of all uniformly continuous maps* from Y into X is not closed whenever $|Y| \geq \kappa$ and $|X| > 1$.

However these deficiencies appear only if we continue to focus on the usual set theoretic reductions of the fundamental notions of operation and relation. We shall show that, assuming a non purely set theoretic foundational framework like the *Ample theory* (theory A) of [2], many of these deficiencies can be partially amended.

It was with this in mind that we assumed urelements in our models. These have played no role up to now. We intend here to *activate* them as *relations, operations* and *qualities*, not withstanding the fact that many of these notions *do not have a corresponding set (graph, extension) in the model.* This method is similar to that of Oberschelp [15].

We assume for the rest of this section that κ is an uncountable strongly inaccessible weakly compact cardinal and that the space U of the urelements is isometric to the universal κ-compact space 2^κ of all κ-sequences of 0's and 1's, endowed with the *first-difference hypermetric* (see section 2).

Since the mapping $x \mapsto x \cup (\{\bar\kappa\} \times \{0, 1\})$ provides an isometry of 2^κ into a closed subspace L of N, we can fix an isometric inner labelling of the atoms by elements of L, say $\ell : U \to L$. We fix also a uniformly isomorphic embedding of $N = N_\kappa(\mathcal{U})$ into U, which will be denoted by j. Note that both ℓ and j, being continuous, belong to N. Further conditions on the embedding j will be specified later.

From now onwards, we assume the reader acquainted with the definitions and the notation of [2], which we shall adopt without further explanation for lack of space. In particular we shall deal freely with the over 200 distinguished objects of the Ample theory, which will be denoted by the names of the corresponding constants. Similarly, we shall refer to any axiom of the seventeen groups constituting the theory A by simply quoting the reference number it received in [2]. [7]

Our goal is to extend the set-theoretic structure $(N, \in)$ to an *Ample structure* with domain N. An ample structure is a first order relational structure capable of accomodating the interpretation of the constants and predicates of the ample theory. The fundamental predicates of the theory A are:

x is a quality, x enjoys the quality y;

x is a relation, x is in the relation z with y;

x is an operation, y is the result of the operation z on x;

z is the pair with first component x and second component y.

If the axioms 1.A-J hold, the ample structure is uniquely determined by the interpretations of the distinguished constants and of the ternary predicate x *is in the relation* z *with* y (see [2, §I Appendix]).

We call *natural* an *ample* structure with domain N if its structure of collections is *standard*, i.e. if the extension of the quality

---

[7] We modify slightly the formulation of the axioms 1.K and 17.F of [2], in order to make it closer to the common use and meaning of the objects involved. Namely, we do not reduce the quality of being a q-r-structure to the simple extensional condition given in [2]; consequently we take in the corresponding axiom only a one-sided implication (as done in [2] for the quality quniv):

1.K - qqrs S implies that $S=(q,r)$, $qqual\ r$ and, if for some $y\ yrx$, then qx. On the other hand, we replace the functional relation rbid by an operation bid which associates to any pair (x,y) and to any relation r the proposition "xry". This seems to represent better the act of bidding some opinion:

17.F - qop bid, if y rval bid then qprop y, and x rdom bid iff $x= ((u,v),r)$ and qrel r. (due to a misprint, the axioms on propositions received in [2] the number 16 instead of 17).

qcoll is $N \cap \mathcal{P}(N)$ and the relation *rcoll* is the *true membership* restricted to N. We proceed now to sketch a first extension of $(N, \in)$ to a natural ample structure with universe N.

In the rest of this section, we shall use *square brackets* to denote the usual set-theoretic codification *à la Kuratowski* of *pairs* and *n-tuples*, namely

$$[x, y] = \{\{x\}, \{x, y\}\} \text{ and } [x_1, ..., x_n, x_{n+1}] = [[x_1, ..., x_n], x_{n+1}].$$

The standard notation (x, y) and $(x_1, ..., x_n)$ will be reserved for denoting *primitive pairs* and *n-tuples*.

Similarly, we shall distinguish between (ordinary) graphs and products, which are collections of *primitive pairs or n-tuples*, and *K-graphs* or *K-products*, which are built up à la Kuratowski and marked by a subscript K. For instance, the K-product of A and B is

$$A \times_K B = \{[a, b] | a \in A \quad \& \quad b \in B\},$$

and is therefore distinct from the ordinary cartesian product

$$A \times B = \{(a, b) | a \in A \ \& \ b \in B\}.$$

First of all, we put

$Coll = N \backslash U$,

$Sys = \{j(x) | x \subseteq N \times_K N \ \& \ |x| < \kappa\}$, and

$Card = \{j(\lambda) | \lambda \text{ a Von Neumann cardinal} < \kappa\} \cup \{j(\bar{\kappa})\}$.

We interpret the elements of *Coll* as *collections* with the *ordinary membership* $\in$. We interpret those of *Sys* as *systems*, with the natural rule stating that, for the system $S = j(x), uSv$ holds iff $[u, v] \in j(x)$. Finally we interpret the elements of *Card* as *cardinals*, with $\varPi = j(\bar{\kappa})$ intended as the size of any *large* collection (the quality *qsmall* meaning having size less than κ). The standard cardinal operations and ordering (on *small* cardinals, i.e. below κ) are transferred by means of j.

In particular the collection of the *natural numbers* is $\mathsf{N} = \hat{j}(\omega)$, with $0 = j(\emptyset), 1 = j(\{\emptyset\})$, $n + 1 = j(j^{-1}(n) \cup \{j^{-1}(n)\})$, and $\aleph_o = j(\omega)$.

Moreover, in accord with the axiom 9.B of [2], we define the *natural pair* of x and y by $(x, y) = j(\{[1, x], [2, y]\})$, and similarly the *n-tuple* $(x_1, ..., x_n) = j(\{[1, x_1], ..., [n, x_n]\})$.

Before proceeding we remark that *Coll* and *Card* are closed, hence are collections themselves, whereas *Sys* is not. We could have chosen as systems the images of *all closed subsets of* $N \times_K N$, thereby obtaining a closed collection, but we prefer to deal only with *small systems*. So doing we allow for the *greatest manageability of systems*, which is a typical feature of the theory A.

In order to encode qualities, relations and operations as suitable definable subsets of U we proceed as follows.

Given a set $A \subseteq \mathcal{P}(N)$, define the *Gödel closure* $\mathcal{C}l(A)$ of A as the least superset of A which is closed under all *Gödel operations* (we can take only the operations $X^{-1}, X^*Y, \hat{X}(Y)$ considered in the proof of Theorem 3.2, together with the complement $N\backslash X$).

Let $Q = \mathcal{C}l(N)$ be the Gödel closure of N, let $\varphi : Q \to_K U$ be an injective K-mapping, and put $\Phi = \{[\varphi(x), y] \mid y \in x \in Q\}$.

Let $R = \mathcal{P}(N \times N) \cap \mathcal{C}l(N \cup \{\Phi\})$, be the set of all binary graphs belonging to the Gödel closure of $N \cup \{\Phi\}$, let $\psi : R \to_K U$ be an injective K-mapping, and put $\Psi = \{[\psi(x), y] \mid y \in x \in R\}$.

Let $F = N^{\subseteq N} \cap \mathcal{C}l(N \cup \{\Phi, \Psi\})$ be the set of all functional graphs belonging to the Gödel closure of $N \cup \{\Phi, \Psi\}$, and let $\chi : F \to_K U$ be an injective K-mapping.

The domains N, Q, R and F being overlapping, one can have *more than one atom* associated to a given set by the encoding functions φ, ψ, χ, and j (e.g. each of them is defined at any closed functional graph). We assume that the ranges of φ, ψ, χ, j are *pairwise disjoint* leaving uncovered a large part of U. On the other hand, we assume that there is some *uniform definable combinatorial rule connecting the labels of atoms which correspond to the same set via different encodings*. Since we set no topology on Q, R and F, we have no topological constraints on φ, ψ, χ; further conditions on them will be specified later on.

Now we put

$$Qual_o = \hat{\varphi}(Q), \quad Op_o = \hat{\chi}(F), \quad Rel_o = \hat{\psi}(R),$$

and we interpret $q = \varphi(X)$ as a *quality whose extension is X* (hence qx holds iff $x \in X$), $f = \chi(Y)$ as an *operation* and $r = \psi(Z)$ as a *relation whose graphs are Y and Z*, respectively (hence $y = fx$ holds iff $(x, y) \in Y$ and xry iff $(x, y) \in Z$).

We have thus determined the interpretations of all basic predicates of the ample theory. More specifically, we have an ample structure

$$\mathcal{N}_o = <N; Qual_o, P_1; Rel_o, P_2; Op_o, P_3; P_4>$$

where $P_1 = \cup_{X \in Q} X \times_K \{\varphi(X)\}$, $P_2 = \cup_{X \in R} X \times_K \{\psi(X)\}$, $P_3 = \cup_{X \in F} X \times_K \{\chi(X)\}$ and $P_4 = \{[x, y, j(\{[1, x], [2, y]\})] | x, y \in N\}$.

In order to complete the definition of the ample structure $\mathcal{N}_o$ we have to give the interpretation of each constant of the ample theory. Since there are more than two hundred constants to interpret, it would be cumbersome to list explicitly all the corresponding assignements in the model $\mathcal{N}_o$. We prefer to explain instead the general idea underlying our definitions, namely that of *interpreting any constant of the theory as the "qualified urelement" associated to the subset of N which naturally codes the corresponding mathematical concept.*

In doing this, we make use of the fact that a quality exists in $\mathcal{N}_o$ iff its extension belongs to Q (i.e. iff it is $\in$-definable). This can be obtained, in many cases, by a suitable choice of the labels of the objects which have to enjoy that quality. E.g. one can settle in this way the interpretations of all "descriptive" qualities like *qqual, qrel, qop, qrefl,* etc., by choosing *a priori* suitable *definable* subsets of U into which φ, ψ, χ have to map the graphs and extensions enjoying the corresponding properties.

Similarly, a relation (an operation) exists iff its graph belongs to R (resp. F); this can again be obtained by imposing suitable connections between the functions j, ℓ, ψ, χ (in this way are easily settled, among others, the operations *invop, invrel*).

In particular any closed subset of N is the extension of a quality, and any closed set of pairs is the graph of a relation and also of an operation when it is functional. It follows that all *small* qualities, operations and relations are present in N.

According to our initial stipulations, we interpret the fundamental structures of collections, systems and cardinals in the natural way, namely:

$qcoll = \varphi(Coll)$ $rcoll = \varphi(E)$;

$qsys = \varphi(Sys)$ $rsys = \psi(\{((x, y), s) \in N^2 \times Sys \,|[x, y] \in j^{-1}(s)\})$;

$qcard = \varphi(Card)$ $rcard = \psi(\{(x, y) \in Card^2 | \ j^{-1}(x) \subseteq j^{-1}(y)\})$.

The intended interpretation of most constants is determined by the corresponding axioms of the ample theory, once the *fundamental structures of qualities, relations, operations, collections, systems,* and *cardinals* are given. E.g.:

$$rnid=\psi(N\backslash I), \quad id=\chi(I), \quad \mathcal{P} = \chi(\{(A,\mathcal{P}(A))|A \in N\backslash U\}),$$

$$psys=\chi(\{((x,j(X)),y)|[x,y] \in X \ \& \ X \ is \ a \ functional \ K\text{-}graph \ \}).$$

Other constants which are not uniquely determined by the fundamental structures, like $qqrs$, $quniv$, $qprop$, will be interpreted below in a natural way.

In fact, all qualities, relations and operations involving only collections, systems and cardinals, like those introduced in Chapters II and IV of [2], exist in $\mathcal{N}_o$ since all the corresponding subsets belong to Q. Moreover, the comprehension properties stated in Theorem 3.2 yield all axioms of the groups 4-8, 13 and 14 of [2]. Similarly there are all relations connecting qualities, for their graphs belong to R, and all operations not involving other operations, for their graphs are elements of F.

Therefore the following axioms of [2] are directly satisfied by the natural interpretation sketched above:

1.ABC, F, HIJ   2.ABCD, FG, IJ

4.ABCDE 5.ABCDE 6.ABCDEF 7.ABCD 8.ABCDE

9.AB 10.A 11.EFG 13.ABCDEFGHI 14.ABCDEF

Being careful in mapping graphs which are reflexive, symmetric, etc. onto previously determined definable subsets of U, we can satisfy also the axioms 3.ABCDE. Paying similar attention in mapping qualities relations and operations of any cardinality $\lambda < \kappa$, we get 12.A. A suitable choice of a set of less that κ q-r-structures, including those which are explicitly postulated by the ample theory, yields 1.K.

We can satisfy also the axioms 16.ABCD by the following interpretation of the qualities $qinac$ and $quniv$:

(i)   a cardinal $j(\lambda) < h$ is *inaccessible* iff λ is a strongly inaccessible Von Neumann ordinal;

(ii)   a collection V is a *universe* iff $\aleph_o < |V| = \lambda < \kappa$ is inaccessible and, for any $X \subseteq V$, if $|X| < \lambda$, then X belongs V as well as $j(X), \varphi(X), \psi(X)$, and $\chi(X)$ (whenever they are defined).

Finally, for sake of simplicity, we trivialize the structure of *propositions* by allowing only two of them, the *true* proposition t and the *false* proposition f (t and f being two new atoms). In this way we easily obtain the validity of the axioms 17.ABCDEFG (recall that we have replaced the relation *rbid* of [2,§V.2] by the corresponding operation *bid*).

Thus we are left with the problem of assigning a graph in R to each of the relations

$$(3.1) \qquad rrel, \quad rop, \quad rdom, \quad rval, \quad rginc, \quad rexteq,$$

and one in F to each of the operations

$$(3.2) \qquad eval, \quad oprest, \quad hat, \quad graph, \quad dom, \quad img.$$

The natural assignement is possible for the operation *graph,* since *only continuous operations with closed domains can have a graph in N,* and we can arrange χ so that the set $\{(x, \chi(x))|x \in N \cap F\}$ belongs to Q, hence to F. We can also choose X in such a way that *the closures of the domain and of the image of each operation are encoded* by suitable κ-subsequences into the *label* of the atom corresponding to the operation itself. Moreover, we can make distinguishable those operations whose *domains and/or images are closed.* In this way one finds in Op_o also the natural interpretation of the operations *dom* and *img;* one can even discover when a *given collection includes the domain of an operation,* thus getting *all trivial restrictions.*

Unfortunately, one cannot find within $Rel_o \cup Op_o$ the natural full interpretation of the remaining constants (3.1-2). One can find instead homologous operations and relations acting on *qualities of pairs.* Therefore, using the correspondence between the images of the same graph under ϕ, ψ and χ, we decide to interpret the constants above as acting in the natural way only on *relations and operations whose graphs belong to Q.*

Having thus completely defined the *natural ample structure* $\mathcal{N}_o$ we see that all axioms of the theory A hold in it, but

$$1.DE \quad 2.EH \quad 10.C \quad 15.CE \qquad (B)$$

Moreover, introducing the qualities *qrequa* and *gopqua* (of being a *relation* and an *operation corresponding to a quality of pairs*), the given interpretation of the constants (3.1-2) satisfies the axioms

$$1.D_oE_o \ , \ 2.E_oH_o \ , \ 10.C_o \ , \ 15.C_oE_o \qquad (B_o)$$

obtained by restricting the correponding axioms of A to relations and operations enjoing *qrequa* and *qopqua*.[8]

Let A_o be the axiomatic theory resulting from A by replacing the axioms (B) by their weakenings (B_o). It is then straightforward to complete the proof of the following theorem (see also [10], where constructions similar to the one sketched above are developed in full details).

Theorem 3.3. *The ample structure $\mathcal{N}_o$ is a model of theory A_o.*

Moreover the following supplementary axioms of extensionality, comprehension and stability hold in $\mathcal{N}_o$:

I *The fundamental structures of qualities, relations and operations are extensional.*

II *Any system has a graph, any collection is the extension of a quality, any quality of pairs is associated to a relation, and any functional relation to an operation. Moreover all qualities and relations have characteristic operations.*

---

(8) Therefore, e.g., the axioms $1.D_o$ and $2.H_o$ are

$1.D_o$ (x,y) *rrel r iff both qrelqua r and xry.*

$2.H_o$ *eval* $\downarrow z$ *iff* $z=(f,x)$, *qopqua f and* $f\downarrow x$. *In this case eval* $(f,x)=fx$.

III *The collections are closed under union, intersection, cartesian product, power-set operator and power, the qualities and relations are closed under negation and disjunction, the relations and the operations under composition, fibred and tensor product and restriction to any collection.* [9]

The self-referential power of the theory A is seriously weakened by the above restrictions of the axioms 1.DE and 2.EH. In fact, the operations cannot have a complete internal description, since the absence of objects such as *rop* and *eval* is provable, as well as that of many other relations and operations which could replace them in describing the actions of all operations. However one can deal freely with relations and relational pairs inside $\mathcal{N}_o$, since the *characteristic operations* of the full relations *rrel*, *rginc* and *resteq* are elements of Op_o.

This lack of self-description is partially balanced by the strong axioms of extensionality, comprehension and stability I, II and III. Actually the wide stability of the model $\mathcal{N}_o$ goes even beyond the properties III above, which are for themselves already inconsistent with the full theory A, the Antinomy II of [2, §VI.6] being derivable from A+ III. [10]

We conclude this section by expanding $\mathcal{N}_o$ to a model $\mathcal{N}'$ of a very highly self-descriptive theory A'. Namely, we will give below an extensive interpretation to the relations and operations (3.1-2), whose domains had been restricted in defining the model $\mathcal{N}_o$. In doing this, we need to qualify only a finite number of new atoms. Any definable operation acting directly on relations (like *domrel*, *invrel*,

---

[9] The axioms I-III are particular cases of the "strong axioms" of [2, §VI]. We list here those which hold in $\mathcal{N}_o$: SA.1 CA.13.2 DC AS.1-3 AO.3 AQ.1 AR.1,2,4,5 NF.1,4-6 R.1,2,4-6 C.1-3,3* SI. Many more axioms could be satisfied by imposing that suitably chosen subsets of Q, R, F have definable images. However the following axioms cannot be valid in $\mathcal{N}_o$: RA.1-2 IA.1-2 CA.4-7,11,12,14,15 AR.3 NF.2,3 C.4,4*.

[10] In fact, it has recently been shown by G.Lenzi (personal communication) that both theories

A+ any two operations have a composition and

A+ any two relations have a composition + there is a diagonal relation rdiag such that x rdiag y iff $y = (x, x)$

are inconsistent.

etc.) will then have a definable extension, which treats correctly all new relations and still belongs to Op_o.

The same argument entails that also the operations *graph*, *dom* and *img* are still available. It cannot work, however, for relations and operations, like *rginc* and both restrictions, which act on *relational* or *functional pairs*.

Going again through the constants of the theory A, we see that we have to reinterpret, together with the relations and operations (3.1-2), only the four operations

(3.3)   *cext*,   *syext*,   *gcard*,   *birest*

Therefore we pick two finite sets of new atoms

$$Rel_1 = \{r_1, \ldots, r_{11}\} \quad \text{and} \quad Op_1 = \{f_1, \ldots, f_8\},$$

which will be used to interpret the constants (3.1-3), and we put

$$rdom = r_1 \quad rval = r_2 \quad rop = r_3 \quad rrel = r_4$$
$$rginc = r_5 \quad rexteq = r_6 \quad \text{and} \quad r_{12-n} = r_n^{-1};$$
$$eval = f_1 \quad bid = f_2 \quad hat = f_3 \quad oprest = f_4$$
$$birest = f_5 \quad cext = f_6 \quad syext = f_7 \quad gcard = f_8.$$

Since self-description in the theory A is mostly obatained by means of relations, it seems appropriate to pick a third set of atoms $Rel_{-1} = \{r_{-1}, \ldots, r_{-11}\}$ to interpret the *negations* of $r_1, \ldots, r_{11}$. [11]

---

[11] We shall obtain at once the operation *notr* providing the negation of any relation and satisfying the axiom AR.1 of [2, §VI], namely

$$notr = \chi(\{(\psi(X), \psi(N^2 \setminus X)) \mid X \in R\} \cup \{(r_n, r_{-n}) \mid -11 \leq n \leq 11\}).$$

We go now to extend $\mathcal{N}_o$ to a new natural ample structure

$$\mathcal{N}' = <N;\ Qual_0,\ P_1;\ Rel',\ P_2';\ Op',\ P_3';\ P_4>$$

where $Rel' = Rel_o \cup Rel_1 \cup Rel_{-1}$, $Op' = Op_o \cup Op_1$, and

$P_2' = P_2 \cup \bigcup_{n=1}^{11} G_n \times_K \{r_n\} \cup \bigcup_{n=1}^{11} (N^2 \backslash G_n) \times_K \{r_{-n}\}$,
$P_3' = P_3 \cup \bigcup_{m=1}^{8} F_m \times_K \{f_m\}$

We interpret the constants (3.1-3) as stipulated above and the remaining ones by extending in the natural way the interpretation given in $\mathcal{N}_o$. Thus we have only to specify the external graphs G_n and F_m of the new operations and relations.

Due to the simultaneous presence of many *large* and many *non-wellfounded small collections*, providing a model of the whole ample theory A would require particular devices, not only of technical nature. Moreover we want to preserve as much as possible of the properties of comprehension and stability of our previous model $\mathcal{N}_o$. Last but not least, we are looking for a honest compromise between easy definability and wide applicability of the fundamental operations and relations.

Therefore we decide to maintain the full self-descriptive power of the most important objects, which are

$$rop,\ rdom,\ rval,\quad \text{and}\quad eval,\ bid,$$

by defining their graphs in such a way that the axioms 1.DE 2.EH 17.FG are satisfied.

We slightly weaken instead the actions of the relations *rginc*, *rexteq* and of the operations *oprest*, *birest*, *hat* on pairs involving themselves or the other objects r_n, f_m.

Let A$'$ be the axiomatic theory resulting from A by replacing the axioms 10.C and 15.CE by

10.C': *qpreo rginc*. If $(r,x)rginc(r',x')$, then $trx \Longrightarrow tr'x'$.
    The converse implication holds whenever r, r' belong to Rel_o.

15.C': *If f is an operation belonging to Op_o, then oprest is defined at (f, C) for any collection C.*

    If r is a relation belonging to Rel_o, then birest is defined at $(r, (C, D))$ for any pair of collections C, D.

15.E': *Like 15.E with the addition: provided $f \neq$ hat, bid, eval.*

Then, by suitably choosing the graphs F_m, G_n, one can prove

Theorem 3.4. *The natural ample structure $\mathcal{N}'$ is a model of the theory* A' *plus the following axioms of extensionality, comprehension and stability:*

I *The fundamental structures of qualities, relations and operations are extensional.*

II' *Any system has a graph, any collection is the extension of a quality, any quality of pairs is associated to a relation and any functional relation to an operation.*

III' *The collections are closed under union, intersection, cartesian product, power-set operator and power, the qualities are closed under negation and disjunction and the relations under negation.* [12]

Sketch of the proof. We only have to define the graphs $F_m(1 \leq m \leq 8)$ and $G_n(1 \leq n \leq 5)$, since G_{12-n} is to be taken equal to G_n^{-1} and $G_6 = G_5 \cap G_7$.

(a) The domains and codomains of all relations r_n, as well as the ranges of all operations f_m are easily determined *a priori* (e.g. the range of *eval* is N, that of *cext* is $N \backslash U$, etc.).

Hence the graph G_2 of *rval* is completely determined.

(b) The operations *cext*, *syext* and *gcard* have to be reconsidered only on relational pairs (r_n, x), since at any other pair the previous definition works. An easy inspection shows that the operation *cext* (hence a fortiori *syext*) can be made *undefined* in all critical cases, while *gcard* takes on at the corresponding arguments only the value Л.

Therefore the graphs F_m are determined for $6 \leq m \leq 8$.

(c) In order to complete the graph of *rdom*, we need only to fix the domains of the operations f_m $(3 \leq m \leq 5)$, since dom *bid* is known

---

and dom $eval = \bigcup_{n<\omega} D_n$ where $D_o = \bigcup_{f \neq eval}\{f\} \times$ dom f and $D_{n+1} = \{eval\} \times D_n$.

According to the axioms 15.C' DE' we put

$$\text{dom } f_5 = Rel_o \times Coll^2 \cup \{(r_n, (C, D)) | C \times D \supseteq \text{ dom } r_n \times \text{cod } r_n\}$$

$$\text{dom } f_4 = Op_o \times Coll \cup \{(f_m, C) | C \supseteq \text{ dom } f_m\}$$

$$\text{dom } f_3 = \{(f, C) | f \neq f_1, f_2, f_3 \ \& \ \hat{f}(C) \in N\} \cup$$
$$\{(f_m, C) | C \supseteq \text{ dom } f_m \quad m = 1, 2, 3\}.$$

Since any collection is closed, $C \supseteq$ dom f holds iff *the closure of the domain* of f is included in C. Therefore the above definitions are wellposed, since the closure of dom f_m is easy to specify also when *the domain itself* is yet unknown, provided the atoms f_m are suitably choosen (e.g. as *limits of everywhere defined operations.*)

(d) Now the graph of *rdom* is completely defined, while those of the operations f_3, f_4 and f_5 have unique natural definitions once the domains are fixed according to (c). Thus G_1 and F_3, F_4, F_5 are settled.

(e) We define G_5 by stating that $(r, s)r_5(r', x')$ holds iff either $(r, x) = (r', x')$ or the extensions of the rational pairs $(r, x), (r', x')$ are already completely determined by the preceding stipulations and $trx \implies tr'x'$.

(f) Finally we define the graphs of the remaining objects by an inductive procedure involving all of them at once. Namely, we put

$$F_m = \bigcup_{i<\omega} F_m^{(i)} \quad (m = 1, 2) \quad \text{and} \quad G_{\pm n} = \bigcup_{i<\omega} G_{\pm n}^{(i)} \quad (n = 3, 4)$$

where the six sequences of graphs $G_{\pm n}^{(i)}(n = 3, 4)$ and $F_m^{(i)}(m = 1, 2)$ are defined by induction on $i \in \omega$ in the following way:

$$G_3^{(0)} = \{((x, y), f) \in N^3 \times Op''' | fx = y\}$$

$$G_{-3}^{(0)} = N^2 \backslash (N^2 \times Op') \cup \bigcup_{f \in Op''} (N \backslash \text{graph } f) \times \{f\} \cup$$
$$\cup \bigcup_{f \in Op'''} (N \backslash (\text{dom } f \times \text{rng } f)) \times \{f\}$$

$$G_4^{(0)} = \{((x,y),r) \in N^2 \times Rel'''|xry\}$$

$$G_{-4}^{(0)} = N^2 \backslash (N^2 \times Rel') \cup \{((x,y),r) \in N^2 \times Rel'''|xr_-y\}$$

$$F_1^{(i)} = \{((f,x),y)|((x,y),f) \in G_3^{(i)}\}$$

$$F_2^{(i)} = \left(G_4^{(i)} \times \{t\}\right) \cup \left(G_{-4}^{(i)} \times \{f\}\right)$$

$$G_3^{(i+1)} = G_3^{(i)} \cup (F_1^{(i)} \times \{f_1\}) \cup (F_2^{(i)} \times \{f_2\})$$

$$G_{-3}^{(i+1)} = G_{-3}^{(i)} \cup (((\mathrm{dom}F_1^{(i)} \times N) \backslash F_1^{(i)}) \times \{f_1\}) \cup$$
$$\cup (((\mathrm{dom}\, F_2^{(i)} \times N) \backslash F_2^{(i)}) \times \{f_2\})$$

$$G_4^{(i+1)} = G_4^{(i)} \cup \bigcup_{n=3}^{4} (G_{\pm n}^{(i)} \times \{r_{\pm n}\} \cup (G_{\pm n}^{(i)})^{-1} \times \{r_{\pm(12-n)}\})$$

$$G_{-4}^{(i+1)} = G_{-4}^{(i)} \cup \bigcup_{n=3}^{4} (G_{\mp n}^{(i)} \times \{r_{\pm n}\} \cup (G_{\mp n}^{(i)})^{-1} \times \{r_{\pm(12-n)}\}).$$

(We have set above $Rel'' = \{r_{\pm n}|n = 3,4,8,9\}$, $Rel''' = Rel'\backslash Rel''$, $Op'' = \{f_1, f_2\}$, $Op''' = Op'\backslash Op''$. We have also denoted by r_- the *negation* of the relation $r \in Rel'''$).

(g) In order that the axioms 1.DE 2.EH 17.FG hold with the assignements (f), it sufficies that $G_n \cup G_{-n} = N^2$ for $n = 3,4$, or equivalently that dom $f_m = $ dom F_m for $m = 1,2$.

To this aim, we assume that j has been choosen in such a way that one can assign to each object x a weight $w(x)$ verifying

$$w(x) = 0 \quad \text{for } x \notin N^2 \cup Rel'' \cup Op'',$$
$$w(x) = 1 \quad \text{for } x \in Rel'' \cup Op'',$$
$$w(x,y) = w(x) + w(y) \quad \text{for } (x,y) \in N^2.$$

Let (x,y) be a pair of least weight not belonging to $G_4 \cup G_{-4}$. Then $x = (u,v)$, $y = r_{\pm k}$ with $k \in \{3,4,8,9\}$ and $(u,v) \notin G_k \cup G_{-k}$.

Therefore $k \neq 4, 8$, since $w(x, y) > w(u, v)$, and we can assume w.l.o.g. that $y = r_3$ and (u, v) is a pair of least weight outside $G_3 \cup G_{-3}$.

Then $u = (s, t)$, $v = f_m$ with $m \in \{1, 2\}$ and $s \in \text{dom } f_m \setminus \text{dom } F_m$.

If $m = 1$, then $s = (f_h, z)$ with $h \in \{1, 2\}$ and $z \in \text{dom } f_h \setminus \text{dom} F_h$. It follows that $((z, t), f) \notin G_3 \cup G_{-3}$, contradicting the minimality of (u, v).

If $m = 2$, then $s = ((a, b), r) \notin G_4 \cup G_{-4}$, contradicting the minimality of (x, y). The sketch of the proof is thus concluded.

$$\text{Q.E.D.}$$

We conjecture that the theory A+I,II′,III′ is consistent, too. Actually, one can give an inductive simultaneous definition of the graphs $F_m (1 \leq m \leq 5)$ and $G_{\pm n}(3 \leq n \leq 9)$, thus expanding the ample structure $\mathcal{N}'$ to one where all fundamental constants (3.1-3) have a natural interpretation. However it is by no means obvious that one can then assume the strong wellfoundedness property of the encoding of pairs which is needed in order to apply to this wider context the concluding argument (g) sketched above.

4. The axiom of choice. It is well known, since a celebrated result of Specker's [17], that the axiom of choice can be inconsistent with set theories admitting large sets: we refer to [6] for a short but exciting review of some negative results.

We shall briefly discuss here how our models behave w.r.t. various kinds of choice principles. It is worth noticing, in view of the above remark, that we obtain *inter alia* the consistency of the well-ordering principle relative to GPK, the generalized positive comprehension principle. As it is done in most classical analyses of universal choice principles, we consider here in particular the axioms studied in [9]. We phrase them below in a form suitable for set theories with a universal set V :

WoV : *The universe V can be well-ordered.*

 E : *The universe V has a choice function.*

 H : *Any equivalence has a set of representatives.*

F : *Any relation with domain V includes a function with domain V.*

DCC : *Let R be a relation and X a set such that, for any subset Y of X, there is some $x \in X$ with YRx; then there is a X-valued function f defined at all ordinals and verifying $\hat{f}(\beta)Rf(\beta)$ for any ordinal β.*

DCC$_\alpha$: the same as DCC for ordinals less than α.

DC : *Let R be a relation and X a set such that for any $x \in X$ there is $y \in X$ such that xRy; then there is a function f verifying $f(n)R\ f(n+1)$ for any positive integer n.*

We shall also consider the ordering principle

LoV : *The universe V can be linearly ordered.*

It is easily seen that the generalized comprehension principle GPK yields the implications

$$\text{WoV} \to \text{E}, \quad \text{H} \to \text{F} \to \text{E} \quad \text{and} \quad \text{LoV} \to \text{AC}_{fin}$$

(AC$_{fin}$ is the axiom of choice for any set of finite sets). E.g. given a relation R with domain V, consider the equivalence

$$Q = \{((x,y),(x,z))|xRy\ \&\ xRz\}.$$

Any set of representatives for Q is clearly a function with domain V which is included in R.

On the other hand, GPK does not yield either WoV $\to$ F or WoV $\to$ DCC (see Theorem 4.3 below); while both implications hold in Gödel-Bernays class theory, even without foundation.[13]

Since in this context urelements are an inessential complication, we shall consider $U = \emptyset$ throughout this section. The models $N_\kappa(\mathcal{U})$ are thus the same as the models N_κ of [7]. Moreover, since we are interested in the connections between strong principles of comprehension and choice, we restrict our attention to κ-compact models. Therefore we assume that $\kappa \to (\kappa)^2_2$ throughout this section.

---

[13] The exact strength of the axiom H is still unknown, even in pure set theory without urelements, when the axiom of foundation fails. Clearly H follows from WoV and implies F, but the converse implications are open. The authors can only prove that H is strictly stronger than both E and DCC$_{Ord}$ (see [9-II]).

Before proceeding we state in the following lemma a topological property of N_κ which will be useful in the sequel.

Lemma 4.0. *Any accumulation point of N_κ is complete, i.e. it has 2^κ α-equivalent points for any $\alpha < \kappa$.*

PROOF. Let x be an accumulation point of $N = N_\kappa$. There is an accumulation point y belonging to x, otherwise x would be a set of size less than κ of isolated points, which would therefore be all α-isolated for some $\alpha < \kappa$; hence x would be $(\alpha + 1)$-isolated.

For fixed $\alpha < \kappa$, let B_α be the set of all points of N which are α-equivalent to y. Pick an injective κ-sequence $< y_\beta >_{\beta<\kappa}$ of elements of B_α converging to y and, for any subset s of κ, put

$$x_s = (x \backslash B_\alpha) \cup \{y\} \cup \{y_\beta | \beta \in s\}.$$

The sets x_s are clearly α- (indeed at least $\alpha + 1$) equivalent to x.

Q.E.D.

We begin by defining in N_κ a linear ordering of the universe.

Lemma 4.1. *There is a closed subset O of N_κ^2 such that*

(i) O is reflexive, antisymmetric and transitive;

(ii) if $x \not\approx_\alpha y$, then either $B(x, \epsilon_\alpha) \times B(y, \epsilon_\alpha)$ or $B(y, \epsilon_\alpha) \times B(x, \epsilon_\alpha)$ is included in O.

PROOF. Let $<_\alpha$ be a linear ordering of the set B_α of all closed ϵ_α-balls of N. Since B_α is a partition of N_κ, it is possible to choose a κ-sequence $<<_\alpha |\alpha < \kappa >$ in such a way that, given $b <_\alpha b'$, if c, c' are any ϵ_β-balls (with $\beta > \alpha$) included in b, b' respectively, then $c <_\beta c'$.

Define $O = \{(x,y) | x = y \text{ or } \exists \beta \ B(x, \epsilon_\beta) <_\beta B(y, \epsilon_\beta)\}$: then O verifies (ii) by costruction and (i) since all $<_\alpha$'s are linear orderings (note that if $B(x, \epsilon_\beta) <_\beta B(y, \epsilon_\beta)$ holds for some β, then it holds for any β for which they are different).

Therefore we only need to prove that O is closed. Let x_α, y_α be κ-sequences such that, for any $\alpha < \kappa$, $x_\alpha \approx_\alpha x$, $y_\alpha \approx_\alpha y$ and

$(x_\alpha, y_\alpha) \in O$. If for all α $x_\alpha \approx_\alpha y_\alpha$, then $x = y$; otherwise for some β $B(x, \epsilon_\beta) <_\beta B(y, \epsilon_\beta)$, hence in any case $(x, y) \in O$.

Q.E.D.

We shall show now that neither O nor other relations on N_κ can be wellorderings. Note that the statement of Lemma 4.2 below refers to *external true* wellorderings. We shall see later that if $\kappa = \omega$ the above defined relation O is a wellordering *in the sense of N_ω*.

Lemma 4.2. *There are no (standard) closed well-orderings of* N_κ.

PROOF. [14]   Assume $N = N_\kappa$ wellordered and fix an (external) indexing of N by ordinals, say $N = \{x_\alpha | \alpha < \lambda\}$. The corresponding graph belongs to N iff it is closed, i.e. iff whenever $\alpha_\delta \leq \beta_\delta$ for any $\delta < \kappa$ and $x = \lim_{\delta \to \kappa} x_{\alpha_\delta}$, $y = \lim_{\delta \to \kappa} x_{\beta_\delta}$, also $x \leq y$.

Assume that the given wellordering is closed. Then $\lim_{\alpha \to \kappa} x_\alpha = x_\kappa$, since any two convergent κ-subsequences of $< x_\alpha >_{\alpha < \kappa}$ have the same limit.

Now pick another κ-sequence converging to x_κ, whose indices are greater than κ, which exists since x_κ is a complete accumulation point of N. Let $< x_{\gamma_\delta} >_{\delta < \kappa}$ be any κ-subsequence of it with increasing indices γ_δ in the fixed indexing: then we would have $x_{\gamma_\delta} \leq x_{\gamma_\nu}$ whenever $\delta \leq \nu$, whence $x_{\gamma_\delta} \leq x_\kappa = \lim x_{\gamma_\nu}$, and simultaneously $x_\kappa < x_{\gamma_\delta}$ for any δ, since $\kappa < \gamma_\delta$ for any $\delta < \kappa$.

Therefore no closed wellordering of N can exist.

Q.E.D.

Since all α-sequences for $\alpha < \kappa$ are elements of N_κ, any internal wellordering of N_κ would be a *real* wellordering when κ is uncountable. Hence WoV fails in N_κ for any uncountable κ.

This is also a consequence of the following theorem, which summarizes the main choice properties of our models:

---

[14] This proof is essentially due to Malitz [13].

Theorem 4.3.

(i) The axioms LoV *and* $\forall \alpha$ DCC$_\alpha$ *(hence also* AC$_{fin}$*) hold in* N_κ, *whereas the axioms* F *and* DCC *(hence also* H*) fail in* N_κ *for any* κ.

(ii) The axiom WoV *(hence also* E*) holds, whereas the axiom* DC *fails in* N_ω.

PROOF. (i) The set O of the above lemma witnesses that $N_\kappa \models$ LoV. Taking into account that the ordinals of N_κ are exactly those which are less than κ, we get immediately $N_\kappa \models \forall \alpha$DCC$_\alpha$, since all functions of size less than κ belong to N_κ.

In order to prove the failure of DCC, let $\bar{\kappa} = \kappa \cup \{\bar{\kappa}\}$ be the closure of κ in N_κ, put $C = \bar{\kappa} \times \{0,1\}$ and consider the closed relation $R \subseteq \mathcal{P}(C) \times C$ defined by

$$(x, (\alpha, i)) \in R \quad \text{iff} \quad \alpha \times \{0,1\} \subseteq x, \ (\alpha, i) \notin x \ \text{and} \ \begin{cases} (\alpha, 1-i) \in x \\ \text{or} \\ i = 0 \end{cases}$$

$$(C, (\bar{\kappa}, i)) \in R \quad \text{for} \quad i = 0, 1.$$

Any function in N_κ which is defined at all ordinals less than κ must be uniformly continuous and defined at $\bar{\kappa}$.

Let $f : A \to C$ verify $\hat{f}(\alpha) R f(\alpha)$ for any $\alpha < \kappa$ (hence A includes $\bar{\kappa}$). Then $f(0) = (0,0)$, and f proceedes by taking alternately all values $(\alpha, 0)$ and $(\alpha, 1)$, with α increasing without any jump.

Therefore, by continuity, both $(\bar{\kappa}, 0)$ and $(\bar{\kappa}, 1)$ have to be taken as values of f at $\bar{\kappa}$, so f is not a function.

Note that the same relation R provides a counterexample also for the axiom F, since any continuous function included in R should associate to C both $(\bar{\kappa}, 0)$ and $(\bar{\kappa}, 1)$.

The fact that dom R is not the whole universe is easily settled, since there is a projection of N_κ onto any closed set. [15] So we

---

[15] Note that we can easily obtain from the existence of projections another choice-like property of N_κ, namely that *the injective ordering of cardinalities is coarser than the surjective one.* In fact, if $j : A \to B$ is an injective continuous mapping and both A and B are closed, then by κ-compactness j is a homeomorphism between A and $A' = \hat{j}(A)$. If p is a continuous projection of N onto A', then $j^{-1} \circ p_{|B}$ is a projection of B onto A. We do not know whether the converse property holds in N_κ.

can use such a projection onto $\mathcal{P}(C)$ and transform R into a relation with universal domain.

In order to find in N a projection of N onto the closed set A, working from outside we associate to each α-ball B_α meeting A a point $\sigma(B_\alpha) \in B_\alpha \cap A$, in such a way that $\sigma(B_\beta) = \sigma(B_\alpha)$ whenever B_β is a β-ball (with $\beta > \alpha$) to which $\sigma(B_\alpha)$ belongs.

For $x \notin A$, let $\alpha + 1$ be the least (necessarily successor) ordinal such that x is $(\alpha + 1)$-inequivalent to each element of A, and put $p(x) = \sigma(B_\alpha(x))$. Since, for $x, y \notin A$ and $z \in A$, $x \approx_\alpha y \approx_\alpha z$ implies $p(x) \approx_\alpha p(y) \approx_\alpha z$, we extend p by the identity on A and obtain a continuous projection of N_κ onto A.

(ii) We get the wellordering principle in N_ω by showing that any closed set has a least element w.r.t. the ordering O of Lemma 4.1.

Let $A \subseteq N_\omega$ be closed and, for any $n < \omega$, let b_n be the least n-ball meeting A (least in the ordering induced by O). By definition of O, the balls B_n are a descending chain under inclusion, which has non-empty intersection by Cauchy completeness.

Again by definition of O, the unique point lying in the intersection of all balls B_n is the least element of A (and belongs to A as limit of a sequence of points of A).

The proof that DC fails in N_ω could be omitted, since DC is equivalent to DCC for $\kappa = \omega$. However it is easy to verify that the closed relation

$$S = \{((n,0),(n,1))|n \in \omega\} \cup \{((n,1),(n+1,0))|n \in \omega\} \cup$$
$$\cup \{\bar{\omega}\} \times \{0,1\}$$

does not admit a continuous function f with closed domain verifying $f(n)Sf(n+1)$ for any $n \in \omega$, since $f(\bar{\omega})$ should be simultaneously 0 and 1.

$$\text{Q.E.D.}$$

Finally N_κ verifies the axiom of *strong extensionality*
Sext - *Transitive $\in$-isomorphic sets are equal.*

This is a consequence of AR and of the fact that N_κ is a transitive set without urelements.

Therefore we can considerably improve the consistency results of [7] by putting together Theorems 3.2 and 4.3, so as to obtain

Corollary 4.4.

(i) Con (ZF) $\implies$ Con (GPK + Sext+Inf+WoV)

(ii) Con (ZFC $+\exists\kappa > \omega$ $\kappa \to (\kappa)^2_2$) $\implies$
Con (GPK + Sext + SInf + LoV + $\forall\alpha$ DCC$_\alpha$)
where Inf *is the usual axiom of infinity*
Inf - $\exists w(\emptyset \in w$ & $(x \in w \to x \cup \{x\} \in w))$,
while Sinf *is some strong axiom of infinity, e.g.*
Sinf - $\forall\alpha\exists\mu > \alpha$ μ *is a strongly (hyper-hyper-...) Mahlo cardinal.*

We conjecture that $\forall\alpha$DCC$_\alpha$ in (ii) above can be replaced by AC, but at present we do not even know whether the axiom of choice holds in N_κ for some uncountable κ.

References

[1] A.Church, *Set theory with a universal set*, in *Proceedings of the Tarski Symposium*, Proc. Symp. Pure Math. XXV, Amer. Math. Soc., Providence, R.I., 1974, 297-308.

[2] M.Clavelli, E.De Giorgi, M.Forti, V.M.Torterelli, *A selfreference oriented theory for the Foundations of Mathematics*, in *Analyse Mathématique et Applications*, Gauthiers-Villars, Paris 1988, 67-115.

[3] E.De Giorgi, M.Forti, *Una teoria-quadro per i fondamenti della matematica*, Atti Accad. Naz. Lincei Rend. Cl. Sci. Fis. Mat. Nat. (8) **79** (1985), 55-67.

[4] E.De Giorgi, M.Forti, V.M.Tortorelli, *Sul problema dell'autoriferimento*, Atti Accad. Naz. Lincei Rend. Cl. Sci. Fis. Mat. Nat. **80** (1986), 363-372.

[5] R.Engelking, *General Topology*, Polish Scientific Publishers, Warszawa 1977.

[6] T.E.Forster, *The status of the axiom of choice in set theory with a universal set*, J. Symb. Logic **50** (1985), 701-707.

[7] M.Forti, R.Hinnion, *The consistency problem for positive comprehension principles*, J. Symb. Logic (to appear).

[8] M.Forti, F.Honsell, *Set theory with free construction principles*, Ann. Scuola Norm. Sup. Pisa **10** (1983), 493-522.

[9] M.Forti, F.Honsell, *Axioms of choice and free construction*

principles, I. Bull. Soc. Math. Belg. **36 B** (1984), 69-79; II.
Ibid. **37 B** (1985), 1-12; III. Ibid. **39 B** (1987), 259-276.

[10] M.Forti, G.Lenzi, *Higher models of the* **A**-*theory* (in preparation).

[11] H.Friedman, *A cumulative hierarchy of predicates,* Z. Math.
Log. Grundlagen Math. **21** (1975), 309-314.

[12] T.Jech, *Set theory,* Academy Press, New York 1978.

[13] R.J.Malitz, *Set theory in which the axiom of foundation fails,*
Ph. D. Thesis, UCLA, Los Angeles 1976 (unpublished).

[14] E.Mitchell, *A model of set theory with a universal set,* Ph. D.
Thesis, University of Wisconsin, Madison 1976 (unpublished).

[15] A.Oberschelp, *Eigentliche Klassen als Urelemente in der Mengenlehre,* Math. Annalen **157** (1964), 234-260.

[16] W.V.O.Quine, *New Foundations for Mathematical Logic,* Amer.
Math. Monthly **44** (1937), 70-80.

[17] E.P.Specker, *The axiom of choice in Quine's* NF, Proc. Nat.
Acad. Sci. U.S.A. **39** (1953), 972-975.

[18] F.W.Stevenson, W.J.Thron, *Results on ω -metric spaces,* Fundam. Math. **65** (1969), 317-324.

Università di Cagliari

Dipart. di Matematica

Via Ospedale 72

I-09100 CAGLIARI

Università di Torino

Dipartimento di Informatica

Corso Svizzera 165

I-10100 TORINO

Progress in Nonlinear Differential Equations and Their Applications

Editor
Haim Brezis
Department of Mathematics
Rutgers University
New Brunswick, NJ 08903
U.S.A.
and
Département de Mathématiques
Université P. et M. Curie
4, Place Jussieu
75252 Paris Cedex 05
France

Progress in Nonlinear Differential Equations and Their Applications is a book series that lies at the interface of pure and applied mathematics. Many differential equations are motivated by problems arising in diversified fields such as Mechanics, Physics, Differential Geometry, Engineering, Control Theory, Biology, and Economics. This series is open to both the theoretical and applied aspects, hopefully stimulating a fruitful interaction between the two sides. It will publish monographs, polished notes arising from lectures and seminars, graduate level texts, and proceedings of focused and refereed conferences.

We encourage preparation of manuscripts in some such form as LaTex or AMS TEX for delivery in camera ready copy, which leads to rapid publication, or in electronic form for interfacing with laser printers or typesetters.

Proposals should be sent directly to the editor or to: Birkhäuser Boston, 675 Massachusetts Avenue, Suite 601, Cambridge, MA 02139.